MICROBIOLOGY RESEARCH ADVANCES

MICROBES IN PROCESS

MICROBIOLOGY RESEARCH ADVANCES

Additional books in this series can be found on Nova's website under the Series tab.

Additional E-books in this series can be found on Nova's website under the E-book tab.

MICROBIOLOGY RESEARCH ADVANCES

MICROBES IN PROCESS

NEELAM GARG
AND
ABHINAV AERON
EDITORS

New York

For permission to use material from this book please contact us:
Telephone 631-231-7269; Fax 631-231-8175
Web Site: http://www.novapublishers.com

Library of Congress Cataloging-in-Publication Data

Library of Congress Control Number: 2014930663

ISBN: 978-1-63117-127-7

Published by Nova Science Publishers, Inc. † New York

Preface

Microorganisms have a very diverse range of metabolism which makes them capable of existing and multiplying at the sites with extreme types of environment where other life forms usually do not exist. The miniscule organisms produce an array of metabolites in various facets of processes. These metabolites and processes find numerous commercial and industrial applications which involve the exploitation of the microorganisms for the production of the enzymes, antibiotics, organic acids, fermented foods, milk products, beverages, biofuels, animal feeds, biofertilizers, biopesticides, biopharmaceuticals, recombinant DNA products etc. Various technologies employed for waste treatment and pollution control also entail the use of microorganisms in various ways. Infact, microorganisms themselves are also used as food, food supplements, prebiotics, probiotics, vaccines etc. For all these purposes, microorganisms are isolated from diverse surroundings where they perform different physiological roles. These microbial strains are improved by employing various techniques of biotechnology to make them more effective as a candidate for commercial production. Many industrial productions involve the use of chemicals that pollute our environment in numerous ways and are precarious for the health of living organisms. Therefore now-a-days, the microbiologists are in great demand to explore the use of microorganisms to make these processes eco-friendly, safe and cost effective. However the selection and use of microorganisms in industry needs proper understanding of their physiology, metabolism and interactions with other organisms along with the recovery of the desired product economically.

With better perception of the biotechnological processes, improvement and utilization of modern techniques, the man is looking forward for a better future, through a sustainable industry based on microorganisms. Henceforth, the editors envisaged that there is a need to gather information from repertoire of scientists working in the field of industrial microbiology and bring together these scientists belonging to different thoughts of process on a single international platform. We have full confidence that this book 'Microbes in Process' with collection of articles on a wide variety of allied topics concerning modern society will be a valuable tool for researchers, teachers, students, policy makers and for those who are interested to work in the field of microbiology, microbial biotechnology, bioengineering and applied microbiology.

We would like to convey our sincere gratitude to all the subject experts and writers for their erudite contributions. Assistance rendered by our research scholars is thankfully acknowledged. We extend our sincere appreciation to Dr. Carra Faegaiga of Nova Scientific Publishers, USA for her valuable support to facilitate the completion of this book.

28 August 2013
Dr. Neelam Garg
Dr. Abhinav Aeron

In: Microbes in Process
Editors: Neelam Garg and Abhinav Aeron

ISBN: 978-1-63117-127-7

Chapter 1

Industrial Microbiology and Biotechnology

Shadia M. Abdel-Aziz*
Microbial Chemistry Department, Genetic Engineering and Biotechnology Division, National Research Center, Dokki, Cairo, Egypt

Abstract

Microorganisms have been used for several decades in large-scale production of a variety of biochemicals ranging from alcohols to antibiotics, and in the processing of foods and feeds. Use of microbes to obtain a product or service of economic importance constitutes "Industrial Microbiology or Microbial Biotechnology". With respect to the scope, objectives, and activities, 'Industrial Microbiology' is synonymous to 'fermentation', as fermentation includes any process mediated by or involving microorganisms in which a product of economic value is obtained. Developing functional foods to improve public health requires contributions from ongoing basic and applied research and modifications to the current regulatory framework to facilitate the review of new functional components and their health claims. Additional biomarkers that signal changes in health status are urgently needed, and the meaning of changes in those biomarkers must be clearly demonstrated relative to a defined health condition. Research is needed to expand the validated biomarkers of health status including assessing how genes and gene products relate to disease risk.

1. Introduction

Microorganisms are a collection of organisms that share the characteristic of being visible only with a microscope. They constitute the subject matter of microbiology. Members of the microbial world are very diverse and include the bacteria, cyanobacteria, rickettsiae, chlamydiae, fungi, unicellular (single-celled) algae, protozoa, and viruses. The majority of

* Corresponding author: abdelaziz.sm@gmail.com.

microorganisms contribute to the quality of human life by maintaining the balance of chemical elements in the natural environment, by breaking down the remains of all that dies, and by recycling carbon, nitrogen, sulfur, phosphorus, and other elements [1]. Some species of microorganisms cause infectious disease. They overwhelm body systems by sheer force of numbers, or they produce powerful toxins that interfere with body physiology. Viruses inflict damage by replicating within tissue cells, thereby causing tissue degeneration.

Microbiology has had a long, rich history, initially centered in the causes of infectious diseases but now including practical applications of the science. Many individuals have made significant contributions to the development of microbiology [2]. In early history of microbiology, Historians are unsure who made the first observations of microorganisms, but the microscope was available during the mid-1600s, and an English scientist named *Robert Hooke* made key observations. In the 1670s and the decades thereafter, a Dutch merchant named *Anton van Leeuwenhoek* made careful observations of microscopic organisms. Until his death in 1723, van Leeuwenhoek revealed the microscopic world to scientists of the day and is regarded as one of the first to provide accurate descriptions of protozoa, fungi, and bacteria.

Microorganisms are used to produce human or animal biologicals such as insulin, growth hormone, and antibodies. Microorganisms may also help in the treatment of diseases such as cancer. Research shows that *Clostridia* can selectively target cancer cells. Various strains of non-pathogenic *Clostridia* have been shown to infiltrate and replicate within solid tumours [3]. *Clostridia* therefore have the potential to deliver therapeutic proteins to tumours. *Lactobacillus* spp. and other lactic acid bacteria possess numerous potential therapeutic properties including anti-inflammatory and anti-cancer activities [4]. Members of the *Streptomyces* genus are among the most prolific microorganisms producing secondary metabolites with wide uses in medicine and in agriculture. Genomic studies, genomic mining and biotechnological approaches are being employed in the search for new antibiotics and other drugs in *Streptomyces* [5]. Microbial vaccines are also essential to combat infectious diseases. Recombinant DNA technology has allowed for the production of vaccines that offer protection without risk of infection [6] and new vaccines are being developed by this technique every day.

2. Industrial Microbiology

Industrial Microbiology or Microbial Biotechnolgy encompasses the use of microorganisms for the production of industrially important substances such as antibiotics, food products, enzymes, amino acids, vaccines, etc.

2.1. Definition

With respect to the scope, objectives, and activities, "Industrial Microbiology" is synonymous to "fermentation", as fermentation includes any process mediated by or involving microorganism in which a product of economic value is obtained [7]. Degradation of organic wastes, detoxification of industrial wasters and toxic compounds, degradation of oil spills, biocontrol agents, biofertilizers, etc., all come under industrial microbiology. Industrial Microbiology is also includes the application of scientific and engineering principles to the processing of materials by microorganisms. Many groups of microorganisms form products that range from organic acids to enzymes used to create various sugars, amino acids [8]. Industrial microbiology has a wide variety of applications. They can be broadly categorized into: 1) metabolite production; 2) biotransformation; 3) production of biofuels; 4) treatment of organic and industrial wastes; 5) recovery of metals; 6) production of microbial biomass (microbial protein or single cell protein) for food and feed; 7) production of biocontrol agents; and 8) fermentation of food products.

2.2. Development of Microbiology

In the late 1800s and for the first decade of the 1900s, scientists seized the opportunity to further develop the germ theory of disease as enunciated by Pasteur and proved by Koch. There emerged a ***Golden Age of Microbiology*** during which many agents of different infectious diseases were identified. Many of the etiologic agents of microbial disease were discovered during that period, leading to the ability to halt epidemics by interrupting the spread of microorganisms. Despite the advances in microbiology, it was rarely possible to render life-saving therapy to an infected patient. Then, after World War II, the **antibiotics** were introduced to medicine. The incidence of pneumonia, tuberculosis, meningitis, syphilis, and many other diseases declined with the use of antibiotics [2]. Work with viruses could not be effectively performed until instruments were developed to help scientists see these disease agents. In the 1940s, the electron microscope was developed and perfected. In that decade, cultivation methods for viruses were also introduced, and the knowledge of viruses developed rapidly. With the development of vaccines in the 1950s and 1960s, such viral diseases as polio, measles, mumps, and rubella came under control.

2.3. Modern Microbiology

Modern microbiology reaches into many fields of human endeavor, including the development of pharmaceutical products, the use of quality-control methods in food and dairy product production, the control of disease-causing microorganisms in consumable waters, and the industrial applications of microorganisms. Microorganisms are used to produce vitamins, amino acids, enzymes, and growth supplements [2]. One of the major areas of applied microbiology is biotechnology. **Biotechnology**: The term biotechnology was introduced in 1917 by Karl Ereky, a Hungarian Engineer [9]. Ereky defined biotechnology as "all lines of work by which products are produced from raw materials with the aid of living things" [10]. Biotechnology is essentially the product of interaction between the science of biology and technology. The products of biotechnology are going to have a major impact on the quality of human life, productivity, trade and economics in the world. Microorganisms are used as living factories to produce pharmaceuticals that otherwise could not be manufactured [10]. Microbial strains for the production of e.g. fine chemicals, biofuels, secondary metabolites and proteins can now be developed by a rational design strategy. Biotechnology is the application of biochemistry, biology, microbiology and chemical engineering [11] to industrial process and products and on environment.

3. Microbial Metabolites

Microbial metabolites are the molecules of interest produced by microorganisms including Enzymes, Amino acids, Vitamins, Ethanol, Organic acid, Antibiotics, Toxins and Biopolymers etc. Among these enzymes are essential in the metabolism of all living organisms and are widely applied as processing aids in the food and beverage industry [12]. Enzymes are biological catalysts used to facilitate and speed up metabolic reactions in living organisms. They are proteins and require a specific substrate on which to work. Their catalysing conditions are set within narrow limits, e.g. optimum temperature, pH conditions and oxygen concentration. Most enzymes are denatured at temperatures above 42°C. However, certain bacterial enzymes are tolerant to a broader temperature range. Bacteria and fungi are exploited and used for the commercial production of a diversity of enzymes. Several strains of microorganisms have been selected or genetically modified to increase the efficiency with which they produce enzymes. Microbial metabolites are broadly divided under two categories as Primary and Secondary metabolites.

3.1. Primary Metabolites

When microbial growth in a culture is in exponential phase (trophophase) many intermediate metabolic products are produced, which are further required for growth or for energy-yielding catabolism. These metabolites produced during trophophase are known as 'primary metabolites'. Amino acids, nucleotides, proteins, carbohydrates, lipids, vitamins, etc. are the primary metabolites needed for the growth of the culture, whereas acetone, ethanol, butanol, organic acids, etc., are those required for deriving energy. The quantity of some of these metabolites exceeds much more than that required by the producers.

3.1.1. Vitamins

Vitamins and other growth stimulating compounds are synthesized by all phototrophic microorganisms, using the chemical constituents from the culture medium. These compounds synthesized in excess of their requirements, accumulate in cultures and are recovered. Such microorganisms are now commercially exploited to produce vitamins on a large scale, under different cultural conditions. Examples of microbiologically produced vitamins are carotene, the precursor of vitamin A from *Blakeslea trispora*, riboflavin from *Ashbya gossrypii*, L-sarbose in vitamin C synthesis from *Gluconobacter oxidans*, and vitamin B12 from *Bacillus coagulans*, *B megaterium*, *Pseudomonas denitrificans* and *Streptonryces olivaceous*.

3.1.2. Amino acids

Microbial fermentation of substrates like glucose, acetate and sucrose by various microorganisms can yield D and L-alanine, L-glutamic acid, L-leucine, L-lysine and L-threonine. *Microbacterium ammoniaphilum* ferments glucose to give D and L-alanine.

Contents

Fermentation of the same substrate by *Micrococcus glutamicus* yields L-glutamic acid, and *Brevibacterium lactofermentum* yields L-leucine. Acetate is fermented by *Brevibacterium flavum* producing L-lysine. *Escherichia coli K12* ferments glucose and give L-threonine.

3.1.3. Ethanol

Manufacture of ethanol for human consumption by alcoholic fermentation dates back to the prehistoric times. In recent years, it is also used for industrial, laboratory and fuel purposes. Ethanol is nowadays considered synonymous with energy and has been used as an important chemical feed stock and as a fuel supplement. Generally, some sugar-rich products are fermented by the yeast *Saccharomyces cerevisiae* or rarely *Kluyveromyces fragilis*. Many strains of bacteria and fungi are also known to produce small quantities of ethanol. However bacteria and fungi are not used for ethanol production, since along with alcohol, some undesirable products are also generated. Anaerobic fermentation converts carbohydrate into ethanol through pyruvate and acetaldehyde. Ethanol is widely used as a solvent, extractant and antifreeze. Ethanol forms the base for many solvents of dyes, pharmaceuticals, lubricants, detergents, pesticides, plasticizers, explosives and resins and for the manufacture of synthetic fibres.

3.1.4. Organic acids

Metabolism of carbohydrate results in the production of organic acids which accumulate in the culture medium. This is then separated and purified. Organic acids are manufactured on a large scale and are marketed as pure chemicals or their respective salts. Thus, they offer a great promise for future development. The organic acids are either the terminal products of EMP pathway (lactic acid, propionic acid) or products of incomplete oxidation of sugars (citric acid, itaconic acid, and gluconic acid). Organic acids find their application in a wide variety of industries, research, etc. For example, acetic acid is used in food industry and research. Fumaric acid is used in resins. Calcium, iron and potassium salts of gluconic acid are used in medicines and also in cleaning bottles. Lactic acid is used in food industry (like preparation of fruit extracts, syrup and pickles), in dye mordant, in tanning, in decalcifying skins, in plastics and in medicine as calcium and iron lactates. Citric acid is used in food industry (e.g. preparation of fruit drinks, confectionary, jams, jellies, preserved fruits, candies, wines, etc.), in pharmacy (e.g. blood transfusion, effervescent products), in cosmetics (e.g. astringent lotions, shampoos and hair setting fluids) and in industries (e.g. electroplating, leather tanning, cleaning of pipes, etc.).

3.2. Secondary Metabolites

A growing microbial culture, after trophophase, enters into idiophase. The cells are now slow-growing or non-growing and produce certain substances other than primary metabolites. These are called secondary metabolites and play no role in the growth of the microorganisms. Toxins, gibberellins, alkaloids and antibiotics are examples of secondary metabolites or idiolites. Secondary metabolites are produced by a limited number of microorganisms, when there is depletion of one or more nutrients in the culture medium. Secondary metabolites include:

3.2.1. Antibiotics

Antibiotics are metabolites having preferential antimicrobial activity. The phenomenon of growth inhibition by antibiotics is known as 'antibiosis'. Antibiotics are defined as 'the complex chemical substances, which are produced by microorganisms as secondary metabolites and act against other microorganisms. The first antibiotic to be isolated was Penicillin from *Penicillium notatum* by Alexander Flemming in 1929. Penicillin was produced on a large scale using culture of *P. notatum* during the 1940s. The high demand for chemotherapeutic agents to treat wound infections during World War II led to the development of a production process for penicillin and this marked the beginning of an era of antibiotic research and continues to be the most fascinating area of microbial biotechnology even today. Biotransformations are now routinely employed for the commercial production of several useful antibiotics, e.g. production of semi-synthetic penicillin and cephalosporins.

3.2.2. Toxins

Microbial toxins are toxins produced by micro-organisms, including bacteria and fungi. Microbial toxins cause infection and disease by directly damaging host tissues and by disabling the immune system. Bacterial toxins have been defined as “soluble substances that alter the normal metabolism of host cells with deleterious effects on the host”. The major symptoms associated with some diseases caused by *Vibrio cholerae* (cholera), *Bacillus anthracis* (anthrax), *Clostridium botulinum* (botulism), *Clostridium tetani* (tetanus), enterohemorrhagic *Escherichia coli* (bloody diarrhea and hemolytic uremic syndrome), and *Staphylococcus aureus* (a common cause of skin infections, respiratory disease, and food poisoning) are all related to the activities of the toxins produced by these organisms. With the recognition of the central role of toxin in these and other diseases has come the application of inactive toxins (toxoids) as vaccines. Such toxoid vaccines have had an important positive impact on public health. On other hand, however, microbial toxins also have important uses in medical science and research. Potential applications of toxin research include combating microbial virulence, the development of novel anticancer drugs and other medicines, and the use of toxins as tools in neurobiology and cellular biology.

3.2.3. Biopolymers

A huge variety of biopolymers, such as polysaccharides, polyesters, and polyamides, are produced by microorganisms. These products range from viscous solutions to plastics. The genetic manipulation of microorganisms has permitted the biotechnological production of biopolymers with tailored material properties suitable for high-value medical application such as tissue engineering and drug delivery. Industrial microbiology can be used for the biosynthesis of xanthan, alginate, cellulose, cyanophycin, poly-glutamic acid, levan, hyaluronic acid, organic acids, oligosaccharides and polysaccharides [13]. On other hand, the production and sensory benefits of exopolysaccharides from lactic acid bacteria is well established. Functional food products that offer health and sensory benefits beyond their nutritional composition are becoming progressively more important to the food industry. There is evidence for the health properties that are attributable to exopolysaccharides from LAB [8].

4. Microbial Products

The fungi: *Trichoderma barzianum, T. ressei*, and *T.viride* effectively produce cellulases, whereas some algae such as *Chlorella sorokiniana, Spirulina maxima,* and *S. platensis* are used for the production of single cell protein (SCP). The yeast *Saccharomyces cerevisiae* is used for the production of SCP as well as ethanol, while the bacterium, *Bacillus polymyxa*, produces antibiotic such as polymixin [secondary metabolite] and the enzyme amylase (Figure 1).

On other hand, some microbial products of realized potential importance are revealed in Figure 2.

Microorganisms	Primary Metabolites	Secondary Metabolite (Antibiotics)	Secondary Metabolites (Enzymes)
Bacteria *Acetobacter aceti* *B.polymyxa* *Gluconobacter suboxidans* *Methylophilus methylotrophus* *Pseudomonas denitrificans*	Acetic acid Vinegar Glutamic acid Vitamin B_{12}	Polymixin B	Amylase
Fungi *Asergillus niger* *A. oryzae* *Candida lipolytica* *Gibberella fujikuroi* *Streptomyces lipolytica* *Saccharomyces cerevisiae*	Citric acid Gibberellin Citric acid Ethanol		Amylase, cellulose Lipase Wine, bakery.
Actinomycetes *Streptomyces tradine* *Nocardia mediterranei* *Micromonosporsa purpurea*		Neomycin Refamycin Gentamicin	

Figure 1. Primary and Secondary Metabolites of Some Microorganisms.

Products/Activities	Examples
Amino acids	L-glutamic acid, L-lysine.
Antibiotics	Streptomycin, penicillin, tetracycline, polymyxin.
Beverages	Wine, beer, distilled beverages.
Biodegradable plastic	Polyhydroxybutyrate
Enzymes	Amylase, proteases, pectinases, invertase, cellulose.
Flavouring agents	Monosodium glutamate, nucleotides.
Foods	Cheese, pickles, yoghurt, bread, vinegar.
Gases	CO_2, H_2, CH_4
Organic acids	Lactic, citric, acetic, butyric, fumaric.
Organic solvents	Acetone, ethanol, butanol, amyl alcohol.
Others	Glycerol, fats, steroids, gibberellins.
Vitamins	B_{12}, riboflavin, A
Recombinant proteins	Insulin, interferon, subunit vaccines.
Substrates	A wide range of compounds used for chemical syntheses of valuable products.
Biomass	Food and feed yeast, other organisms used as single cell protein.
Vaccines	A variety of viral and bacterial vaccines.
Biotransformation	Steroids, antibiotics, D-sorbitol.
Degradation	Disposal of biological and industrial compounds, petroleum.
Solubilization/ accumulation	Improved recovery of oil and metals, discovery of new oil reserves, removal of toxic metals.

Figure 2. Microbial Products of Realized Potential Importance.

4.1. Microbes Used in Food Industry

Microorganisms are widely used in the food industry to produce various types of foods that are both nutritious and preserved from spoilage because of their acid content. Food industry is the complex, global collective of diverse businesses that together supply much of the food energy consumed by the world. Many studies have been performed to assure the safety of foods [14].

4.1.1. Biotechnology in Food Fermentation

Microorganisms are an integral part of the processing system during the production of fermented foods. Microbial cultures can be genetically improved using both traditional and molecular approaches, and improvement of bacteria, yeasts and moulds is the subject of much academic and industrial research. Traits which have been considered for commercial food applications in both developed and developing countries include sensory quality (flavour, aroma, visual appearance, texture and consistency), virus (bacteriophage) resistance in the case of dairy fermentations, and the ability to produce antimicrobial compounds (e.g. bacteriocins, hydrogen peroxide) for the inhibition of undesirable microorganisms [12]. In many developing countries, the focus is on the degradation or inactivation of natural toxins, mycotoxins (in cereal fermentations) and anti-nutritional factors (e.g. phytates).

4.1.2. Biotechnology in the Production of Food Ingredients

Flavouring agents, organic acids, food additives and amino acids are all metabolites of microorganisms during fermentation processes. Microbial fermentation processes are therefore commercially exploited for production of these food ingredients. Metabolic engineering, a new approach involving the targeted and purposeful manipulation of the metabolic pathways of an organism, is being widely researched to improve the quality and yields of these food ingredients [12]. It typically involves alteration of cellular activities by the manipulation of the enzymatic, transport and regulatory functions of the cell using recombinant DNA and other genetic techniques. Understanding the metabolic pathways associated with these fermentation processes, and the ability to redirect metabolic pathways, can increase production of these metabolites and lead to production of novel metabolites and a diversified product base.

4.1.3. Nutrition and Food Safety

Fermentation processes enhance the nutritional value of foods through the biosynthesis of vitamins, essential amino acids and proteins, through improving protein and fibre digestibility; enhancing micronutrient bioavailability and degrading anti-nutritional factors [12]. Many bacteria in fermented foods also exhibit functional properties (probiotics). The safety of fermented food products is enhanced through reduction of toxic compounds, such as mycotoxins and production of antimicrobial factors, such as bacteriocins, carbon dioxide, hydrogen peroxide and ethanol, which facilitate inhibition or elimination of food-borne pathogens.

Development and improvement of fermentation processes is currently taking place worldwide. Moreover, many of the microorganisms associated with traditional fermentation processes in developing countries are unique. Biotechnological innovations have greatly assisted in industrializing production of certain indigenous fermented foods. The results of biotechnology research led to fermented foods of improved quality, safety and consistency. Some of the most important microbes used in food industry are represented in figure 3 (scribed.com/scribed).

4.2. Functional Foods

While food has long been used to improve health, our knowledge of the relationship between food components and health is now being used to improve food. Strictly speaking, all food is functional, in that it provides energy and nutrients necessary for survival. But the term "functional food" in use today conveys health benefits that extend far beyond mere survival. Most of the functional foods are fortified with "Probiotic Bacteria" such as *Lactobacillus casei, Streptococcus thermophilus* and *Lactobacillus bulgaricus.* Food and nutrition science has moved from identifying and correcting nutritional deficiencies to designing foods that promote optimal health and reduce the risk of disease [15]. The functional food industry, consisting of food, beverage and supplement sectors, is one of the several areas of the food industry that is experiencing fast growth in recent years. It is estimated by BCC Research that the global market of functional food industry will reach 176.7 billion in 2013 [16]. The definition of "disease or health-related condition" reveals that a "disease" is "damage to an organ, part, structure, or system of the body does not function properly (e.g., cardiovascular disease), or a state of health leading to such defective function (e.g., hypertension); except that diseases resulting from essential nutrient deficiencies (e.g., scurvy, pellagra) are not included. Sound science must underlie the development, marketing and regulation of these new functional foods to protect and inform consumers. Regulatory policies must ensure the safety and efficacy of products and the accuracy of their marketing claims.

The Food and Drug Administration (FDA) regulations (in USA) allow certain disease-risk-reduction claims, known as "health claims," to appear in food labeling. Health claims suggest a functional food will has a preventive effect with respect to a disease. Health claims that FDA has approved generally have been claims to reduce the risk of a named disease, e.g., calcium osteoporosis; dietary lipids and cancer; sodium and hypertension; dietary saturated fat and cholesterol and coronary heart disease. The functional foods include several definitions that help clarify the issues: Functional foods, Dietary supplements, Fortified foods, Fortified foods, Nutraceuticals, Medical foods.

4.2.1. Cholesterol-lowering breakfast cereals and Margarines

Several companies have marketed breakfast cereals as functional foods when enriched with inulin (an oligosaccharide) or psyllium (soluble fiber). Plasma total cholesterol was reduced by 7.9 percent and total triglycerides by 21.2 percent in young men consuming a 50 g serving of a breakfast cereal fortified with 18 percent inulin for four weeks. *Bifidobacteria* in the stools also increased and appeared to explain the lipid-lowering effect. A study of a breakfast cereal containing 50 g of soluble fiber as psyllium, oat and barley showed that it significantly reduced total cholesterol and low density lipoprotein concentrations in the blood as compared to a wheat bran cereal for middle-aged mildly hypercholesterolemic men who were already eating a diet low in saturated fat [17]. Plant sterol-enriched margarine, originally positioned as dietary supplements, was shown to reduce blood cholesterol levels in mildly hyper-cholesterolemic subjects in a human trial [18].

4.2.2. Probiotic Beverages

A dietary supplement beverage containing *Lactobacillus casei, Streptococcus thermophilus* and *Lactobacillus bulgaricus* has been marketed as a dietary supplement. This product, with its unique combination of cultures, including *L. casei,* has been clinically proven to fortify human natural defenses. A patented probiotic bacterium, *L. casei,* has been shown in human trials to produce an optimized humoral response in the immune system [19] and a heightened immune response to rotavirus-induced diarrhea in children [20].

4.3. Functional Dairy Foods

Milk is sterile at secretion in the udder but contamination of the milk by microorganisms can take place during milking, handling, storage, and other pre-processing activities. Milk is potentially susceptible to contamination with pathogenic microorganisms. Information on the microbial content of milk can be used to judge its quality and the conditions of production. Certain microorganisms produce desirable chemical changes in the production of dairy products such as cheese, yogurt. Microorganisms in milk include lactic acid bacteria which are normally present in milk and are also used as starter cultures in the production of cultured dairy products such as yogurt. These groups of bacteria, classified as *Lactococci*; *Lactobacilli*; or *Leuconostoc*, are able to ferment lactose to lactic acid. Coliforms, facultative anaerobes with an optimum growth at 37° C, are indicator organisms and can cause rapid spoilage of milk because they are able to ferment lactose with the production of acid and gas, and are able to degrade milk proteins. Spoilage is a term used to describe the deterioration of a food's texture, colour, odour or flavour where it becomes unsuitable and hazardous for human consumption. Microbial spoilage of food often involves the degradation of protein, carbohydrates, and fats by the microorganisms or their enzymes. In milk, microorganisms that are principally involved in spoilage are psychrotrophic organisms where most of them are destroyed by high temperature (pasteurization), however, some bacteria such as *Pseudomonas fluorescens, Pseudomonas fragi* can produce proteolytic and lipolytic extracellular enzymes which are heat stable and capable of causing spoilage. Some species and strains of *Bacillus, Clostridium, Cornebacterium, Arthrobacter, Microbacterium, Micrococcus*, and *Streptococcus* can survive pasteurization and grow at refrigeration temperatures which can cause spoilage problems. Other pathogenic microorganisms such as *Bacillus cereus, Listeria monocytogenes, Yersinia enterocolitica, Salmonella* spp., *Escherichia coli* O157, are still of concern today in raw milk and other dairy products. It should also be noted that moulds, mainly of species of *Aspergillus, Fusarium,* and *Penicillium* can grow in milk and dairy products. If the conditions permit, these moulds may produce mycotoxins which can be a health hazard.

Product	**Microbes**	**Description**
Acidophilus milk	*Lactobacillus acidophilus*	Produced by adding microbesto fluid whole, low fat, or skim milk under controlled conditions.
Butter	*Streptococcus cremoris; Lactobacillus acetylactics* to give it its characteristic flavour and odour.	Made from pasteurized cream, to which a lactic acid starter has been added.
Cheese	*Lactococcus lactis* and *Leuconostoc cremoris* are most often use.	Made by the lactic acid fermentation of milk.
Monosodium glutamate	*Corynebacterium glutamicum*	Produced by fermentation of starch, sugar beets, sugar cane, or molasses.
Baker's yeast	*Saccharomyces cerevisiae*	Provides a variety of enzymes that enable carbohydrates to be broken down producing sufficient carbon dioxide to give bread its characteristic texture.
Olives	*Lactobacillus plantarum* and *Lactobacillus mesenteroides*	Edible only after fermentation.
Tempeh	*Rhizopus oligosporus*	Fermentation process that binds soybeans into a cake form.
Soy sauce	*Lactobacillus delbrueckii, Aspergillus oryzae, Aspergillus soyae,* and *Saccharomyces rouxii.*	Mixture of soy beans and rice fermented by a variety of bacteria and fungi.
Coffee and chocolate	*Erwinia dissolvens, leuconostoc* , and *lactobacillus* species plus the yeasts of the genus *Saccharomyces* to remove the tough outer coats.	The microbes do not affect the taste of coffee but are necessary to confer the characteristic taste to cocoa and chocolate, through fermentation.

Figure 3. Microbes Used in Food Industry.

Microorganism	Antibiotic	Applications
Bacteria		
Bacillus brevis	Tyrothricin, Gramicidin	Mouth and throat infection.
B. polymyxa	Polymyxin B	Gastroenteritis.
B. subtilis	Bacitracin	Dematitis, superfical pyogenic infection, dtsentery.
Streptocous cremoris	Nisin	In cheese, food preservation.
Actionomycetes		
Micromonospora purpurea	Gentamycin	Abscess.
Nocardia mediterranei	Rifamycin	Meningitis
Streptomyces griseus	Streptomycin	Tuberculosis
S.aureofaciens	Tetracyclines	Cholera, tetanus
S. erytbreusa	Erythromycin	Choilera, tetanus, arthritis
S. noursei	Nystatin	Skin lesions
S. spheroids	Novobiocin	Abscess
Fungi		
Cephalosporium acremonium	Cephalosporins	Pneumonia, meningitis
Pencillium crysogenum	Penicillin	Pneumonia, pharyngitis
P. griseofulum	Griseofulin	Skin and hair lesions
P. notatum	Pencillin	Fever, pneumonia, genital infection.

Figure 4. Some Antibiotics Produced by Microorganisms.

Current commercial development in dairy functional foods is large and growing rapidly. Milk is a major provider of important nutrients for humans, including calcium, protein and riboflavin, however, it is now recognized that dairy products and ingredients derived from milk and whey can provide functional food products with beneficial effects on human health as well as nutritional benefits. The role of traditional dairy products [milk, yogurt, fermented dairy drinks, infant milk formulae, cheese and ice-cream] as functional foods is established. Probiotic bacteria and prebiotic carbohydrates are commonly used in yogurts and fermented drinks. The clinical evidence for a wide range of commercial strains of *lactobacilli* and *bifidobacteria* is extensively reviewed and their ability to improve health is assessed [21].

Future trends facing the emerging functional dairy foods and ingredients sector include: 1) development of synergistic probiotic and prebiotic products (synbiotics); 2) new prebiotic carbohydrates more able to mimic human milk oligosaccharides; 3) novel prebiotic proteins and peptides; 4) closer alliance with the glycoscience industry; 5) the emergence of genetically modified milk product; probiotics as carriers of biologicals and vaccines; and 6) the regulatory issues surrounding the introduction of health claims for new functional dairy foods [22].

5. Antibiotics Produced by Microrganisms

Biotechnology has wide range of applications in animal and human health care both directly as well as indirectly. Its direct use involves controlling the spread of a variety of diseases using gene manipulation at different levels. The indirect use is in manufacturing of nutritional medical foods, single cell proteins, and in detecting food contaminants through molecular probes. Following approaches are used to control the diseases and thereby taking care of the human and animal health.

a) Providing immunity against the disease through the development of vaccine and antibiotics.
b) Early diagnosis of a disease.
c) Successful treatment of the disease.

Industrial Microbiology is perhaps best known for its development of antibiotics and pharmaceutical agents. Penicillin, streptomycin, and a host of other antimicrobial agents originated from industrial microbiology in the 1950's and 1960's [10]. New research directions are aimed at discovering microbial metabolites (with pharmacological activities) useful in the treatment of hypertension, obesity, coronary heart disease, cancer, and inflammation. Some of the antibiotics produced by microorganisms are represented in Figure 4.

6. Biotechnology of Enzymes

Enzymes are biological catalysts which accelerate the biochemical reactions in living cells. A purified enzyme can also carry on a specific biochemical reaction outside the cell. Due to this property of the enzymes, they are widely used in laboratory and in the commercial production of important biochemical compounds, drugs and industrial products. This has made enzyme research an important area of biotechnology. Various techniques are being used to modify the enzymes through enzyme modeling and gene manipulation for various specific purposes [10]. Industrial applications of enzymes include the production of cheese, the clarification of apple juice, the development of more efficient laundry detergents, pulp and paper production, and the treatment of sewage. These processes have been dramatically enhanced by the use of recombinant DNA techniques to design enzymes of increased activity, stability, and specificity. Commercial enzymes are used in textile industry as softening agents (amylases), as detergents to remove proteinaceous stains (proteases). Enzymes isolated from certain bacteria are used commercially to produce acetone, butanol, lactic acid, citric acid etc. In meat and beer processing, the proteases "Papain" (from papaya) and "bromelain" (from pineapple) are used to tenderize meat. These enzymes hydrolyze the peptide bonds.

7. Genetically Engineered Bacteria

The origins of biotechnology culminated with the birth of genetic engineering. There were two key events that have come to be seen as scientific breakthroughs beginning the era that would unite genetics with biotechnology. One was the 1953 discovery of the structure of DNA, by Watson and Crick, and the other was the 1973 discovery by Cohen and Boyer of a recombinant DNA technique by which a section of DNA was cut from the plasmid of an *E. coli* bacterium and transferred into the DNA of another [23]. This approach could, in principle, enable bacteria to adopt the genes and produce proteins of other organisms, including humans.

7.1. Metabolic Engineering

Metabolic engineering is the improvement of cellular activities by manipulation of enzymatic, transport and regulatory functions of the cell with the use of recombinant DNA technology. In order to get the overproduction of the desired metabolite, the metabolic pathways and bioprocesses of the cell are modified by the genetic manipulations in the cell. The overproduction of metabolite can also be achieved by mutation followed by selection of the improved strains of living organisms exhibiting the desired trait. However, metabolic engineering involves the introduction of heterologous genes and regulatory elements which leads to targeted alterations in normal cellular activities. The process has several components such as biosynthesis of a metabolite, manipulation of protein processing pathways etc. However, this field needs greater efforts to understand the complex interactions of metabolic networks.

7.2. Protein Engineering

The protein engineering is an important tool to produce superior enzymes and storage proteins. There are several important uses of protein engineering such as: a) to create superior enzymes with the capacity to catalyze the production of high value specific chemicals; b) to produce enzymes for large scale use in the chemical industry;c) to produce biological compounds including synthetic peptides, storage proteins and specific drugs that are superior to the natural ones. Protein engineering is also being used to create immunotoxins, which are the conjugates of cell binding antibody or antigens, covalently bound to a plant or a bacterial toxin. The procedure of fusion of genes using sequences coding for antibodies and toxic peptides is used to create these immunotoxins. The detailed studies of antimetabolites or inhibitors of DNA synthesis has led to the development of drugs for treatment of cancer, gout, malaria and some viral infections like Herpes. Drugs that inhibit the DNA synthesis and the cell division are very effective in cancer chemotherapy. Based on the same principle of inhibition of nucleic acid synthesis, drugs used for treatment of malaria, bacterial infections, and 'acyclovir' for herpes virus infection were also developed. The genetically engineered bacteria have found applications in various areas including:

7.2.1. Crop Production and Protection through Biological Control

Genetically engineered bacterial strains have been developed to control insects and fungal diseases. Engineered bacterial strains such as *Rhizobium melilotii, Pseudomonas fluorescens, Bacillus thuringiensis*, *Clavibacter xyli* are used for increasing the efficiency of N_2 fixation, biocontrol of fungal diseases, assessment of movement of bacteria for biological control and control of corn ear worm.

7.2.2. Production of Chemicals and Fuels like Antibiotics, Enzymes, Diagnostics

Genetically engineered strains of *Bacillus amyloliquefaciens* and *Lactobacillus casei* have been prepared for production of amino acids on a large scale. Industrially useful bacteria utilize cheaper feed stocks (substrates) like D-xylulose, cellulose, e.g. *Zymomonas mobilis* [normally incapable of using lactose] carrying cellulase gene from *Cellulomonas uda* has six fold increasing in cellulose activity. *E.coli* and *Klebsiella planticola* carrying genes from *Z. mobilis* could utilize glucose and xylose to give maximum yield of ethanol. Concerns about the Genetically Modified Organisms (GMOs): there are concerns regarding the biosafety, ethics and issues related to the release of GMOs in the environment.

In the past, time and again, there has been public outrage against the use of genetically modified or transgenic plants and other organisms. In 1999, a British medical journal published the adverse effect of genetically modified (GM) potato. This potato was found to contain snow drop lectin which affected the small intestine of the rats and stunted the growth and damaged their immune system. This led to worldwide public concern about this issue and created a lot of controversy about the safety of GM foods. The transgenic Bt-plants such as cotton, corn, soybean, and potato were approved for cultivation in USA. However, some countries did not allow Bt-plants in their fields e.g. Br-rice was not allowed in Philippines, Bt-cotton in France. Many Governments are also suspicious of the use of GMOs due to various reason-risks, societal beliefs, and economic concerns.

8. Microorganisms and Metals

Bacterial cells can tolerate high levels of metal concentrations due to the presence of specific mechanisms for survival [24]. Metal-microbe interactions have an important role in several biotechnological applications, including the fields of bioremediation, biomineralization, bioleaching, and microbial corrosion. These processes are commercially exploited in bacterial leaching operations such as the recovery of gold from arsenopyrite-pyrite ores [25]. The bacterial oxidation of minerals is important in the formation of acid mine drainage and in the extraction of gold, copper, and uranium from ores. Recent advances in the understanding of the role and application of microorganisms in the remediation of toxic metal include the use of natural and genetically engineered bacteria that are capable of mobilization and immobilization of metal ions. Effective techniques will require bacterial bioremediation coupled with effective chemical and engineering strategies [26]. Microbial-based technology presents an economic alternative for wastewater treatment industries [27]. It is well known that biological systems can provide a number of metal or metal-containing particles in the nanometer size range [28]. For example, many multicellular organisms use inorganic materials (such as calcium carbonate or silica) in combination with an organic matrix (proteins, lipids, or polysaccharides) to produce hard materials such as skeletal units, tooth, and bones.

9. Biotechnology and Environment

The term "Environment" is defined as our surroundings which includes the abiotic component (the non living) and biotic component (the living) around us. The abiotic environment includes water, air and soil while the biotic environment consists of all living organisms – plants, animals and microorganismsn [10]. Environmental pollution broadly refers to the presence of undesirable substances in the environment which are harmful to man and other organisms. In the past decade or two, there has been a significant increase in the levels of environmental pollution mostly due to direct or indirect human activities. Controlling the environmental pollution are the major focus areas of all the countries around the world. The efforts are not only on to use biotechnology to protect the environment from pollution but also to use it to conserve the natural resources. As we all know that microorganisms are known natural scavengers so the microbial preparations (both natural as well as genetically engineered) can be used to clean up the environmental hazards.

9.1. Development of Alternate Technologies Using Biotechnology

Biotechnology is being used to provide alternative cleaner technologies which will help to further reduce the hazardous environmental implications of the traditional technologies. E.g. some fermentation technologies have some serious environmental implications [10]. Various biotechnological processes have been devised in which all nutrients introduced for fermentation are retained in the final product, which ensures high conversion efficiency and low environmental impact.

9.1.1. In paper industry

The pulp bleaching technologies are being replaced by more environmentally friendly technologies involving biotechnology. The pulp processing helps to remove the lignin without damaging valuable cellulosic fibres but the available techniques suffer from the disadvantages of high costs, high energy use and corrosion [10]. A lignin degrading and modifying enzyme was isolated from *Phanerochaete chrysosporum* and was used, which on one hand, helped to reduce the energy costs and corrosion and on the other hand, increased the life of the system. This approach helped in reducing the environmental hazards associated with bleach plant effluents.

9.1.2. In Plastic industry

The conventional technologies use oil based raw materials to extract ethylene and propylene which are converted to alkene oxides and then polymerized to form plastics such as polypropylene and polyethylene [10]. There is always the risk of these raw materials escaping into the atmosphere thereby causing pollution. Using biotechnology, more safely raw materials like sugars [glucose] are being used which are enzymatically or through the direct use of microbes converted into alkene oxides. e.g. *Methylococcus capsulatus* has been used for converting alkene into alkene oxides.

9.2. Biological Procedures Achieved by Microorganisms

Biological processes such as Bioremediation, Biotransformation, Bioaccumulation are achieved by microorganisms using their inbuilt capability, which can be utilized for benefit of mankind.

9.2.1. Bioremediation

It is defined as 'the process of using microorganisms to remove the environmental pollutants where microbes serve as scavengers. The removal of organic wastes by microbes leads to environmental cleanup. The other names/terms used for bioremediation are biotreatment, bioreclamation, and biorestoration. The term "Xenobiotics" (xenos means foreign) refers to the unnatural, foreign and synthetic chemicals such as pesticides, herbicides, refrigerants, solvents and other organic compounds [10]. The microbial degradation of xenobiotics also helps in reducing the environmental pollution. Pseudomonas which is a soil microorganism effectively degrades xenobiotics. Different strains of *Pseudomonas* that are capable of detoxifying more than 100 organic compounds (e.g. phenols, biphenyls, organophosphates, naphthalene etc.) have been identified. Some other microbial strains are also known to have the capacity to degrade xenobiotics such as *Mycobacterium*, *Alcaligenes*, *Norcardia*.

9.2.2. Biotransformation

Biotransformation is the process of modification of an organic compound into a recoverable product by simple chemical reactions catalyzed by cellular enzymes. Transformations were practiced as early as 5000 BC in Babylonia to convert wine to vinegar. But it was only in 1864 that Pasteur showed that the process was mediated by microbes. By 1900, several bioconversions were known. Examples are: 1) conversion of ethanol to acetic acid and glucose, then to luconic acid; 2) conversion of Fructose to mannitol and tannin, then to gallic acid, etc.; this biotransformation is now more efficiently achieved by microbial process. Plant cells, animal cells as well as microbes are capable of performing biotransformations. But due to the rapid microbial growth and high metabolic rates of microorganisms, microbial process of biotransformation is far more efficient and economical. Biotransformation requires much lower energy needs since it occurs at ambient temperature (20-40°C) and so reduces cost. The most important advantage is that the entire process leads to much lower pollution problems.

9.2.3. Bioaccumulation and Bioconcentration

It refers to the accumulation of substances, such as pesticides, or other organic chemicals in an organism. Bioaccumulation occurs when an organism absorbs a toxic substance at a rate greater than that at which the substance is lost. Thus, the longer the biological half-life of the substance the greater the risk of chronic poisoning, even if environmental levels of the toxin are not very high [29]. On other hand, ***Bioconcentration*** is a more specific term, referring to uptake and accumulation of a substance from water alone. By contrast, bioaccumulation refers to uptake from all sources combined (e.g. water, food, air, etc.). Bioconcentration is occurred when the uptake from water is greater than excretion [30]. Absorption of a toxic substance by a microorganism is referred to as *Bioaccumulation*. Bioconcentration and bioaccumulation occur within an organism. For example, though mercury is only present in small amounts in seawater, it is absorbed by algae (generally as methylmercury). It is efficiently absorbed, but only very slowly excreted by organisms [31].

9.2.4. Treatment of sewage using microorganisms

The sewage is defined as the waste water resulting from the various human activities, agriculture and industries and mainly contains organic and inorganic compounds, toxic substances, heavy metals and pathogenic organisms. The sewage is treated to get rid of these undesirable substances by subjecting the organic matter to biodegradation by microorganisms. The biodegradation involves the degradation of organic matter to smaller molecules (CO_2, NH_3, PO^4 etc.) and requires constant supply of oxygen [10]. The process of supplying oxygen is expensive, tedious, and requires a lot of expertise and manpower. These problems are overcome by growing microalgae in the ponds and tanks where sewage treatment is carried out. The algae are capable of adsorbing certain heavy toxic metals due to the negative charges on the algal cell surface which can take up the positively charged metals.

10. Biotechnological Applications in Food Processing

Biotechnology includes a wide range of diverse technologies and they may be applied in each of the different food and agriculture sectors. It includes technologies such as gene modification (manipulation) and transfer; the use of molecular markers; development of recombinant vaccines and DNA-based methods of disease characterization/diagnosis; in-vitro vegetative propagation of plants; embryo transfer and other reproductive technologies in animals or triploidisation in fish. It also includes a range of technologies used to process the raw food materials produced by the crop, fishery and livestock sectors.

During fermentation processes, microbial growth and metabolism result in the production of a diversity of metabolites. These metabolites include enzymes (which are capable of breaking down carbohydrates, proteins and lipids present within the substrate and/or fermentation medium), vitamins, antimicrobial compounds (e.g. bacteriocins and lysozyme), texture-forming agents (e.g. xanthan gum), amino acids, organic acids (e.g. citric acid, lactic acid) and flavor compounds (e.g. esters and aldehydes). Many of these microbial metabolites are produced at the industrial level in both developed and developing countries for use in food processing applications [12].

Fermentation is globally applied in the preservation of a range of raw agricultural materials (cereals, roots, tubers, fruit and vegetables, milk, meat, fish etc.). Commercially produced fermented foods which are marketed globally include dairy products (cheese, yogurt, and fermented milks), sausages and soy sauce. Certain microorganisms associated with fermented foods, in particular strains of the *Lactobacillus* species, are probiotics which influencing the composition and metabolic activity of the flora of the gastrointestinal tract. Microorganisms and metabolic pathways associated with the production of fermented foods are the subject of considerable research, targeting strain isolation and identification; improvement of the efficiency of fermentation processes and the quality, safety and consistency of fermented foods. While microorganisms are beneficial in most fermentation processes, some may pose the risk of food contamination and can cause food-borne illness.

11. New Challenges of Industrial Microbiology

Although the development of the biotechnology in 20th century was, in fact, one of the first technologies developed, and the earliest example of these technologies, fermentation, is still the most economically important [32]. In spite of the length of time that has passed since biotechnology came into use, it is only in the past century that we have begun to understand how these innovations work can be improved [33]. Microbiology is the technological and scientific field that has been used to improve living conditions and to increase survival opportunities by transforming risks into challenges [34].

Microorganisms used to grow and adapt in the external environment (pH, temperature, pressure, feed flow, etc.) that do not affect the microbial machine itself [35]. Furthermore, all the processes focus on trying to avoid variations that might affect the yield. Therefore, mutation, degradation and contamination of the working microbial species are risks must be avoided.

Microbial processes are involved in food, pharmaceutical, fine-chemical, cosmetic, energy and new-material industries. Microbiological processes with industrial applications are usually much more advantageous than any other industrial processes, accounting for the strong continued interest in this field [36]. At first, the specific ability of a given microorganism to produce a particular molecule can be exploited. This molecule can be a secondary metabolite or, more directly, a byproduct of microbial metabolism. While chemical industrial processes may yield the same product, the process is often more complex, involving more pathway steps, and/or mixed with some undesirable byproducts. Secondly, the molecules obtained from microorganisms are synthesized under mild conditions, resulting in less energy and equipment costs compared with chemical synthetic processes.

In general, it is now possible to: (i) convert secondary reactions into main metabolic pathways; (ii) optimize productions and yields; (iii) change the original metabolic pathways to allow the use of less expensive raw materials, or to obtain previously unknown molecules; and (iv) use the enantiomeric properties of enzymes to obtain new chiral molecules (34). The long list of products obtained by these methods increases every year, and many industrial areas are involved. Products most commonly obtained in industrial microbiology include:

- Antibiotics: more than 20% of antibiotics, including biosynthetic penicillin V and natural penicillin G, can be produced by filamentous fungi. Antitumoral agents such as taxol, first discovered in plants but later transferred to and produced by *Taxomyces andreanae*. Mycotoxins including adrenalin inhibitors, estrogens and anabolic agents for cattle and sheep, or gibberellins can be used in brewing and malting industries.
- Pigments, including the carotenoid as taxanthin from *Phaffia rhodozyma*, and β-carotenoid from *Blakeslea trispora*, used in food and textile industries.
- Polyunsaturated-fatty-acids, including γ-linoleic acid produced by *Mucor circinelloides*, and arachidonic acid from *Mortierella isabellina*.

The field of biotechnology has had a lot of beneficial contribution in the area of healthcare, agriculture, food production, manufacture of industrial enzymes, and appropriate environmental management. However, the advancement in this field has also lead to some concerns and controversies raised by a number of groups. The main areas of consideration for safety aspects in biotechnology [10] are the following:

a) How to dispose off spent microbial biomass and purify the effluents from biotechnological processes?
b) The toxicity of the allergy associated with microbial production.
c) How to deal with the increase in the number of antibiotic resistant pathogenic microorganisms?
d) How to evaluate the pathogenicity of the genetically engineered microorganisms to infect humans, plants and animals?
e) How to prevent contamination, infection or mutation of the processed strains?
f) The evaluation of the interaction of the genetically engineered microbes with the elements of natural environment.

Conclusion

Mankind has learned over many millennia which of nature's products must be eaten and which not to eat. Mankind also discovered the value of drying, freezing, smoking, fermentation and acidification as means of preserving the food supply. The introduction of commercially processed foods and sterilized canned foods allowed the foods to be stored for long time periods. However, the heat-resistant spores of, for example, *Clostridium botulinum* cause food poisoning in canned foods. The microbiological risks associated with the consumption of any food depend on the types and number of microorganisms, or amounts of the microbial toxin[s] produced. The biotechnology and development of functional foods, such as milk products, to improve public health requires contributions from applied research and modifications to the current regulatory framework to facilitate the review of new functional components and their health claims. Development in industrial microbiology and biotechnology must be controlled by International Laws and healthcare must be ensured by the so-called Health claims.

References

[1] Brock, T. Madigan, M. Martinko, J. & Parker, J. (1994). *Biology of Microorganisms*. Prentice Hall, New Jersey

[2] Tyson, J. (2012). *Industrial Microbiology*. (www.jtyson.com)

[3] Sails, A. D. (2009). Applications in Clinical Microbiology. *Real-Time PCR:* Curr Technol Applications, Caister Academic Press. ISBN 978-1-904455-39-4

[4] Ljungh, A. & Wadstrom, T. (2009). *Lactobacillus Molecular Biology: From Genomics to Probiotics*. Ljungh A and Wadstrom T (eds): Caister Academic Press. ISBN 978-1-904455-41-7

[5] Dyson, P. (2011). *Streptomyces: Molecular Biology and Biotechnology*, Dyson, P (ed). Caister Academic Press. ISBN 978-1-904455-77-6
[6] Rappuoli, R. & Bagnoli, F. (2011). *Vaccine Design: Innovative Approaches and Novel Strategies*. Rappuoli R and Bagnoli F (eds) Caister Academic Press. ISBN, 978-1-904455-74-5
[7] Thackray, A. (1998). *Biotechnology and the Rise of the Molecular Sciences*. Philadelphia: University of Pennsylvania Press, ISBN 0-8122-3428-6
[8] Sonomoto, K. & Yokota, A. (2011). *Lactic Acid Bacteria and Bifidobacteria: Current Progress in Advanced Research*. Sonomoto and Yokota (eds) Caister Academic Press. ISBN 978-1-904455-82-0
[9] Fiechter, A. (2000). *History of Modern Biotechnology*. Fiechter A. (ed) p. 153. New York: Springer-Verlag Berlin Heidelberg ISBN 3-540-67793-3
[10] Biotechnology Society. (2009). *Biotechnology 4u*. (www.blogs.amlaan.com)
[11] Park I. (2010). *Microbes and Industrial Biotechnology*. A conference, Germany: (Microbes and Industrial Biotechnology 2010).
[12] FAO. Biotechnology Forum. (2004). *Electronic Forum on Biotechnology in Food and Agriculture: Conference 11. Food and Agriculture Organization of the United Nations:* (biotech-admin@ fao.org).
[13] Ullrich, M. (2009). *Bacterial Polysaccharides: Current Innovations and Future Trends*. Caister Academic Press. ISBN 978-1-904455-45-5.
[14] Baird-Parker, A. (1995). Development of industrial procedures to ensure the microbiological safety of foods. *Food Control*, *6*, 29-36.
[15] Mary, M. (2000). *Functional Foods and Nutraceuticals*. Nutrition Science News.
[16] Roberts, W. (2009). *Benefiting Beverages*. Prepared Foods August 2009.
[17] Roberts, D. Truswell, A. Bencke, A. Dewar H. & Farmakalidis, E. (1994).The cholesterol-lowering effect of a breakfast cereal containing psyllium fibre. *Med J Aust*, *161*, 600-64
[18] Law, M. (2000). Plant sterol and stanol margarines and health. *Br Med J*, *320*, 861-864
[19] Kaila, M. Isolauri E, Saxelin M, Arvilommi H, & Vesikari T. (1995). Viable versus inactivated *lactobacillus* strain GG in acute rotavirus diarrhoea. *Arch Dis Child*, *72*, 51-53.
[20] Spanhaak, S. Havenaar, R. & Schaafsma, G. (1998). The effect of consumption of milk fermented by *Lactobacillus casei* Strain Shirota on the intestinal microflora and immune parameters in humans. *Eur J Clin Nutr*, *53*, 899-907.
[21] Playne, M., Bennett L. & Smithers, G. (2003). Functional dairy foods and ingredients. *Aust J Dairy Technol*, *58*, 242-264.
[22] John, Wiley. (2000-2012). Foods from microorganisms.
[23] Grace, E. S. (1997). *Biotechnology Unzipped: Promises and Realities*. Washington, D.C.: Joseph Henry Press, ISBN 0-309-09621-9.
[24] Silver, S. (1996). Bacterial resistance to toxi metal ions – a review. *Gene*, *179*, 9-19.
[25] Dopson, M. & Lindstrom, E. (1999). Potential role of *Thiobacillus caldus* in arsenopyrite bioleaching. *Appl Env Microbiol*, *65*, 36-40.
[26] Stephen, J. & Macnaughton S. (1999). Developments in terrestrial bacterial remediation of metals. *Curr Opin Biotechnol*, *10, 230-233.*

[27] Tanja, K. Joerger, R, Olsson, E. & Granqvist, C. (2001). Bacteria as workers in the living factory: metal-accumulating bacteria and their potential for materials science. *Tren Biotechnol, 19, 15-20.*

[28] Dickson, DP. (1999). Nanostructured magnetism in living systems. *J Magn Mater*, *203*, 46-49.

[29] Bryan, G. Waldichuk, M., Pentreath, R. & Ann, D. (1979). *Bioaccumulation of marine pollutants*. Transactions of the Royal Society of London. Series B, Biological Sciences.

[30] Landrum, P. & Fisher, S. (1999). *Influence of lipids on the bioaccumulation and trophic transfer of organic contaminants in aquatic organisms: Lipids in fresh water ecosystems*. Springer Verlag New York.

[31] Croteau, M. Luoma, S. & Stewart, A. (2005). Trophic transfer of metals along freshwater food webs: Evidence of cadmium biomagnification in nature. *Limnol Oceanogr*, *50*, 1511-1519.

[32] Brown, C., Campbell, I. & Priest, F. (1987). Introduction to Biotechnology. Blackwell Scientific Publications, Oxford.

[33] Joseph, M. (2003). *The unacknowledged workers of the biotechnology industry: microbes*. (db@mail.Ubc. Ca)

[34] Jaume, M., Jacquet, S., Pierre, C. Mace, G., Lionel, S., Planbied. C., Demur, C., Pierre B. & Gilles, D. (2004). *Hepatol*, *40*, 1136-1143

[35] Demain, A. L. (1996). Fungal secondary metabolism: regulation and functions. In: Sutton B (ed.) *A century of mycology*, Cambridge Univ. Press, Cambridge MA, 233-254.

[36] Adrio, J. & Demain A. (2003). Fungal biotechnology. *Int Microbiol*, *6*, 191-199.

In: Microbes in Process
Editors: Neelam Garg and Abhinav Aeron

ISBN: 978-1-63117-127-7

Chapter 2

Valuable Products from Microbes

Sergio Sanchez[1,*] and Arnold L. Demain[2,†]
[1]Departamento de Biología Molecular y Biotecnología, Instituto de Investigaciones Biomédicas, Universidad Nacional Autónoma de México (UNAM), México D.F. México
[2]Research Institute for Scientists Emeriti (R.I.S.E.), Drew University, Madison, NJ, USA

Abstract

Microorganisms are amazing producers of natural products. They are crucial to the health and well-being of people throughout the world. They are excellent producers of primary metabolites such as amino acids, vitamins and nucleotides, and make secondary metabolites which constitute half of the pharmaceuticals on the market today and provide agriculture and livestock with many essential products. This chapter centers on these beneficial secondary metabolites, the discovery of which goes back eighty five years when penicillin was discovered by Alexander Fleming. They include antibiotics, hypocholesterolemics, anticancer drugs, immunosuppressants, stimulators of gastrointestinal motility, enzyme inhibitors, products acting against insects, antiparasitics, ruminant growth stimulants, and herbicides. The recent decrease in the introduction of new antibiotics and other secondary metabolites is mainly due to the loss of interest in this area by the major pharmaceutical companies. However, small biotechnology companies and academic laboratories are pursuing this goal and some new technologies offer hope for a solution; these include metagenomics and genome mining.

1. Introduction

A major contribution of microbes to the health and well-being of people began back in 1928, when Alexander Fleming discovered in a Petri dish seeded with *Staphylococcus aureus* that a compound produced by a mold killed the bacterium. The mold, *Penicillium notatum*, produced an active agent which was named penicillin. Fleming´s discovery began the

---

* E-mail: sersan@biomedicas.unam.mx.
† E-mail: ademain@drew.edu.

microbial drug era. By using the same method, other naturally occurring substances, like chloramphenicol and streptomycin, were later isolated. Naturally occurring antibiotics are produced by fermentation, an old technique which can be traced back almost 8000 years, initially for beer and wine production and recorded in the written history of ancient Egypt and Mesopotamia [www.britannica.com/eb/article-9106004/beer]. During the last 4000 years, *Penicillium roqueforti* has been utilized for cheese production and for the past 3000 years, soy sauce in Asia and bread in Egypt represented examples of traditional fermentations [1].

Natural products [NPs] with high commercial value can be produced from primary or secondary metabolism. Due to technical improvements in screening programs and separation and isolation techniques, the number of natural compounds discovered exceeds one million [2]. Among them, 50 to 60% are produced by plants (alkaloids, flavonoids, terpenoids, steroids, carbohydrates etc.) and 5% have a microbial origin. From all the reported natural products, about 20-25% show biological activity and of these, approximately 10% have been obtained from microbes. Microorganisms produce many compounds with biological activity. From the 22,500 biologically active compounds so far obtained from microbes, 45% are produced by actinomycetes, 38% by fungi and 17% by unicellular bacteria [2]. The role of microorganisms in the production of antibiotics and other drugs for treatment of non-infective diseases has been dramatic. Currently, with less than 1% of the microbial world having been cultured, there have been significant advances in microbial techniques for growth of uncultured organisms as a potential source of new chemicals [3]. Furthermore, metagenomics, i.e., the extraction of DNA from soil plants and marine habitats and incorporation into known organisms, is allowing access to a vast untapped reservoir of genetic and metabolic diversity [4, 5].

2. Natural Products and Chemically-Synthesized Drugs Derived from Them

Microbes have been always our friends. They have given us thousands of valuable natural drugs with novel structures and activities. In addition, by chemical modification, these structures have given rise to drugs with new attractive properties. Drugs of natural origin have been classified as (i) original NPs, (ii) products derived or chemically-synthesized from NPs, or (iii) synthetic products based on NP structures. The importance of NPs in the discovery of leads for the development of drugs for the treatment of human diseases is evidenced by the fact that close to half of the best-selling pharmaceuticals in 1991 either were NPs or their derivatives, especially in the oncological and antihypertensive therapeutic areas (6). From these drugs, 42% were NPs or their derivatives, and 67% of these were antibiotics. In this regard, NP derived drugs were well represented in the 35 top selling drugs reported between 2000-2002 and in 2007, about 37 new drugs were in clinical trials and evaluated for the treatment of cancer (51%) and bacterial infections (27%) [7]. Today, around 140,000 secondary metabolites have been structurally elucidated. A total of 13 NPs and NPs-derived drugs were approved for marketing worldwide between 2005 to 2007.

Through the control of infectious diseases and other medical conditions, secondary metabolites have exerted a major impact on the development of pharmaceutical industry. Their use has significantly contributed to the increase in the average life expectancy in the

USA, which went from 47.3 years in 1900 to 76.0 years (males) and 80.9 years (females) in the year 2009 [8]. Probably the most important use of secondary metabolites has been as anti-infective drugs. In the year 2000, the market for such anti-infectives was 55 billion dollars and in 2009, it was 79 billion dollars.

It should be pointed out that many chemically synthesized drugs owe their origin to natural sources. Figure 1 shows that among the anti-infective drugs, antivirals represented more than 20% of the market. Two antivirals, which are chemically synthesized today, were originally isolated from marine organisms. One is acyclovir (effective against the herpes virus by causing inhibition and inactivation of DNA polymerase) and the other is cytarabine (active against non-Hodgkin´s lymphoma). Both metabolites are nucleoside analogue drugs, originally isolated from sponges [9]. Additional antiviral applications of natural compounds are concerned with human immunodeficiency virus (HIV) treatment. In its pathogenesis, HIV-1, like other retroviruses, depends on its stable integration into the host genome to allow efficient replication of the viral RNA and therefore, maintain the infected state. Thus, the new viral DNA synthesized during reverse transcription is rapidly integrated into the host cell DNA (via a step referred to as "integration"), allowing transcription of viral RNA. In the late phase of HIV viral replication, the large precursor polyprotein (gag-pol precursor, Pr 160 of HIV) is appropriately cleaved by a viral protease to allow maturation and infectivity of the viral particle. If the precursor gag polyprotein is not adequately cleaved, the viral particles produced are not infective. To face this problem, a tremendous effort has been conducted at the U.S. National Cancer Institute (NCI), searching for natural metabolites capable of inhibiting both HIV reverse transcriptase and HIV protease. Such products are the main current agents used to control HIV. Furthermore, there have been reports published on NP inhibitors of HIV integrase obtained among the marine ascidian alkaloids, i.e., the lamellarins produced by terrestrial plants (*Baccharis genistelloides* and *Achyrocline satureioides*) and by a mollusc (*Lamellaria* sp.).

Compounds	Market (US dollars; billions)
Aminoglycosides	1.8
Antifungals and antiparasitics	4.2
Augmentin (amoxicillin + clavulanic acid)	2.1
Antivirals excluding vaccines	16.0
Avermectins	1.0
Azithromycin	2.0
Beta-lactams other than penicillins & cephalosporins	1.5
Cephalosporins	11.0
Ciprofloxacin	1.8
Clarithromycin	1.2
Clavulanic acid	1.0
Erythromycin	4.0
Penicillins	8.2
Quinolones	6.4
Tetracyclines	1.4
Tylosin	2.6
Vaccines	21.0
Vancomycin	1.0
Total	88.2

Figure 1. Anti-Infective Market.

The most reliable anti-HIV activity was observed with extracts prepared from several *Baccharis* species [10]. In addition, NCI has been evaluating the HIV-1 inhibitory activity of pepstatin A, a small pentapeptide produced by several *Streptomyces* species. This compound contains a statine (4-amino-3-hydroxy-6-methylheptanoic acid), which sterically blocks the active site of HIV-1 protease [11, 12].

Additional applications of chemically-synthesized natural metabolites include the use of a NP derived from plant salicylic acid derivatives present in meadowsweet, white willow, and wintergreen to relieve pain and suffering. Concoctions of these plants were administered by Hippocrates back in the year 500 BC, and even earlier in Egypt and Babilonia for pain, fever and childbirth. Synthetic salicylates were produced initially by Bayer in 1874, and later in 1897, Arthur Eichengrun at Bayer discovered that an acetyl derivative (aspirin) reduced acidity, bad taste and stomach irritation. These plant-based systems continue to play an essential role in health care, and the World Health Organization (WHO) has estimated that approximately 80% of the world's inhabitants rely mainly on traditional medicines for their primary health care [13].

The active constituent of the "fever tree" *Cinchona succiruba*, is the alkaloid quinine. This compound has been known for centuries by South American Indians to control malaria. During the 20th century, massive programs to synthesize quinoline derivatives, based on the quinine prototype, were accomplished. The first of the new quinolones to be used clinically as an antibacterial agent was nalidixic acid, which emerged as part of a large chemical synthesis program developed at the Sterling Winthrop Research Institute [14, 15]. The program was begun when 7-chloro-1,4-dihydro-1-ethyl-4-oxoquinolone-3-carboxylic acid was obtained as a side-product during purification of chloroquine and found to have antibacterial activity. During this program, the best compound found was nalidixic acid which had remarkable activity against Gram-negative bacteria and was shown to be an inhibitor of DNA gyrase. Its discovery led to synthesis of a whole series of quinolone and fluoroquinolone antibiotics (pefloxacin, norfloxacin, ciprofloxacin, levofloxacin, ofloxacin, lomefloxacin, sparfloxacin, etc.), which have been very successful in medicine and have achieved major commercial success (Figure 1). Although synthetic, it is important to appreciate that all quinolones are based on the structure of the plant natural product quinine.

Other synthesized compounds originating from NPs include teprotide, a nonapeptide, isolated from the venom of the Brazilian pit viper *Bothrops jaracaca* [16]. This led to the design and synthesis of angiotensin-converting enzyme (ACE) inhibitors such as captopril. Captopril was the first marketed, orally active ACE inhibitor [17]. Another ACE inhibitor, i.e., enalapril, was approved for marketing by the FDA in 1985 for the treatment of cardiovascular diseases [16].

3. Why the Necessity for Developing New Antibiotics?

New antibiotics are continuously required to combat antibiotic- resistant bacteria [18, 19]. Pathogenic yeasts are also a problem. *Candida* infections in the USA amount to 60,000-70,000 cases per year [20].

Bacteria have lived on Earth for several billion years. During this time, they encountered in nature a wide range of naturally occurring antibiotics. To survive, bacteria developed mechanisms for resistance to different antibiotics. Therefore, it is not surprising that resistance exists to most of the natural antimicrobial agents that have been developed over the last 50 years [21]. This resistance increasingly limits the effectiveness of current antimicrobial drugs. In 2004, more than 70% of pathogenic bacteria were estimated to be resistant to at least one of the currently available antibiotics [22]. The problem is not just antibiotic resistance but also multidrug resistance. The so-called 'superbugs' (organisms that are resistant to most of the clinically used antibiotics) are emerging at a rapid rate.

The Infectious Disease Society of America (IDSA) reported in July 2004 that in US hospitals alone, around 2 million people acquire bacterial infections each year [23]. *S. aureus* resistant to methicillin (MRSA) is responsible for many cases of infections each year. This microbe causes half of hospital-associated infections and takes the lives of about 100,000 patients each year in the USA alone [24]. There was an increase in MRSA-resistant organisms from 38% in 1992 to 65% in 2004. Sadly, antibiotic approvals of antibiotics by FDA decreased from 17 in 1980-1984 to 2 in 2005-2007.

Bacteria often produce a biofilm in which they are encased and protected from the environment. Biofilms can grow on wounds, scar tissue and medical implants or devices such as joint prostheses, spinal instrumentations, vascular prosthetic grafts, catheters, and heart valves. Bacteria forming biofilms, including staphylococci, are very resistant to antibiotics. They are also important in native infections such as cystic fibrosis, otitis media, endocarditis, urinary tract infections, etc. The communities of bacteria present in a biofilm are relatively resistant to antibiotics that normally would inhibit the individual species in the free state. More than 70% of the bacterial species producing such biofilms are likely to be resistant to at least one of the drugs commonly used in anti-infectious therapy [14]. There are relatively few antibiotics active against biofilms but new compounds are being pursued [25].

In hospitals, there are additional examples of Gram-positive (*Enterococcus, Streptococcus*) and Gram-negative pathogens (*Klebsiella, Escherichia, Enterobacter, Serratia, Citrobacter, Salmonella, Pseudomonas*); these hospital-inhabiting microbes are called "nosocomial bacteria". More than 60% of sepsis cases observed in hospitals are caused by Gram-negative bacteria [14]. Among them, *Pseudomonas aeruginosa* accounts for almost 80% of these opportunistic infections. They are a serious problem in patients hospitalized with cancer, cystic fibrosis, and burns, causing death in 50 percent of them. Other infections caused by *Pseudomonas* species include pneumonia, endocarditis, and infections of the urinary tract, central nervous system, wounds, eyes, ears, skin, and musculoskeletal system. The pseudomonads are another example of a natural multidrug-resistant microorganism. Although many strains are susceptible to gentamicin, tobramycin and amikacin, resistant forms have also developed. The incidence of multidrug-resistant pathogenic bacteria is increasing. These multi-drug resistant bacteria make hospitals "dangerous places to be, especially if you are sick, but even if not" [26].

Although there is a steady increase in resistance to most of the current antibiotics in almost every pathogen over time, not all the antibacterial agents show the same rate of resistance development. For instance, antimicrobials such as rifampicin, which target single enzymes, are the most susceptible to the development of resistance, whereas resistance to agents such as the penicillins, which inactivate several targets irreversibly, develops more slowly.

Besides the problem of antibiotic-resistance, new families of antibiotics are needed to enter the marketplace at regular intervals to face new diseases caused by evolving pathogens. At least 30 new diseases emerged in the 1980s and 1990s and they are growing in incidence. Emerging organisms with infective capacity often encounter hosts with no prior exposure to them and thus represent a novel challenge to the host's immune system. Several viruses responsible for human epidemics have been transferred from animal hosts to humans and are now transmitted from human to human. HIV, responsible for the AIDS epidemic, is one example. Although it has not been proven, it is suspected that severe acute respiratory syndrome (SARS), caused by the SARS coronavirus, also has evolved from a different species [27].

During recent decades, sexually transmitted diseases have also increased, especially in young people (aged 15 to 24). The human papiloma virus, genital herpes, chlamydia, gonorrhea, and HIV/AIDS are examples. HIV/AIDS has infected more than 40 million people in the world. Together with other diseases like tuberculosis and malaria, HIV/AIDS accounts for over 300 million illnesses and more than 5 million deaths each year.

The incidence of tuberculosis began to increase, after decades of decline, in the early 1990s. The epidemic took place due to inadequate treatment regimens, a diminished public-health system, and the onset of the HIV/AIDS epidemic. The WHO has predicted that between 2000 and 2020, nearly one billion people will become infected with *Mycobacterium tuberculosis* and this disease will cost the lives of 35 million people.

Another problem has been the emergence of evolving pathogens including the (i) Ebola virus, causing the viral hemorrhagic fever syndrome with a resultant 88% mortality rate; (ii) Legionaire's disease, a pulmonary infection mainly caused by *Legionella pneumophila*, an ubiquitous aquatic bacterium that lives in warm environments; (iii) Hantavirus, which can infect humans with two serious illnesses: hemorrhagic fever with renal syndrome (HFRS), and Hantavirus pulmonary syndrome (HPS); (iv) Lyme disease, an emerging infection caused by at least three species of bacteria from the genus *Borrelia*. In this case, the infection is acquired from the bite of ticks belonging to several species of the genus *Ixodes*. *Borrelia burgdorferi* is the predominant cause of Lyme disease in the U.S., whereas *Borrelia afzelii* and *Borrelia garinii* are implicated in most European cases. The disease presentation varies widely, and may include rash and flu-like symptoms in its initial stage, then musculoskeletal, arthritic, neurologic, psychiatric and cardiac manifestations. In a majority of cases, symptoms can be eliminated with the use of antibiotics, especially if the treatment begins early in the course of illness. However, late or inadequate treatment can lead to "late stage" Lyme disease that can be disabling and difficult to treat [28]. (v) Other evolving pathogens include the *Escherichia coli* 0157:H7 (enterohemorrhagic *E. coli* or EHEC), a strain that causes bloody diarrhea and colitis by producing a toxin called Shiga toxin, which damages the intestines. It is estimated that in the USA alone, this bacterium causes infection in more than 70,000 patients a year. Another example is (vi) *Cryptosporidium*, an obligate intracellular parasite commonly found in lakes and rivers. *Cryptosporidium parvum* is one of the common species affecting respiratory and digestive organs. Severely watery diarrhea characterizes intestinal cryptosporidiosis. Pulmonary and tracheal cryptosporidiosis in humans is associated with coughing and frequently a low-grade fever. People with severely weakened immune systems are likely to have more severe and more persistent symptoms than healthy individuals.

Nearly 90% of infectious disease deaths in the developing world are usually due to six diseases or disease processes: acute respiratory infections, diarrhea, tuberculosis, HIV, measles, and malaria. In both developing and developed nations, the leading cause of death by a wide margin is acute respiratory disease. In the developing world, acute respiratory infections are attributed primarily to the following bacteria: *Streptococcus pneumonia, Bordetella pertussis, Haemophilus influenzae, S. aureus, Mycoplasma pneumonia, Chlamydophila pneumonia,* and *Chlamydia trachomatis.* In addition, major viral causes of respiratory infections include respiratory syncytial virus, human parainfluenza viruses 1 and 3, influenza viruses A and B, as well as some adenoviruses. These diseases are highly destructive in economic and social as well as in human terms and cause approximately 17 million deaths per year, innumerable serious illnesses, and affect economic growth, development and prosperity of human societies [29]. Morse [30] identified six general factors in emergence of infectious diseases: international travel, ecological changes, human demographics and behavior, technology and industry, microbial adaptation and change, and breakdown in public health measures [31]. Toxicity of the antibiotics is another reason for developing new antibiotics. As with other therapeutic agents, use of antibiotics may also cause side-effects in patients. These include mild reactions like upset stomach, vomiting and diarrhea (cephalosporins, macrolides, penicillins, tetracyclines), rash and other mild and severe allergic reactions (cephalosporins, penicillins), sensitivity to sunlight (tetracyclines), nervousness, tremors and seizures (quinolones). Some of these side effects are more severe, and depending on the antibiotic, may disrupt the hearing function (aminoglycosides), kidneys (aminoglycosides, polypeptides), or liver (rifampin).

4. Actions Taken against Resistance to Antimicrobials

We have seen an increase in the number of reports highlighting the development of bacterial resistance to almost all antimicrobial agents. In the 1970's, the multidrug-resistance of Gram-negative bacteria was a major problem but later in the 1980's, the Gram-positive bacteria became also important, including methicillin-resistant *staphylococci* (MRSA), penicillin-resistant *pneumococci*, and vancomycin-resistant *enterococci* [32]. Recently, the solution to the problem has depended primarily on the development of novel antimicrobial agents. However, for different reasons, the number of new classes of antimicrobial agents being developed has decreased dramatically in recent years.

The occurrence of resistant Gram-positive bacteria has been noticed by the pharmaceutical, biotechnology, and academic communities. Some of these groups are making concerted efforts to find novel antimicrobial drugs to combat this problem. A novel glycopeptide antibiotic, teicoplanin, was developed against infections with resistant Gram-positive bacteria, especially bacteria resistant to the glycopeptide vancomycin. In another example, the approach involved redesign of a mixture of two compounds called streptogramin into a new mixture called pristinamycin, to allow administration of the drug parenterally and in higher doses than the earlier oral preparation [33]. The two components of streptogramin, quinupristin and dalfopristin, were chemically modified to allow intravenous administration. The new combination, pristinamycin (Synercid) was approved by the FDA for use against infections caused by a vancomycin-resistant bacterium, *Enterococcus faecium*.

Additional novel antimicrobial agents used in treating resistance to Gram-positive infections include the cyclic lipopeptide antibiotic daptomycin, produced by *Streptomyces roseosporus*. This compound was approved in 2003 by the FDA for skin infections resulting from complication of surgery, burns, and diabetic foot ulcers. Its mode of action is distinct from any other approved antibiotic: it rapidly kills Gram-positive bacteria by disrupting multiple aspects of bacterial membrane function (by binding irreversibly to bacterial cell membrane, causing membrane depolarization, destroying the ion concentration gradient and provoking efflux of K+). It represented the first new natural antibiotic approved in many years. It acts against most clinically relevant Gram-positive bacteria (*S. aureus*, *Enterococcus faecalis, Streptococcus pyogenes*, *Streptococcus agalactiae* and *Streptococcus dysgalactiae* subsp. *equisimilis*) and retains *in vitro* potency against isolates resistant to methicillin, vancomycin and linezolid. Traditionally, these infections were treated with penicillin and cephalosporins, but resistance to these agents became widespread [34, 35, 36, 37]. Daptomycin seems to have a favorable side-effect profile, and it might be used to treat patients who cannot tolerate other antibiotics.

Further moves against resistant microorganisms are the glycylcyclines, developed to treat tetracycline-resistant bacteria. These modified tetracyclines show potent activity against a broad spectrum of Gram-positive and Gram-negative bacteria, including strains that carry the two major tetracycline-resistance determinants, involving efflux and ribosomal protection. Two of the glycylcycline derivatives, DMG-DMDOT and DMG-MINO, have been tested against a large number of clinical pathogens isolated from various sources. The spectrum of activity of these derivatives includes organisms with resistance to antibiotics other than tetracyclines, e.g., methicillin-resistant staphylococci, penicillin-resistant *S. pneumoniae*, and vancomycin-resistant *enterococci* [38]. Tigecycline (tygacil™), was approved by the FDA in 2005 as an injectable antibiotic [39].

The macrolide antibiotic telithromycin is the first orally-active compound of a new family of antibacterials named the ketolides. It shows potent activity against pathogens implicated in community-acquired respiratory tract infections, irrespective of their ß-lactam, macrolide or fluoroquinolone susceptibility. Some of the microorganisms susceptible to this antibiotic are pneumococci, *H. influenzae* and *Moraxella catarrhalis*, including ß-lactamase-positive strains. In addition, telithromycin has a very low potential for selection of resistant isolates or induction of cross-resistance found with other macrolides [40].

Streptomyces clavuligerus produces clavulanic acid which contains a bicyclic β-lactam ring fused to an oxazolidine ring with an oxygen in place of a sulfur, a β-hydroxyethylidene substituent at C-2 and no acylamino group at C-6. First described in 1976, this antibiotic showed potent inhibition of the β-lactamases produced by staphylococci and plasmid-mediated β-lactamases of *E. coli*, *Klebsiella*, *Shigella*, *Pseudomonas, Proteus* and *Hemophilus*. Although it is a broad spectrum antibiotic, clavulanic acid possesses only very low antibacterial activity. Therefore, as a β-lactamase inhibitor, the molecule has been combined with a variety of broad-spectrum semi-synthetic penicillins. For example, when administered with amoxicillin, it is marketed under the trade name Augmentin for the treatment of infections caused by β-lactamase-producing pathogenic bacteria [41]. Augmentin has world sales of over $1 billion and in 1995 was the second largest selling antibacterial. Clavulanic acid can be also combined with ticarcillin which is a penicillin effective against organisms like *E. coli*, *Salmonella*, *Proteus*, *Haemophilus*, *Pseudomonas* and *S. aureus*. It is frequently used in hospitals for treating severe infections affecting blood or internal organs, upper or lower airways, bones and joints or skin and soft tissue. By inhibiting the action of the β-lactamases produced by certain bacteria, the combination of these compounds (Timentin) extends ticarcillin antimicrobial activity.

5. Resistance of Antifungals

Fungi are harmless most of the time, but sometimes they can cause infections. Mycosis is a condition in which fungi surpass the resistance barriers of the human or animal body and establish infections. Usually, these infections are not life-threatening. However, when they are deeply invasive and disseminated, they lead to more serious infections, such as in critically ill patients, and people treated with drugs that affect the immune system. In addition, the use of antineoplastic and broad-spectrum antibiotics, prosthetic devices and grafts, and more aggressive surgery has increased invasive fungal infections. Patients with burns, pancreatitis, neutropenia, or after organ transplantation (40% of liver transplants, 15-35% of heart transplants and 5% of kidney transplants) are also predisposed to fungal infection [42]. Around 40% of deaths from nosocomial infections are caused by fungi and 80% of these are caused by species of *Aspergillus and Candida*, although *Cryptococcus* spp., *Fusarium* spp., *Scedosporium* spp., *Penicillium* spp., and zygomycetes are increasingly involved [43]. *Pneumocystis carinii* is the leading cause of death of AIDS patients from Europe and North America [44]. In death of recipients of bone marrow transplants, the main factor involved is pulmonary aspergillosis.

The emergence of broader fungal resistance and the increasing incidence of invasive fungal infections have led to the search for novel antifungal agents. Although amphotericin B is the first-line therapy for systemic infection because of its broad-spectrum and fungicidal activity, considerable side-effects limit its clinical utility. Additional compounds are the echinocandins, large lipopeptide molecules that inhibit the synthesis of 1,3-β-D-glucan, a key component of the fungal cell wall. Three echinocandins have reached the market, i.e., caspofungin, micafungin and anidulafungin. Caspofungin, also known as pneumocandin or MK-0991, was the first cell-wall active antifungal approved in 2000 as a new injectable antifungal [45]. It irreversibly inhibits 1,3-β-D-glucan synthase, preventing the formation of glucan polymers and disrupting the integrity of fungal cell walls [46]. It is more active and less toxic than amphotericin B and shows a broad spectrum of activity against *Candida* (including fluconozole-resistant strains), *Aspergillus*, *P. carinii and Histoplasma*, the major causes of HIV death. Micafungin is licensed for clinical use in Asian countries and the US. This compound exhibits extremely potent antifungal activity against clinically important fungi, including *Aspergillus* and azole-resistant strains of *Candida*. In animal studies, micafungin is as efficacious as amphotericin B with respect to improvement of survival rate. It is characterized by a linear pharmacokinetic profile and substantially has fewer toxic effects. Anidulafungin has proven efficacy against esophageal candidiasis and invasive infections by *Candida* and *Aspergillus*. It is currently licensed in the USA [47].

Some patients remain resistant to treatments, despite the development of several new antifungal drugs in the last six years. This is mainly due to intrinsic or acquired antifungal resistance, organ dysfunction preventing the use of some agents, and drug interactions. In addition, some drugs penetrate poorly into sanctuary sites including eye and urine, and others are associated with considerable adverse events. However, there has been some progress in the development of novel compounds. Posaconazole is a new member of the triazole class of antifungals. It has shown clinical efficacy in the treatment of oropharyngeal candidiasis and has potential as a salvage therapy for invasive aspergillosis, zygomycosis, cryptococcal meningitis and a variety of other fungal infections. This compound is available as an oral suspension and has a favorable toxicity profile. An important addition to the armamentarium against fungal infections is posaconazole [48] with its wide spectrum of activity in *in vitro* studies, animal models, and preliminary clinical studies.

6. Additional Applications of Microbial Natural Products

Over the years, the pharmaceutical industry extended their antibiotic screening programs to other areas [49, 50]. Since microorganisms are such a prolific source of structurally diverse bioactive metabolites, the industry extended the screening programs in order to look for microbes with activity in other disease areas. As a result of this move, some of the most important products of the pharmaceutical industry were obtained. For example, the immunosuppressants have revolutionized medicine by facilitating organ transplantation [51]. Other applications include antitumor drugs, hypocholesterolemic drugs, enzyme inhibitors, gastrointestinal motor stimulator agents, ruminant growth stimulants, insecticides, herbicides, antiparasitics vs. coccidia and helminths and other pharmacological activities (Figure 2). Catalyzed by the use of simple enzyme assays for screening prior to testing in intact animals or in the field, further applications are emerging in various areas of pharmacology and agriculture.

6.1. Products with Hypocholesterolemic Effects

Atherosclerosis is a chronic, progressive disease characterized by continuous accumulation of atheromatous plaque within the arterial wall causing stenosis and ischemia. The last two decades have witnessed the introduction of a variety of anti-atherosclerotic therapies.

The statins form a class of hypolipidemic drugs used to lower cholesterol by inhibiting the rate-limiting enzyme of the mevalonate pathway of cholesterol biosynthesis, i.e., 3-hydroxymethyl glutaryl-CoA (HMG-CoA) reductase. Inhibition of this enzyme in the liver stimulates low-density lipoprotein (LDL) receptors, resulting in an increased clearance of LDL from the bloodstream and a decrease in blood cholesterol levels. Through their cholesterol- lowering effect, they reduce risk of cardiovascular disease, prevent stroke and reduce development of peripheral vascular disease [52]. In addition, statins have anti-thrombotic and anti-inflammatory effect.

Anticancer agents	Immunosuppressants
Antiparasitics	Insecticides
Enzyme inhibitors	Ruminant growth stimulants
Herbicides	Stimulants of gastrointestinal motility
Hypocholesterolemic drugs	

Figure 2. Additional Applications of Natural Products Produced by Microbes.

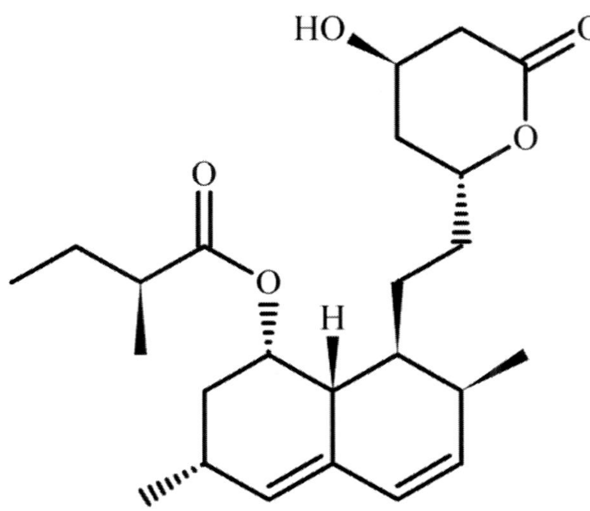

Figure 3. Chemical Structure of Lovastatin.

Currently there are a number of statins in clinical use. They reached an annual market of nearly $30 billion before one became a generic pharmaceutical. The history of the statins has been described by Akira Endo, the discoverer of the first statin, compactin [53]. The first member of the group (compactin; mevastatin) was isolated as an antibiotic product of *Penicillium brevicompactum* and later from *Penicillium citrinum*. Although compactin was not of commercial importance, its derivatives achieved strong medical and commercial success.

The lactone form of mevalonic acid, known as lovastatin (monacolin K; mevinolin; MevacorTM), was isolated in the 1970s in broths of *Monascus rubra* and *Aspergillus terreus* [54]. Lovastatin, developed by Merck & Co. and approved by the FDA in 1987, was the first commercially marketed statin. In its chemical structure, lovastatin has a hexahydronaphthalene skeleton substituted with a p-hydroxy-lactone moiety (Figure 3). A semisynthetic derivative of lovastatin is Zocor® (simvastatin), one of the main hypocholesterolemic drugs, selling for 7 billion dollars per year before becoming generic. Although simvastatin is usually made from lovastatin chemically in a multistep process, an enzymatic/bioconversion process using recombinant *E. coli* has been developed [55]. Another statin, pravastatin (3.6 billion dollars per year), is made via different biotransformation processes from compactin by *Streptomyces carbophilus* [56] and *Actinomadura* sp. [57]. Other genera involved in production of statins are *Doratomyces, Eupenicillium, Gymnoascus,*

Hypomyces, Paecilomyces, Phoma, Trichoderma, and *Pleurotus* [58]. A synthetic compound, modeled from the structure of the natural statins, is Lipitor, which was the leading drug of the entire pharmaceutical industry in terms of market (about 14 billion dollars per year) for many years.

6.2. Anticancer Drugs

Over 12 million new cases of cancer were diagnosed in the world in 2008. 6.6 million cases were in men and 6.0 million in women, resulting in 7.6 million cancer-related deaths. The tumor types with the highest incidence were lung (12.7%), breast (10.9%), and colorectal (9.8%) [59]. Over 60% of the anticancer drugs in clinical use are NPs derived from microorganisms. They started to appear around 1940 with the discovery of actinomycin and since then, many compounds with anticancer properties were isolated from natural sources. Among the approved products deserving special attention are actinomycin D, anthracyclines (daunorubicin, doxorubicin, epirubicin, pirirubicin and valrubicin), bleomycin, mitosanes (mitomycin C), anthracenones (mithramycin, streptozotocin, pentostatin), endiynes (calcheamycin), taxol and epothilones.

The oldest antibiotic used in cancer therapy is actinomycin D. Its relative, actinomycin A, was the first antibiotic isolated from actinomycetes. It was obtained from *Actinomyces antibioticus* (now *Streptomyces antibioticus*) by Waksman and Woodruff [60] and from *Micromonospora*. Since it binds DNA at the transcription initiation complex, it prevents elongation by RNA polymerase. This property, however, confers some human toxicity and it has been used primarily as an investigative tool in the development of molecular biology. Despite the toxicity, however, it has served well against Wilms tumor in children. It is also used for rhabdomyosarcoma.

Some of the most effective antitumor agents ever developed are the anthracyclines. These are effective against more types of cancer than any other class of chemotherapy agents [61]. They are used to treat a wide range of cancers, including leukemias, lymphomas, and breast, ovarian, uterine, and lung cancers. Anthracyclines act by intercalating DNA strands, which results in complex formation, inhibiting synthesis of DNA and RNA in the target cells. It also triggers DNA cleavage by topoisomerase II, resulting in mechanisms that lead to cell death. In their cytotoxic effects, the binding to cell membranes and plasma proteins plays an important role. Their main adverse effects are heart damage (cardiotoxicity), which considerably limits their usefulness, and vomiting. The first anthracycline discovered was daunorubicin (daunomycin) in 1966, which is produced naturally by *Streptomyces peucetius*. Doxorubicin (adriamycin), a 14-hydroxylated version of daunorubicin, was developed in 1967. This compound showed better activity against solid tumors and a higher therapeutic index.

Another anthracycline is epirubicin. This compound, approved by FDA in 1999, is favored over doxorubicin as it appears to cause fewer side effects. Epirubicin has a different spatial orientation of the hydroxyl group at the 4' carbon of the sugar, which may account for its faster elimination and reduced toxicity. Epirubicin is primarily used against breast and ovarian cancer, lung cancer, gastric cancer, and lymphomas. However, the use of this compound may cause serious heart problems during the cancer treatment or after treatment has ended. Valrubicin (Valstar) is a semisynthetic analog of doxorubicin approved as a chemotherapeutic drug in 1999 and used by intravesical administration to treat bladder cancer. Valstar was removed from the market in 2002 due to manufacturing problems. However, in 2007, Valstar was placed back on the FDA Drug Shortages List, after it was considered to be a medical necessity. Many new anthracyclines are in clinical trials against prostate cancer and leukemia.

The non-ribosomal glycopeptide antibiotic bleomycin is produced as a family of structurally-related compounds by the bacterium *Streptomyces verticillus*. First reported by Umezawa and coworkers [62], bleomycin obtained FDA approval in 1973. When used as an anti-cancer agent (inducing DNA strand breaks), the chemotherapeutic forms are primarily bleomycins A2 and B2. The drug is used in the treatment of Hodgkin's lymphoma, squamous cell carcinomas, and testicular cancer. Mitosanes are composed of several mitomycins, which are formed during the cultivation of *Streptomyces caespitosus*. Although the mitosanes are excellent antitumor agents, they have limited utility due to their toxicity. Mitomycin C was approved by FDA in 1974, showing activity against several types of cancer (lung, breast, bladder, anal, colorectal, head and neck), including melanomas and gastric or pancreatic neoplasms [63]. Mitomycin has shown to be effective in killing bacteria by producing DNA crosslinks. However, the prolonged use of this drug may result in permanent bone-marrow damage. Mitomycin dimers have been explored as potential alternatives for lowering toxicity and increasing efficiency [64].

An antitumor aromatic polyketide produced by *Streptomyces argillaceus* is mithramycin (plicamycin) which shows both antibacterial and antitumor activity [65]. It is one of the older chemotherapy drugs used in the treatment of testicular cancer, disseminated neoplasms and hypercalcemia. It binds to G-C rich DNA sequences inhibiting the binding of transcription factors like Sp1, which is believed to affect neuronal survival/death pathways. It may also indirectly regulate gene transcription by altering histone methylation. With its repeated use, organotoxicity (kidney, liver, haematopoietic system) can become a problem. However, this drug prevents neurodegeneration by selective DNA-binding inhibitors [66].

Another antibiotic with antitumor properties is streptozotocin which is produced by *Streptomyces achromogenes*. Chemically, it is a glucosamine-nitrosourea compound. As with other alkylating agents in the nitrosourea class, it is toxic to the β-cells of the pancreatic islets by causing damage to DNA, though other mechanisms may also contribute. It is similar enough to glucose to be transported into the cell by the glucose transport protein of these cells, but is not recognized by the other glucose transporters. Since beta-cells have relatively high levels of glucose permease, the relative streptozotocin toxicity for these β-cells can be explained [67]. In 1982, FDA granted streptozotocin approval as a treatment for pancreatic islet cell cancer. Its use is generally limited to patients whose cancer cannot be removed by surgery.

Pentostatin (deoxycoformycin) is an anticancer chemotherapeutic drug produced by *Streptomyces antibioticus*. It is classified as a purine analog, which mimics the nucleoside adenosine and thus tightly binds and inhibits adenosine deaminase (Ki of 2.5 x 10-12 M). It interferes with the cell's ability to process DNA [68]. Pentostatin is commonly used to treat hairy cell leukemia, acute lymphocytic leukemia, prolymphocytic leukemia (B-cell and T-cell origin), T-cell leukemia, and lymphoma. However, it can cause kidney, liver, lung, and neurological toxicity [69]. The FDA granted pentostatin approval in 1993.

Calicheamicins are highly potent antitumor antibiotics of the enediyne family produced by *Micromonospora echinospora*. Their antitumor activity is apparently due to cleavage of double-stranded DNA [70]. They are highly toxic but it was possible to introduce one such compound into the clinic by ingeniously attaching it to an antibody, which delivered it to certain cancer types selectively. In this regard, Gentuzumab (marketed as Mylotarg) is effective against acute myelogenous leukemia (AML). The calicheamicin is bound to a monoclonal antibody against a transmembrane receptor (CD33) expressed on cells of monocytic/myeloid lineage. CD33 is expressed in most leukemic blast cells, but in normal hematopoietic cells, the intensity diminishes with maturation. A few years ago, it was approved by the FDA for use in patients over the age of 60 with relapsed AML who are not considered candidates for standard chemotherapy [71]. Recently, the drug was conjugated to a humanized antibody (CMC-544) and shown to kill primary acute lymphoblastic leukemia cells [72]. The activity of the immunosuppressant rapamycin and its derivatives against cancer are discussed in section 6.3.

Camptothecin is a modified monoterpene indole alkaloid produced by certain plants (angiosperms) [73]. It also is produced by the endophytic fungus, *Entrophospora infrequens*, from the plant *Nathapodytes foetida*. In view of the low concentration of camptothecin in tree roots and poor yield from chemical synthesis, the fungal fermentation is very promising for industrial production of camptothecin. It is used for recurrent colon cancer and has unusual activity against lung, ovarian, and uterine cancer (74). Colon cancer is the second-leading cause of cancer fatalities in the USA and the third most common cancer among US citizens. Camptothecin is known commercially as Camptosar and Campto and achieved sales of $1 billion in 2003 [75]. Camptothecin's water-soluble derivatives irinotecan and topotecan have been approved and are used clinically. Metastatic colorectal cancer is treated by irinotecan whereas topotecan has use for ovarian cancer, cervical cancer and small-cell lung cancer. A review of the activities of camptothecin and its many small and macromolecular derivatives has been published by Venditto and Simanek [76].

The cellular target of camptothecin is type I DNA topoisomerase. When patients become resistant to irenotecan, its use can be prolonged by combining it with the monoclonal antibody Erbitux (Cetuximab). Erbitux blocks a protein that stimulates tumor growth and the combination helps metastatic colorectal cancer patients expressing epidermal growth factor receptor (EGFR). This protein is expressed in 80% of advanced metastatic colorectal cancers. The drug combination reduces invasion of normal tissues by tumor cells and the spread of tumors to new areas.

The most successful non-actinomycete cancer-fighting molecule is taxol (Paclitaxel), which was first isolated from the Pacific yew tree, *Taxus brevifolia.* It is a steroidal alkaloid diterpenoid that has a characteristic N-benzoylphenyl isoserine side chain and a tetracycline ring (Figure 4). This compound inhibits rapidly dividing mammalian cancer cells by promoting tubulin polymerization and interfering with normal microtubule breakdown during cell division.

The benzoyl group of the molecule is particularly crucial for maintaining the strong bioactivity of taxol. The drug also inhibits several fungi (*Pythium, Phytophthora, Aphanomyces*) by the same mechanism. In 1992, taxol was approved for refractory ovarian cancer and today is used against breast cancer and advanced forms of Kaposi's sarcoma [77].

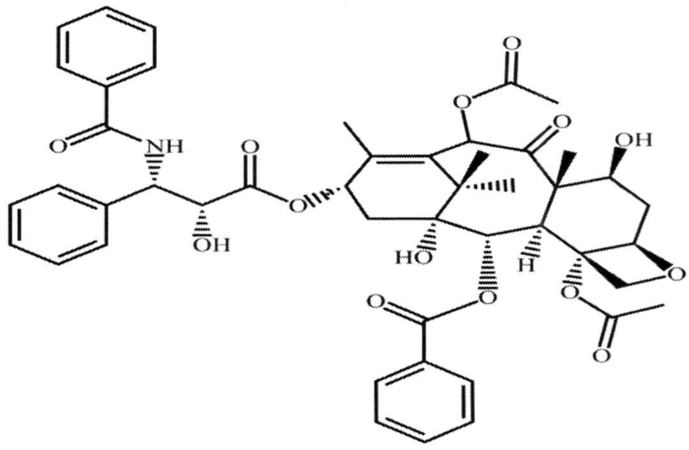

Figure 4. Chemical Structure of Taxol (The Benzoyl Group is Located in the Left Side of the Structure).

A new formulation in which paclitaxel is bound to albumin is sold under the trademark Abraxane. Taxol sales amounted to $1.6 billion in 2006 for Bristol Myers-Squibb, representing 10% of the company's pharmaceutical sales and its third largest selling product. Paclitaxel has reached $3.7 billion annual sales in international markets.

Although synthetic methods for taxol production have been tried, the chemical molecular structure is so complex that commercial synthetic production is unfeasible. Taxol production by cell culture of *Taxus* sp. amounted to 67 mg/L [78]. Currently, Italy, the United Kingdom, the Netherlands and other Western countries are engaged in the production of taxol by plant cell fermentation technology.

Although originally discovered in plants, taxol has also been found to be a fungal metabolite [79, 80]. Fungi such as *Taxomyces andreanae, Pestalotiopsis microspora, Tubercularia* sp., *Phyllosticta citricarpa* and *Nodulisporium sylviforme* produce taxol [79, 81, 82, 83]. Production by *P. citricarpa* amounted to 265 μg/L [84]. Production was reported at 417 μg/L by submerged fermentation with an engineered strain of the endophytic fungus *Ozonium* sp. (EFY-21). The transformed strain overproduced the taxol rate-limiting enzyme taxadiene synthase [85]. Another endophytic fungus, *Phoma betae*, isolated from the medicinal tree *Ginkgo bilova*, produced taxol at 795 μg/L [86]. *Cladosporium cladosporoides*, an endophyte of the *Taxus media* tree, produces 800 μg/L of taxol [87]. *Metarhizium anisopiliae* H-27, isolated from the tree *Taxus chinensis* yielded 846 μg/L [88]. Although a review of taxol production by endophytic fungi indicated that strain improvement had resulted in levels of only 0.4-1.0 mg/L [89], it was reported that another fungus, *Alternaria alternate* var. *monosporus*, from the bark of *Taxus yunanensis*, after ultraviolet and nitrosoguanidine mutagenesis, could produce taxol at 227 mg/L [90].

The epothilones are macrolides originally isolated from the broth of the soil myxobacterium *Sorangium cellulosum* as weak agents against rust fungi [91]. They were later identified as microtubule-stabilizing drugs, acting in a similar manner to taxol [92, 93]. However, they are generally 5 to 25 times more potent than taxol in inhibiting cell growth in cultures. Five analogs are now undergoing investigation as candidate anticancer drugs, and their preclinical studies have indicated a broad spectrum of antitumor activity including taxol-resistant tumor cells. With the best currently available therapies, the median survival time for patients with metastatic breast cancer is only 2–3 years, and many patients develop resistance to taxanes or other chemotherapy drugs. One epothilone, ixabepilone, was approved in October 2007 by the FDA for use in the treatment of aggressive metastatic or locally advanced breast cancer no longer responding to available chemotherapies [94]. In tumor cells, p-glycoprotein reduces intracellular antitumor drug concentrations, thereby limiting access of chemotherapeutic substrates to the site of action. The epothilones are attractive because they are active against p-glycoprotein-producing tumors and have good solubility [95]. Epothilone B is a 16-membered polyketide macrolactone with a methylthiazole group connected to the macrocycle by an olefinic bond. Five additional epothilones, including patupilone (for ovarian cancer) and sagopilone are in clinical trials.

Romidepsin (Istodax, NSC630176, FR 901228, FK228) is a depsipeptide produced by *Chromobacterium violaceum*, a Gram-negative bacterium. It was isolated from soil by Ueda et al. [96] of Fujisawa Corp. (now Astellas). They found it to have antitumor activity. In 1998, they agreed to have it further investigated by the US National Cancer Institute (NCI). Fujisawa carried out clinical trials in 2002 after NCI confirmed its antitumor activity. It was licensed to Gloucester Pharmaceuticals in the USA in 2004, later acquired by Celgene Corp. It was found active against cutaneous T-cell lymphoma (CTCL) and was FDA-approved in 2009. In the USA, there are 1500 new cases and 500 deaths per year from CTCL. The history of romidepsin has been reviewed by VanderMolen et al. [97]. It is an inhibitor of histone deacetylase.

The most common cancer diagnosis in men between the ages of 15 and 35 years is testicular cancer, with approximately 8000 cases detected in the United States annually [98]. The majority [95%] of testicular neoplasms are germ cell tumors, which are relatively uncommon carcinomas, accounting for only 1 % of all male malignancies. Remarkable progress has been made in the medical treatment of advanced testicular cancer with a substantial increase in cure rates from approximately 5% in the early 1970s to almost 90% [99, 100]. This cure rate is the highest of any solid tumor and improved survival is primarily due to effective chemotherapy. Two chemotherapy regimens are effective for patients with a testicular germ cell tumor prognosis: four cycles of etoposide and cisplatin or three cycles of bleomycin, etoposide, and cisplatin [101]. Of the latter three agents, bleomycin and etoposide are NPs. A major advance in chemotherapy for testicular germ cell tumors was the introduction of cisplatin in the mid-1970s. Cisplatin is synthetic and also shows activity against microorganisms [102].

6.3. Immunosuppressant Drugs

An individual's immune system is capable of distinguishing between native and foreign antigens and to mount a response only against the latter. Suppressor cells are critical in the regulation of the normal immune response. The suppression of the immune response either by drugs or radiation, in order to prevent the rejection of grafts or transplants or to control autoimmune diseases, is called immunosupression.

Microbial compounds capable of suppressing the immune response have been discovered. Cyclosporin A was originally introduced as a narrow spectrum antifungal peptide produced by the mold, *Tolypocladium nivenum* by aerobic fermentation. Cyclosporins are a family of neutral, highly lipophilic, cyclic undecapeptides containing some unusual amino acids, synthesized by a nonribosomal peptide synthetase, cyclosporin synthetase. Discovery of the immunosuppressive activity led to use in heart, liver, and kidney transplants and to the overwhelming success of the organ transplant field [103]. Cyclosporin was approved for use in 1983. It is thought to bind to the cytosolic protein cyclophilin (immunophilin) of immunocompetent lymphocytes, especially T-lymphocytes. This complex of cyclosporin and cyclophylin inhibits calcineurin, which under normal circumstances, is responsible for activating the transcription of interleukin-2. It also inhibits lymphokine production and interleukin release and therefore leads to a reduced function of effector T-cells. Annual world sales of cyclosporin A are approximately $2 billion.

Sirolimus (rapamycin) and tacrolimus (FK506, Fujimycin) are important organ transplant agents produced by actinomycetes. Rapamycin is especially useful in kidney transplants since it lacks the nephrotoxicity seen with cyclosporin A and tacrolimus. This macrolide, discovered in 1975 as a product of *S. hygroscopicus*, was initially proposed as an antifungal agent. However, this application was abandoned when it was discovered that it had potent immunosuppressive and antiproliferative properties. Rapamycin not only has immunosuppressant and antifungal activities, it also exhibits antitumor, neuroprotective, autoimmune and anti-aging properties. It binds to the immunophilin FK506-binding protein (FKBP12) and this binary complex interacts with the rapamycin-binding domain (FRB) and inactivates a serine-threonine kinase termed the mammalian target of rapamycin (mTOR) [104]. The latter is known to control proteins that regulate mRNA translation initiation and G1 progression [105]. The anti-proliferative effect of rapamycin has also been used in conjunction with coronary stents to prevent restenosis, which usually occurs after treatment of coronary artery disease by balloon angioplasty. A rapamycin-eluting coronary stent is marketed under the trade name of Cypher.

The biosynthesis, regulation and mutagenic improvement of rapamycin biosynthesis has been reviewed [106]. The importance of various factors such as carbon source regulation [107], nitrogen source regulation [108], and the precursor effects of acetate, propionate, methionine [109], pipecolic acid [110] and shikimic acid [111] were all considered for strain improvement of *Streptomyces hygroscopicus* for rapamycin production by Zhu et al. [112]. Starting with a strain producing 150 mg/L, a mutant resistant to lysine inhibition produced 260 mg/L. Further mutation to high glucose resistance yielded a strain making 350 mg/L. Mutation to resistance to high shikimate concentrations and then to auxotrophy to tryptophan and phenylalanine resulted in production of 450 mg/L. Fed-batch fermentation with glycerol and potassium phosphate supplementation yielded 812 mg/L.

Two genes of the rapamycin biosynthetic cluster in the rapamycin-producing *S. hygroscopicus*, i.e. *rap G* and *rap H*, encode positive regulatory proteins for rapamycin production [113]. Overexpression of either gene increases rapamycin formation whereas their deletions eliminate rapamycin biosynthesis. They act by affecting the promoter of the operon.

The antitumor activity of rapamycin [114] is due to its ability to interfere with angiogenesis [115] and induce apoptosis. This finding was taken advantage of by chemical modification, yielded important derivatives such as temsirolimus (CCI-779; Torisel®), everolimus (RAD001), and deforolimus (A23573). Temsirolimus, was approved by FDA for renal cell carcinoma [116], while everolimus and deferolimus, are in clinical trials against cancer [117]. Ridaforolimus (deforolimus) of Merck and ARIAD is in phase III clinical trials vs sarcoma.

Rapamycin also shows promise in treating tuberous sclerosis complex (TSC), a congenital disorder that leaves sufferers prone to benign tumor growth in the brain, heart, kidneys, skin and other organs. In a study of rapamycin as a treatment for TSC, UCLA researchers observed a major improvement in mice regarding retardation related to autism [118].

Poor aqueous solubility of rapamycin has been improved in its analogues, i.e., everolimus, tensirolimus and AP23573. Everolimus (marketed as Certican) is currently used as an immunosuppressant to prevent rejection of organ transplants. It is marketed by Novartis under the tradenames Zortress (USA) and Certican (Europe and other countries) in transplantation medicine, and Afinitor in oncology. Everolimus may have a role in heart transplantation as it has been shown to reduce chronic allograft vasculopathy in such transplants [119]. In 2010, phase III trials were under way in gastric cancer, hepatocellular carcinoma and lymphoma [120]. Everolimus is also used in drug-eluting coronary stents as an immunosuppressant to prevent rejection. Deforolimus is a novel non-prodrug rapamycin analogue with a non-linear pharmacokinetic behavior, which has demonstrated anti-proliferative activity against several human tumor cell lines *in vitro* and against experimental tumors *in vivo* [121]. This agent is currently under evaluation in Phase I-II trials including patients with different tumors. Two additional small-molecule rapamycin analogues, AP23841 and AP23675, are in preclinical development for treatment of bone metastases and primary bone cancer [122].

Tacrolimus (FK-506) was discovered in 1987 [123]. It is produced by *Streptomyces tsukubaensis* and is approximately 100 times more active as an immunosuppressive than cyclosporin A [124]. It was introduced in Japan in 1993, and in 1994, was approved by FDA for use as an immunosuppressant in liver transplantation. Tacrolimus had a market of $2 billion in 2007. Its use has been extended to include bone marrow, cornea, heart, intestines, kidney, lung, pancreas, trachea, small bowel, skin and limb transplants, and for prevention of graft-vs-host disease. Topically, it is used also against atopic dermatitis, a widespread skin disease. In the laboratory, tacrolimus inhibits the mixed-lymphocyte reaction, the formation of interleukin-2 by T lymphocytes, and the formation of other soluble mediators, including interleukin-3 and interferon gamma. Recently, it has been reported that tacrolimus inhibits tumor growth factor-β-induced signaling and collagen synthesis in human lung fibroblastic cells. This factor plays a pivotal role in tissue fibrosis, including pulmonary fibrosis. Therefore, tacrolimus may be useful for the treatment of pulmonary fibrosis, although its use in the acute inflammatory phase may exacerbate lung injury [125].

6.4. Stimulators of Gastrointestinal Motility

The most effective route for feeding is via the gastrointestinal tract. Critically ill patients that accept early feeding improve their health. In some postoperative patients, gastric stasis and excessive volumes in the stomach increase the risk of reflux and aspiration with subsequent pneumonia. Because of the importance of achieving early and adequate nutritional intake, it is common practice in many intensive care units to use drugs to improve gastrointestinal motility.

Erythromycin is a macrolide antibiotic with a broad spectrum of activity. It is well recognized that when prescribed, either intravenously or orally, it causes side-effects such as diarrhea, nausea and vomiting. These side-effects are in part due to the action of erythromycin at motilin receptors in the gut. This makes this antibiotic very attractive to be used in ill patients with gastrointestinal motility problems. There have been some developments on erythromycin analogs which lack antibiotic action but retain action at motilin receptors. These have been named `motilides' [126, 127]. An orally active erythromycin-derived motilin receptor agonist [mitemcinal] was tested in patients with idiopathic and diabetic gastroparesis. In both cases, an improvement of gastroparetic symptoms were observed [128].

6.5. Inhibitors of Enzyme Activity

Enzyme inhibitors are compounds that interact with enzymes and decrease their activity. They have received increasing attention as useful tools, not only for the study of enzyme structures and reaction mechanisms, but also for potential utilization in medicine and agriculture. Several enzyme inhibitors with various industrial uses have been isolated from microbes [129]. The most important are clavulanic acid, the inhibitor of β-lactamases discussed above in section 4, and the statins, hypocholesterolemic drugs presented previously in section 6.1. Some of the common targets for other inhibitors are lipases, proteases, amylases, glucosidase, and xanthine oxidase (XO). A number of enzyme inhibitors are produced by species of *Streptomyces* (Figure 5).

Lipstatin is a pancreatic lipase inhibitor produced by *Streptomyces toxitricini* and used to combat obesity and diabetes. It interferes with the gastrointestinal absorption of fat [130]. The commercial product is tetahydrolipstatin which is called Orlistat.

In the pathogenic processes of some diseases like emphysema, arthritis, pancreatitis, cancer, and AIDS, protease inhibitors are potentially powerful tools for inactivating target proteases. Examples of microbial products include antipain, produced by *Streptomyces yokosukaensis*, leupeptin from *Streptomyces roseochromogenes,* and chymostatin from *Streptomyces hygroscopicus* [131]. Leupeptin is produced by more than 17 species of actinomycetes [129].

Amylase inhibitors are useful for the control of carbohydrate-dependent diseases such as diabetes, obesity and hyperlipemia [132, 131]. Amylase inhibitors are also known as starch blockers because they contain substances that prevent dietary starches from being absorbed by the body. The inhibitors may also be useful for weight loss since some versions of amylase inhibitors do show potential for reducing carbohydrate absorption in humans [133, 134]. The use of amylase inhibitors for the treatment of rumen acidosis has been also reported [135]. Examples of microbial α-amylase inhibitors are Paim, obtained from culture filtrates of *Streptomyces corchorushii* [136], and TAI-A and TAI-B, oligosaccharide compounds from *Streptomyces calvus* TM-521 [137].

Acarbose is a pseudotetrasaccharide made by *Actinoplanes* sp. SE50. It contains an aminocyclitol moiety, valienamine, which inhibits intestinal α-glucosidase and sucrase. This results in a decrease in starch breakdown in the intestine which is useful in combating diabetes in humans [138].

Xanthine oxidase (XO) catalyzes the oxidation of hypoxanthine to uric acid via xanthine. An excessive accumulation of uric acid in the blood, called hyperuricacidemia, causes gout [139]. The inhibitors of XO decrease the uric acid levels and result in an antihyperuicemic effect. A potent inhibitor of XO, hydroxyakalone, was purified from the fermentation broth of *Agrobacterium aurantiacum* sp. nov., a marine bacterial strain [140].

Fungal products are also used as enzyme inhibitors against cancer, diabetes, poisoning, Alzheimer's disease, etc. The enzymes inhibited include acetylcholinesterase, protein kinase, tyrosine kinase, glycosidases and others [141].

Inhibitor	Target	Application
Acarbose	α-Amylase	Type 2 diabetes
Ancovenine	Angiotensin-converting enxyme	Hypertension
Bialaphos, Fosfinotricine	Glutamine synthetase	Herbicide
Clavulanic acid	β-Lactamase	β-Lactam resistance
Fibrostatin C	Proline hydroxylase	Fibrosis
Phthoxazoline	Cellulose synthetase	Herbicide
Nikkomycin	Chitin synmthetase	Insectide, fungicide
Validoxylamine	Trehalase	Stress tolerance enhancer in plants vs. cold amd draught

Figure 5. Enzyme Inhibitors from *Streptomyces* spp. [171].

6.6. Products Acting Against Insects

Insecticides are compounds used to kill insects in all developmental forms. They are used in agriculture, medicine, industry and the household. It is believed that their use represented one of the major factors behind the increase in agricultural productivity in the 20th century. Synthetic insecticides pose some hazards whereas the natural insecticides offer adequate levels of pest control and pose fewer hazards. Microbially produced insecticides are especially valuable because their toxicity to non-target animals and humans is extremely low. Compared to other commonly used insecticides, they are safe for both the pesticide user and consumers of treated crops. The action of microbial insecticides is often specific to a single group or species of insects and this specificity means that most microbial insecticides do not naturally affect beneficial insects [including predators or parasites of pests] in treated areas. The macrotetrolide tetranactin [142] has been in use since 1974 as a mitocide for plants. A screen for microbial insecticides was developed by Fabre et al. [143].

Among the most important NPs with insecticidal activity are the spinosyns (A83543 group), a group of non-toxic compounds produced by *Saccharopolyspora spinosa*, which were discovered at Eli Lilly in 1989 and sold today by Dow AgroSciences. These are environmentally friendly tetracyclic macrolides with activity against insects of the orders Coleoptera, Diptera, Hymenoptera, Isoptera, Lepidoptera, Siphonoptera and Thysanoptera. The researchers isolated spinosyns A and D as well as 21 minor analogues. Spinosad is a mixture of spinosyns A and D, i.e., two 21-carbon atom tetracyclic macrolides (polyketides) containing the deoxysugars forosamine and tri-O-methyl rhamnose with different levels of methylation on the polyketide moiety. They differ by a methyl group at position 6 of the polyketide. Spinosad is used for protection of crops and feedstock animals [144]. The spinosyns are active on a wide variety of insect pests, especially lepidopterans and dipterans but they do not have antibiotic activity [145]. The compounds attack the nervous system of insects by targeting two key neurotransmitter receptors, i.e., nicotinic acetylcholine receptors, with no cross resistance to other known insecticides. Spinosad, the mixture of 85% spinosyn A and 15% spinosin D, is produced via fermentation and was introduced to the market in 1997 under the trademark Naturalyte® for the control of chewing insects on a variety of crops. Genome shuffling has been used for strain improvement to reach a titer of 547 mg/L [146].

Spinosyn formulations were recently approved for use on organic crops (Entrust®) and for animal health applications (Elector®). Spinosad has improved activity over target insects as compared to older insecticides, is less of a risk against beneficial insects, and lacks carcinogenic, mutagenic or teratogenic activity in humans and animals.

A semi-synthetic second generation spinosyn is spinetoram which is more active, has an earlier onset of activity and is longer acting. Novel spinosyns have also been prepared by biotransformation, using a genetically engineered strain of *Saccharopolyspora erythraea* [147]. A new naturally occurring series of insect-active compounds was discovered in recent years from a novel soil isolate, *Saccharopolyspora pogona* NRRL30141 [148]. The culture produces a unique family of over 30 new spinosyns. They have a butenyl substitution at the 21 position on the spinosyn lactone and are named butenyl-spinosyns or pogonins.

Bioinsecticides include *Bacillus thuringiensis* (BT) spores which contain the crystalline δ-endotoxin, active against insects such as Lepidoptera and Diptera which occur in grain silos and forestry, and also against tropical mosquitos.

Nikkomycins are substrate analogs of UDP-acetylglucosamine which inhibit chitin synthetase Chs3 and fungal cell wall formation. They are structurally related to the polyoxins which are used as agricultural antifungal agents. Despite their excellent target, the nikkomycins have not been commercialized as antifungals due to the existence of alternate chitin synthetases (e.g., Chs2), their narrow spectrum of activity, and poor permeability. The nikkomycin antibiotics [149] are of considerable interest as bioinsecticides however. Since chitin is an important structural material for insects, the nikkomycins have potent insecticidal activity.

6.7. Antiparastic Drugs and Ruminant Growth Stimulants

In 2010, the global animal health market was valued at US $20.1 billion, of which an estimated 30% was represented by parasiticides. Parasites are organisms which inhabit the body and benefit from a prolonged, close association with the host. Antiparasitics are compounds that inhibit growth or reproduction of a parasite; some antiparasitics directly destroy parasites. In general, parasites are much smaller than their hosts, show a high degree of specialization for their mode of life and reproduce more quickly and in greater numbers than their hosts. Classic examples of parasitism include the interactions between vertebrate hosts and such diverse animals as tapeworms, flukes, *Plasmodium* species and fleas. Parasitic infections can cause potentially serious health problems and even kill the host. They mainly enter the body via the mouth, usually through ingestion of tainted food or drink. This is a very common problem in tropical areas of the world, but is not limited to those regions. There are 3200 varieties of parasites in four major categories: Protozoa, Trematoda, Cestoda and Nematoda.The major groups include protozoans (organisms having only one cell) and parasitic worms (helminthes). Each of these can infect the digestive tract, and sometimes two or more can cause infection at the same time. The WHO reported that about 25% of the world's population is infected with roundworms. In addition, a major agricultural problem has been the infection of farm animals by worms.

Testing of synthetic compounds against nematodes has been the predominant type of antiparasitic screening effort over the years and commercial products did result. Certain antibiotics were also shown to possess antihelmintic activity against nematodes or cestodes, but these failed to compete with the synthetic compounds. Although Merck & Co. had earlier developed a commercially useful synthetic product, thiobenzole, they had enough foresight to examine microbial broths for antihelmintic activity, and found a nontoxic fermentation broth which killed the intestinal nematode *Nematosporoides dubius* in mice. The *Streptomyces avermitilis* culture, isolated by Omura and coworkers at the Kitasato Institute in Japan [150], was found by Merck scientists to produce a family of secondary metabolites (eight compounds) with both antihelmintic and insecticidal activities. These compounds, named "avermectins" are pentacyclic, 16-membered macrocyclic lactones, which harbor a disaccharide of the methylated sugar oleandrose [Figure 6] with exceptional activity against parasites especially Nemathelminthes (nematodes) and arthropod parasites, i.e., ten times higher than any known synthetic antihelmintic agent. Surprisingly, avermectins lack activity against bacteria and fungi, do not inhibit protein synthesis, and are not ionophores. Instead, they interfere with neurotransmission in many invertebrates causing paralysis and death by neuromuscular attack [151].

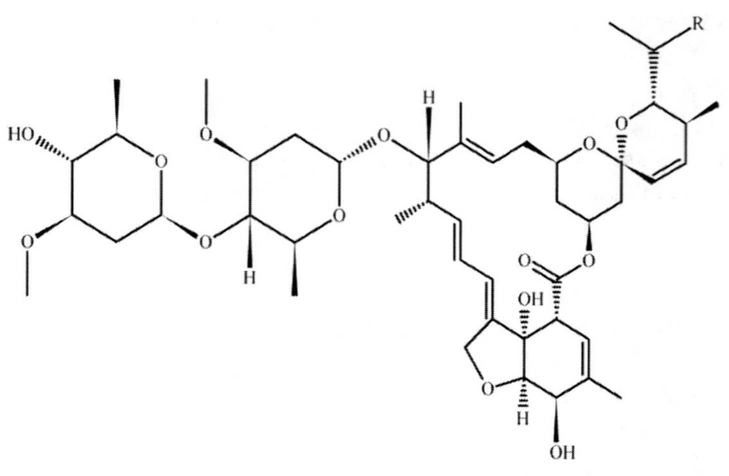

Figure 6. Chemical Structure of Avermectins.

Annual sales of avermectins have surpassed one billion dollars. They are used against both nematode and arthropod parasites in sheep, cattle, dogs, horses, and swine. A semisynthetic derivative, 22,23-dihydro-avermectin B1 (ivermectin) is one thousand times more active than thiobenzole and is a commercial veterinary product. The efficacy of ivermectin has made it a powerful drug for the control of human onchocerciasis (river blindness) and human strongyloidiasis [152]. Another avermectin, called doramectin (=cyclohexylavermectin B1), produced by "mutational biosynthesis" was commercialized for use by food animals [153].

DNA shuffling of the *ave C* gene of *S. avermitilis* gave an improved ratio of the undesirable CHC-B2 to the desirable CHC-B1 of 0:07:1. This was an improvement of 21-fold over the ratio with the starting gene [154]. The effort was carried out by personnel of Pfizer, Maxygen and Codexis.

A semisynthetic monosaccharide derivative of doramectin called selamectin is the most recently commercialized avermectin, and is the first from the class to be active against heartworms (*Dirofilaria immitis*) and fleas in companion animals. Although the macrocyclic backbone of each of these molecules (ivermectin, doramectin, and selamectin) is identical, there are different substitutions at pharmacologically relevant sites such as C-5, C-13, C-22, 23, and C-25 [155]. The avermectins are closely related to the milbemycins, a group of non-glycosilated macrolides produced by *S. hygroscopicus* subsp. *aureolacrimosus* [156]. These compounds possess activity against worms and insects.

In both extensively and intensively reared poultry, coccidiostats are used for the prevention of coccidiosis. It is a common intestinal disease caused by the invading protozoan parasites of the genus *Eimeria* that affects several different animal species (cattle, dogs, cats, poultry, etc.). The major damage is due to the rapid multiplication of the parasite in the intestinal wall, and the subsequent rupture of the cells of the intestinal lining, leading to high mortality and severe loss of productivity. Coccidia are obligate intracellular parasites, which show host specificity. Only cattle coccidia will cause disease in cattle, whereas other species-specific coccidia will not.

For a long time, synthetic compounds were used to combat coccidiosis in poultry; however resistance developed rapidly. A solution came with the discovery of the narrow spectrum polyether antibiotic monensin, which had extreme potency against the coccidian [157]. Made by *Streptomyces cinnamonensis*, monensin led the way for additional microbial ionophoric antibiotics such as lasalocid, narasin and salinomycin. All are produced by various *Streptomyces* species. They form complexes with the polar cations K+, Na+, Ca2+ and Mg2+, severely affecting the osmotic balance in the parasitic cells, thus causing their death [158]. The widespread use of anticoccidials has revolutionized the poultry industry by reducing mortality and production losses caused by coccidiosis.

Of great interest was another extremely valuable application of monensin, i.e., growth promotion in ruminants. Previously synthetic chemicals were tested for years to inhibit wasteful methane production by cattle and sheep and increase fatty acid formation (especially propionate) in order to improve feed efficiency; however, they failed. The solution was monensin which became a major success as a ruminant growth enhancer.

Certain antibiotics have been used in food-animal production for more than 40 years, to enhance feed utilization and weight gain [159]. From a production standpoint, feed antibiotics have been consistently shown to improve animal weight gain and feed efficiency, especially in younger animals. These responses are probably derived from an inhibitory effect on the normal microbiota, which can lead to reduced intestinal inflammation and improved nutrient utilization [160]. Pigs in the USA are exposed to a great variety of antibiotics. These include β-lactam antibiotics (including penicillins), lincosamides and macrolides (including erythromycin and tetracyclines). All these groups have members that are used to treat infections in humans. In addition, bacitracin, flavophospholipol, pleuromutilins, quinoxalines and virginiamycin are utilized as growth stimulants. Flavophospholipol and virginiamycin, are also used as growth promoters in poultry. As described above, cattle are also exposed to ionophores such as monensin to promote growth. The Animal Health Institute of America 161] has estimated that, without the use of growth-promoting antibiotics, the USA would require an additional 452 million chickens, 23 million more cattle and 12 million more pigs to reach the levels of production attained by the current practices.

Animal health research and development of new anti-infective products have decreased over the past 15 years, resulting in few new drug approvals [162]. Therefore, it will be incumbent upon veterinary practitioners to use the existing products in a responsible manner to ensure their longevity. It remains to be seen what effects the dearth of new antibiotics for veterinary medicine will have on the future practice of veterinary medicine, production agriculture, food safety and public health [163]. Since the 1999 European Union decision to prohibit antibiotic use for food-animal growth promotion, four antibiotic growth promoters have been banned, including the macrolide drugs tylosin and spiramycin [164]. Although macrolides are no longer formally used as 'growth promoters', their use under veterinary prescription rose from 23 tons in 1998 to 55 tons in 2001, which suggests that more is now being used than before the prohibition.

6.8. Herbicidal Drugs

These are compounds marketed to inhibit or interrupt plant growth and development. They are applied in agriculture, industry, and urban areas for weed management. About 30,000 kinds of weeds are widely distributed in the world; yield losses caused by 1800 kinds of weeds are about 9.7% of total crop production every year [165]. With a minimum of labor, herbicides provide cost-effective weed control. Herbicides are mainly used on crops planted in large acreages, such as soy, cotton, corn, and canola [166]. There are many classes of herbicides with different modes of action as well as different potentials for adverse effects on health and the environment. The use of chemical herbicides employed to control various weeds may have caused many serious side-effects such as injured crops, threat to the applicator and others exposed to the chemicals, herbicide-resistant weed populations, reduction of soil and water quality, herbicide residues, and detrimental effects on non-target organisms [167]. In this regard, alachlor (Lasso) and atrazine (AAtrex) were reported to cause cancer in animal tests. On the other hand, with increasing global environmental consciousness, bio-herbicides, which are highly effective for weed control and environmentally friendly as well, are very attractive both for research and for application. Microbial herbicides can be divided into microbial preparations [microorganisms that control weeds] and microbially-derived herbicides.

Groups in Germany and Japan independently discovered the first microbial herbicide. In 1972, the Zähner group in Germany isolated phosphinothricin tripeptide. They isolated it from *Streptomyces viridochromogenes* as a broad-spectrum antibacterial including activity against *Boyritis cinerea* [168]. Chemically, it is a peptide antibiotic consisting of two molecules of L-alanine and one molecule of the unusual amino acid L-phosphinothricin, i.e., N[4[hydroxyl[methyl]phosphinoyl] homoalanyl]alanylalanine. In Japan, it was discovered at the Meiji Seiki laboratories in 1973 from *S. hygroscopicus* and called bialaphos [169]. The bioactive L-phosphinothricin is a structural analogue of glutamic acid, acting as a competitive inhibitor of glutamine synthetase, and has bactericidal (Gram-positive and Gram-negative), fungicidal (*Botrytis cinerea*), and herbicidal properties [170]. Glufosinate, (DL-phosphinothricin without Ala-Ala), was developed as a herbicide. Therefore, the group acts as herbicides with or without Ala-Ala. Bialaphos has no influence on microorganisms in the soil and is easily degraded in the environment, having a half-life of only two hours. This low level of environmental impact is of great interest to environmentalists.

Bioherbicides acting against weeds include spores of the mold *Colletotrichum gloeosporoides* (Collego®) which controls northern jointvetch in rice and soybean fields [171]. Spores of *Alternaria cassiae* are used against the weed sickelpod which can grow in soybean fields and on peanut plants.

Conclusion

Microorganisms have greatly contributed for about 85 years to the development of medicine and agriculture. However, due to different situations, microbes have become resistant to many antibiotics creating a dangerous situation and therefore the need for new antibiotics is imperative. Unfortunately, most of the large pharmaceutical companies have abandoned the search for new antimicrobial compounds. Due to economics, they have concluded that drugs directed against chronic diseases offer a better revenue stream than do antimicrobial agents, for which the length of treatment is short and government restriction is likely. Some small pharmaceutical and biotechnology companies are still developing antibiotics but most depend on venture capital rather than sales income, and with the present regulations, face huge barriers to enter into the market. These barriers were raised with the best intentions of ensuring public safety but they are having the opposite effect, i.e., termination of antibiotic development while resistance continues to increase [172]. However, there are some new bright possibilities. One of the most promising is the utilization of uncultivated microorganisms. Considering that 99% of bacteria and 95% of fungi have not been cultivated in the laboratory, efforts into finding means to grow such uncultured microorganisms is proceeding and succeeding [3]. Furthermore, researchers are now extracting bacterial DNA from soil samples, cloning large fragments into, for example, bacterial artificial chromosomes, expressing in a host bacterium and screening the library for new antibiotics. This metagenomic effort could open up the exciting possibility of a large untapped pool from which new NPs could be discovered [173]. Another exciting possibility is that of genome mining [174]. In addition to these three relatively new techniques, chemical and biological modification of old antibiotics could still supply new and powerful drugs. These comments also apply to non-antibiotics such as antitumor agents and other microbial products.

In addition, NPs must continue to be tested for desirable therapeutic activities. We believe that significant progress in identifying new antibiotics, oncology therapeutics and other useful medicines will be made, probably not by the big pharmaceutical companies, but by small research groups from institutes and universities.

Acknowledgments

The authors acknowledge the excellent assistance of Sana Siddiqui in preparation of the revised article.

Conflict of Interest

None.

References

[1] Hölker U, Höfer M, Lenz J (2004). Biotechnological advantages of laboratory-scale solid-state fermentation with fungi. *Appl. Microbiol. Biotechnol.* 64:175-186.

[2] Berdy J (2005) Bioactive microbial metabolites. A personal view. *J. Antibiot.* 58:1-26.

[3] Kaeberlein T, Lewis K, Epstein SS (2002) Isolating "uncultivable" microorganisms in pure culture in a simulated natural environment. *Science* 296:1127-1129.

[4] Colwell RR (2002) Fulfilling the promise of biotechnology. *Biotechnol. Adv.* 20:215-228.

[5] Gaudilliere B, Bernardelli P, Berna P (2001) To market, to market-2000. In: Doherty AM (ed). *Annual Reports in Medicinal Chemistry*, Chapter 28, Vol 36, Academic Press: Amsterdam, pp 293-318.

[6] Cragg GM, Newman DJ, Snader KM (1997) Natural products in drug discovery and development. *J. Nat. Prod.* 60:52-60.

[7] Butler MS (2008) Natural products to drugs: natural product derived compounds in clinical trials. *Nat. Prod. Rep.* 25:475-516.

[8] Health United States 2011. *National Center for Health Statistics.* http://www.cdc.gov/nchs/hus/contents2011.htm#022.

[9] Rayl AJS (1999) Oceans: medicine chests of the future? *The Scientist* 13:1-5.

[10] Robinson WE, Reinecke MG, Abdel-Malek S, Jia O, Chow A (1996) Inhibitors of HIV-1 replication that inhibit HIV integrase. *Proc. Natl. Acad. Sci. USA* 93:6326-6331.

[11] Cragg GM, Newman DJ (2001) Medicinals for the millennia. The historical record. *Ann. NY Acad. Sci.* 953a:3-25.

[12] Yang SS, Cragg GM, Newman DJ, Bader JP (2001) Natural products based anti-HIV drug discovery and development facilitated by the NCI developmental therapeutics program. *J. Nat. Prods.* 64:265-277.

[13] Farnsworth NR, Akerele O, Bingel AS, Soejarto DD, Guo Z (1985) Medicinal plants in therapy. Bull. WHO 63:965-981.

[14] Overbye KM, Barrett JF (2005) Antibiotics: where did we go wrong? *Drug Discov. Today* 10:45-52.

[15] Topliss JG, Clark AM, Ernst E, Hufford CD, Johnston GAR, Rimoldi JM, Weimann BJ (2002) Natural and synthetic substances related to human health. *Pure Appl. Chem.* 74:1957-1985.

[16] Patchett AA (2002) Natural products and design: interrelated approaches in drug discovery. *J. Med. Chem.* 45:5609-5616.

[17] Cushman DW, Ondetti MA (1999) Design of angiotensin converting enzyme inhibitors. *Nature Med.* 5:1110-1112.

[18] Spizek J, Novotna J, Rezanka T, Demain AL (2010) Do we need new antibiotics? The search for new tyargets and new compounds. *J. Ind. Microbiol. Biotechnol.* 37:1241-1248.

[19] Demain AL, Spizek J (2012) The antibiotic crisis. In G. Tegos & E. Mylonakis, eds., Antimicrobial Drug Discovery: Emerging Strategies. *CAB International*, pp. 26-43.

[20] Perlroth J, Choi B, Spellberg B (2007) Nosocomial fungal infections: epidemiology, diagnosis and treatment. *Med. Mycol.* 45:321-346.

[21] Hancock REW (2007) The end of an era? *Nature Rev. Drug Discov.* 6:26.

[22] Katz ML, Mueller LV, Polyakov M & Weinstock SF (2006) Where have all the antibiotic patents gone? *Nature Biotechnol.* 24:1529-1531.

[23] Infection Disease Society of America (2004) http://www.idsociety.org/pa/IDSA-_Paper4_final_web_.pdf.

[24] Balaban N, Dell'Acqua G (2005) Barriers on the road to new antibiotics. *The Scientist* 19:42-43.

[25] Lynch AS, Abbanat D (2010) New antibiotic agents and approaches to treat biofilm-associated infections. *Expert Opin. Ther. Patents* 20:1373-1387.

[26] Levin BR, Bonten MJM (2004) Cycling may be bad for your health. *Proc. Natl. Acad. Sci. USA* 101:13101-13102.

[27] Ecker DJ, Sampath R, Willett P, Wyatt JR, Samant V, Massire C, Hall TA, Hari K, McNeil JA, Büchen-Osmond C, Budowle B (2005) *The Microbial Rosetta Stone Database: A compilation of global and emerging infectious microorganisms and bioterrorist threat agents. BMC Microbiol.* 5:19.

[28] Ryan KJ, Ray CG (2004) *Sherris Medical Microbiology*, 4th ed., McGraw Hill, New York, pp 434-437.

[29] OECD forum (2004) Health of Nations: Combating infectious diseases. organization for economic co-operation and development. *Centre de Conférences Internationales* 19 avenue Kléber 75116 Paris, France. 12-13 May, 2004.

[30] Morse SS (1997) The public health threat of emerging viral disease. *J. Nutr.* 127:951S-957S.

[31] Coates A, Hu Y, Bax R, Page C (2002) The future challenges facing the development of new antimicrobial drugs. *Nature Rev. Drug Discov.* 1:895-910.

[32] Moellering RC Jr (1998) Problems with antimicrobial resistance in gram-positive cocci. *Clin. Infect. Dis.* 26:1177-1178.

[33] Bacque E, Barriere JC, Berthand N (2005). Recent progress in the field of antibacterial pristinamycins. *Curr. Med. Chem. Anti-infect. Agents* 4:185-217.

[34] Tally FP, DeBruin MF (2000) Development of daptomycin for Gram-positive infections. *J. Antimicrob. Chemother.* 46:523–526.

[35] Raja A, LaBonte J, Lebbos J, Kirkpatrick P (2003) Daptomycin. *Nature Rev. Drug Discov.* 2:943-944.

[36] Eisenstein BI (2004) Lipopeptides, focusing on daptomycin, for the treatment of Gram-positive infections. *Expert Opin. Investig. Drugs* 13:1159-1169.

[37] LaPlante KL, Rybak MJ (2004) Daptomycin - a novel antibiotic against Gram-positive pathogens. *Expert Opin. Pharmacother.* 5:2321-2331.

[38] Sum PE, Sum FW, Projan SW (1998) Recent developments in tetracycline antibiotics. *Curr. Pharm. Des.* 4:119-132.

[39] Sum PE (2006) Case studies in current drug development: 'glycylcyclines'. *Curr. Opin. Chem. Biol.* 10:374-379.

[40] Leclercq R (2001) Overcoming antimicrobial resistance: profile of a new ketolide antibacterial, telithromycin. *J. Antimicrob. Chemother.* 48:9-23.

[41] Jensen SE, Paradkar AS (1999) Biosynthesis and molecular genetics of clavulanic acid. *Ant. V. Leeuwenhoek*. 75:125-133.

[42] Singh N (2005) Invasive aspergillosis in organ transplant recipients: New issues in epidemiologic characteristics, diagnosis, and management. *Med. Mycol*. 43 (Suppl. 1) S267-S270.

[43] Enoch DA, Ludlam HA, Brown NM (2006) Invasive fungal infections: A review of epidemiology and management options. *J. Med. Microbiol*. 55:809-818.

[44] Alexander BD, Perfect JR (1997) Antifungal resistance trends towards the year 2000. Implications for therapy and new approaches. *Drugs* 54:657–678.

[45] Hoang A (2001) Caspofungin acetate: an antifungal agent. *Am. J. Health Syst. Pharm.* 58:1206-1217.

[46] Georgopapadakou NH (2001) Antifungals targeted to cell wall: focus on β-1,3-glucan synthase. *Expert Opin. Invest. Drugs* 10:269-280.

[47] Ikeda F, Tanaka S, Ohki H, Matsumoto S, Maki K, Katashima M, Barrett D, Aoki Y (2007) Role of micafungin in the antifungal armamentarium *Curr. Med. Chem.* 14:1263-1275.

[48] Kwon DS, Mylonakis E (2007) Posaconazole: A new broad-spectrum antifungal agent. *Expert Opin. Pharmacother.* 8:1167-1178.

[49] Cardenas ME, Sanfridson A, Cutler NS, Heitman J (1998) Signal-transduction cascades as targets for therapeutic intervention by natural products. *Trends Biotechnol.* 16:427-433.

[50] Kremer L, Douglas JD, Baulard AR, Morehouse C, Guy MR (2000) Thiolactomycin and related analogues as novel anti-mycobacterial agents targeting KasA and KasB condensing enzymes in Mycobacterium tuberculosis. *J. Biol. Chem.* 275:16857-16864.

[51] Verdine GL (1996) The combinatorial chemistry of nature. *Nature* 384:11-13.

[52] Nicholls SJ, Tuzcu EM, Sipahi I, Grasso AW, Schoenhagen P, Hu T, Wolski K, Crowe T, Desai MY, Hazen SL, Kapadia SR, Nissen SE (2007) Statins, high-density lipoprotein cholesterol, and regression of coronary atherosclerosis. *JAMA* 297:499-508.

[53] Endo A (2010) A historical perspective on the discovery of statins. *Proc. Jpn. Acad. Ser. B* 86:484-492.

[54] Alberts AW, Chen J, Kuron G, Hunt V, Huff J, Hoffman C, Rothrock J, Lopez M, Joshua H, Harris, E et al. (1980) Mevinolin, a highly potent competitive inhibitor of hydroxylmethylglutaryl-coenzyme A reductase and a cholesterol-lowering agent. *Proc. Natl. Acad. Sci. USA* 77:397-3961

[55] Xie X, Tang Y(2007) Efficient synthesis of simvastatin by use of whole-cell biocatalysts. *Appl. Environ. Microbiol.* 73:2054-2060.

[56] Serizawa N, Matsuoka T (1991) A two-component-type cytochrome P-450 monooxygenase system in a prokaryote that catalyzes hydroxylation of ML-236B to pravastatin, a tissue-selective inhibitor of 3-hydroxy-3-methylglutaryl coenzyme A reductase. *Biochim. Biophys. Acta* 1084:35-40.

[57] Peng Y, Demain AL (1998) A new hydroxylase system in Actinomadura sp. cells converting compactin to pravastatin. *J. Indust. Microbiol. Biotechnol.* 20:373-375.

[58] Alarcon J, Aguila S, Arancibia-Avila P, Fuentes O, Zamorano-Ponce E, Hernandez M (2003) Production and purification of statins from Pleurotus ostreatus (Basidiomycetes) strains. *Z. Naturforsch*. 58:62-64.

[59] GLOBOCOAN database (2008) http://globocan.iarc.fr.

[60] Waksman SA, Woodruff HB (1941) Actinomyces antibioticus, a new soil organism antagonistic to pathogenic and non-pathogenic bacteria. *J. Bacteriol.* 42:231-249.

[61] Minotti G, Menna P, Salvatorelli E, Cairo G, Gianni L (2004) Anthracyclines: molecular advances and pharmacologic developments in antitumor activity and cardiotoxicity. *Pharmacol. Rev.* 56:185-229.

[62] Umezawa H, Maeda K, Takeuchi T, Okami Y (1966) New antibiotics, bleomycin A and B. *J. Antibiot.* 19:200-209.

[63] Schein PS, Macdonald JS, Hoth DF, Wooley PV (1979) The FAM (5-fluorouracil, adriamycin, mitomycin C) and SMF (streptozotocin, mitomycin C, 5-fluorouracil) chemotherapy regimens. In. Carter et al. (eds) Mitomycin C: *Current Status and New Developments.* Academic Press, New York pp. 133-143.

[64] Paz MM, Kumar GS, Glover M, Waring MJ, Tomasz M (2004) Mitomycin dimers: polyfunctional cross-linkers of DNA. *J. Med. Chem.* 47:3308-3319.

[65] Fernández E, Weißbach U, Sánchez Reillo C, Braña AF, Méndez C, Rohr J and Salas JA (1998) Identification of two genes from Streptomyces argillaceus encoding glycosyltransferases involved in transfer of a disaccharide during biosynthesis of the antitumor drug mithramycin. *J. Bacteriol.* 180:4929–4937.

[66] Ratan RR (2011) Mithramycin is a gene-selective Sp1 inhibitor that identifies a biological intersection between cancer and neurodegeneration. *J. Neurosci.* 31:6858-70.

[67] Wang Z, Gleichmann H (1998) GLUT2 in pancreatic islets: crucial target molecule in diabetes induced with multiple low doses of streptozotocin in mice. *Diabetes* 47:50-56.

[68] Schowalter DH, Bunge RH, French JC, Hurley TR, Leeds RL, Leja B, McDonnell PD, Edmunds CR (1992) Improved production of pentostatin and identification of fermentation cometabolites. *J. Antibiot.* 45:1914-1918.

[69] Dang NH, Hagemesiter FB, Duvic M, Romaguera JE, Younes A, Jones D, Samuels B, Fayad LE, Pro B, Samaniego F, Sarris A, Goy A, McLaughlin P, Tong AT, Walker PL, Tiongson LP, Smith TL, Huh YO, Morimoto C, Rodriguez MA (2003) Pentostatin in T-non-Hodgkin's lymphomas: efficacy and effect on CD26+ T lymphocytes. *Oncol. Rep.* 10:1513-1518.

[70] Walker S, Landovitz R, Ding WD, Ellestad GA, Kahne D (1992) Cleavage behavior of calicheamicin gamma 1 and calicheamicin T. *Proc. Natl. Acad. Sci. USA* 89:4608-4612.

[71] Bross PF, Beitz J, Chen G, Chen XH, Duffy E, Kieffer L, Roy S, Sridhara R, Rahman A, Williams G, Pazdur R (2001) Approval summary: gemtuzumab ozogamicin in relapsed acute myeloid leukemia. *Clin. Cancer Res.* 7:1490-1496.

[72] de Vries JF, Zwaan CM, De Bie M, Voerman JS, den Boer ML, van Dongen JJ, van der Velden VH (2012) The novel calicheamicin-conjugated CD22 antibody inotuzumab ozogamicin (CMC-544) effectively kills primary pediatric acute lymphoblastic leukemia cells. *Leukemia* 26:255-264.

[73] Wall ME, Wani MC (1996) Camptothecin and taxol: from discovery to clinic. *J. Ethnopharmacol.* 51:239-254.

[74] Amna T, Puri SC, Verma V, Sharma, JP, Khajuria RK, Spiteller M, Qasi GN (2006) Bioreactor studies on the endophytic fungus Entrophospora for the production of an anticancer alkaloid camptothecin. *Can. J. Microbiol.* 52:189-196.

[75] Lorence A, Nessler CL (2004) Camptothecin, over four decades of surprising findings. *Phytochemistry* 65:2735-2749.

[76] Venditto VJ, Simanek EE (2010) Cancer therapies utilizing the camptothecins: a review of the in vivo literature. *Molec. Pharmaceut.* 7:307-349.

[77] Newman DJ and Cragg GM (2007) Natural products as sources of new drugs over the last 25 Years. *Nat. Prod.* 70:461-477.

[78] Sabater-Jara AB, Tudela LR, Lopez-Perez AJ (2010) In vitro cultura of Taxus sp.: strategies to increase cell growth and taxoid production. *Phytochem. Res.* 9:343-356.

[79] Stierle A, Strobel G, Stierle D (1993) Taxol and taxane production by Taxomyces andreanae, an endophytic fungus of Pacific yew. *Science* 260:214-216.

[80] Flores-Bustamante ZR, Rivera-Orduna FN, Martinez-Cardenas A, Flores-Cotera LB (2010) Microbial paclitaxel: advances and perspectives. *J. Antibiot.* 63:460-467.

[81] Li J-Y, Strobel G, Sidhu R, Hess WM, Ford EJ (1996) Endophytic taxol-producing fungi from bald cypress, Taxodium distichum. *Microbiology* 142:2223-2226.

[82] Wang JF, Li GL, Lu HY, Zhang ZH, Huang YJ, Su WJ (2000) Taxol from Tubercularia sp. strain TF5, an endophytic fungus of Taxus mairei. *FEMS Microbiol. Lett.* 193:249-253.

[83] Zhao K, Zhou D, Ping W and Ge J (2004) Study on the preparation and regeneration of protoplast from taxol-producing fungus Nodulisporium sylviforme. *Nat. Sci.* 2:52-59.

[84] Kumaran RS, Muthumary JP, Hur B-K (2008) Taxol from Phyllosticta citricarpa, a leaf spot fungus of the angiosperm Citrus medica. References and further reading may be available for this article. To view references and further reading you must purchase this article *J. Biosci. Bioeng.* 106:103-106.

[85] Wei Y, Liu L, Zhou X, Lin J, Sun X, Tang K (2012) Engineering taxol biosynthetic pathway for improving taxol yield in taxol-producing endophytic fungus EFY-21 (Ozonium sp.). *African J. Biotechnol.* 11:9094-9101.

[86] Kumaran RS, Choi Y-K, Lee S, Jeon H J Jung H, Kim HJ (2012) Isolation of taxol, an anticancer drug produced by the endophytic fungus, Phoma betae. *African J. Biotechnol.* 11: 950-960.

[87] Zhang P, Zhou P-P, Yu L-J (2009) An endophytic taxol-producing fungus from Taxus media, Cladosporium cladosporoides MD2. *Curr. Microbiol.* 59:227-232.

[88] Liu K, Ding X, Deng B, Chen W (2009) Isolation and characterization of endophytic taxol-producing fungi from Taxus chinensis. *J. Ind. Microbiol. Biotechnol.* 36: 1171-1177.

[89] Zhou X, Zhu H, Liu L, Lin J, Tang K (2010) A review: recent advances and future prospects of taxol-producing endophytic fungi. *Appl. Microbiol. Biotechnol.* 86: 1707-1717

[90] Duan L-L, Chen H-R, Chen J-P, Li, W-P, Hong L (2008) Screening the high-yield paclitaxel producing strain Alternaria alternate var monosporus. *Chi. J. Antibiot.* 33: 650-652

[91] Gerth K, Bedorf N, Hofle G (1996) Epothilons A and B: Antifungal and cytotoxic compounds from Sorangium cellulosum (Myxobacteria): Production, physico-chemical and biological properties. *J. Antibiot.* 49:560-563.

[92] Bollag DM, McQueney PA, Zhu J (1995) Epothilones, a new class of microtubule-stabilizing agents with a taxol-like mechanism of action. *Cancer Res.* 55:2325-2333.

[93] Kowalski RJ, Giannakakou P, Hamel E (1997) Activities of the microtubule-stabilizing agents epothilones A and B with purified tubulin and in cells resistant to paclitaxel (Taxol). *J. Biol. Chem.* 272:2534-2541.

[94] Goodin S (2008) Novel cytotoxic agents: epothilones. *Am. J. Health Syst. Pharm.* 65 (10 Suppl 3): S10-5.

[95] Goodin S, Kane MP, Rubin EH (2004) Epothilones: mechanism of action and biologic activity. *J. Clin. Oncol.* 22:2015-2025.

[96] Ueda H, Nakajima H, Hori Y, Fujita T, Nishimura M, Goto T, Okuhara M (1994) FR901228, a novel antitumor bycyclic depsipeptide produced by Chromobacterium violaceum No. 968. I. Taxonomy, fermentation, isolation, physico-chemical and biological properties, and antitumor activity. *J. Antibiot.* 47:301-310.

[97] VanderMolen KM, McCulloch W, Pearce CJ, Oberlies NH (2011) Romidepsin (Istodax, NSC 630176, FR901228, FK228, depsipeptide): a natural product recently approved for cutaneous T-cell lymphoma. *J. Antibiot.* 64:525-531.

[98] Jemal A, Siegel R, Ward E, Murray T, Xu J, Thun MJ (2007) Cancer statistics, 2007. *CA- Cancer J. Clin.* 57:43-66.

[99] Einhorn LH (2002) Curing metastatic testicular cancer. *Proc. Natl. Acad. Sci. USA* 99:4592-4595.

[100] Feldman DR, Bosl GJ, Sheinfeld J, Motzer RJ (2008) Medical treatment of advanced testicular cancer. *JAMA* 299:672-684.

[101] Culine S, Kerbrat P, Kramar A, Théodore C, Chevreau C, Geoffrois L, Bui NB, Pény J, Caty A, Delva R, Biron P, Fizazi K, Bouzy J, Droz JP (2007) Refining the optimal chemotherapy regimen for good-risk metastatic nonseminomatous germ-cell tumors: a randomized trial of the Genito-Urinary Group of the French Federation of Cancer Centers (GETUG T93BP). *Ann. Oncol.* 18:917-924.

[102] Joyce C, Saxena S, Williams A, Damurjian C, Auricchio N, Aluotto S, Tynan H, Demain AL (2010) Antimicrobial spectrum of the antitumor agent, cisplatin. *J. Antibiot.* 63:530-532.

[103] Borel JF (2002) History of the discovery of cyclosporin and of its early pharmacological development. *Wien Klin. Wochenschr.* 114:433–437.

[104] Huang S, Houghton PJ. (2003) Targeting mTOR signaling for cancer therapy. *Curr. Opin. Pharmacol.* 3:371-377.

[105] Aggarwala D, Fernandez ML, Solimanb GA (2006) Rapamycin, an mTOR inhibitor, disrupts triglyceride metabolism in guinea pigs. *Metabolism* 55:794-802.

[106] Park SR, Yoo YJ, Ban Y-H, Yoon YJ (2010) Biosynthesis of rapamycin and its regulation: past achievements and recent progress. *J. Antibiot.* 63:434-441.

[107] Kojima I, Cheng YR, Mohan V, Demain AL (1995) Carbon source nutrition of rapamycin biosynthesis in Streptomyces hygroscopicus. *J. Ind. Microbiol.* 14:436-439.

[108] Lee MS, Kojima I, Demain AL (1997) Effect of nitrogen source on biosynthesis of rapamycin by Streptomyces hygroscopicus. *J. Ind. Microbiol. Biotechnol.* 19:83-86.

[109] Paiva N, Demain AL, Roberts MF (1991) Incorporation of acetate, propionate and methionine into rapamycin by Streptpmyces hygroscopicus. *J. Nat. Prods.* 54:167-177.

[110] Paiva N, Demain AL, Roberts MF (1993a) The immediate precursor of the nitrogen-containing ring of rapamycin is free pipecolic acid. *Enzyme Microb. Technol.* 15:581-585.

[111] Paiva N, Demain AL, Roberts MF (1993b) The cyclohexane moiety of rapamycin is derived from shikimic acid in Streptomyces hygroscopicus. *J. Ind. Microbiol.* 12:423-425.

[112] Zhu X, Zhang W, Chen X, Wu H, Duan Y, Xu Z (2010) Generation of high rapamycin producing strain via rational metabolic pathway-based mutagenesis and further titer improvement with fed-batch bioprocess optimization. *Biotechnol. Bioeng.* 107:506-515.

[113] Kuscer et al. (2007) Roles of rapH and rapG in positive regulation of rapamycin biosynthesis in Streptomyces hygroscopicus. *J. Bacteriol.* 189:4766-4763.

[114] Brown VI, Fang J, Alcorn K, Barr R, Kim JM, Wasserman R, Grupp SA (2003) Rapamycin is active against B-precursor leukemia in vitro and in vivo, an effect that is modulated by IL-7-mediated signaling. *Proc. Natl. Acad. Sci. USA* 100: 15113-15118.

[115] Guba M, von Breitenbuch P, Steinbauer M, Koehl G, Flegel S, Hornung M, Bruns CJ, Zuelke C, Farkas S, Anthuber M, Jauch KW, Geissler EK (2002) Rapamycin inhibits primary and metastatic tumor growth by antiangiogenesis: involvement of vascular endothelial growth factor. *Nature Med.* 8:128-135.

[116] Rini B, Kar S, Kirkpatrick, P (2007) Temsirolimus. *Nature Rev./Drug Discov.* 6: 599-600.

[117] Bailly C (2009) Ready for a comeback of natural products in oncology. *Biochem. Pharmacol.* 77:1447-1457.

[118] *Nature Medicine* June 22, 2008 online edition.

[119] Eisen HJ, Tuzcu EM, Dorent R, Kobashigawa J, Manzini D, Valantine-von Kaeppler HA, Starling RC, Sørensen K, Hummel M, Lind JM, Abeywickrama KH, Bernhardt P (2003) Everolimus for the prevention of allograft rejection and vasculopathy in cardiac-transplant recipients. *New Eng. J. Med.* 349:847-858.

[120] http://www.genengnews.com/gen-news-highlights/novartis-afinitor-cleared-by-fda-for-treating-sega-tumors-in-tuberous-sclerosis/81244159/.

[121] Dancey JE (2006) Therapeutic targets: MTOR and related pathways. *Cancer Biol. Ther.* 5:1065-1073.

[122] Chen Y-J (2007) Targeted mTOR in human gynecologic cancers. *J. Cancer Molec.* 3: 101-106.

[123] Kino T, Hatanaka H, Hashimoto M, Nishiyama M, Goto T, Okuhara M, Kohsaka M, Aoki H, Imanaka H (1987) FK-506, a novel immunosuppressant isolated from a Streptomyces. 1. Fermentation, isolation, and physico-chemical and biological characteristics. *J. Antibiot.* 40:1249-1255.

[124] Jain AB, Todo S, Fung JJ, Venkataramanan R, Day R, Bryant J, Abu-Elmagd KM, Alessiani M, Takaya S, Tzakis A, Starzl TE (1991) Correlation of rejection episodes with FK 506 dosage, FK 506 level and steroid following primary orthotopic liver transplant. *Transplant. Proc.* 23:3023-3025.

[125] Nagano J, Iyonaga K, Kawamura K, Yamashita A, Ichiyasu H, Okamoto T, Suga M, Sasaki Y, Kohrogi H (2006) Use of tacrolimus, a potent antifibrotic agent, in bleomycin-induced lung fibrosis. *Eur. Respir. J.* 27:460-469.

[126] Funabashi Y, Maeshiba Y, Inatomi N, Tanayama S, Harada S, Itoh Z, Omura S (1996) Bioactive metabolites of EM574 and EM523, erythromycin derivatives having strong gastrointestinal motor stimulating activity. *J. Antibiot.* 49:794-801.

[127] Bradley C (2001) Erythromycin as a gastrointestinal prokinetic agent. *Int. & Crit. Care Nursing* 17:117-119.

[128] McCallum RW, Cynshi O (2007) Clinical trial: effect of mitemcinal (a motilin agonist) on gastric emptying in patients with gastroparesis - a randomized, multicentre, placebo-controlled study. *Aliment. Pharmacol. Ther.* 26:1121-1130.

[129] Umezawa H (1972) *Enzyme inhibitors of microbial origin*, University of Tokyo Press, Tokyo.

[130] Weibel EK, Hadvary P, Hochuli E, Kupfer E, Lengsfeld H (1987) Lipstatin, an inhibitor of pancreatic lipase, produced by Streptomyces toxytricini. *J. Antibiot.* 40:1081-1085.

[131] Imada C (2004) Enzyme inhibitors of marine microbial origin with pharmaceutical importance. *Mar. Biotechnol*. 6:193-198.

[132] Boivin M, Flourie B, Rizza RA, Go VL, DiMagno EP (1988) Gastrointestinal and metabolic effects of amylase inhibition in diabetics. *Gastroenterology* 94:387-394.

[133] Díaz E, Aguirre C, Gotteland M (2004) Effect of an α-amylase inhibitor on body weight reduction in obese women. *Rev. Chil. Nutr.* 31:306-317.

[134] Boivin M, Zinsmeister AR, Go VL, DiMagno EP (1987) Effect of a purified amylase inhibitor on carbohydrate metabolism after a mixed meal in healthy humans. *Mayo. Clin. Proc*. 62:249-255.

[135] Banks BJ, Haxell MA, Lunn G, Pacey MS, Roberts LR (2006) Treatment of rumen acidosis with alpha-amylase inhibitors. *European Patent* EP1157696.

[136] Arai M, Oouchi N, Murao S (1985) Inhibitory properties of an α-amylase inhibitor, paim, from Streptomyces corchorushii. *Agric Biol. Chem*. 49:987-991.

[137] Kangouri K, Namiki S, Nagate T, Sugita K, Omura S (1980) Novel amylase inhibitors. *US Patent* 4197292.

[138] Truscheit E, Frommer W, Junge B, Müller L, Schmidt DD, Wingender W (1981) Chemistry and biochemistry of microbial α-glucosidase inhibitors. *Angew. Chem. Intl. Ed. Engl.* 20:744-761.

[139] Borges F, Fernandes E, Roleira F (2002) Progress towards the discovery of xanthine oxidase inhibitors. *Curr. Med. Chem.* 9:195-217.

[140] Izumida, H, Miki W, Sano H, Endo M (1995) Agar plate method, a new assay for chitinase inhibitors using a chitin-degrading bacterium. *J. Mar. Biotechnol*. 2:163-166.

[141] Paterson RRM (2008) Fungal enzyme inhibitors as pharmaceuticals, toxins and scourge of PCR. *Curr. Enzyme Inhib*. 4:46-59.

[142] Oishi H, Sugawa T, Okutomi T, Suzuki K, Hayashi T, Sawada M, Ando K (1970) Insecticidal activity of macrotetrolide antibiotics. *J. Antibiot.* 23:105-106.

[143] Fabre B, Armau E, Etienne G, Legendre F, Tiraby G (1988) A simple screening method for insecticidal substances from actinomycetes. *J. Antibiot.* 41:212-219.

[144] Kirst HA (2010) The spinosyn family of insecticides: realizing the potential of natural products research. *J. Antibiot*. 63:101-111.

[145] Kirst HA, Creemer LC, Naylor SA, Pugh PT, Snyder DE, Winkle JR, Lowe L, Rothwell JT, Sparks TC, Worden TV (2002) Evaluation and development of spinosyns to control ectoparasites on cattle and sheep. *Curr. Topics Med. Chem.* 2:675-699.

[146] Jin ZH, Xu B, Lin SZ, Jin QC, Cen PL (2009) Enhanced production of spinosad in Saccharopolyspora spinosa by genome shuffling. *Appl. Biochem. Biotechnol.* 2159: 655-663.

[147] Gaisser S, Martin CJ, Wilkinson B, Sheridan RM, Lill RE, Weston AJ, Ready SJ, Waldron C, Crouse GD, Leadlay PF, Staunton J (2002) Engineered biosynthesis of novel spinosyns bearing altered deoxyhexose substituents. *Chem. Commun.* 21:618-619.

[148] Hahn DR, Gustafson G, Waldron C, Bullard B, Jackson JD, Mitchell J (2006) Butenyl-spinosyns, a natural example of genetic engineering of antibiotic biosynthetic genes. *J. Ind. Microbiol. Biotechnol.* 33:94-104.

[149] Brillinger GU (1979) Metabolic products of microorganisms. 181. Chitin synthase from fungi, a test model for substances with insecticidal propertries. *Arch. Microbiol.* 121: 71-74.

[150] Stapley EO (1982) Avermectins, antiparasitic lactones produced by Streptomyces avermitilis isolated from a soil in Japan. In: Umezawa H, Demain AL, Hata R, Hutchinson CR (eds) Trends in antibiotic research. *Japan Antibiotic Research Association*, Tokyo, pp 154-170.

[151] Tooth JA, Davis, MW, Avery LA (1997) avr-15 encodes a chloride channel subunit that mediates inhibitory glutamatergic neurotransmission and ivermectin sensitivity in Caenorhabditis elegans. *EMBO J.* 16:5867-5879.

[152] Shikiya K, Kinjo N, Uehara T, Uechi H, Ohshiro J, Arakaki T, Kinjo F, Saito A, Iju M, Kobari M (1992) Efficacy of ivermectin against Strongyloides stercoralis in humans. *Intern. Med.* 31:310-312.

[153] Goudie AC, Evans NA, Gration KAF, Bishop BF, Gibson SP, Holdom KS, Kaye B, Wicks SR, Lewis D, Weatherly AJ, Bruce CI, Herbert A, Seymour DJ (1993) Doramectin-A potent novel endectocide. *Vet. Parasitol.* 49:5-15.

[154] Stutzman-Engwall K, Conlon S, Fedechko R, McArthur H, Pekrun K, Chen Y, Jenne S, La C, Trinh N, Kim S et al. (2005) Semi-synthetic DNA shuffling of aveC leads to improved industrial scale production of doramectin by Streptomyces avermitilis. *Metab. Eng.* 7:27-37.

[155] Michael B, Meinke PT, Shoop WL (2002) Comparison of ivermectin, doramectin, selamectin, and eleven intermediates in a nematode larval development assay. *J. Parasitol.* 87:692-696.

[156] Mishima H, Ide J, Muramatsu S, Ono M (1983) Milbemycins, a new family of macrolide antibiotics. Structure determination of milbemycins D, E, F, G, H, J and K. *J. Antibiot.* 36:980-990.

[157] Westley JW (1977) Polyether antibiotics: versatile carboxylic acid ionophores produced by Streptomyces. *Adv. Appl. Microbiol.* 22:177-223.

[158] Matabudul DK, Lumley ID, Points JS (2002) The determination of 5 anticoccidial drugs (nicarbazin, lasalocid, monensin, salinomycin and narasin) in animal livers and eggs by liquid chromatography linked with tandem mass spectrometry (LC-MS-MS). *Analyst.* 127:760-768.

[159] Page SW (2003) The role of enteric antibiotics in livestock production. *National Association for Crop Production and Animal Health*, Avcare, Australia, 2003.

[160] Gaskins HR, Collier CT, Anderson DB (2002) Antibiotics as growth promotants: mode of action. *Anim. Biotechnol.* 13:29-42.

[161] Animal Health Institute USA (1998) Antibiotic resistance back in the news. *AHI Quarterly* 19:1-4.

[162] Nelson R (2003) Antibiotic development pipeline runs dry. *Lancet* 362:1726-1727.

[163] Shryock TR (2004) The future of anti-infective products in animal health. *Nat. Rev. Microbiol.* 2:425-430.

[164] Casewell M, Friis C, Marco E, McMullin P, Phillips I (2003) The European ban on growth-promoting antibiotics and emerging consequences for human and animal health. *J. Antimicrob. Chemother.* 52:159-161.

[165] Li Y, Sun Z, Zhuang X, Xu L, Chen S, Li M (2003) Research progress on microbial herbicides. *Crop. Prot.* 22:247:252.

[166] Randall JM (1996) Weed control for the preservation of biological diversity. *Weed Technol.* 10:370-383.

[167] Kline WN and Duquesnel JG (1996) Control of problem vegetation: a key to ecosystem management. *Down to Earth* 51:20-28.

[168] Bayer E, Gugel KH, Hägele K, Hagenmaier H, Jessipow S, König WA, Zähner H (1972) Metabolic products of microorganisms. 98. Phosphinothricin and phosphinothricyl-alanyl-analine. *Helv. Chim. Acta* 55:224-239.

[169] Kondo Y, Shomura T, Ogawa Y, Tsuruoka T, Watanabe H, Totsukawa K, Suzuki T, Moriyama C, Yoshida J, Inouye S, Niida T (1973) Studies on a new antibiotic, SF-1293. I. Isolation and physico-chemical and biological characterization of SF-1293 substances. *Sci. Rep. Meiji Seika* 13:34-41.

[170] Schwartz D, Berger S, Heinzelmann E, Muschko K, Welzel K, Wohlleben W (2004) Biosynthetic gene cluster of the herbicide phosphinothricin tripeptide from Streptomyces viridochromogenes Tu494. *Appl. Environ. Microbiol.* 70:7093-7102.

[171] Vandamme EJ (2007) Microbial gems: Microorganisms without frontiers. *SIM-News* 57:81-91.

[172] Livermore DM (2004) The need for new antibiotics. *Clin. Microbiol. Infect.* 10:1-9.

[173] Clardy J, Fischbach MA, Walsh CT (2006) New antibiotics from bacterial natural products. *Nature Biotechnol.* 24:1541-1550.

[174] Scheffler R, Colmer S, Tynan H, Demain AL, Gullo VP (2013) Antimicrobials, drug discovery, and genome mining. *Appl. Microbial. Biotechnol.* 97:969-978.

In: Microbes in Process
Editors: Neelam Garg and Abhinav Aeron
ISBN: 978-1-63117-127-7

Chapter 3

Biotechnological Potential of Citrus Peels

Diomi Mamma[1] and Paul Christakopoulos[2,*]
[1]Biotechnology Laboratory, School of Chemical Engineering,
National Technical University of Athens, Zografou Campus,
Zografou, Attica, Greece
[2]Biochemical and Chemical Process Engineering,
Division of Sustainable Process Engineering, Department of Civil,
Environmental and Natural Resources Engineering,
Luleå University of Technology,
Luleå, Sweden

Abstract

Citrus by-products are the processing wastes generated after citrus juice extraction and constitute about 50% of fresh fruit weight. This solid residue comprised of peel (flavedo and albedo), pulp (juice sac residue), rag (membranes and cores) and seeds. The disposal of the fresh peels is becoming a major problem to many factories. Usually, citrus juice industries dry the residue and it is either sold as raw material for pectin extraction or pelletized for animal feeding, though none of these processes is very profitable. This residual material is a poor animal feed supplement because of its extremely low protein content and high amount of sugar. The application of agroindustrial by-products in bioprocesses offers a wide range of alternative substrates, thus helping solve pollution problems related to their disposal. Attempts have been made to use citrus by-products to generate several value-added products, such as enzymes, single cell protein, natural antioxidants, ethanol, organic acids, polysaccharides and prebiotics. This chapter reviews developments regarding processes and products that employed citrus peels as a substrate.

* E-mail: paul.christakopoulos@ltu.se.

1. Introduction

The genus *Citrus* includes several important fruits [1], with the most important on a worldwide basis being sweet orange (*C. sinensis*: 61.1% of world citrus production), tangerine (*C. reticulata*: 19.9%), lemon and lime (*C. limon* and *C. aurantifolia*: 12.1%) and grapefruit (*C. paradisi*: 5.0%). Minor citrus genuses that comprise the bulk of the remaining 2.0% include sour orange (*C. Quarantium*), shaddock (*C. Grandis*), citron (*C. Medica*). According to FAO estimates the world citrus production for the year 2010 was 111930847 tonnes (Figure 1) [2]. About 16% of world production of citrus is in the Mediterranean countries of Spain, Italy, Greece, Egypt, Turkey and Morocco, with Brazil (19%), China (17%) and the USA (9%) being major individual citrus producing countries (Figure 1). About 30% of the production of citrus fruits(and 40% of orange production) is processed, principally to make juice, and results large quantities of by-products.

Exocarp, is the outermost layer of the fruit. The exocarp forms the tough outer skin of the fruit, which bears oil glands and pigments. The exocarp in citruses is called the *flavedo*. *Flavedo* is mostly composed of cellulosic material but also contains other components, such as essential oils, paraffin waxes, steroids and triterpenoids, fatty acids, pigments (carotenoids, chlorophylls, flavonoids), bitter principles (limonene), and enzymes. When ripe, the flavedo cells contain carotenoids (mostly xanthophyll) inside chromoplasts, which, in a previous developmental stage, contained chlorophyll. This hormonally controlled progression in development is responsible for the fruit's change of color from green to yellow upon ripening. The internal region of the flavedo is rich in multicellular bodies with spherical or pyriform shapes, which are full of essential oils. The mesocarp is also referred to as *albedo*. It is the inner part of the peel which is commonly removed before eating. The endocarp is separated into sections, which are most commonly called segments. The juicy pulp filling the segments is usually referred to as juice vesicles.

Citrus by-products is the principal solid by-product of the citrus processing industry and constitutes about 50% of fresh fruit weight [3]. It contains the peel (60-65%), internal tissues (30-35%) and seeds (0-10%). Until comparatively recently these waste products led to significant disposal problems, since there was no satisfactory means of disposal other than dumping on land adjacent to the production sites. This led in some regions to the generation of large tracts of land containing significant quantities of putrefying waste which presents a significant risk to local water courses and in some cases leads to uncontrolled methane production. Traditionally, the only other disposal routes for waste orange peels were to utilize it as raw material in the manufacture of cattle feed or simply to burn it [4, 5]. These routes would also generate global warming gases, although some of the carbon dioxide generated from combustion would not contribute significantly to a net increase, since it would be subsequently assimilated by plants. This disposal problem has stimulated significant recent interest in developing more responsible ways of dealing with waste products, ideally with the added benefit of yielding high value products and establishing more environmentally responsible approaches.

	Total	Oranges	Tangerines	Lemon and limes	Grapefruits
WORLD	**111930847**	**69416336**	**14244782**	**21311892**	**6957837**
Africa	**9932561**	**6749760**	**883310**	**1678423**	**621068**
Algeria	998100	740000	71000	185800	1300
Egypt	3518235	2401020	318111	796867	2237
Ghana	602100	556100	46000		
Morocco	1345529	849197	22498	472834	1000
South Africa	2116750	1415090	215980	142500	343180
Tunisia	296000	118500	42900	45400	89200
Others	1055847	669853	166821	35022	184151
Americas	**45931097**	**34898652**	**5680628**	**3192911**	**2158906**
Argentina	2559423	833486	1113380	423737	188820
Belize	273200	237200			36000
Brazil	21327480	19112300	1020350	1122730	72100
Chile	289000	134000	155000		
Colombia	527839	330000	87474	110365	
Costa Rica	356400	312000	26300		18100
Cuba	345000	178263	6060	23017	137660
Ecuador	88794	47494	5900	31400	4000
El Salvador	69694	45699	16076	7919	
Guatemala	249600	139000	110600		
Honduras	286492	266221	9853		10418
Mexico	6753406	4051630	1891400	409442	400934
Paraguay	327179	229898	9057	44224	44000
Peru	853147	394573	233032	221324	4218
United States of America	9941830	7478830	800140	539770	1123090
Uruguay	315214	154211	37656	121450	1897
Venezuela	533800	400000	51600	75000	7200
Others	833599	553847	106750	62533	110469
Asia	**45564491**	**20868872**	**6452399**	**14142136**	**4101084**
China	19051144	5003289	1058105	10121000	2868750
India	9627300	6268100	3098900		260300
Indonesia	2032670	2032670			
Iran (Islamic Republic of)	2532258	1502820	706800	276138	46500
Israel	556523	134829	65079	152207	204408
Japan	843500	53000	4500	786000	
Lebanon	402600	241100	113100	38500	9900
Pakistan	2203000	1542100	88120	572780	
Republic of Korea	614871			614871	
Syrian Arab Republic	866100	668900	142200	26000	29000
Thailand	1118913	372700	171074	280190	294949
Turkey	3570030	1710500	787063	858699	213768
Viet Nam	752000	729400			22600
Others	1393582	609464	217458	415751	150909
Europe	**9951774**	**6495029**	**1192649**	**2199197**	**64899**
Greece	981700	770000	75200	130000	6500
Italy	3163790	2393660	522377	240628	7125
Portugal	244600	193900	13800	34400	2500
Spain	5449600	3120000	578200	1708200	43200
Others	112084	17469	3072	85969	5574
Oceania	**550924**	**404023**	**35796**	**99225**	**11880**
Australia	521645	391343	28900	91002	10400
New Zealand	24450	10500	4700	8200	1050
Others	4829	2180	2196	23	430

Figure 1. World Citrus Production (tonnes) [2].

2. Production and Composition of Citrus by-Products

Citrus fruits are principally consumed by humans as fresh fruit or processed juice, either fresh, chilled or concentrated. After juice is extracted from the fruit, there remains a residue (Figure 2) comprised of peel (flavedo and albedo), pulp (juice sac residue), rag (membranes and cores) and seeds. These components, either individually or in various combinations, are the source materials from which citrus by-product feedstuffs (BPF) are produced [6, 7]. The main citrus BPF from citrus processing are fresh citrus pulp which is the whole residue after extraction of juice, representing between 492 and 692 g/kg of fresh citrus fruit with 600–650 g dry matter (DM)/kg peel, 300–350 g/kg pulp and 0–100 g/kg seeds [8], and dried citrus pulp (DCP) which is formed by shedding, liming, pressing and drying the peel, pulp and seed residues to about 80 g/kg moisture, and citrus meal and fines which is formed and separated during the drying process. A typical processing plant produces these BPF in a ratio of about 850 g/kg DCP, 140 g/kg citrus meal and 10 g/kg citrus fines. Other citrus BPF include citrus molasses, made by concentrating the press liquor from the citrus peel residue, which has a bitter taste and contains about 100–150 g/kg solubles of which 500–700 g/kg consists of sugar [7], citrus peel liquor, which is similar to citrus molasses, but not as concentrated, and citrus activated sludge which is produced from liquid wastes from citrus processing plants. Other minor BPF from citrus include cull or excess fruit [9].

The composition of citrus fruit is affected by factors such as growing conditions, maturity, rootstock, variety and climate [1]. Citrus fruits contain N (1–2 g/kg on a wet basis), lipids (oleic, linoleic, linolenic, palmitic, stearic acids, glycerol, and a phytosterol), sugars (glucose, fructose, sucrose), acids (primarily citric and malic, but also tartaric, benzoic, oxalic, and succinic), insoluble carbohydrates (cellulose, pectin), enzymes (pectinesterase, phosphatase, peroxidise), flavonoids (hesperidin, naringin), bitter principles (limonin, isolimonin), peel oil (*d*-limonene), volatile constituents (alcohols, aldehydes, ketones, esters, hydrocarbons, acids), pigments (carotenes, xanthophylls), vitamins (ascorbic acid, Vitamin B complex, carotenoids), and minerals (primarily calcium and potassium) [10]. The nutrient content of citrus BPF is influenced by factors that include the source of the fruit (growing conditions, maturity, rootstock, variety, and climate) and type of processing [11].

Bampidis and Robinson [10] investigated the composition of several citrus by-products (Figure 3). It is worthy of note that the pH of orange peel can be as low as 3.64. This means neutralization of pH may be required for some applications. The carbohydrate content of different citrus by-products is presented in Figure 4.

3. Bioprocesses Involving Citrus Peels

Major strategies for citrus waste utilization include: production of enzymes (Figure 5), bioethanol (Figure 6), heteropolysaccharides, citric acid, single cell protein, natural antioxidants, prebiotics (Figure 7).

Whole peel or rind (pericarp)	Consists of flavedo (exterior yellow peel, epicarp) and albedo (interior whity spongy peel, mesocarp). Albedo is rich in pectin. The whole peel combined with the pulp residue (rag) and/or molasses can becomes a feed for animals. It is also used for production of human foods and food supplements
Pulp (principal edible portion, endocarp)	Used mainly to produce raw juice for human nutrition, after mechanical extraction and screening. The material screened from the raw juice is also called pulp and is usually combined with other residues to produce by-products used in animal nutrition
Pulp residue (called rag in the industry)	Consists of the fraction screened from the pulp, being cores, segment walls or membranes, juice vesicles and seeds. The pulp residue is usually combined with peel residue to manufacture by-products feeds. From the lime-treated mass peel and pulp residues, citrus processors produce such by-products as press liquor, citrus molasses, citrus pulp, citrus meal and feed yeast. It is also used for production of human foods and food supplements
Seeds	Sometimes separated from the rag to produce seed oils, seed meals and dried seed pressed cake
Waste waters (aqueous effluent emulsions from processing plants)	Have potential uses for production of such products as activated sludge and yeasts. It is also used for production of human foods and food supplements

Figure 2. Products and by-Products of Various Tissues of Citrus Fruits [6, 10].

	Orange peels (fresh)	Lemon peels (fresh)	Bergamot peels	Grapefruit pulp (fresh)	Mandarin pulp fresh
DM[a]	233	202	163	197	197
OM	975	967	957	962	960
CP	58	66	-	71	80
Crude Fat	-	16	13	63	61
NDF	200	330	101	-	-
ADF	129	-	73	168	174
Lignin	-	-	27	19	21
Sugar	-	191	-	-	-

[a]ADF, acid detergent fibre; CP, crude protein; DM, dry matter; NDF, neutral detergent fibre; NE, net energy; OM, organic matter.

Figure 3. Chemical Composition of Several Citrus by-Products (g/kg DM) [10].

	Ash	Sugar	Fat	Protein	Pectin	Lignin	Cellulose	Hemicellulose
Lemon peels[a]	2.5	6.5	1.5	7.0	13.0	7.6	23.1	8.09
Lemon pulp[a]	2.5	9.0	3.1	8.7	22.5	7.6	36.2	11.1
Orange peels[a]	2.6	9.6	4.0	9.1	23.0	7.5	37.1	11.0
Orange pulp[a]	2.6	6.0	1.5	6.6	12.1	7.5	24.5	7.6
Orange peels[b]	1.7	41.4	3.9	7.9	14.4	1.0	16.2	13.8
Kinnow mandarin waste[c]	3.2	31.6	n.d.*	5.8	22.6	0.6	10.1	4.3
Mandarin peel waste[d]	3.0	21.6	n.d.	n.d.	14.2	8.9	20.8	17.2
Grapeftuit Peels[e]	3.3	n.d.	n.d.	n.d.	16.1	8.2	19.8	18.3

[a]From [(12)]for citrus wastes from Spanish citrus juice factories.
[c] From [(13)]for kinnow waste (composed of peel and pulp) from a fruit processing facility in India
[d] From [(14)]for mandarin citrus peel waste (MCPW), from citrus juice factories in Spain
[b]From [(15)] for orange peels from a fruit processing facility in Greece
[e] From [(16)]
*n.d. not determined

Figure 4. Composition of Different Citrus Wastes (% of dry weight).

Enzyme	Microorganism	Substrate	Process	Reference
Pectinases	*Aspergillus foetidus* ATCC 16878	Citrus waste	SSF[(a)]	[3]
	A. niger A-20	Orange peel	SmF	[17]
	A. flavipes FP-500, A. terreus FP-370	Lemon peel	SmF	[18]
	A.niger	Lemon peel pomace	SSF	[19]
Alkaline pectinase	*Bacillus* sp. MG-cp-2	Orange peel	SmF	[20]
Polygalacturonase	*Aspergillus* sp.	Decaying lemons	SmF	[21]
	Fusarium oxysporum F3, *A. niger* BTL, *Neurospora crassa* DSM 1129, *Penicillium decumbens*	Orange peel	SSF	[15]
	A. niger, Trichoderma viride	Citrus Pulp	SLSF[(c)]	[22]
	Tubercularia vulgaris	Orange-pulp pellets	SmF	[23]
	Thermoascus aurantiacus 179-5	Orange bagasse	SSF	[24]
	P. vericatum Rfc 3	Orange bagasse	SSF	[25]
Endo-polygalacturonase	*Aspergillus* sp. CH-Y-1043	Lemon peel	SmF	[26]
	Aureobasidium pullulans	Orange peel	SmF	[27]
	A. niger MTCC 281	Citrus peel	SmF[(b)]	[28]
	Sporotrichum thermophile	Citrus peel	SmF	[29]
Pectinesterase	*Aspergillus* sp.	Decaying lemons	SmF	[21]
Polygalacturonase lyase	*B. subtilis* strain 11089	Orange waste	SmF-CC[(d)]	[30]
Pectate lyase	*F. oxysporum* F3, *A. niger* BTL, *N. crassa* DSM 1129, *P. decumbens*	Orange peels	SSF	[15]
Pectin lyase	*P. vericatum* Rfc 3	Orange bagasse	SSF	[25]
	Aspergillus giganteus	Orange waste	SmF	[31]
α-Amylase	*A. niger* A-20	Orange peels	SmF	[17]
	B. subtilis strain 11089	Orange waste	SmF-CC	[30]
Xylanase	*A. niger* A-20	Orange peels	SmF	[17]
	Trichoderma harzianum	Orange pomace, orange peel, lemon pomace, lemon peel	SSF	[32]
	F. oxysporum F3, *A. niger* BTL, *N. crassa* DSM 1129, *P. decumbens*	Orange peels	SSF	[15]
Cellulase	*A. niger* A-20	Orange peels	SmF	[17]
	F. oxysporum F3, *A. niger* BTL, *N. crassa* DSM 1129, *P. decumbens*	Orange peels	SSF	[15]
Protease	*B. subtilis* strain 11089	Orange waste	SmF-CC	[30]

[(a)]SSF: Solid State Fermentation, [(b)]SmF: Submerged Fermentation, [(c)]SLSF: Slurry State Fermentation, [(d)]CC: Continuous Culture

Figure 5. Citrus Wastes as a Substrate for Enzyme Production of Industrial Interest.

Substrate	Process	Microorganism	Ethanol concentration (g/L)	Ethanol yield	Reference
Orange peel hydrolysate	SHF	*Saccharomyces cerevisiae*	40.0-45.0	-	[33]
Orange peel hydrolysate	SHF	*Escherichia coli KO11*	35.0-38.0	-	[34]
Orange peel hydrolysate	SHF	*S. cerevisiae* and *Kluyveromyces marxianus*	37.1-40.9	-	[35]
Citrus peel waste	SSF	*S. cerevisiae*	39.0	0.210**	[36]
Kinnow waste and banana peel	SSF	*S. cerevisiae*	26.8	0.426*	[37]
Orange peel	Two stage hydrolysis followed by fermentation	*S. cerevisiae*	30.4	0.250*	[38]
Kinnow mandarin waste	SSF	*S. cerevisiae*	42.0	0.280**	[13]
Pretreated mandarin peel	SHF	*S. cerevisiae*	46.2	0.462*	[39]
Steam exploded manadarin citrus peel waste	SSF	*S. cerevisiae*	59.3***	0.220**	[14]
Steam exploded lemon peel waste	SSF	*S. cerevisiae*	67.8***	0.250**	[16]
Steam pretreated citrus peels waste	SSF	*S. cerevisiae*	40.0	-	[40]

* expressed in g of ethanol per g of total biomass
** expressed in g of ethanol per g of consumed sugar
***expressed in L of ethanol per 1000 kg of fresh mandarin citrus peel or lemon peel waste

Figure 6. Citrus Wastes as a Substrate for Bioethanol Production.

Application	Microorganism or enzyme used	Process	Reference
Xanthan production			
	Xanthomonas campestris	SmF(a)	[41, 42]
	X. campestris	SSF(b)	[43]
Citric acid production			
	A. niger	SmF	[44, 45]
	A. niger	SSF	[46, 47, 48, 49]
Single Cell Protein			
	Penicillium spp	SSF	[50]
	A. niger, T. viride	SLSF(c)	[22]
	Geotrichum candidum	SSF	[51, 52]
Enzymatic pectin extraction			
	Endopolygalacturonase, pectin methyl esterase	-	[53, 54, 55, 56]
	Protopectinase	-	[57, 58, 59]
	Pectinase 62L, Pectinase 690L, Cellulase CO13P	-	[60]
	Viscozyme® L		[61]
	Promod 24L+ Cellulyve TR400,+PME-A		[62]
Prebiotic oligosaccharides			
	Pectinase 62L	-	[63]
	Endo-polygalacturonase	EMR(d)	[64]
	Rapidase Liq+		[62]
	Celluclast 1.5L ,Viscozyme® L		[65]
Enzymatic extraction of flavonoids			
	Celluzyme MX	-	[66]
	Pectinase 690L, Pectinase 62L, Cellulase C013P	-	[60]

(a)SSF: Solid State Fermentation, (b)SmF: Submerged Fermentation, (c)SLSF: Slurry State Fermentation, (d)EMR: Enzyme Membrane Reactor
Cellulase CO13P: cellulolytic preparation (Biocatalysts Ltd. Wales, UK)
Celluclast 1.5L: cellulase from *Trichoderma reesei* ATCC 26921 (Novozymes, A/S, Denmark)
Celluzyme MX: cellulolytic preparation (Zymus International Limited, Auckland, New Zealand)
Cellulyve TR400: cellulolytic preparation produced by *Trichoderma reesei* (Lyven, Colombelle, France)
Pectinase 62L and Pectinase 690L: pectinolytic preparations, produced by *Aspergillus* sp. (Biocatalysts Ltd. Wales, UK)
PME-A : Pectin Methyl Esterase from *A. aculeatus* expressed in *A. oryzae* (Novozymes, A/S, Denmark)
Promod 24L: broad spectrum protease produced by *Bacillus* sp. (Biocatalysts Ltd. Wales, UK)
Rapidase Liq+: cell wall degrading enzymes from a mixture of *T. longibrachiatum* and *A. niger* (DSM, Delft, The Netherlands)
Viscozyme® L: multi-enzyme complex containing a wide range of carbohydrases, including arabanase, cellulase, β-glucanase, hemicellulase, and xylanase from *A. aculeatus* (Novozymes, A/S, Denmark)

Figure 7. Other Bioprocesses Using Citrus Peels as Substrate.

3.1. Enzyme Production

The most important area of citrus peels utilization is the production of enzymes, especially pectinolytic. Pectinases are a heterogeneous group of related enzymes that hydrolyze the pectic substances. Pectinases are classified under three headings according to the following criteria: whether pectin, pectic acid or oligo-D-galacturonate is the preferred substrate, whether pectinases act by trans-elimination or hydrolysis and whether the cleavage is random (endo-, liquefying of depolymerizing enzymes) or endwise (exo- or saccharifying enzymes). The three major types of pectinases are: pectinesterases, depolymerising enzymes (enzymes that either hydrolyze glycosidic linkages or cleave α-1, 4-glycosidic linkages by trans-elimination) and protopectinase [67]. Pectinolytic enzymes are of significant importance in the current biotechnological era with their all embracing applications in fruit juice extraction and its clarification, scouring of cotton, degumming of plant fibers, waste water treatment, vegetable oil extraction, tea and coffee fermentations, bleaching of paper, in poultry feed additives and in the alcoholic beverages and food industries [67, 68, 69].

Due to the chemical composition of cirtus peels other enzymes could be produced resulted in multienzyme complexes. Cellulases and hemicellulases are among these enzymes. Cellulases are a complex enzyme system, comprising endo-1,4-β-D-glucanase, exo-1,4-β-glucanase and β-D-glucosidase [70]. These enzymes are employed in feed, fuel and chemical industries for the processing of lignocellulosic materials [71, 72]. The hemicellulolytic system carrying out the xylan hydrolysis is usually composed of a repertoire of hydrolytic enzymes: β-1,4-endoxylanase, β-xylosidase, α-L-arabinofuranosidase, α-glucuronidase, acetyl xylan esterase, and phenolic acid [ferulic and *p*-coumaric acid] esterase. Xylanolytic enzymes have attracted a great deal of attention in the last decade, particularly because of their biotechnological potential in various industrial processes such as food, feed, and pulp and paper industries [73]. Bacteria, yeasts, and fungi under both submerged (SmF) and solid state fermentation (SSF) conditions are able to produce these enzymes.

Maldonado et al. [21] tested *Aspergillus* sp. from decaying lemons for extracellular pectinase production using differently pretreated lemon peel as the carbon source instead of pectin. It was found that the production of extracellular polygalacturonase was about the same and that of pectinesterase substantially higher when unwashed fresh lemon peel was used instead of pectin. Pectinase production by *Tubercularia vulgaris* using orange-pulp pellets or citrus pectin as carbohydrate sources was investigated by Fonseca and Said [23]. The highest levels of extracellular polygalacturonase were detected with OPP as inducing substrate. High levels of endo-polygalacturonase were produced by *Aureobasidiurn pullulans* on orange-peel waste [27]. Polygalacturonase and pectin lyase production by *P. viridicatum* strain Rfc3 was carried out by means of solid state fermentation using orange bagasse, corn tegument, wheat bran and mango and banana peels as carbon sources. The maximal activity value of polygalacturonase (30U/g) was obtained using wheat bran as carbon source while maximal pectin lyase (2000 U/g) activity value was obtained in medium composed of orange bagasse. The mixture of orange bagasse and wheat bran (50%) increased the production of polygalacturonase and pectin lyase to 55 U/g and 3540 U/g respectively [25]. Larios et al. [26] studied the endo-polygalacturonase production by *Aspergillus sp.* CH-Y-1043 using untreated lemon peel and citrus pectin as carbon sources. Untreated lemon peel proved to be a better substrate, while *Aspergillus* sp. CH-Y-1043 produced more endo-polygalacturonase at 37°C than at 29°C. Endo-polygalacturonase production was very sensitive to pH changes.

Optimization of the culture medium as far as the nitrogen source, concentration of phosphates and initial culture pH resulted 65.2 U/ml endopolygalactorunase activity. The production of endo-polygalacturonase by the fungus *A. niger* MTCC 281 using powdered citrus peel was studied by Dhillon et al. [28]. Maximum enzyme activity (0.940 μM/ml/min) was observed with 15% substrate semi-solid substrate, when incubated at 30°C for 120 h, using 5% inoculum.

Mahmood et al. [30] found that *Bacillus* species 11089 was capable of growth in continuous culture on orange substrate as carbon-energy source in a mineral salts basal medium and produced α-amylase, neutral and alkaline proteases, and polygalacturonate-lyase. The activities obtained were 19.0 U/ml α-amylase, 20 U/ml neutral protease, 7.5 U/ml alkaline protease and 2.0 Units/ml polygalacturonate lyase. The enzyme production was similar better than that produced by glucose used at the equivalent weight-volume concentrations. *A. foetidus* ATCC 16878 when grown under solid-state fermentation using citrus waste as carbon source produced pectic enzyme activities up to 1600-1700 U/g after 36 h of culture, as reported by Garzón and Hours [3]. Yield of pectinases was 25% higher than that achieved with the same fungal strain and culture conditions using apple pomace as a substrate. *Bacillus* sp. MG-cp-2, isolated from the outer covering of seeds of *Celastrus paniculatus*, produced 140.1 U/ml of an alkaline and thermostable polygalacturonase when grown on orange peels as carbon source. The partially purified polygalacturonase was optimally active at 60°C at pH 10.0 with half-lives of 120, 118 and 20 min at 60, 70 and 80°C, respectively. The enzyme was 100% stable at 50°C for more than 12 h. Polygalacturonase was stable in a broad alkaline pH range 7.0–12.0 for more than 24 h at room temperature, retaining more than 80% of its activity [20].

Martins et al. [24] reported the pectin lyase and polygalacturonase production by the thermophilic fungus *Thermoascus aurantiacus* 179-5 under solid state fermentation using orange bagasse as carbon source. *T. aurantiacus* cultivation provided 43 and 19320 U/g polygalacturonase and pectin lyase activity respectively, while time course of enzymes production revealed sequential induction of them. Polygalacturonase and pectin lyase had optimum activity at pH 5.0 and 10.5–11.0, respectively. Maximal activity of the enzymes were determined at 65 °C. Seyis and Aksoz [32] investigated the use of apple pomace, orange pomace, orange peel, lemon pomace, lemon peel, pear peel, banana peel, melon peel, and hazelnut shell as substrate for xylanase production using *Trichoderma harzianum*. The maximum enzyme activity was observed when melon peel was used as the substrate for SSF, followed by the apple pomace and hazelnut shell. *Sporotrichum thermophile* produces a thermostable polygalacturonase under submerged culture and citrus peel as carbon source [29].

Single cell protein (SCP) and crude pectinolytic enzymes production from citrus pulps was reported by De Gregorio et al. [22]. SCP and enzymes were produced by slurry-state flask cultivation of *A. niger* and *T. viride*. *A. niger* showed a significant enzyme production starting with the 7th incubation day while *T. viride* showed later production. For the former mould the maximum polygalacturonase production was recorded about the 14th incubation day while in *T. viride* it was detected later, at 25th incubation day, when the observed Units (9.01 U/ml) were about seven times more than in *A. niger* (1.27 U/ml). Ismail [17] investigated the production of multienzyme preparations containing pectinase, cellulase and xylanase using six fungal isolates, namely *Aspergillus niger* 2, *A. niger* A-20, *A. oryzae* 1911, *Memmoniella* sp. 6, *Penicillium chrysogenum* 3486 and *P. oxalicum* 7, all grown on orange peels as the sole carbon source in SmF. Of the fungal isolates tested, *A. niger* A-20 proved to be the most potent and produced highly active multienzyme systems (56.0 U/ml pectinase, 4.39 U/ml cellulase, 3.33 U/ml xylanase and 0.01 U/ml α-amylase) after 5 days at 30°C. Production of multienzyme preparations containing pectinolytic, cellulolytic, and xylanolytic enzymes, by the mesophilic fungi *Aspergillus niger* BTL, *Fusarium oxysporum* F3, *Neurospora crassa* DSM 1129 and *Penicillium decumbens*, under solid-state fermentation of dry orange peel was also investigated. The selected fungi were also compared for their abilities to secret specific enzymes when grown on orange peel. Under optimal conditions *A. niger* BTL was found to produce the greatest yield of polygalacturonase, pectate lyase, xylanase, β-xylosidase, and invertase. *N. crassa* DSM 1129 produced the highest yield of endoglucanase [15].

Aspergillus giganteus grown on orange waste under SmF produces pectin lyase at 26.0 U/ml [31]. Co-culture of *A. niger* GJ-2 and *S. cerevisiae* J-1 using dry orange peels resulted in nearly twofold higher enzyme production than that of the culture without the presence of *S. cerevisiae* J-1 [74]. Lemon peel can be used as a substrate in submerged cultures to obtain high pectinase titres by *A. flavipes* FP-500 and *A. terreus* FP-370 [18]. *A. niger* when grown on lemon peels pomace in a solid state bioreactor produces approximately 2181.23 U/L pectinase [19].

3.2. Bioethanol Production

The spiraling crude oil prices and increasing concerns over greenhouse gas (GHG) emissions inflicted by the use of gasoline and diesel have led to an increased global interest in alternative fuels. Ethanol production by microbial fermentation of starch- and sugar-based raw materials is being practiced (1st generation biofuels), and the recent turbulence in the oil market has triggered the worldwide demand for bioethanol to such an extent that the conventional starch and sugar sources are proving to be insufficient. Lignocellulose feedstocks, such as agricultural and forest residues, industrial and municipal wastes, and dedicated energy crops, by virtue of their high carbohydrate content, hold tremendous potential for large-scale bioethanol production, and need to be exploited for commercial production (2nd generation biofuels) [75, 76].

Overall fuel ethanol production from lignocellulosic biomass includes five main steps: biomass pretreatment, cellulose hydrolysis, fermentation of hexoses, separation and effluent treatment Furthermore, detoxification and fermentation of pentoses released during the pre-treatment step can be carried out [77]. The sequential configuration employed to obtain cellulosic ethanol implies that the solid fraction of pretreated lignocellulosic material undergoes hydrolysis (saccharification); this fraction contains the cellulose in an accessible form to enzymes. Once hydrolysis is completed, the resulting cellulose hydrolyzate is fermented and converted into ethanol. This process is called separate hydrolysis and fermentation (SHF) [78]. The above mentioned two process steps can be performed together in so-called simultaneous saccharification and co-fermentation (SSCF), which has been shown to have several advantages over performing the steps separately [77,78]. The SSCF process alleviates end-product inhibition of the enzymes, and is also less capital intensive than SHF [79]. Furthermore, SSCF has been shown to be superior to SHF in terms of overall ethanol yield [80].

Several researchers have successfully hydrolyzed both orange and grapefruit peel waste using commercial cellulase and pectinase enzymes to glucose, galactose, fructose, arabinose, xylose, rhamnose, and galacturonic acid [81, 82, 83, 84]. According to Grohmann et al. [82], glucose, fructose and galactose from hydrolyzed citrus peel waste can be fermented to ethanol by *Saccharomyces cerevisiae* yeast. Galacturonic acid from pectin hydrolysis, arabinose, and xylose as well as the sugars mentioned above can be fermented by *Escherichia coli* K011 to produce ethanol and acetic acid [33, 34]. *E. coli* KO11 is a recombinant bacterial strain developed to ferment arabinose and xylose as well as hexoses to ethanol [85]. However in order to ferment these sugars, orange peel oil concentration in the hydrolysate must be reduced prior to fermentation [82]. The inhibitory effect on yeast growth due to orange peel oil and/or D-limonene, a monoterpene that makes up more than 90% of orange and grapefruit peel oils has been observed by several researchers [35, 36]. The mechanisms by which limonene and other monoterpenes similar in structure to limonene inhibit yeast function and growth have been the subject of several studies [86].

Wilkins et al. [35] reported that ethanol produced by *S. cerevisiae* and *Kluyveromyces marxianus* during fermentation of a solution modelling hydrolyzed orange peels waste was 37.1 g/l and 40.9 g/l respectively (80% and 88.3% theoretical yield, respectively) in the absence of limonene, while in the presence of 0.2% limonene ethanol production reduced at 23.3 g/l and 13.1 g/l respectively (50.3% and 28.3% theoretical yield, respectively). It also should be noted that limonene concentrations tested by Wilkins et al. [35] were less than concentrations observed in commercial citrus peel, which have been reported as 1.8% (w/w) for orange peel waste. Grohman et al. [33] reported that the recombinant bacterium *E. coli* KO11 produced 27.6 g/l ethanol from approximately 66.6 g/l sugars in orange peels hydrolysate. The hydrolysate contained 18.6 g/l galacturonic acid which could efficiently fermented to ethanol by the bacterium. A two step acid hydrolysis was applied by Oberoi et al. [38] for the hydrolysis of orange peels. Fermentation of the hydrolyzate by *S. cerevisiae* resulted in ethanol yield of 0.25 g/g on a biomass basis (0.46 g/g on a substrate-consumed basis). The ethanologenic bacterium *Zymomonas mobilis* can achieve greater ethanol yields than *S. cerevisiae* and other yeasts with productivities that can be as much 5 times greater than yeast fermentation [87]. Wilkins[88] investigated the inhibition caused by orange peel oil on ethanol production by *Z. mobilis* and reported that the bacterium is inhibited by concentrations similar to a previous study using S. *cerevisiae* and *K. marxianus*, but the ethanol yield was greater for *Z. mobilis* [35]. Since limonene inhibits ethanol production and keeping in mind that pretreatment is also required to improve the accessibility of substrates to cellulose-degrading enzymes [89], several researchers investigated the effect of different pretreatment methods on ethanol production from citrus wastes.

Choi et al. [39] applying popping pretreatment in mandarin peel [MP] waste, reduced limonene concentration from 0.21% in raw MP to 0.01%. Following enzymatic hydrolysis using commercial cellulase, xylanase, β-glucosidase and pectinase preparations, ethanol fermentation was conducted applying a modified SHF process by the yeast *S. cerevisiae* KCTC 7906. Ethanol production reached 46.2 g/L (0.462 g/g sugar, 90.6% of the theoretical) while the respected value for the raw MP was 39.8 g/L (73% of the theoretical) while productivity was tripled. Steam explosion was applying to citrus peels waste (CPW) followed by SSF of the pretreated material. Steam explosion removed most of the limonene present in CPW. In the SSF commercial enzyme were used, while the fermented microorganism was *S. cerevisiae*. Ethanol concentrations as high as 39.03 g/l (0.21 g/g on biomass basis) (in the presence of 0.08%, v/w, limonene) were achieved at high pectinase loading [36]. Sharma et al. [37] obtained an ethanol concentration of 26.84 g/L (0.426 g/g sugar, 83.52% of the theoretical) using crude cellulase enzyme and a combination of hexose (*S. cerevisiae*) and pentose (*Pacysolen tannophilus*) fermenting yeasts from a mixture of kinnow waste and banana peel pretreated by steam explosion.

In pilot plant scale Zhou et al. [40] applying steam pretreatment for the removal of limonene followed by SSF of the pretreated citrus peels waste (using commercial enzymes and the yeast *S. cerevisiae*) obtained 40 g/l ethanol (approximately 90% of the theoretical) with limonene as co-product. Hydrothermally treated Kinnow mandarin waste was used for ethanol production applying SSF process. Approximately 89% of the initial D-limonene was removed during pretreatment while 42 g/l (0.28 g/g on biomass basis) ethanol under optimal conditions were produced [13]. Mandarin (*Citrus reticulata* L.) peel wastes (MCPW) and lemon (*Citrus limon* L.) peel wastes were evaluated for bioethanol production, obtaining also as co-products: limonene, galacturonic acid, and citrus pulp pellets (CPP). Both wastes were pretreated by steam explosion. SSF was applied for ethanol production using commercially available cellulolytic and pectinolytic preparations and the yeast *S. cerevisiae.* Steam explosion reduces the concentration of the initial limonene content of both wastes to non-inhibitory levels (36). Ethanol production from pretreated MCPW reached 60 L/1000 kg fresh MCPW (approximately 0.22 g/g on dry biomass basis) [14], while the respected value from the lemon waste was 67.83 L/1000 kg fresh lemon CPW (approximately 0.25 g/g on dry biomass basis) [16]. A process for production of ethanol and biogas was developed by Pourbafrani et al. [90]. In this process, the CW is mixed with dilute sulfuric acid and then steam-exploded to hydrolyze the CW and also get rid of limonene. The resultant slurry is then centrifuged, where the liquid part is fermented to ethanol and distilled. The stillage from the distillation and solids from the centrifuge were mixed and digested to biogas (methane). This biogas can partly be utilized to produce steam required for the process, such as steam explosion and distillation. Zhou et al. [91] performed an economic analysis of ethanol production from citrus peels. They that concluded citrus ethanol has many advantages over starch or cellulosic ethanol, including low feedstock price, lower maintenance and equipment costs of the pretreatement facility and limonene which has a higher value than distiller's dried grain or lignin.

3.3. Heteropolysaccharides

Long-chain, high-molecular-mass polymers that dissolve or disperse in water to give thickening or gelling properties are indispensable tools in food product formulation. Such food polymers are also used for secondary effects, which include emulsification, stabilization, suspension of particulates, control of crystallization, inhibition of syneresis, encapsulation, and film formation. Xanthan gum is the most important microbial polysaccharide from the commercial point of view, produced by bacteria of the genus *Xanthomonas*, with a worldwide production of about 30000 tons per year. It has widespread commercial applications as a viscosity enhancer and stabilizer in the food, pharmaceutical and petrochemical industries [92, 93]. The utilization of agroindustrial byproducts for the production of polysaccharides by microorganisms has many advantages, such as reducing production costs and recycling natural resources. In studies of Bilanovic et al. [41] and Green et al. [42] four different fractions of citrus waste were compared as substrates for xanthan fermentation: whole citrus waste, pectic, hemicellulosic and cellulosic extracts, in SmF. The whole waste was found to be a good substitute for glucose media for xanthan production, since the production was 37% higher than that of on standard glucose medium. *X. campestris* ATCC 13951 utilized both simple and complex carbon compounds originating from citrus wastes. Substrate utilization in the medium based on pectin extract was similar to that in the medium based on a whole citrus waste and the pectic extract yielded the same amount of xanthan as the whole waste. This indicated that water soluble substances in citrus waste such as pectins, organic acids and simple carbohydrates were readily converted into xanthan and that they were the main contributor to xanthan production from the whole waste. The biodegradability of the hemicellulose and cellulose extracts of citrus waste was found to be much lower than that of the pectic extract. Substrate utilization and its conversion to xanthan in the hemicellulosic extract was 36% lower, and in the cellulosic extract 60% lower, than those in the whole citrus waste.

In submerged culture, the excretion of the polysaccharide during fermentation results in a highly viscous and shear thinning broth. The rheological behaviour of the fermentation broth causes serious problems of mixing, heat transfer, and oxygen supply, thus limiting the maximum gum concentration achievable as well as the product quality [94, 95]. Stredansky and Conti [43] studied the solid state fermentation as an alternative strategy for the production of xanthan by *X. campestris*. The choice was based on the observation that solid substrates reproduce the natural habitat of this phytopathogenic bacterium. This technique allows problems connected with broth viscosity to be overcome and utilizes cheap substrates. Citrus peels and apple pomace based substrates were employed. The conversion efficiency of citrus peel into xanthan was 24.5%, lower than that obtained with apple pomace [28–32%] but comparable to that reported in the aforementioned study [41] for xanthan production in submerged cultivation.

3.4. Citric Acid Production

Among the various products obtained through microbial cultivation on agroindustrial residues, organic acids are particularly important. The ratio of carboxylic acids manufactured microbiologically in the bulk of biotechnological products is very high. These compounds are valuable building blocks for the synthesis of different chemicals, which can be used in several applications.

Among organic acids, citric acid production has been well studied and reported. Citric acid is widely used in the food, beverage, pharmaceutical, and cosmetic industries and finds applications in a variety of other industries, from textiles to electroplating. In addition, production of citric acid could offset the disposal costs of the wastes [96]. Aravantinos-Zafiris et al. [44] reported citric acid fermentation from orange processing wastes having sugar content of 55 g/l, in submerged culture, using *A. niger* and 40g/l methanol. Yields obtained were citric acid concentration 30 g/L, productivity (QP=0.104 g/[L·h]) and yield YP/S=0.63 g/g. *A. niger* does not assimilate methanol, and, although its exact role in the stimulation of citric acid production is not yet known, it is believed that methanol increases the permeability of the microorganism cell membrane, thereby making the excretion of citric acid easier [97, 98].

Rivas et al. [45] reported that a treatment of autohydrolysis at 130°C and liquid/solid ratio of 8.0 g/g, a novel technology for orange peel, had a beneficial effect on its hydrolysis, producing liquors rich in soluble sugars, mainly sucrose, glucose, and fructose, which could be utilized for citric acid production by *A. niger* ATCC 9142. The highest values of citric acid concentration (9.2 g/L), product yield on consumed sugars (*YP/S*=0.53 g/g), and productivity (*Q*P=0.128 g/[L·h]) were achieved within 3 days from initial sugar content of 25.5 g/l and in the presence of CaCO3 and 40 ml/kg methanol. Usually less methanol is required to stimulate citric acid release in solid-state fermentation. Hang et al. [97] reported optimal methanol concentration of only 20 ml/kg in solid-state fermentation of kiwifruit peel by *A. niger* ATCC 9142, obtaining 82 g/L of citric acid after 5 days from 168 g/l of initial sugars (*Q*P=0.683 g/[L·h], *YP/S*=0.60 g/g). Similar results were reported by Zhang [46] for the solid residue of an orange juice factory and by Kang et al.[47] for tangerine peel. Kuforiji et al. [48] applying solid state fermentation, reported that *A. niger* strains NRRL 567 and 328 produced 57.6% and 55.4% [g citric acid/100 g of glucose used] respectively at a moisture content of 38.9% in orange waste. The addition of methanol resulted in reduction of citric acid yields. Dhillon et al. [49] reported citric acid concentration of 63.6±2.9 g/kg dry substrate by *A. niger* NRRL 2001 in 72 h incubation period using citrus waste as solid substrate.

3.5. Production of SCP

Citrus by-products are very high in cellulosic materials (cellulose and hemicelluloses), but low in lignin, making them potentially good feed sources for ruminants and promising substrates for the production of microbial protein. In countries with inadequate supplies of conventional ruminant feeds the use of citrus waste can impact quite positively on the supply of feed for animal nutrition while reducing environmental pollution [99]. Unfortunately, citrus wastes have only minimal protein content which limits their value in animal nutrition. So, exploitation of these wastes in animal nutrition will depend on the deployment of processes for their protein enrichment by biotechnological means. Fermentative processes in both SSF and slurries employing both filamentous and unicellular microorganisms have been employed for the protein enrichment of a citrus wastes. Taiwo et al. [100] reported that unfermented citrus pulp (910 g/kg DM) had relatively high levels of glucose and low levels of other nutrients. However, fermentation of citrus pulp without or with 100 g/kg molasses for 61 days resulted in production of primarily lactic and acetic acid [100], which enhanced citrus pulp ammonia holding capacity from 0.1 g NH3 N/kg DM in unfermented citrus pulp to 10.6 and 16.4 g NH3 N/kg DM in fermented citrus pulp without and with molasses, respectively. This suggests that the N content of citrus pulp can be enhanced by trapping excess ammonia generated from, for example, urea treated barley straw [100].

Scerra et al. (50) found that colonization of bergamot fruit peel with 10 strains of *Penicillium* spp. improved its nutritional value by increasing levels of crude protein, crude fat and structural carbohydrates *versus* untreated bergamot fruit peel. Furthermore, solid-state fermentation with *Penicillium roqueforti* Pr2 could represent an economical system for recycling, as animal feed, orange and lemon pulps, increasing their protein and lipid content by microbial bioconversion [101]. De Gregorio et al. [22] showed that by utilizing *A. niger*, but preferably *T. viride*, it was possible to extensively hydrolyse the lemon pulps, producing useful amounts of single cell protein (SCP) and a high-activity crude pectinases. The highest protein level was reached with *A. niger* after 14 days growth and later with *T. viride*, though the final amount of N was higher in *T. viride* (31.9%) than in *A. niger* (25.6%). Pectinases could be utilized in the citrus processing factories producing the wastes as well as in other fruit processing industries, while protein residues could be utilized in ruminant feeding. Productions from solid-state cultures of *Geotrichum candidum* using acid pretreated orange peel has been successfully demonstrated [51, 52]. This typically contains 35–40% crude protein characterized by a very high in vitro digestibility (73–88%). From the solid residue obtained after an acidic pre-treatment of the peel, hesperidin, a high-value by-product (vitamin P), was recovered, reaching yields ranging between 3.7% and 4.5% DM [51]. This by-product thus increases the cost-efficacy of the whole process.

3.6. Enzymatic Pectin Extraction

Pectin is an important component of dicotyledonous plant cell walls, besides cellulose and hemicelluloses and it is probably the most complex macromolecule in nature, because it can be composed of as many as 17 different monosaccharides [102]. Rather than making all possible combinations with these monosaccharides, mother nature has provided us a number of distinct polysaccharides which together form pectin. Three pectic polysaccharides (homogalacturonan, rhamnogalacturonan-I, and rhamnogalacturonan-II) have been isolated from primary cell walls and structurally characterized [103]. Homogalacturonan (HG) is a linear chain of 1,4-linked α-D-galactopyranosyluronic acid (Gal*p*A) residues. The Gal*p*A residues can be methyl-esterified at *C*-6, and carry acetyl groups on *O*-2 and *O*-3. "Smooth" regions are mainly composed of HG [102,103]. Rhamnogalacturonan-I (RG-I) is a family of pectic polysaccharides that contain a backbone composed of as many as 100 repeats of the disaccharide [→4]-α-D-Gal*p*A-[1→2]-α-L-Rha*p*-[1→]. The ramnosyl residues can be sustitued at *O*-4 with neutral sugars. The side chains are mainly composed of galactosyl and/or arabinosyl residues. They can be single unit (β-D-Gal*p*-[1→4]), but also polymeric such as arabinogalactan I [(AG-I) and arabinan (50 glycosyl residues or more) [103]. AG-I is composed of a 1,4-linked β-D-Gal*p* backbone; α-L-Ara*f* residues can be attached to the *O*-3 of the galactosyl residues [102]. The arabinans consist of a 1,5-linked α-L-Ara*f* backbone, which can be substitueted with α-L-Ara*f*-[1→2]-, α-L-Ara*f*-[1→3]-, and/or α-L-Ara*f*-[1→3]-α-L-Ara*f*-[1→3]- side chains [102,103]. Complexes of RG-I and AG-I and arabinan are often referred to as pectic "hairy regions" (HR), in which AG-I and arabinan are the "hairs". The abundance of HR can differ from species to species. RG-II is not structurally related to RG-I since its backbone is composed of 1,4-linked α-D-galactopyranosyluronic acid (Gal*p*A) residues rather than the repeating disaccharide [→4]-α-D-Gal*p*A-[1→2]-α-L-Rha*p*-[1→]. A nonasaccharide and an octasaccharide are attached to *C*-2 of some of the backbone GalA residues and two structurally different disaccharides are attached to *C*-3 of the backbone [102].

In food industries, pectin is used as a jellifying agent in jams and jellies. Gelation by pectin in fruit drink concentrates provides stabilization of emulsions, suspensions and foam [104]. Extracted pectin is widely used as functional food ingredient and it (or its EU code, E440) is listed among the ingredients of innumerable food products. Worldwide annual consumption is estimated at around 45 million kilograms, with a global market value of at least 400 million Euros [105]. Pectins are used in the pharmaceutical sector as detoxifying agents, and are well known for their anti-diarrheal effects. Pectin is also used for the production of single cell protein in a modified 'symba' process [106], as well as in cosmetics as gels and pastes [104].The chief raw materials for the industrial production of pectin are the residues from the manufacture of fruit juices, apple pomace and citrus fruits [107]. Classically, pectin extraction is carried out by acid hydrolysis at a pH range of 2–3 for 5 h at high temperature (70–100°C). The solid to liquid ratio is normally about 1:18. The pectin extract is separated from the pomace using a hydraulic press or by centrifugation. The extract is filtered, and finally concentrated to a standard setting strength. For powdered pectin preparation, the concentrated liquor is treated with organic solvents or certain metallic salts to precipitate the polymers. This process usually results in corrosion of equipment and water pollution problems. This uncontrolled hydrolysis may also result in a reduced degree of pectin polymerization, an important characteristic of commercial pectins [104]. Alternative treatments have been therefore taken into account for minimizing the use of detrimental chemicals during pectin extraction. Thus, several thermal and/or mechanical treatments have been applied to extract pectin including ultrasound [108], autoclaving [109], extrusion-cooking or microwave-assisted extraction [110, 111, 112, 113, 114] and subcritical water extraction [115].

Nowadays, the consumer demand for "green" products stimulates the search for alternative means of extraction. Therefore, enzymes could represent, despite their potential cost, an alternative and environmentally friendly way to extract "green labeled" pectins. Two different approaches can be considered. The first one involves enzymes to degrade the pectins and isolate pectin fragments. Enzymes degrading the pectin backbone (endo-polygalacturonase together with pectin methyl esterase, or endo-pectin lyase) are able to solubilize all the galacturonic acid [116]. These enzymes extract high molecular mass fragments of "hairy regions", carrying many side chains, but also galacturonic acid oligomers from "smooth regions" [53]. The yield of extraction obtained with pectinolytic enzymes is higher than that with chemical means [53, 55]. Some studies have also been done with various "protopectinases" [57, 58, 59]. According to their authors, these enzymes are a heterogeneous group that solubilizes pectins from the insoluble pectin in plant tissues (the so-called "protopectin") by restricted depolymerization. They can be active against the pectin backbone or against the pectin side chains. Their action is strongly dependent on the type of enzyme and the nature of substrate [58]. Some "protopectinases" can be more efficient to extract pectin than acid, depending on the substrate. This is the case for protopectinases having a pectin lyase or an endo-polygalacturonase activity, but also an arabinanase activity [59]. However, pure arabinanase and galactanase were found to be inefficient to extract galacturonic acid-containing polymers and oligomers [117, 118].

The second approach consists of using enzymes able to deconstruct the plant cell wall and isolate pectins. The primary plant cell wall of dicotyledons is composed of various polysaccharides [cellulose, xyloglucan, pectin] and proteins, which form entangled networks [119]. Xyloglucan is known to associate to cellulose microfibrils, probably via hydrogen bonding. The cellulose/xyloglucan network is embedded in a pectic matrix, together with a protein network. However, studies have shown the possibility to form interactions between cellulose and pectin side chains [120].

Commercial enzyme preparations [usually multienzyme complexes] have been used by several researchers for the extraction of pectin from citrus by-products. Pectinolytic, cellulolytic, hemicellulolytic, proteolytic, mixtures of the abov e[56, 60, 61, 62] have been used so far.

3.7. Enzymatic Production of Prebiotic Oligosaccharides

Prebiotics have been defined as "selectively fermented ingredients that allow specific changes, in both the composition and/or activity in the gastrointestinal microbiota that confers benefits upon host well-being and health". Several oligosaccharides with prebiotic properties (such as fructooligosaccharides, galactooligosaccharides, or lactulose) are commercially available, but there is an increasing interest in the identification and development of new prebiotic ingredients with added functionality [63]. Pectic oligosaccharides (POS) are good candidates as prebiotics [121, 122]. Recent studies reported healthy effects for POS, including regulation of lipid and glucose metabolism with decreased glycemic response and blood cholesterol levels, anticancer and immunological properties, anti-obesity effects, antibacterial and antioxidant properties [123, 124]. The prebiotic potential of POS is also under evaluation by means of in vitro fermentation assays using both individual cells and faecal inocula, and, even, some studies were also carried out in humans [125, 126]. Experiments carried out using POS derived from high methoxy citrus pectin, low-methoxy apple pectin and orange peel enhanced the growth of *Bifidobacteria* and *Lactobacilli*, while limiting the growth of pathogenic genera [63, 127, 128]. The prebiotic effect of pectins and POS depend on the physicochemical characteristics of substrates, and particularly, on the molecular weight distribution and degree of esterification [124]. For instance, it was demonstrated that POS with low degree of methylation show higher prebiotic index than high degree of methylation POS [127]. Olano-Martín et al. [64] reported on the continuous production of POS in an enzyme membrane reactor using pectin from citrus and apple. Pectinolytic enzymes were used to obtain POS from bergamot peel [63] and while pectinases and cellulases were used during POS production from orange peels waste [65]. The enzymatic treatment of the solid residue, after extraction of "green labeled pectins" from citrus peels, led to the extraction of a soluble fraction, containing pectic oligosaccharides of specific structure, known as modified hairy regions (MHR) [62]. Orange peel albedo (white part) was also a good source of pectic oligosaccharides with prebiotic properties produced by a microwave and autoclave extraction [129]. Martínez Sabajanes et al. [130] reported the manufacture of pectic oligosaccharides by nonisothermal processing with hot, compressed water (autohydrolysis or hydrothermal treatments) of the orange peel wastes.

3.8. Natural Antioxidants in Citrus by-Products

Phenolic compounds are important for their sensory and nutritional qualities that impart the colours, flavours and tastes of many plants. Of them, flavonoids have been found to have health-related properties, including anticancer, antiviral, and anti-inflammatory activities [131, 132]. It is believed that they act as scavengers of free radicals, as well as modulate the activities of metabolic enzymes involved in the initiation of low-density lipoprotein oxidation (e.g. xanthine oxidase, glutathione reductase, lipoxygenase, and NADPH-oxidase) and inhibit cellular proliferation [133]. Citrus processing by-products potentially represent a rich source of natural flavonoids (e.g., hesperidin, diosmin, naringin, and tangeretin), owing to the large amount of peel produced, and that citrus peels contain a high concentration of phenolic compounds [132, 134]. Moreover, while flavonoids are abundant elsewhere in the plant kingdom, there are several compounds (e.g. flavanones, flavanone glycosides and polymethoxylated flavones) unique to citrus, which are relatively rare in other plants [134].Up to now, several extraction techniques have been reported for the extraction of phenols from citrus peels like solvent extraction [135, 136, 137, 138], hot water extraction [139], resin-based extraction [140], electron beam- and c-irradiation-based extractions [141], supercritical fluid extraction [142], ultrasound assisted extraction [143] and enzyme-assisted extraction [60, 66].

3.8.1. Enzymatic Extraction of Flavonoids

Traditionally the extraction process of citrus flavonoids is based on organic solvents. The method worked well and a high percentage extraction was possible. An alternative process, enzyme-assisted extraction was used with similar results using food-grade enzymes, containing glucanase and pectinase activities such as Cellulase® MX, Cellulase® CL, Kleerase® AFP [66] and Pectinase 690L, Pectinase 62L, Cellulase C013P [60]. One of the main advantages of the enzymatic extraction of phenolics is that during this process 90% of the flavonoid glycosides present were cleaved to their aglycones [60]. Many of the ingested dietary flavonoids (e.g., rutin, rutinosides, neohesperosides, and complex acylated flavonoid glycosides) reach the colon without degradation. In the colon, they are substrates for the complex indigenous microflora which are known to contain species and genera able to exert varying effects on the health of the host [144, 145]. While the majority of dietary phytochemicals are hydrolyzed by the colonic microflora, small percentages are taken up during transit through the small intestine, and in some cases, the presence of a glucose moiety may enhance absorption [146]. It has also been demonstrated that the bioavailability of some xenobiotics is dependent mainly on small intestinal uptake [147]. Polyphenol glycosides, however, are relatively hydrophilic and do not diffuse passively across biological membranes. Simple flavonoid glucosides can be taken up into cells via SGLT1 (sodium-dependent glucose transporter 1) and other hexose transporters, and the aglycones are readily absorbed by passive diffusion. Rutinosides and some other glycosides are not absorbed in the small intestine.

Thus, adeglycosylation step is critical for the absorption of dietary flavonoids [148]. The low solubility of the released flavonoid aglycones could be increased by addition of food/pharmaceutical grade cyclodextrins as has been recently demonstrated for naringenin and hesperetin [149, 150].

3.8.2. Enzymatic Esterification of Flavonoids

The use of flavonoids in several domains is limited by their low stability and solubility in the fatty and aqueous phases. To improve their properties, several authors have studied the modification of their structure by chemical, enzymatic or chemo-enzymatic reactions. Two reactions (glycosylation and acylation) have received particular attention. The first allowed flavonoids to reinforce their hydrophilic character by adding sugars, whereas the second reaction make them more hydrophobic by fatty acid linkage. The chemical acylation of flavonoids by various fatty acids has been patented [151], but this process is not regioselective and leads to an unwanted functionalisation of phenolic hydroxyl groups which are responsible for the antioxidant activity of flavonoids [152]. However, the enzymatic acylation of flavonoids by lipases with phenolic acids is more regioselective than chemical acylation and may enhance not only their solubility in various media, but also their stability [153] and their antioxidant activity [154].

Enzymatic modification of flavonoids by lipases is described in quite a lot of studies. Flavonoid glycosides were acylated by butanoic acid in the presence of subtilisin [155] or by vinyl esters of fatty acids as well as phenolic acids using *Candida antarctica* as biocatalyst [149, 156]. More specifically, several groups have already performed enzymatic acylation of rutin and naringin [151].

3.8.3. Enzymatic Halogenation of Flavonoids

The effects of flavonoids, on the central nervous system have been considered. They process anxiolytic activity and low sedative or myorelaxant effects [157]. Among the most active compounds, a number of halogenated flavones have been reported; in particular, 6-bromoflavone and 6-bromo-3-nitroflavone showed activities close to or higher than that of diazepam, a benzodiazepine derivative which is a classical anxiolytic, anticonvulsant, sedative and skeletal muscle relaxant drug. In order to show these activities, the presence of electro-donating or withdrawing substituents on the aromatic ring of the flavonoids seems to be essential. In the literature, several methods for halogenating aromatic compounds are reported. Direct bromination, e.g. with elemental bromine, is a highly polluting method which, in addition, involves serious difficulties connected with the handling of a highly corrosive agent. Other methods, including NBS-amberlyst, [158] metal-oxo-catalysed KBr–H2O2 [159] and KBr–NaBO3 [160] suffer from harsh conditions or require complex or laborious work-up.

The use of microbial enzymes for the transformation of organic compounds has been employed as a powerful method in metabolism studies and in modern synthetic organic chemistry for decades. One of the enzymatic reactions that has been widely studied is a chloroperoxidase-catalyzed halogenations [161]. Chloroperoxidase from *Caldariomyces fumago* [CPO; EC 1.11.1.10] is a well-known enzyme, capable of halogenating a great variety of organic compounds such as b-ketoacids, cyclic b-diketones, steroids, alkenes, activated aromatic compounds and heterocylcic compounds [162]. The whole cells and the chloroperoxidase enzyme of Caldariomyces fumago were capable of halogenating the flavanones, naringenin and hesperetin, at C-6 and C-8 in the presence of either Cl_{-} or Br_{-} [163].

The biohalogenated products of naringenin and hesperetin were isolated and found to be identical to those obtained from chemical reactions using molecular halogen and hypohalous acid.

4. Other Uses

Citrus by-products represent a great potential for use as substrates in biotechnological processes as mentioned in previous paragraphs. This valuable by-product can also be used as a source for dietary fibres and for the removal of heavy metals from contaminated wastewaters.

4.1. Citrus By-Products as Sources for Dietary Fibres

Dietary fibre is often classified as soluble dietary fibre (SDF) and insoluble dietary fibre (IDF) [164]. It consists of a variety of non-starch polysaccharides which include cellulose, hemicellulose, pectin, b-glucans, gums and lignin [165]. Dietary fibre is composed mainly of remnants of edible plant cells; parenchymatous tissues are known to be the most important source of vegetable fibre [166]. Cell walls of fruits, vegetables and cereals make up most of the dietary fibre intake. Increasing attention has been given to their beneficial physiological effects of dietary fibre on humans and animals. High dietary fibre diets are associated with the prevention, reduction and treatment of some diseases, such as diverticular and coronary heart diseases [167]. It is widely known that dietary fibre obtained by different methods and from different sources, behave differently during their transit through the gastrointestinal tract, depending on their chemical composition and physicochemical characteristics and on the processing that food undergo [168].

Currently, there is a great variety of raw materials, mainly processing by-products, from which dietary fibre powders are obtained [169]. Residues from orange juice extraction are potentially an excellent source of DF compared to other alternative sources, such as cereals, due to its higher proportion of soluble dietary fibre and the availability in large quantities [170]. This is important, considering that the requirement for dietary fibre intake must be balanced, i.e. the water-soluble fraction should represent between 30% and 50% of the total dietary fibre. Moreover, fibers from citrus fruits have an additional advantage over dietary fibers from other sources due to the presence of associated bioactive compounds (i.e. flavonoids and vitamin C) with antioxidant properties [12, 171, 172]. Citrus fibre, added to meat products (bologna sausages, chicken hamburgers), are effective inhibitors of lipid oxidation, thereby improving their oxidative stability and prolonging their shelf life [171]. Citrus fibre added to bologna sausage can also decrease residual nitrite levels [171]. Garcia et al. [173] showed that the addition of 1.5% orange fibre, to dry fermented sausages gives an organoleptic characteristic similar to the conventional high fat product.

4.2. Citrus by-Products as Biosorbens for Heavy Metal Removal

With the rapid development of industries such as metal plating facilities, mining operations, fertilizer industries, tanneries, batteries, paper industries and pesticides, etc., heavy metals wastewaters are directly or indirectly discharged into the environment increasingly, especially in developing countries. Unlike organic contaminants, heavy metals are not biodegradable and tend to accumulate in living organisms and many heavy metal ions are known to be toxic or carcinogenic. Toxic heavy metals of particular concern in treatment of industrial wastewaters include zinc, copper, nickel, mercury, cadmium, lead and chromium. Faced with more and more stringent regulations, nowadays heavy metals are the environmental priority pollutants and are becoming one of the most serious environmental problems. So these toxic heavy metals should be removed from the wastewater to protect the people and the environment. Many methods that are being used to remove heavy metal ions include chemical precipitation, ion-exchange, adsorption, membrane filtration, electrochemical treatment technologies, etc. [174]. Research interest into the production of cheaper adsorbents to replace costly wastewater treatment methods such as such the above are attracting attention of scientists. Adsorption is one the physico-chemical treatment processes found to be effective in removing heavy metals from aqueous solutions. The emerging process of 'biosorption' uses nonviable or viable biological materials to bind contaminants *via* physico-chemical mechanisms, whereby factors like pH, size of biosorbent, ionic strength and temperature influence metal biosorption [175]. Plant wastes are inexpensive as they have no or very low economic value and thereby most of the adsorption studies have been focused on untreated plant wastes.

Some of the advantages of using plant wastes for wastewater treatment include simple technique, requires little processing, good adsorption capacity, selective adsorption of heavy metal ions, low cost, free availability and easy regeneration. However, the application of untreated plant wastes as adsorbents can also bring several problems such as low adsorption capacity, high chemical oxygen demand (COD) and biological chemical demand (BOD) as well as total organic carbon (TOC) due to release of soluble organic compounds contained in the plant materials. The increase of the COD, BOD and TOC can cause depletion of oxygen content in water and can threaten the aquatic life. Therefore, plant wastes need to be modified or treated before being applied for the decontamination of heavy metals [176]. Citrus peel principally consists of cellulose, pectin, hemicellulose, lignin, chlorophyll pigments and other low molecular weight hydro-carbons. These components contain various functional groups, such as carboxyl and hydroxyl, which make citrus peel to be a potential adsorbent material for removing metal ions from aqueous solutions. Several studies have been proposed in literatures about using modified or raw citrus waste in relation to heavy metal adsorption from water [177, 178, 179, 180, 181, 182, 183, 184, 185, 186].

5. The Biorefinery Concept

A biorefinery is most commonly defined as a facility that integrates biomass conversion processes and equipment to produce fuels, power, and chemicals ideally from waste biomass. The biorefinery concept is analogous to today's petroleum refineries, which produce multiple fuels and products from petroleum and importantly utilizing the entire gross starting product. By producing multiple products, a biorefinery can take advantage of the differences in biomass components and intermediates, and maximize the value derived from the biomass feedstock while leaving little waste. López et al. [187] proposed a general scheme for intergrated utilization of orange peels. The first step would consist of the removal of essential oils. This would be followed by an assessment of how the impact of oil removal influenced subsequent ethanol and methane production. Other high value products such as pectin, industrial enzymes, and single cell protein would also been removed before the final step of utilizing the remaining lignin as an energy source.

Pourbafrani et al. [90] presented a method for commercial treatment of citrus wastes [CWs] to yield different value- added products. The process focused on ethanol, biogas and limonene production. Depending on the market and profitability of the process, pectin can be recovered as a by-product from the process. Briefly, citrus wastes were hydrolyzed by dilute-acid process in a reactor equipped with an explosive drainage. At this stage, high solubilization of pectin present in the CWs was obtained, and 77.6% of total pectin content of CWs could be recovered by solvent recovery. The limonene of the CWs was effectively removed through flashing of the hydrolyzates into an expansion tank. The sugars present in the hydrolyzates were converted to ethanol using baker's yeast while the stillage and the remaining solid materials of the hydrolyzed CWs were anaerobically digested to obtain biogas. By applying this process, 39.64 l ethanol, almost 45 m3 pure methane, 8.9 l limonene, and up to maximum 38.8 kg pectin can be produced per ton of the wet CW. Furthermore the same authors were performed an economic analysis of the proposed process and concluded that the total cost of ethanol for base case process with 100,000 tons/year CW capacity was calculated as 0.91 USD/L, assuming 10 USD/ton handling and transportation cost of CW to the plant. This price is sensitive to the plant capacity. With constant price of methane and limonene, changing the plant capacity from 25.000 to 400.000 tons CW per year results in reducing ethanol costs from 2.55 to 0.46 USD/L in an economically feasible process. In addition, the ethanol production cost is sensitive to the transportation cost of CW. Increasing this cost from 10 to 30 USD/ton for the base case results in increasing the ethanol costs from 0.91 to 1.42 USD/L [188].

Conclusion

Citrus by-products represent a great potential for use as substrates in biotechnological processes. Several studies have been described regarding the employment of this residue for the production of value-added compounds, such as enzymes, biofuels, biopolymers, SCP, organic acids, prebiotic compounds and natural antioxidants among others. Biotechnological applications of the citrus by-products are interesting not only from the point of view of low-cost substrate, but also in solving problems related to their disposal.

However, more effort is still required in the following areas: (a) a greater knowledge of the chemical constituents of orange peel, with the aim of detecting more high value products, (b) design of chemical engineering processes that can adapt to a variety of feedstocks, (c) development of microbial strains more efficient at converting orange peel into high value products using classic mutagenesis or metabolic engineering, (d) integrated study of the orange peel treatment steps, in which the residues obtained in one step will be employed as raw material in the following one, according to the definition of a biorefinery facility.

Conflict of Interest

The Authors declare that there is no conflict of interest.

References

[1] Kale PN, Adsule PG (1995) Citrus. In: Salunkhe DK, Kadam SS (eds) Handbook of Fruit Science and Technology: *Production, Composition, Storage, and Processing.* Marcel Dekker Inc., New York, NY, USA, pp 39-65.

[2] *FAO* (Food and Agriculture Organization) available at www.fao.org.

[3] Garzón CG, Hours RA (1992) Citrus waste: an alternative substrate for pectinase production in solid-state culture. *Bioresource Technol*. 39:93-95.

[4] Lapuerta M, Hernandez JJ, Pazo A, Lopez J (2008) Gasification and co-gasification of biomass wastes: Effect of the biomass origin and the gasifier operating conditions. *Fuel Process Technol.* 89(9):828–837.

[5] Ghani WA, Alias AB, Savory RM, Cliffe KR (2009) Co-combustion of agricultural residues with coal in a fluidized bed combustor. *Waste Manage* 29:767–773.

[6] Sinclair WB (1984) The Biochemistry and Physiology of the Lemon and Other Citrus Fruits, *Division of Agriculture and Natural Resources*, University of California, Oakland, CA, USA.

[7] Ensminger ME, Oldfield JE, Heinemann WW (1990) *Feeds and Nutrition* (2nd Edn) The Ensminger Publishing Company, Clovis, CA, USA.

[8] Martínez-Pascual J, Fernández-Carmona J (1980) Composition of citrus pulp. *Anim. Feed Sci. Technol.* 5:1-10.

[9] Madrid J, Hernández F, Pulgar MA, Cid JM (1996) Dried lemon as energetic supplement of diet based on urea-treated barley straw: effects on intake and digestibility in goats. *Anim. Feed Sci. Technol*. 63:89-98.

[10] Bampidis VA, Robinson PH (2006) Citrus by-products as ruminant feeds: A review. *Anim. Feed Sci. Technol.* 128:175-217.

[11] Ammerman CB, Henry PR (1991) Citrus and vegetable products for ruminant animals. In: *Proceedings of the Alternative Feeds for Dairy and Beef Cattle Symposium*, St. Louis, MO, USA, pp 103-110.

[12] Marín FR, Soler-Rivas C, Benavente-García O, Castillo J, Pérez-Alvarez JA (2007) By-products from different citrus processes as a source of customized functional fibres. *Food Chem.* 100:736–741.

[13] Oberoi HS, Vadlani PV, Nanjundaswamy A, Bansal S, Singh S, Kaur S, Babbar N (2011) Enhanced ethanol production from Kinnow mandarin (*Citrus reticulata*) waste *via* a statistically optimized simultaneous saccharification and fermentation process. *Bioresource Technol.* 102:1593–1601.

[14] Boluda-Aguilar M, García-Vidal L, González-Castañeda FP, López-Gómez A (2010) Mandarin peel wastes pretreatment with steam explosion for bioethanol production. *Bioresource Technol.* 101:3506–3513.

[15] Mamma D, Kourtoglou E, Christakopoulos P (2008) Fungal multienzyme production on industrial byproducts of the citrus-processing industry. *Bioresource Technol.* 99:2373-2383.

[16] Boluda-Aguilar M, López-Gómez A (2013) Production of bioethanol by fermentation of lemon (*Citrus limon* L.) peel wastes pretreated with steam explosion. *Ind. Crop. Prod.* 41:188-197.

[17] Ismail AS (1996) Utilization of orange peels for the production of multi-enzyme complexes by some fungal strains. *Process Biochem*. 1:645-650.

[18] Martínez-Trujillo A, Arreguín-Range L, García-Rivero M, Aguilar-Osorio G (2011). Use of fruit residues for pectinase production by *Aspergillus flavipes* FP-500 and *Aspergillus terreus* FP-370. *Lett. Appl. Microbiol*. 53:202–209.

[19] Ruiz HA, Rodríguez-Jassoa RM, Rodríguez R, Contreras-Esquivel JC, Aguilar CN (2012) Pectinase production from lemon peel pomace as support and carbon source in solid-state fermentation column-tray bioreactor. *Biochemical Eng. J.* 65:90-95.

[20] Kapoor M, Beg QK, Bhushan B, Dadhich KS, Hoondal GS (2000) Production and partial purification and characterization of a thermo-alkali stable polygalacturonase from *Bacillus* sp. MG-cp-2. *Process Biochem.* 36:467-473.

[21] Maldonado MC, Navarro A, Callieri DAS (1986) Production of pectinases by *Aspergillus* sp. using differently pretreated lemon peel as the carbon source. *Biotechnol. Lett.* 8:501-504.

[22] De Gregorio A, Mandalari G, Arena N, Nucita F, Tripodo MM, Lo Curto RB (2002) SCP and crude pectinase production by slurry-state fermentation of lemon pulps. *Bioresource Technol.* 83:89-94.

[23] Fonseca MJV, Said S (1994) The pectinase produced by *Tubercularia vulgaris* in submerged culture using pectin or orange-pulp pellets as inducer. *Appl. Microbiol. Biot.* 42:32-35.

[24] Martins ES, Silva D, Da Silva R, Gomes E (2002) Solid state production of thermostable pectinases from thermophilic *Thermoascus aurantiacus*. *Process Biochem.* 37:949-954.

[25] Silva D, Martins ES, Silva R, Gomes E (2002) Pectinase production by *Penicillium viridicatum* Rfc3 by solid state fermentation using agricultural wastes and agro-industrial by-products. *Braz. J. Microbiol.* 33:318-324.

[26] Larios G, Garcia JM, Huitrdn C (1989) Endopolygalacturonase production from untreated lemon peel by *Aspergillus* sp. CHY- 1043. *Biotechnol. Lett.* 11:729-734.

[27] Federici F, Petruccioli M (1985) Growth and polygalacturonase production by *Aureobasidium pullulans* on orange peel waste. *Microbiologie Aliments Nutrition* 3:39-46.

[28] Dhillon SS, Gill RK, Gill SS, Singh M (2004) Studies on the utilization of citrus peel for pectinase production using fungus *Aspergillus niger*. *Int. J. Environ. Stud.* 61(2):199-210.

[29] Kaur G, Kumar S, Satyanarayana T (2004) Production, characterization and application of a thermostable polygalacturonase of a thermophilic mould *Sporotrichum thermophile* Apinis. *Bioresource Technol.* 94:239-243.

[30] Mahmood AU, Greenman J, Scragg AH (1998) Orange and potato peel extracts: analysis and use as *Bacillus* substrates for the production of extracellular enzymes in continuous culture. *Enzyme Microb. Tech.* 22:130-137.

[31] Pedrolli DB, Carmona EC (2009) Pectin Lyase from *Aspergillus giganteus*: Comparative study of productivity of submerged fermentation on citrus pectin and orange waste. *Appl. Biochem. Microbio.* 45(6):610-616.

[32] Seyis I, Aksoz N (2005) Xylanase production from *Trichoderma harzianum* 1073 D3 with alternative carbon and nitrogen sources. *Food Technol. Biotech.* 43:37-40.

[33] Grohmann K, Baldwin EA, Buslig BS, Ingram LO (1994) Fermentation of galacturonic acid and other sugars in orange peel hydrolysates by the ethanolgenic strain of *Escherichia coli*. *Biotechnol. Lett.* 16:281-286.

[34] Grohmann K, Cameron RG, Buslig BS (1996) Fermentation of sugars in orange peel hydrolysates to ethanol by recombinant *Escherichia coli* K011. *Appl. Biochem. Biotech.* 51/52:423-435.

[35] Wilkins MR, Suryawati L, Maness NO, Chrz D (2007) Ethanol production by *Saccharomyces cerevisiae* and *Kluyveromyces marxianus* in the presence of orange-peel oil. *World J. Microb. Biot.* 23:1161-1168.

[36] Wilkins MR, Widmer WW, Grohmann K (2007) Simultaneous saccharification and fermentation of citrus peel waste by *Saccharomyces cerevisiae* to produce ethanol *Process Biochem*. 42:1614-1619.

[37] Sharma N, Kalra KL, Oberoi HS, Bansal S (2007) Optimization of fermentation parameters for production of ethanol from kinnow waste and banana peels by simultaneous saccharification and fermentation. *Indian J. Microbiol*. 47:310–316.

[38] Oberoi HS, Vadlani PV, Madl RL, Saida L, Abeykoon JP, (2010) Ethanol production from orange peels: two stage hydrolysis and fermentation studies using optimized parameters through experimental design. *J. Agr. Food Chem*. 58:3422–3429.

[39] Choi IS, Kim J-H, Wi SG, Kim K-H, Bae HY (2012) Bioethanol production from mandarin (*Citrus unshiu*) peel waste using popping pretreatment. *Applied Energy*, (in press, corrected proof).

[40] Zhou W, Widmer W, Grohmann K (2008) Developments in ethanol production from citrus peel waste. *P Fl St Hortic Soc*. 121:307-310.

[41] Bilanovic D, Shelet G, Green M (1994) Xanthan fermentation of citrus waste. *Bioresource Technol*. 48:169-172.

[42] Green M, Shelef G, Bilanovic D (1994) The effect of various citrus waste fractions on xanthan fermentation. *Chem. Eng. J*. 56:B37-B41.

[43] Stredansky M, Conti E (1999) Xanthan production by solid state fermentation. *Process Biochem*. 34:581-587.

[44] Aravantinos-Zafiris G, Tzia C, Oreopoulou V, Thomopoulos CD (1994) Fermentation of orange processing wastes for citric acid production. *J. Sci. Food Agr*. 65:117-120.

[45] Rivas B, Torrado A, Torre P, Converti A, Domínguez JM (2008) Submerged citric acid fermentation on orange peel autohydrolysate. *J. Agr. Food Chem*. 56:2380-2387.

[46] Zhang Q (1988) Utilization of citrus wastes in production of citric acid. *Shipin Kexue* (Beijing, China) 104:21-24.

[47] Kang SK, Park HH, Lee JH, LeeYS, Kwon IB, Sung NK (1989) Citric acid fermentation from mandarin orange peel by Aspergillus niger. *Sanop. Misaengmul. Hakhoechi* 17:510-518.

[48] Kuforiji OO, Kuboye AO, Odunfa SA (2010) Orange and pineapple wastes as potential as potential substrates for citric acid production. *Int. J. Plant. Biol*. 1e4:19-21.

[49] Dhillon GS, Brar SK, Vermab M, Tyagia RD (2011) Utilization of different agro-industrial wastes for sustainable bioproduction of citric acid by Aspergillus niger. *Biochemical Engineering Journal* 54:83–92.

[50] Scerra V, Caridi A, Foti F, Sinatra MC (1999) Influence of dairy Penicillium spp. on nutrient content of citrus fruit peel. *Anim. Feed Sci. Technol*. 78:169-176.

[51] Locurto R, Tripodo MM, Leuzzi U, Giuffre D, Vaccarino C (1992) Flavonoids recovery and SCP production from orange peel. *Bioresource Technol*. 42(2):83-87.

[52] Vaccarino C, Locurto R, Tripodo MM, Patane R, Lagana G, Ragno A (1989) SCP from orange peel by fermentation with fungi-acid-treated peel. *Biol. Waste* 30(1):1-10.

[53] Renard CMGC, Voragen AGJ, Thibault J-F, Pilnik W (1991) Studies on apple protopectin. V: Structural studies on enzymatically extracted pectins. *Carbohyd. Polym*. 16:137-154.

[54] Massiot P, Thibault J-F (1989) Enzymic analysis of carrot cell-wall polysaccharides. *Carbohyd. Res*. 190:121-136.

[55] Donaghy JA, McKay AM (1994) Pectin extraction from citrus peel by polygalacturonase produced on whey. *Bioresource Technol.* 47:25-28.

[56] Contreras-Esquivel JC, Voget CE, Vita CE, Espinoza-Perez JD, Renard CMGC (2006) Enzymatic extraction of lemon pectin by endo-polygalacturonase from Aspergillus niger. *Food Sci. Biotechnol.* 15:163-167.

[57] Sakai T, Ozaki Y (1988) Protopectin solubilizing enzyme that does not catalyse the degradation of polygalacturonic acid. *Agr. Biol. Chem. Tokyo* 52(4):1091-1093.

[58] Nakamura T, Hours RA, Sakai T (1995) Enzymatic maceration of vegetables with protopectinases. *J. Food Sci.* 60:468- 472.

[59] Sakamoto T, Hours RA, Sakai T (1995) Enzymic pectin extraction from protopectins using microbial protopectinases. *Process Biochem.* 30:403-409.

[60] Mandalari G, Bennett RN, Kirby AR, Lo Curto RB, Bisignano G, Waldron KW, Faulds GB (2006) Enzymatic hydrolysis of flavonoids and pectic oligosaccharides from Bergamot (Citrus bergamia Risso) peel. *J. Agr. Food Chem.* 54:8307-8313.

[61] Lim J, Yoo J, Ko S, Lee S (2012) Extraction and characterization of pectin from Yuza (Citrus junos) pomace: A comparison of conventional-chemical and combined physical-enzymatic extractions. *Food Hydrocolloid* 29:160-165.

[62] Zykwinska A, Boiffard M-H, Kontkanen H, Buchert J, Thibault J-F, Bonnin E (2008) Extraction of green labeled pectins and pectic oligosaccharides from plant byproducts. *J. Agr. Food Chem.* 56:8926–8935.

[63] Mandalari G, Nueno Palop C, Tuohy K, Gibson GR, Bennett RN, Waldron KW, Bisignano G, Narbad A, Faulds CB (2007) In vitro evaluation of the prebiotic activity of a pectic oligosaccharide-rich extract enzymatically derived from bergamot peel. *Appl. Microbiol. Biot.* 73:1173-1179.

[64] Olano-Martin E, Mountzouris KC, Gibson GR, Rastall RA (2001) Continuous production of oligosaccharides from pectin in an enzyme membrane reactor. *J. Food Sci.* 66:966-971.

[65] Martínez Sabajanes M, Yáñez R, Alonso JL, Parajó JC (2012) Pectic oligosaccharides production from orange peel waste by enzymatic hydrolysis. *Int. J. Food Sci. Tech.* 47:747–754.

[66] Li BB, Smith B, Hossain MdM (2006) Extraction of phenolics from citrus peels II. Enzyme-assisted extraction method. *Sep. Purif. Technol.* 48: 189-196.

[67] Jayani RS, Saxena S, Gupta R (2005) Microbial pectinolytic enzymes: A review. *Process Biochem.* 40:2931-2944.

[68] Kashyap DR, Vohra PK, Chopra S, Tewari R (2001) Applications of pectinases in the commercial sector: A review. *Bioresource Technol.* 77:215-227.

[69] Hoondal GS, Tiwari RP, Tewari R, Dahiya N, Beg QK (2002). Microbial alkaline pectinases and their industrial applications: A review. *Appl. Microbiol. Biotechnol.* 59:409-418.

[70] Hildén L, Johansson G (2004) Recent developments on cellulase and carbohydrate-binding modules with cellulose affinity. *Biotechnol. Lett.* 26:1683-1693.

[71] Pandey A, Selvakumar P, Soccol CR, Nigam P (1999) Solid state fermentation for the production of industrial enzymes. *Curr. Sci.* 77 (1):149-162.

[72] Cherry JR, Findatsef AL (2003) Directed evolution of industrial enzymes: an update. *Curr. Opin. Biotech.* 14:438-443.

[73] Beg QK, Kapoor M, Mahajan L, Hoonda GS (2001) Microbial xylanases and their industrial applications: a review. *Appl. Microbiol. Biot.* 56:326-338.

[74] Zhou J-M, Ge XY, Zhang WG (2011) Improvement of polygalacturonase production at high temperature by mixed culture of Aspergillus niger and Saccharomyces cerevisiae. *Bioresource Technol.* 102:10085–10088.

[75] Banerjee S, Mudliar S, Sen R, Giri B (2010) Commercializing lignocellulosic bioethanol: technology bottlenecks and possible remedies. *Biofuel Bioprod. Bior*. 4:77–93.

[76] Balat M (2011) Production of bioethanol from lignocellulosic materials via the biochemical pathway: A review. *Energ. Convers. Manage* 52:858-875.

[77] Cardona CA, Sánchez OJ (2007) Fuel ethanol production: Process design trends and integration opportunities. *Bioresource Technol.* 98:2415-2457.

[78] Hahn-Hägerdal B, Gable M, Gorwa-Grauslund MF, Lidén G, Zacchi G (2006) Bio-ethanol- the fuel of tomorrow from the residues of today. *Trends Biotechnol.* 24:549-556.

[79] Wingren A, Galbe M, Zacchi G (2003) Techno-economic evaluation of producing ethanol from softwood-a comparison of SSF and SHF and identification of bottlenecks. *Biotechnology Progr*. 19:1109-1117.

[80] Söderström J, Galbe M, Zacchi G (2005) Separate versus simultaneous saccharification and fermentation of two-step steam pretreated softwood for ethanol production. *J. Wood Chem. Technol.* 25:187-202.

[81] Grohmann K, Baldwin EA (1992) Hydrolysis of orange peel with pectinase and cellulase enzymes. *Biotechnol. Lett.* 14:1169-1174.

[82] Grohmann K, Baldwin EA, Buslig BS (1994) Production of ethanol from enzymatically hydrolyzed orange peel by the yeast Saccharomyces cerevisiae. *Appl. Biochem. Biotech.* 45/46:315-327.

[83] Grohmann K, Cameron RG, Buslig BS (1995) Fractionation and pre-treatment of orange peel by dilute acid hydrolysis. *Bioresource Technol.* 54:129-141.

[84] Wilkins MR, Widmer WW, Grohmann K, Cameron RG (2007) Hydrolysis of grapefruit peel waste with cellulase and pectinase enzymes. *Bioresource Technol*. 98:1596-1601.

[85] Beall DS, Ohta K, Ingram LO (1991) Parametric studies of ethanol production from xylose and other sugars by recombinant Escherichia coli. *Biotechnol. Bioeng.* 38:296-303.

[86] Uribe S, Pena A (1990) Toxicity of allelopathic monoterpene suspensions on yeast. *J. Chem. Ecol.* 16:1399-1408.

[87] Panesar PS, Marwaha SS, Kennedy JF (2006) Zymomonas mobilis: an alternative ethanol producer. *J. Chem. Technol. Biot*. 81:623-635.

[88] Wilkins MR (2009) Effect of orange peel oil on ethanol production by Zymomonas mobilis. *Biomass Bioenerg.* 33:538-541.

[89] Alvira P, Tomαs-Pejo E, Ballesteros M, Negro MJ (2010) Pretreatment technologies for an efficient bioethanol production process based on enzymatic hydrolysis: A review. *Bioresource Technol.* 101(13):4851-4861.

[90] Pourbafrani M, Forgács G, Sárvári Horváth I, Niklasson C, Taherzadeh MJ (2010) Production of biofuels, limonene and pectin from citrus wastes. *Bioresource Technol.* 101:4246-4250.

[91] Zhou W, Widmer W, Grohmann K (2007) Economic analysis of ethanol production from citrus peel waste. *P Fl St Hortic Soc.* 120:310-315.

[92] Margaritis A, Pace GW (1985) Microbial polysaccharides. In: Moo-Young M (ed) *Comprehensive biotechnology*, vol. 3, Oxford: Pergamon Press, pp 1005-1041.

[93] Galindo E (1994) Aspects of the process for xanthan production. *T I Chem. Eng.-Lond* 72:227-237.

[94] Wecker A, Onken V (1991). Influence of dissolved oxygen concentration and shear rate on the production of pullulan by Aureobasidium pullulans. *Biotechnol. Lett.* 13:155-160.

[95] Peters HV, Herbst H, Hesselink PGM, Lunsdorf H, Schumpe A, Deckwer WD (1989) The influence of agitation rate on xanthan production by Xanthomonas campestris. *Biotechnol. Bioeng.* 34:1391-1397.

[96] Sauer M, Porro D, Mattanovich D, Branduardi P (2008) Microbial production of organic acids: expanding the markets. *Trends Biotechnol.* 26(2):100-108.

[97] Hang YD, Luh BS, Woodams EE (1987) Microbial production of citric acid by solid state fermentation of kiwifruit peel. *J. Food Sci.* 52:226-227.

[98] Navaratnam P, Arasaratnam V, Balasubramaniam K (1998) Channelling of glucose by methanol for citric acid production from Aspergillus niger. *World J. Microb. Biot.* 14:559-563.

[99] Obeta Ugwuanyi J, McNeil B, Harvey LM (2009) Production of Protein-Enriched Feed Using Agro-Industrial Residues as Substrates In: P. Singh nee' Nigam, A. Pandey (eds), *Biotechnology for Agro-Industrial Residues*, Springer Science+Business Media B.V. pp 77-100.

[100] Taiwo AA, Adebowale EA, Greenhalgh JFD, Akinsoyinu AO (1995) Techniques for trapping ammonia generated from urea treatment of barley straw. *Anim. Feed Sci. Technol.* 56:133-141.

[101] Scerra V, Caridi A, Foti F, Sinatra MC, Caparra P (2000) Changes in chemical composition during the colonisation of citrus pulps by a dairy Penicillium roqueforti strain. *Bioresource Technol.* 72:197-198.

[102] Ridley BL, O'Neill MA, Mohnen D (2001) Pectins: Structure, biosynthesis, and oligogalacturonide-related signaling. *Phytochem.* 57:929-967.

[103] Vincken JP, Schols HA, Oomen RJFJ, Beldman G, Visser, RGF, Voragen AGJ (2003) Pectin-The Hairy thing. In: Voragen F, Schols H and Visser R (eds), *Advances in Pectin and Pectinase Research,* Kluwer Academic Publishers, The Netherlands, pp 47-59.

[104] Sakai T, Sakamoto T, Hallaert E, Vandamme EJ (1993) Pectin, pectinase and protopectinase: production, properties, and applications. *Adv. Appl. Microbiol.* 39:213-294.

[105] Savary BJ, Hotchkiss AT, Fishman ML, Cameron RG, Shatters RG (2003) Development of a Valencia orange pectin methyl esterase for generating novel pectin products. In: Voragen F, Schols H and Visser R (eds), *Advances in Pectin and Pectinase Research,* Kluwer Academic Publishers, The Netherlands, pp 345-361.

[106] Fellows PJ, Worgan JT (1986) Studies on the growth of Candida utilis on D-galacturonic acid and the products of pectin-hydrolysis. *Enzyme Microb. Tech.* 9:537-540.

[107] Willats WGT, Knox JP, Mikkelsen JD (2006) Pectin: new insights into an old polymer are starting to gel. *Trends Food Sci. Tech.* 17:97-104.

[108] Panchev I, Kirchev N, Kratchanov C (1988). Improving pectin technology. *Int. J. Food Sci. Tech.* 23:337–341.

[109] Oosterveld A, Beldman G, Schols HA, Voragen AGJ (2000) Characterization of arabinose and ferulic acid rich pectic polysaccharides and hemicelluloses from sugar beet pulp. *Carbohyd. Res.* 328:185–197.

[110] Fishman ML, Chau HK, Hoagland P, Ayyad K (2000) Characterization of pectin, flash-extracted from orange albedo by microwave heating, under pressure. *Carbohyd. Res.* 323:126-138.

[111] Fishman ML, Walker PN, Chau HK, Hotchkis AT (2003) Flash extraction of pectin from orange albedo by steam injection. *Biomacromolecules* 4:880-889.

[112] Kratchanova M, Pavlovaa E, Panchev I (2004) The effect of microwave heating of fresh orange peels on the fruit tissue and quality of extracted pectin. *Carbohyd. Polym.* 56:181–185.

[113] Shin H-H, Kim C-T, Cho Y-J, Hwang J-K (2005) Analysis of extruded pectin extraction from apple pomace by response surface methodology. *Food Sci. Biotechnol.* 14:28-31.

[114] Liu Y, Shi J, Langrish TAG (2006) Water-based extraction of pectin from flavedo and albedo of orange peels. *Chem. Eng. J.* 120:203–209.

[115] Ueno H, Tanaka M, Hosino M, Sasaki M, Goto M (2008) Extraction of valuable compounds from the flavedo of Citrus junos using subcritical water. *Sep. Purif. Technol.* 62:513-516.

[116] Voragen AGJ, Heutink R, Pilnik W (1980) Solubilization of apple cell walls with polysaccharide-degrading enzymes. *J. Appl. Biochem.* 2:452-468.

[117] Thibault JF, De Dreu R, Geraeds CJ, Rombouts FM (1988) Studies on extraction of pectins from citrus peels, apple marks and sugar pulps with arabinase and galactanase. *Carbohyd. Polym.* 9:119-131.

[118] Renard CMG, Voragen AGJ, Thibault JE, Pilnik W (1991) Comparison between enzymatically and chemically extracted pectins from apple cell walls. *Anim. Feed Sci. Technol.* 32:69-75.

[119] Carpita N, McCann MC (2000) The cell wall. In: Buchanan B, Gruissel W, Jones R (Eds) *Biochemistry and Molecular Biology of Plants American Society of Plant Physiologists*: Rockville, MD, pp 52-108.

[120] Zykwinska AW, Ralet M-CJ, Garnier CD, Thibault J-FJ (2005) Evidence for in vitro binding of pectin side chains to cellulose. *Plant. Physiol.* 139:397-407.

[121] Olano-Martin E, Williams MR, Gibson GR, Rastall RA (2003) Pectins and pectic-oligosaccharides inhibit Escherichia coli O157:H7 Shiga toxin as directed towards the human colonic cell line HT29. *FEMS Microbiol. Lett.* 218:101–105.

[122] Olano-Martin E, Rimbach GH, Gibson GR, Rastall RA (2003) Pectin and pectic-oligosaccharides induce apoptosis in in vitro human colonic adenocarcinoma cells. *Anticancer Res.* 23:341–346.

[123] Rastall RA, Gibson GR, Gill HS, Guarner F, Klaenhammer TR, Pot B, Reid G, Rowland IR, Sanders ME (2005) Modulation of the microbial ecology of the human colon by probiotics, prebiotics, prebiotics and synbiotics to enhance human health: an overview of enabling science and potential applications. *FEMS Microbiol. Ecol.* 52:145–152.

[124] Gullón P, Gullón B, Moure A, Alonso JL, Domínguez H, Parajó JC (2009) Manufacture of prebiotics from biomass sources. In: Charalampopoulos D, Rastall RA (eds) *Prebiotics and Probiotics Science and Technology*, New York, NY: Springer, pp. 535–589.

[125] Fanaro S, Jelinek J, Stahl B, Boehm G, Kock R, Vigi V (2005) Acidic oligosaccharides from pectin hydrolysate as new component for infant formulae: effect on intestinal flora, stool characteristics, and pH. *J. Pediatr. Gastr. Nutr.* 41:186-190.

[126] Magne F, Hachelaf W, Suau A, Boudraa G, Bouziane-Nedjadi K, Rigottier-Gois L, Touhami M, Desjeux J-F, Pochart P, (2008) Effects on faecal microbiota of dietary and acidic oligosaccharides in children during partial formula feeding. *J. Pediatr. Gastr. Nutr.* 46(5):580–588.

[127] Olano-Martin E, Gibson GR, Rastall RA (2002) Comparison of the in vitro bifidogenic properties of pectins and pectic-oligosaccharides. *J. Appl. Microbiol.* 93:505–511.

[128] Manderson K, Pinart M, Tuohy KM, Grace WE, Hotchkiss AT, Widmer W, Yadhav MP, Gibson GR, Rastall RA (2005) In vitro determination of prebiotic properties of oligosaccharides derived from an orange juice manufacturing by-product stream. *Appl. Environ. Microbiol.* 71:8383-8389.

[129] Hotchkiss AT, Olano-Martin E, Grace WE, Williams MR, Gibson GR, Rastall RA (2003) Pectic oligosaccharides as prebiotics. In: Eggleston G, Côté GL (Eds) *Oligosaccharides in food and agriculture*, ACS Symposium series 849, Oxford University Press, USA, pp 54-62.

[130] Martínez Sabajanes M, Yáñez R, Alonso JL, Parajó JC (2010) Chemical production of pectic oligosaccharides from orange peel wastes. *Ind. Eng. Chem. Res.* 49:8470-8476.

[131] Benavente-Garcia O, Castillo J, Marin FR, Ortuno A, Del Rio JA (1997) Uses and properties of Citrus flavonoids. *J. Agr. Food Chem.* 45:4505-4515.

[132] Moure A, Cruz JM, Franco D, Domínguez M, Sineiro J, Domínguez H, Núñez MJ, Parajó JC (2001) Natural antioxidants from residual sources. *Food Chem.* 72:145-171.

[133] Duthie G, Crozier A (2000) Plant-derived phenolic antioxidants. *Curr. Opin. Lipidol.* 11:43-47.

[134] Manthey JA, Grohmann K (2001) Phenols in citrus peel byproducts: concentrations of hydroxycinnamates and polymethoxylated flavones in citrus peel molasses. *J. Agr. Food Chem.* 49:3268-3273.

[135] Anagnostopoulou MA, Kefalas P, Papageorgiou VP, Assimopoulou AN, Boskou D (2006) Radical scavenging activity of various extracts and fractions of sweet orange peel (Citrus sinensis). *Food Chem.* 94:19–25.

[136] Xu G, Ye X, Chen J, Liu D (2007) Effect of heat treatment on the phenolic compounds and antioxidant capacity of citrus peel extract. *J. Agr. Food Chem.* 55:330–335.

[137] Zia-ur-Rehman (2006) Citrus peel extract – A natural source of antioxidant. *Food Chem.* 99:450–454.

[138] Loizzo ML, Tundis R, Bonesi M, Menichini F, De Luca D, Colica C, Menichinia F, (2012) Evaluation of Citrus aurantifolia peel and leaves extracts for their chemical composition, antioxidant and anti-cholinesterase activities. *J. Sci. Food Agr.* (doi: 10.1002/jsfa.5708).

[139] Xu GH, Chen JC, Liu DH, Zhang YH, Jiang P, Ye XQ (2008) Minerals, phenolic compounds, and antioxidant capacity of citrus peel extract by hot water. *J. Food Sci.* 73:C11–C18.

[140] Kim MR, Kim WC, Lee DY, Kim CW (2007) Recovery of narirutin by adsorption on a non-ionic polar resin from a water-extract of Citrus unshiu peels. *J. Food Eng.* 78:27–32.

[141] Kim JW, Lee BC, Lee JH, Nam KC, Lee SC (2008) Effect of electron-beam irradiation on the antioxidant activity of extracts from Citrus unshiu pomaces. *Radiat. Phys. Chem.* 77:87-91.

[142] Giannuzzo AN, Boggetti HJ, Nazareno MA, Mishima HT (2003) Supercritical fluid extraction of naringin from the peel of Citrus paradise. *Phytochem. Analysis* 14:221–223.

[143] Khan MK, Abert-Vian M, Fabiano-Tixier A-S, Dangles O, Chemat F (2010) Ultrasound-assisted extraction of polyphenols (flavanone glycosides) from orange (Citrus sinensis L.) peel. *Food Chem.* 119:851-858.

[144] Aura AM, O'Leary KA, Williamson G, Ojala M, Bailey M, Puupponen-Pimia R, Nuutila AM, Oksman-Caldentey KM, Poutanen K (2002) Quercetin derivatives are deconjugated and converted to hydroxyphenylacetic acids but not methylated by human fecal flora in vitro. *J. Agr. Food Chem.* 50:1725- 1730.

[145] Aura AM, Martin-Lopez P, O'Leary KA, Williamson G, Oksman-Caldentey KM, Poutanen K, Santos-Buelga C (2005) In Vitro metabolism of anthocyanins by human gut microflora. *Eur. J. Nutr.* 44, 133-142.

[146] Hollman PCH, Arts ICW (2000) Flavonols, flavones and flavonols-nature, occurrence and dietary burden. *J. Sci. Food Agr.* 80:1081-1093.

[147] Morand C, Manach C, Crespy V, Rémésy C (2000) Respective bioavailability of quercetin aglycone and its glycosides in a rat model. *Biofactors* 12:169-174.

[148] Nemeth K, Plumb GW, Berrin JG, Juge N, Jacob R, Naim HY, Williamson G, Swallow DM, Kroon PA (2003) Deglycosylation by small intestinal epithelial cell β-glucosidases is a critical step in the absorption and metabolism of dietary flavonoid glycosides in humans. *Eur. J. Nutr.* 42:29-42.

[149] Tommasini S, Raneri D, Ficarra R, Calabro ML, Stancanelli R, Ficcara P (2004) Improvement in solubility and dissolution rate of flavonoids by complexation with (beta)-cyclodextrin. *J. Pharmaceut. Biomed.* 35:379-387.

[150] Tommasini S, Calabro ML, Raneri D, Ficarra P, Ficarra R (2004) Combined effect of pH and polysorbates with cyclodextrins on solubilization of naringenin. *J. Pharmaceut. Biomed.* 36:327-333.

[151] Bok S-H, Jeong T-S, Lee S-K, Kim J-R, Moon S-S, Choi M-S (2001) *Flavanone derivatives and composition for preventing or treating blood lipid level-related diseases comprising same*. US 20010006978A1.

[152] Rice-Evans CA, Miller NJ, Paganga G (1996) Structure–antioxidant activity relationships of flavonoids and phenolic acids. *Free Radical. Bio. Med.* 20:933-956.

[153] Ishihara K, Nakajima N (2003) Structural aspects of acylated plant pigments: stabilization of flavonoid glucosides and interpretation of their functions. *J. Mol. Catal. B-Enzym.* 23:411-417.

[154] Tamura H, Yamagami A (1994) Antioxidative activity of monoacylated anthocyanins isolated from Muscat Bailey A grape. J. Agr. Food Chem. 42:1612-1615.

[155] Danieli B, De Bellis P, Carrea G, Riva S (1990) Enzyme-mediated regioselective acylations of flavonoid disaccharide monoglycerides. *Helv. Chim. Acta* 73:1837-1844.

[156] Sakai M, Suzuki M, Nanjo F, Hara Y (1994) *3-O-acylated catechins and methods of producing same*. EP 0618203.
[157] Medina JH, Vioila H, Wolfman C, Marder M, Wasowski C, Calvo D, Paladini AC (1997) Overview-Flavonoids: A new family of benzodiazepine receptor ligands. *Neurochem. Res.* 22:419-425.
[158] Goldberg Y, Alper H (1994) Electrophilic halogenation of aromatics and heteroaromatics with N-halosuccinimides in a solid/liquid system using an H+ ion exchanger or ultrasonic irradiation. *J. Mol. Catal.* 88:377-383.
[159] Clague MH, Butler A (1995) On the mechanism of cis-dioxovanadium(V)-catalyzed oxidation of bromide by hydrogen peroxide: evidence for a reactive, binuclear vanadium(V) peroxo complex. *J. Am. Chem. Soc.* 117: 3475-3484.
[160] Roche D, Prasad K, Repic O, Blacklock TJ (2000) Mild and regioselective oxidative bromination of anilines using potassium bromide and sodium perborate. *Tetrahedron. Lett.* 41:2083–2086.
[161] Franssen MCR, van Boven HG, van der Plas HC (1987) Enzymatic halogenation of pyrazoles and pyridine derivatives. *J. Heterocyclic. Ch*em. 24:1313-1316.
[162] Franssen MCR (1994) Halogenation and oxidation reactions with haloperoxidases. *Biocatalysis.* 10:87-111.
[163] Yaipakdeea P, Robertsonb LW (2001) Enzymatic halogenation of flavanones and flavones. *Phytochem*. 57:341–347.
[164] Gorinstein S, Zachwieja Z, Folta M, Barton H, Piotrowicz J, Zember M, Weisz M, Trakhtenberg S, Martín-Belloso O (2001) Comparative content of dietary fiber, total phenolics, and minerals in persimmons and apples. *J. Agr. Food Chem*. 49:952-957.
[165] Gallaher D, Schneeman BO (2001) Dietary fiber. In: Bowman B, Russel R (Eds) *Present knowledge in nutrition* (8th ed., 805p) Washington, DC: ILSI.
[166] De Vries JW, Faubion JM (1999) Defining dietary fiber: A report on the AACC/ILSI NA consensus workshop. *Cereal Food World* 44:506-507.
[167] Villanueva-Suarez MJ, Redondo-Cuenca A, Rodríguez-Sevilla MD, de las Heras M (2003) Characterization of nonstarch polysaccharides content from different edible organs of some vegetables, determined by GC and HPLC: Comparative study. *J. Agr. Food Chem.* 51:5950-5955.
[168] Chau CF, Huang YL (2003) Comparison of the chemical composition and physicochemical properties of different fibers prepared from the peel of Citrus sinensis L. Cv. Liucheng. *J. Agr. Food Chem.* 51:2615-2618.
[169] Elleuch M, Bedigian D, Roiseux O, Besbes S, Blecker C, Attia H (2011) Dietary fibre and fibre-rich by-products of food processing: Characterisation, technological functionality and commercial applications: A review. *Food Chem.* 124:411-421.
[170] Grigelmo-Miguel N, Martín-Belloso O (1999) Characterization of dietary fiber from orange juice extraction. *Food Res. Int.* 131:355-361.
[171] Fernández-Ginés JM, Fernández-López J, Sayas-Barberá E, Pérez-Alvarez JA (2003) Effects of storage conditions on quality characteristics of bologna sausages made with citrus fiber. *J. Food Sci.* 68:710-715.
[172] Figuerola F, Hurtado ML, Estévez AN, Chiffelle I, Asenjo F (2005) Fibre concentrates from apple pomace and citrus peel as potential fibre sources for food enrichment. *Food Chem.* 91:395-401.

[173] Garcia ML, Dominguez R, Galvez M, Gavlez D, Casas C, Selgas MD (2002) Utilisation of cereal and fruit fibres in low fat dry fermented sausage. *Meat Sci.* 60:227–236.

[174] Fu F, Wang Q (2011) Removal of heavy metal ions from wastewaters: A review. *J. Environ. Manage* 92:407-418.

[175] Volesky B, Schiewer S (1999) Biosorption of metals. In: Flickinger MC, Drew SW (eds) *Encyclopedia of Bioprocess Technology: Fermentation, Biocatalysis, and Bioseparation*, John Wiley & Sons, New York, pp 433-453.

[176] Ngah WSW, Hanafiah MAKM (2008) Removal of heavy metal ions from wastewater by chemically modified plant wastes as adsorbents: A review. *Bioresource Technol.* 99:3935-3948.

[177] Ajmal M, Rao RAK, Ahmad R, Ahmad J (2000) Adsorption studies of Citrus reticulata (fruit peel of orange): removal of Ni(II) from electroplating wastewater. *J. Hazard. Mater.* 79:117-131.

[178] Annadurai G, Juang HS, Lee DJ (2002) Adsorption of heavy metal from water using banana and orange peels. *Water Sci. Technol.* 47:185-190.

[179] Dhakal RP, Ghimire KN, Inoue K (2005). Adsorptive separation of heavy metals from an aquatic environment using orange. *Hydrometallurgy* 79:182-190.

[180] Schiewer S, Patil SB (2008) Pectin-rich fruit wastes as biosorbents for heavy metal removal: Equilibrium and kinetics. *Bioresource Technol.* 99:1896-1903.

[181] Pérez-Marín AB, Meseguer Zapata V, Ortuño JF, Aguilar M, Sáez J, Lloréns M (2007) Removal of cadmium from aqueous solutions by adsorption onto orange waste. *J. Hazard. Mater.* 139:122-131.

[182] Liang S, Guo X, Feng N, Tian Q (2010) Isotherms, kinetics and thermodynamic studies of adsorption of Cu2+ from aqueous solutions by Mg2+/K+ type orange peel adsorbents. *J. Hazard. Matter.* 174:756–762.

[183] Bhatnagar A, Minocha AK, Sillanpää M (2010) Adsorptive removal of cobalt from aqueous solution by utilizing lemon peel as biosorbent. *Biochem. Eng. J.* 48:181-186.

[184] Feng N, Guoa X, Lianga S, Zhub Y, Liub J (2011) Biosorption of heavy metals from aqueous solutions by chemically modified orange peel. *J. Hazard. Mater.* 185:49-54.

[185] Bayo J (2012) Kinetic studies for Cd(II) biosorption from treated urban effluents by native grapefruit biomass (Citrus paradisi L.): The competitive effect of Pb(II), Cu(II) and Ni(II). *Chem. Eng. J.* 191:278-287.

[186] Shan W, Fang D, Zhao Z, Shuang Y, Ning L, Xing Z, Xiong Y (2012) Application of orange peel for adsorption separation of molybdenum(VI) from Re-containing industrial effluent. *Biomass. Bioenerg.* 37:289-297.

[187] López JAS, Li Q, Thompson IP (2010) Biorefinery of waste orange peel. *Crit. Rev. Biotechnol.* 30(1):63–69.

[188] Lohrasbi M, Pourbafrani M, Niklasson C, Taherzadeh MJ (2010) Process design and economic analysis of a citrus waste biorefinery with biofuels and limonene as products. *Bioresource Technol.* 101:7382–7388.

In: Microbes in Process
Editors: Neelam Garg and Abhinav Aeron
ISBN: 978-1-63117-127-7

Chapter 4

Production of Fuel Ethanol from Cane Molasses

Mohammed Fadel [*]
Microbial Chemistry Dept., Genetic Engineering and Biotechnology Division,
National Research Center, Dokki, Cairo, Egypt

Abstract

Molasses, is a thick dark syrup that is a by-product of sugar refining through repeated crystallization of sugar syrup obtained by crushing sugar cane. Molasses is used mainly in manufacture of industrial/ potable alcohol, yeast and cattle feed. Alcohol in turn is used to produce ethanol, rectified spirit and various value added chemicals. Ethanol is consumed by chemical industry and is also used in blending with petroleum to produce Ethanol Blended Petroleum (EBP). The yield of molasses per ton of sugar cane crushed varies in the range of 4.5% and 5%.Nearly 90% of molasses produced is consumed by industrial alcohol manufacturers and remaining 10% for various other uses like potable liquor. The components of molasses include Major components (water, sugar, non-sugars) and minor components (trace elements, vitamins, significant amounts of several minerals and growth substances. As well as molasses is a source of calcium, magnesium, potassium, and iron;

1. Introduction

The molasses is defined as the end product of sugar manufacture from which no more sugar may be crystallized by conventional methods. Molasses therefore, is a solution of sugars, organic matter and inorganic material in water. Molasses cane has a wide variety of uses and is of importance both in temperate and tropical economies. The total production of cane molasses from world sugar production is 15,000,000 tons per annum or more, and of this approximately 4,000,000 tons is the extent of the international molasses trade. The

[*] E-mail: mfadel54@yahoo.com.

international movement of molasses from producing areas to consuming areas is a complex operation involving the collection of supplies from factories by road and rail transport to storage at deep sea ports followed by ocean freight in bulk tanker vessels to ports in consumer countries and, finally, the distribution from port storage to individual uses. Generally, molasses obtained after sugar recovery is a cheap source of fermentable sugars that in addition to sugars also contains a number of other important nutrients. Other advantages of molasses are by-products, thereby, prices are not politically controlled and sugars in molasses can be easily utilized by most of the microorganisms.

2. Composition of Cane Molasses

Cane molasses composition is affected by the variety and soil conditions. In addition, processing conditions in the sugar factory may affect the composition of molasses. For this reason it is not possible to establish a typical analysis of molasses but a broad range of composition may be given as in Figure 1.

The total sugar content of cane molasses varies from below 50% to 60% and in addition there is a variable split between sucrose and reducing sugars. The reducing sugars in cane molasses are largely present in the original cane stalk and are concentrated in processing although some inversion may take place under poor processing conditions in the sugar factory. The solubility of sucrose is affected by the presence of reducing sugars and inorganic ash. In general, reducing sugars decrease the solubility of sucrose and inorganic ash increases its solubility.

For sugar factory control various formulae have been devised to predict the composition of a well exhausted molasses taking into account the above two factors; one very simple formula is as follows:

$$\text{Sucrose \% dry matter} = 40 - 4 \times I / A,$$

where I, is reducing sugars (% Dry matter); A, is ash (% dry matter). This may be modified to predict total sugars i.e. sucrose and reducing sugars as follows: Total Sugars (sucrose + reducing sugars) = $40 - 4 \times I / A + I$.

The latter formula for total sugar has been applied to a number of mixed shipments of molasses from various ports. It is emphasized that these results apply to commercial shipments of molasses and are not necessarily representative of the efficiency of the sugar industry in the area named. The fermentation of sugars into ethanol is one of the earliest organic reactions employed by humanity. The intoxicating effects of ethanol consumption have been known since ancient times. In modern times, ethanol has widespread use as a solvent of substances intended for human contact or consumption, including scents, flavorings, colorings, and medicines.

In chemistry, it is both an essential solvent and a feedstock for the synthesis of other products. It has a long history as a fuel for heat and light, and more recently as a fuel for internal combustion engines. The molasses discussed in the following pages is a product of the sugar industry which is based upon sugar cane. Sugar cane is a grass like plant grown in tropical areas and is the most prolific converter of sunlight into carbohydrates and producing some of the highest dry matter yields per area of any plant.

Sr. no	Analytical Parameters	Units	Normally observed values
A. Chemical Analysis:			
1	Total solids	% by wt.	65 to 80
2	Total Reducing Sugars	% bywt.	45 to 60
3	Unfermentable Sugars	% bywt.	4to5
4	Fermentable Sugars	% by wt.	40 to 46
5	F/N ratio		1 to 2.5
6	Total inorganics	% by wt.	8 to 12
7	Calcium as CaO	% by wt.	1.5 to 3.5
8	Total settlable sludge at 4.5 PH	% by wt.	5 and below
9	Specific gravity		1.38 to 1.55
10	Titrable volatile acidity in terms of acetic acid & acetate salts	Ml of 0.1 N NaOH	29ml
11	pH at 40° Bx dilution		4.5 to 5.2
12	Suspended particles >100 microns	% by wt.	Up to 0.05
13	Color in terms of Optical Density (OD) at 375 nm with 0.01% w/v dilution	OD	0.15 to 0.4
14	Soluble sulphates as SO4		1.5 to 5
B. Microbiological:			
1	Total bacterial count	CFU/gm	10 exp 3 and below
2	Wild yeast and fungal count	CFU/gm	10 exp 2 and below

Figure 1. Typical Composition of Cane Molasses.

2.1. Sugar Content in Cane Molasses

Sugars are an important part of molasses and before discussing the general composition it might be as well to refer to the more important classes of sugar, i.e. sucrose. Sucrose is a disaccharide which when hydrolyzed splits into two monosaccharides, glucose and fructose (sometimes referred to as dextrose and laevulose). Sucrose does not reduce cupric ions in an alkaline solution such as Fehling solution but the reducing sugars do, hence their name. Molasses contains both reducing sugars and sucrose so that the original reducing sugars are determined first and then the total reducing sugars after hydrolysis.

The difference between the two determinations will give the reducing sugars derived from sucrose.

($Cn\ H_{22}\ On + H_20$)		($C_6\ H_{12}\ O_6 + C_6\ H_{J2}\ O_6$)
(Sucrose + water)		(Glucose + Fructose)
(Sucrose+ water)	Hydrolysis	(Reducing (Invert) Sugar
(342 parts + 18 parts)		(360 parts)

It should be noticed that 342 parts of sucrose will produce 360 parts of reducing sugars after hydrolysis, or to put it in simpler terms, the original sucrose is equivalent to 0.95 times the reducing sugars formed.

A second method for the determination of sucrose is based upon the optical activity of sugar solutions. Sucrose is a combination of glucose and fructose rotates the plane of polarization to the left. For this reason hydrolysis of sucrose is often referred to as inversion and the product as invert sugar. The change in polarization measured on a saccharimeter or polarimeter under standard conditions can be used to determine the original sucrose content. These standard conditions include the elimination of the effect of interfering substances by the use of lead acetate and then the elimination of lead by potassium oxalate. It is considered that the reduction method using Fehlings solution will give more easily reproducible results in different laboratories and is therefore the preferred method for the determination of sucrose in molasses. Both methods, however, rely upon an inversion procedure and it is known that under conditions of high acidity and high temperature sugars may be destroyed. It has been a matter of experience that disagreements between laboratories have very often been due to differing methods of inversion and it is necessary therefore that inversion conditions are chosen with care and can be and are standardized.

2.2. Reducing (Invert) Sugars

Both methods chosen in the handbook for the determination of reducing sugars rely upon the reduction of cupric salts, one facing a volumetric method and the other a gravimetric one. The volumetric method is well known in the sugar industry and is simple and capable of giving reproducible results under standard conditions. Some non-sugars, in particular calcium, can influence the results of this determination possibly forming a complex with glucose and fructose. This effect is eliminated by sequestering the calcium with sodium ethylene di- amine tetra acetate. Raffinose is a trisaccharide which is present in small proportions and consists of three monosaccharide units of sugar joined together. For its more precise determination it is necessary to use chromatographic methods.

2.3. Unfermentable Reducing Substances

As well as the reducing sugars themselves, there are present in molasses other compounds that have a reducing effect on Fehlings solution. Consequently the results of sugar determination using reduction methods include these non-sugar organic reducing compounds. Whereas these compounds are of value in animal feeding they are not capable of being utilised in fermentation processes and are therefore determined by difference following fermentation.

2.3.1. Reporting Sugar Content

Very often in molasses trading the sugar content is reported as the total of sucrose plus invert 95 (or reducing sugars). It is however, possible to relate the two sugars to a single basis such as "total sugars as sucrose", or "total sugars as invert" (TSAI). If sucrose is represented by A% and the invert content by B% then the method of reporting may be as follows:

a) Total Sugars A% + B %
b) Total Sugars as Sucrose A% -f (B % x0.95)
c) Total Sugars as Invert A% 0.95 + B %

2.3.2. General Observations

The analysis of sugars depends upon the measurement of various properties such as the rotation of plane polarized light for sucrose and reducing sugars under which the determinations are carried out must be very carefully standardized. The situation with regard to molasses is even more critical because molasses is a mixture of other materials, in addition to sugars, and these may also affect the properties being measured, both in polarimetric and chemical determinations. This has already been pointed out above in the paragraph dealing with unfermentable reducing substances. It is therefore of great importance that the conditions of analysis for molasses are standardized in details in order that reproducible results could be obtained in different laboratories.

2.3.3. Viscosity

The viscosity of cane molasses is influenced by constituents other than sugars. Beet molasses is usually of lower viscosity, particularly at ambient temperatures than cane molasses. Molasses is affected by the dry matter content. The viscosity of cane molasses is very positively influenced by the non sugar organic matter and there are wide variations. The variations are a function of climate, soil and factory processing and also the conditions under which cane is harvested. The age of cane reaching the sugar factory, particularly when mechanically harvested, has an effect on the viscosity of the final molasses.

Molasses exhibits the phenomenon called "critical viscosity" which means that above a certain dry matter content the viscosity increases at a greater rate than might be expected from the increased dry matter content. The critical viscosity for cane molasses lies between 81 and 85 hydrometer Brix. Brix is a measure of the specific gravity of a liquid. The molasses trade commonly use the term Brix as an indicator of specific gravity and it represents an approximation of total solids content. Brix is originally initiated for pure sucrose solutions to indicate the percentage of sucrose in solution on a weight basis. However, in addition to sucrose, molasses contains glucose, fructose, raffinose and numerous non-sugar organic materials. Brix is tested by diluting a representative sample of the molasses with an equal weight of water, letting the mixture stand for about 20 minutes to allow entrained air to escape, and observing the level of the liquid on the stem of the hydrometer. The reading is then doubled to allow for the dilution with water. Brix is used in the official definition of beet, citrus, starch and cane molasses (AAFCO, 1982). The viscosity of molasses is affected both by dry matter content and a temperature, for example a rise in temperature of 10°C may reduce the viscosity to a half or less and a reduction in the dry matter content will also decreases the viscosity. It can be seen that for molasses at 75% dry matter and at a temperature of 40°C not only is the viscosity reduced but the differences between various origins tend to disappear. It is for this reason that for handling molasses in a feed mill or factory a temperature of 40°C is recommended because at this temperature little or no destruction takes place and molasses may be handled efficiently both for distribution and measurement.

3. Heating of Molasses

At a temperature of 40°C molasses is relatively stable but as the temperature is raised sugar may be lost by thermal decomposition. In Figure 2 the effect of temperature over a number of days is shown on two samples of cane molasses.

It can be seen that at a temperature of 60°C there is a significant loss of sugar within a few days and when the temperature is further increased to 120°C and 130°C, which corresponds to steam at 1 kg. per sq. cm. and 1.75 kg per sq.cm., the loss becomes rapid. After 1 hour at 120 °C a b < of the sugar has been destroyed. Under extreme conditions of heating, molasses may decompose completely.

	Trinidad Molasses		Jamaican Molasses	
Temperature	**40°C**	**60°C**	**40°C**	**60°C**
Time Elapsed				
3 days	**Nil**	**-0.83**	**-0.07**	**-1.-13**
6 days	**+0.04**	**-1.12**	**-0.10**	**-1.84**
9 days	**+0,16**	**-1.66**	**-0.09**	**-2.46**
12 days	**+0,12**	**-1.97**	**-O.O9**	**-2.98**
15 days	**+0.16**	**-2 .56**	**-0.08**	**-3.46**

Figure 2. Change in Sugar Content at Low Temperature.

	Trinidad Molasses		Jamaican Molasses	
Temperature	120°C	131°C	120°C	131°C
Time Elapsed				
1 hour	-5.93	-14.26	-7.70	-11.42
3 hour	-9.84	-26.78	-11.42	-34.03

120°C = 1 Kg/cm² steam 131°C = 1.75kg/cm² steam

Figure 3. Change in Sugar Content at Higher Temperature.

From time to time instances occur, particularly on tropical sugar estates, which result in the complete destruction of large quantities of cane molasses. When molasses is heated the decomposition that occurs is exothermic and therefore heat is generated. If the heat generated by the decomposition can't escape from the sides of the storage tank fast enough then the temperature increases still further and the rate of reaction becomes faster until in the end molasses may either be ejected from the tank with the formation of steam or in extreme cases will catch alight to leave charred molasses.

Although it is not possible to be precise about the level of heat required to start destructive decomposition, at any temperature over 60°C there is always the possibility of thermal decomposition. Earlier it was mentioned that the products produced by the reaction of reducing sugars and amino acids may play some part in thermal decomposition but whatever the mechanisms of decomposition it is most important to keep the temperature of cane molasses below 60°C for safe storage. It is a useful precaution to fit molasses storage tanks with thermometers so that a rise in temperature may be seen if it occurs and then action may be taken before a complete disaster happens.

4. Fermentation Nutrients

In many of traditional fermentations, the use of yeast nutrients is not required, however, in some instances the supply of such nutrients may be limiting. For instance, the supply of nitrogen is often inadequate. In batch fermentations there is usually a 5- to 10-fold multiplication of yeast cell. With an inoculum of 6-8 million cells per ml or about 0.24 kg (2 Ib) of yeast solids per kl (1000 gal.) one would expect the growth of about 1.2 to 2.2 kg (10 to 18 Ib) of yeast solids in 1 kl (1000 gal) of wort or mash. Newer processes of continuous fermentation with some aeration during the fermentation period may result in the production of larger quantities of yeast.

Yeast requires the following nutrients for growth per 45 kg (100 Ib) of yeast solids: About 3-3.6 kg of nitrogen from an assailable nitrogen source such as ammonia, ammonium salts, urea, or amino acids; about 1.1 kg of phosphorus (as P_2O_5), usually in the form of orthophosphoric acid; and minor quantities magnesium continuously pumped into the fermenter and an equal volume of the fermenter liquid is continuously pumped out for recovery of ethanol and yeast. It is assumed that the concentration of all substances is uniform in all parts of the fermenter and that the same concentration prevails in the withdrawn (harvested) liquid. The rate at which medium is added (or at which fermenter liquid is withdrawn) is usually expressed as the dilution rate. This is the ratio of withdrawn liquid to the volume of total liquid in the fermenter. Hence, a dilution rate of $D = 0.1$ indicates that one-tenth of the fermenter liquid is withdrawn per hour (or that the "residence time" in the fermenter and the yield of ethanol declines.

5. Wild Yeast and Bacterial Contamination

The proccess of fermentation to yield ethanol is often affected by cross contamination of wild yeasts and bacteria which may cause loses in the final yield and product formation.

5.1. What Is Wild Yeast?

Yeasts, other than the elliptical culture yeast, are called as wild yeast, and occurs naturally in the raw material and produce small amounts of alcohol [1]. They can be harder to be detected and controlled than bacteria.

5.2. Sources of Contamination

Wild yeast is considered to be indigenous and found in the air and on surfaces [2]. Wild yeast primarily comes through molasses in the fermentation process because of the presence of fermentable sugars, solid addition as well as liquid streams entering the process can be a source of wild yeast contamination [3].

5.3. Problems Caused by Wild Yeast Contamination

These include situations like scum formation and reduced efficiency.

5.3.1. Scum Formation

In the fermentation process, yeast multiplies in the form of heavy branched structure. This branching yeast comes at the top of the fermenter. The sparging of air in the fermenter helps the branched yeast to rise to the top and form a thick layer called the scum. There is washout of culture yeast and a gradual drop in the alcohol concentration in the fermentation process.

According to Lorenz et al. [4] the morphological changes in the culture yeast (*S. cerevisiae*) leads to filament formation under unfavorable fermentation conditions. This causes scum formation and foaming. The budding yeast *S. cerevisiae*, starved for nitrogen, differentiates into a filamentous growth form. In nitrogen poor conditions leucine, the precursor of iso-amyl alcohol (a chief constituent of fuel oil) can also induce elongation of cells.

5.3.2. Reduced Efficiency

According to Lorenz et al. [4], the reduction in efficiency is attributed to the presence of wild yeast and bacterial contamination in the input to the fermenter. Due to heavy foaming, juice losses result in the loss of alcohol. The efficiency reduction depends upon the extent of contamination.

6. Wild Yeast Management

Many of the fermentative processes use non-pasteurized medium, which can enhance the predominance of wild yeast. These contaminants can contribute to the fermentation rate decrease or blockage, increases in acidity, fuel oil production, and an overall decrease in ethanol production.

6.1. Killer Yeast Strains

There are widespread occurrences of killer phenotypes in yeasts in alcoholic fermentation. Isolation of killer yeast strains from the ethanol process is imperative for good yields. The presence of cultured yeast with "killer" tendencies against wild yeast will help overcome the problem of excess growth of wild yeast contamination during the process [5]. The availability of good quality molasses depends on the efficiencies' on fit specific sugar refineries, weather conditions, and harvesting techniques. Unfortunately, these variables lead to an inconsistent supply of good quality sugar cane molasses and a constant dilemma for sugar cane molasses distillers [6]. It is not altogether necessary to sterilize molasses. However, there are advantages in terms of alcohol yield and purity when pre-treating the molasses. Treatment of the molasses solution by means of heat and sulfuric acid will precipitate undesirable salts and help to improve the purity of alcohol obtained and reduce the amount of scaling.

6.2. Sulfur Dioxide and Sulfite Solutions

The principle source of SO_2 is sodium bisulfite or potassium meta-bisulfite. Sulfite is available in powdered form but because of its elevated level of toxicity, its use is limited. Potassium metabisulfite contains approximately 55% SO_2 by weight. This free SCX, kills the microorganisms [7]. Culture yeast is generally tolerant to SO_2, but at higher concentrations it shows loss of viability. SO_2 is toxic and requires appropriate handling.

6.3. Process Parameters

For control of infection, as discussed earlier, pasteurization of molasses can help control wild yeast and bacteria. Contaminants generally occupy surfaces covered with deposits of starch, sugar, protein or mineral rich materials. Contaminants thrive in places that are hard to clean such as, porous surfaces, cracks, sharp angles, corners, gaskets, valves, pressure gauges, in-place thermometers, and pump packing. Deposits in these areas protect the organisms in them from heat and sterilizing solutions.

6.3.1. Adequate Yeast Cell Population

According to Abbott et al. [8] the growth rate of *Dekker*a WM yeast is lower than culture yeast *S. cerevisiae* and hence their ability to compete with *S. cerevisiae* is hindered in batch fermentation. Due to different reaction rates *Dekkera* will produce different end product as they ferment the juice sugars to alcohol [9].

Considering this fact, the cell concentration of the culture yeast can be higher but suppressed if wild yeast outnumbers the culture yeast. Recycle of yeast builds up higher population of yeast in the fermenting mash, thus giving higher productivity as well as robustness of operation [10]. There are processes that would allow the yeast cell concentration in the fermentation process to be increased. Yeast recycling is one of the most common methods. This process maintains desirable cell concentration for healthy fermentation. However, maintaining the population of these contaminants at lower levels and the use of proper operating procedures is key to the fermentation process.

6.3.2. Temperature

Online cooling and mixing of molasses is important to avoid the caramelization of sugars. Molasses, if not cooled properly, helps the growth of heat stable microbes that affects fermentation. At higher temperatures, generally in the summer, wild yeast is produced in high quantities.

The processing of fresh molasses usually results in a high microbial count in distillery fermentation. Sulfurous gases can inhibit yeast. Similarly fresh molasses has a high foaming tendency as well as a high buffering capacity. It also contains high levels of suspended sludge. This is why fresh molasses is stored for at least a month before use in a distillery. Cooling and frequent mixing by recirculation during storage helps to avoid internal combustion and caramelization that can deteriorate quality. Excessive storage time (more than six month) on the other hand can reduce the fermentable sugar content slowly. Molasses is stored in steel tank to prevent contamination. In order to control the effects of aging, recording the production span, life, and operating with "first in first out" principle are normally standard procedure.

The sugar factory process along with handling, storage and transportation of molasses is vital to managing the contamination of wild yeast. As discussed earlier, the presence of sugar in the medium favors growth of wild yeast along with nutrients. Contact with water and pockets of dilution in bulk stock can give rise to high microbial flora. Contact between soil with molasses also should be avoided. It is necessary to handle and store via a protective environment [11, 12].

7. Inhibition of Bacterial Contamination

Bacterial contamination is a major cause of reduced yields in ethanol production [13]. Contaminants constantly utilize carbon available for conversion to ethanol and compete for growth factors needed by yeast [14]. The contaminating agents produce deleterious end products such as lactic and acetic acids that inhibit the growth of *S. cerevisiae* [15, 16]. Fermentation tanks and yeast propagation systems can act as reservoirs of bacteria that can continually reintroduce contaminants [17].

A number of antimicrobial agents to control bacterial contamination in ethanol fermentations under laboratory conditions have been described. Urea hydrogen peroxide reduced the numbers of *Lactobacillus* while providing nutrients to aid performance of the yeast [18]. Various agents have been tested for control of bacterial contaminants under laboratory conditions, such as hydrogen peroxide, potassium metabisulfite and 3, 4, 4-trichlorocarbanilide [18-21]. Hop acids are reported as replacing agents of antibiotics [22]. Penicillin and antibiotic monensin have shown effective results against the strains of *Bacillus* and *Lactobacillus* isolated from Brazilian alcoholic fermentation units [23]. Different antibiotics including penicillin, virginiamycin and tetracycline have been reported to control contamination by lactic acid bacteria in experimentally infected alcoholic fermentations [24, 25, 26].

Although, there are different designs and operating practices in distilleries, numerous chances exist for contaminants to persist or thrive in the system. To clean and sterilize the propagation and fermentation tanks is common in distilleries and as there are batch process so much time is available. Yeast propagation is a potential point of contamination which increases during fermentation. The contamination load comes continuously with molasses and this causes a great problem in fermentation. Sugar cane molasses produced by sugar mills in

Pakistan are stored in very poor conditions and so when they reach the distilleries, they are found to be loaded with contaminating microorganisms. The profitability of ethanol production is dependent on favorable sugar cane molasses price and the quality of molasses (sugar % and contamination level). After the year 2000, alcohol business in Pakistan expanded rapidly. Bacterial Contamination can be controlled by:

7.1. Pure Culture

Pure culture is always required for fermentation to get inoculums free of any kind of bacteria to compact the continuous load of contamination of molasses in distilleries.

7.2. Reducing pH

Bacterial contamination can be avoided by reducing the pH value with sulfuric acid, as low pH is not suitable for most bacterial growth.

7.3. Use of Antibiotics

To avoid bacterial contamination, antibiotics are also used at different levels. When virginiamycin was used, maximum achievable results were obtained keeping all other factors constant [27]. Comparing the control with antibiotic treated at 0.5 ppm, there was an increase of 0.3% ethanol (v/v). When bacterial population decreased to the range of 10^4, more than 0.4% ethanol was produced. When the antibiotic dose increased to 1.5 ppm, ethanol increased by 0.8 % (v/v) and optimally 8.5 % ethanol was obtained at 2ppm concentration of the antibiotic used. Occurrence of 10 x 10^6 lactobacilli / ml mash resulted in 1% reduction in the final ethanol produced by the yeast, depending on the strain of the contaminant bacteria.

There is a need to optimize the concentration of the antibiotic used. If antibiotics are not administered correctly, the development of antibiotic resistant strains can become a reality [28]. High level of contamination will not make any impact to yeast growth nor have affect on the recovery of ethanol. In the terms of viability, there is heavy lost in yeast viability when culture is contaminated [29]. Virginiamycin has certain advantages over other antibiotics and factors such as temperature and pH stability. Use of Virginamycin at industrial scale ethanol fermentation is a reliable source to avoid losses due to contamination. Its use should be carefully done as per requirement. It can save more $4000 per day for the distillery producing 100, 000 L / day [27]. On other hand, use of lactrol and sodium fluoride successfully resulted in high ethanol-producing levels [26, 27]. Moreover, successful results and high fuel ethanol fermentation were got when virginiamycin and sodium fluoride were used to treat the bacterial contamination.

8. Modern Ethanol Production

Prior to selection a yeast strain, formulation type, or yeast management protocol, a review of plant objectives and priorities should be conducted.

8.1. Yeast Selection

To facilitate selection of appropriate yeast-related options for a specific ethanol plant, choose one of three main operating objectives i.e., efficiency, throughput, and consistency. If each ethanol production plant was designed to achieve one of these main objectives, the remaining two objectives will also be addressed by good plant designs. However, when analyzed in depth, the overall them of each plant design can be associated with one main objective. Additionally, each ethanol production plant is managed by an individual or team with a unique operating philosophy and approach to achieving plant objectives such as profitability or revenue growth. Selected operating philosophies - focused on parameters that impact yeast choices, – include process control, fast and complete fermentation as well as minimal bacterial and wild yeast contamination.

Production and formulation technologies have led to valuable specialty yeast products suited for today's ethanol production facilities. As a result, the ethanol producer has a number of viable commercial yeast product options to choose from, including a variety of yeast strains and formulations. Additionally, in-plant yeast management options abound. The optimal yeast choices - strain, formulation and management for each facility depend on a number of interrelated factors. Decisions are getting more complicated and more critical as ethanol production technologies and markets mature.

8.2. Yeast Strains

Hundreds of specialty strains have been commercialized for ethanol or CO_2 production, including specific strains for fuel ethanol. Specialty *S. cerevisiae* strains for ethanol production can be categorized based on the criteria used to select them. There are strains selected for continuous fermentation processes and strains for batch fermentation processes. Specific ethanol production strains for sugarcane and beet, separately for juice or molasses derived from each, have been marketed for years. Recently, high performance yeast strains have been selected and commercialize for ethanol production utilizing batch fermentation processes. Some yeast strains are able to convert substrate to ethanol with increased yields. Genetically- enhanced microorganisms for ethanol production are in various stages of development.

Whatever yeast strains can be selected or developed using a number of proven techniques comprise isolation from fermentations, screening from culture collections, mutation and hybridization as well as spheroplast fusion and genetic enhancement and cloning. Today's commercial yeast strains selected using the first five listed techniques differ across a broad range of selection and performance criteria such as temperature, pH value, sugar, organic acid and ethanol tolerance as well as flocculation, foam production, and respiratory growth rate.

The yield, stability of strains, carbohydrate utilization, and activity of enzymes used are important. Killer factor production and killer factor resistance are also critical.

9. Commercial Yeast Production

The main objective of commercial yeast production is to propagate a single yeast cell to a very large number of genetically identical yeast cells. Propagation is certainly necessary for economic ethanol production, but an equally important objective of modern yeast production is to ensure consistently excellent performance of the yeast in end-user applications. Both of these objectives can be accomplished under meticulously controlled conditions. Many precautions are taken to avoid contamination by other microbes; clean fermentation is critical to achieving consistent high performance. The first steps of the process, from vial to seed flask propagation, are carried out in a sterile laboratory environment to avoid contamination from other yeast or bacteria. Extraordinary precautions are taken to ensure scale-up steps, are conducted without introducing contaminating microorganisms. Contaminants (usually bacteria) in commercial yeast products are typically present at less than 0.001% of the number of desired yeast colony-forming units (CFU). Parameters such as temperature, pH, nutrient concentration, air purity and concentration and agitation are thoroughly controlled from seed to final formulation to ensure consistency of the final product. Process control and cleanliness are also critical during optional separation and drying steps. Yeast performance in end-user processes can be impacted by yeast growth rate during yeast production. In general, if yeast cells grow slowly, more mature cells with higher carbohydrate are obtained; in ethanol production plants, yeast cells grown in this manner give relatively low initial activity, but high ethanol yield, and the resulting commercial yeast products exhibit good stability. Conversely, if yeast cells are grown rapidly, less mature cells with higher protein and enzyme levels are obtained, resulting in higher initial activity, but lower ethanol yield and less stability in handling and use. A typical commercial yeast production process is depicted Three types of commercial products can be derived from this yeast production process.

The first available commercial product format is liquid yeast, also called yeast cream, which is harvested just after broth has been centrifuged to remove most of the water from the yeast cells. The second product format is fresh cake yeast, which is typically harvested after a vacuum filtration step in which more water is removed from the yeast solids. Fresh cake yeast can be sold as bricks or extruded and bagged; typical water content is 50-70%. Finally, the extruded cake yeast can be dried further, typically in an air lift dryer, to give a free-flowing product that is dry to the touch, called active dry yeast (ADY). In *a* modern, sophisticated yeast production facility, final product quality is assessed in a dedicated analytical laboratory using techniques such as DNA karyotyping to ensure trueness to type and purity of the strain. Other parameters monitored include pH value, protein, solids and water content. Yeast performance is normally assessed with a quick test (CO_2 production) and a fermentation performance test.

10. Commercial Yeast Formats

Before selecting a yeast product format, one must consider how the yeast will be handled prior to ethanol fermentation: What is the distance from the yeast production site to ethanol plant? Is refrigerated yeast storage available? Are automated dosing systems preferred? Are they available? Will in-plant yeast propagation or metabolic acclimatization be employed? If so, what is the main goal of this step? Will daily yeast usage be steady or inconsistent? These questions are important because there is significant variation in the transportation, storage, handling and performance characteristics of the three main commercial yeast formats, which are active dry yeast (ADY); fresh cake yeast; and fresh liquid yeast.

10.1. Active Dry Yeast

ADY products are widely accepted in industry today. For those yeast strains that are amenable to drying, ADY formulations are convenient because of the long shelf life. However, ethanol fermentation with ADY is characterized by a long lag phase because of the time needed for rehydration and acclimatization to compensate for the stress of the drying process. Fresh cake or liquid yeasts start fermenting relatively quickly and have a higher vitality, which result in superior fermentation performance, optimized for stability, acceptable fermentation, and unrefrigerated storage. ADY products does not require refrigeration performance, can be stable for 2 years, require rehydration, and need lowest cost storage. ADY products are good for small volume strains and acclimatization recommended must handle solid product (bags or boxes).

10.2. Fresh Cake Yeast

Fresh yeast products typically have shorter shelf life than dry yeast products. Stabilized liquid yeast, a new product with up to three months shelf life (refrigerated), can be directly pitched into the fermenter. Fresh yeast products are characterized with minimal acclimatization requirements, shorter lag phase in fermenter, and stable for at least 6 weeks. High viability must handle solid product in bags or boxes. Fresh yeast products are sealed and refrigerated.

10.3. Fresh Liquid Yeast

Liquid yeast is characterized by superior performance achieved with fresh yeast but without the need for a propagating or separate metabolic acclimatization step. Stability of cake yeast and cream yeast vary by strain and composition, but are generally similar if strain and composition are similar. Liquid yeast products are characterized with better stability and higher activity than fresh or dry products, suitable for direct pitch to fermenter, and no need for rehydration. Liquid yeast products are unstable (shelf life 1-2 weeks maximum). All aspects involved in the choice of yeast and fermentation management demonstrate that ethanol yeasts are specialty products. Each ethanol plant has unique objectives and constraints. Similarly, each yeast 'strain + formulation + management' combination has different attributes. Finding the best yeast choice for an ethanol plant is not trivial; analysis and selection requires expertise in yeast technology and experience in the industry. Additional ethanol plant considerations such as workforce cost and capabilities, operating philosophy and commodity prices can impact how a plant is operated, and therefore which yeast approach - a combination of strain, format and management is optimal. Responses to these considerations can impact yeast in a number of ways. Reputable yeast suppliers support their products with regular technical support visits and plant audits to optimize use of their products. They develop an understanding of plant management objectives and offer viable product and process alternatives in the context of the plant's needs and problems. Additionally, they provide both practical and theoretical operator and staff training on fermentation and the use of their products. The choice of an overall yeast strategy will depend mainly on the objectives of ethanol plant management. An example to illustrate this point: One objective might be to optimize efficiency of a batch fermentation plant. Efficiency can be measured as the yield of ethanol produced per unit of feedstock. Analysis shows plant efficiency can be impacted by a number of operating philosophies; plant management could select from several viable approaches to achieve optimal plant efficiency. They could, for this example, choose to pursue the approach to take extraordinary steps to avoid and control microbial contamination. A yeast solution for our example could be:

- strain = optimized for batch fermentation;
- format = stabilized liquid yeast;
- management = direct pitch.

This overall yeast approach matches the plant design and operating philosophy with the positive attributes of each selection. The strain was designed specifically for the process and should perform as expected. The product format and management options chosen will result in excellent fermentation performance and, of all options available, the lowest number of contaminants being added to the production fermenters. This is accomplished mainly through the elimination of in-plant yeast propagation or acclimatization steps that also propagate contaminants. The yeast management solutions are therefore matched to the specific objective and operating philosophy of plant management.

11. Cell Recycle

Cell recycle is a method for maintaining a high yeast cell population in a continuous fermentation. Yeast cells are separated from the withdrawn returned to the fermenter while the supernatant essentially free from yeast cell is sent to the still. A decline in yeast activity (and presumably viability) has also been reported for batch fermentations with recycle of yeast and maintenance of high cell populations [30]. This could be alleviated by working with lower sugar concentrations by diluting sugar cane juice from 16 to 13% of glucose. Most authors provide information on the concentration of yeast in terms of cell numbers per ml or in terms of dry cell weight per liter, rarely in both. This makes comparison of results often difficult, since the ratio of cell number to cell weight varies with the size of the cells, fermentation of molasses and cell recycle used between 2 and 3x10" cells per ml.

Other workers indicate cell mass values of 30-60 g/liter for continuous fermentations with cell recycle [31, 32, 33]. If the cell mass is given in g per liter, one can calculate the specific productivity of ethanol (g ethanol/g yeast solids/hr). Maximum values in the range of 0.5 to 1 g/g/hr have been reported [31, 34]. It is clear that the continuous fermentations with cell recycle show high productivities because of the very much larger number of yeast cells, and not because of an improved performance: of the individual cells.

11.1. Inoculum Dose and Thermo-Tolerant Srains

The yeast dose rate also has effect on performance. A higher dose rate will result in a faster start of fermentation, which helps control contamination. Dose rates must be optimized for cost effective performance. Nutrition is another important parameter to take into consideration.

Tmperature for optimal fermentation, for growth and ethanol fermentation1, is 30°C but somewhat higher temperatures (35°-38°C) are tolerated. However, at such higher temperatures growth rate, yield of ethanol, and the death rate may be affected. It would be very desirable to obtain strains which perform well at higher temperatures because it would reduce the energy required for cooling of the fermenters, and it might increase productivity [35]. At this time it is doubtful whether this can actually be achieved with *S. cerevisiae* strains.

11.2. Batch Fermentation

Batch fermentations are carried out without the need for establishing pure culture conditions, that is, without the need for complete sterilization of media (although pasteurization is advantageous) and without maintenance of complete sterility of equipment. However, this presupposes a rapid start of the yeast fermentation. This fermentation inhibits the growth of other microbes by depleting the available nutrients, by a lowering of the pH, and most importantly by the formation of ethanol. Industrial alcohol fermentations require cooling either through internal cooling coils or through external heat exchangers. The temperature is usually maintained between 30 and 35°C and the pH between 4.5 and 6. Fermenters are not stirred and not aerated.

The rapid development of carbon dioxide gas during the fermentation provides some mixing. Most of the currently practiced alcohol fermentations are based on the fermentional processes which have been outlined above, and so are many of the projected installations. However, the need for reducing capital investment, better utilization of energy, and increased productivity has led to the development of advanced methods. Such advances are the use of continuous fermentations, the increase of the yeast population by recycling, and the removal of ethanol during the fermentation.

11.3. Fed Batch Fermentation

Yeast growth and ethanol formation are inhibited by solutions of high osmotic pressure and by high concentrations of ethanol. Many strains of *S. cerevisiae* will readily produce 10-13% of ethanol (by vol.) in batch fermentations of molasses solutions with initial fermentable sugar concentrations of 20-25 %. Higher concentrations up to 16-18 % can be achieved by incremental feeding. However, productivity of ethanol declines rapidly at alcohol concentrations above 5-7 % by vol. Therefore, traditional batch fermentations for the production of ethanol from molasses, for instance, have been carried out with an initial sugar concentration of about 12g/l (20° Brix molasses). Fed batch fermentation has been carried out by charging the fermenter through the needed molasses quantity on two or three batches with interval between batches,the method more suitable for yeast strains can produce high ethanol under low diluted molasses.

11.4. Continuous Fermentation

Most of the experimental work has been carried out with homogeneous fermentations, that is, with a single stirred fermenter. Fresh medium is a non-sterile process. The presence of undesirable microorganisms in the process can't be completely avoided. The raw material used for fermentation itself shows the presence of contaminating microorganisms. Normally 60-70 % of the production cost is contributed by the feedstock in alcohol production. Owing to the increasing demand and cost of molasses, improving and maintaining the quality of the fermentation process has become a crucial factor. Additionally, the distillery must consistently achieve the maximum possible alcohol yields from a given feedstock. It therefore becomes important to monitor and achieve the proper quality of raw material and fermentation parameters.

12. Addition of Enzymes

Enhancement of ethanol yield from molasses fermentation by addition of enzymes is reported. With the dramatic increase in demand of ethanol across the globe and the increase in the cost of molasses, there is an immediate need to improve the efficiency of ethanol production from molasses. By converting unfermented sugars and oligosaccharides into fermentable sugars during molasses fermentation, ethanol yield can be enhanced by about 5% or by 5000 million liters worldwide. The value of this increased ethanol in India alone will be approximately 402 crores rupees annually.

Production of ethanol from molasses by industrial microorganisms such as yeast *S. cerevisiae* is of great commercial importance. Most of the readily available sugars (sucrose, glucose and fructose) in molasses can be consumed by the yeast during fermentation. In addition to these sugars, there are some oligosaccharides and other pentose sugars in molasses, such as xylose, which are not readily fermented by yeasts or other contaminating microorganisms. Efficiency of the ethanol production from molasses by *S. cerevisiae* can be enhanced to a significant level by enzymatic processes, using a mix of specific enzymes and an antibiotic during molasses fermentation process [37]. Un-fermented sugars and oligosaccharides are converted into fermentable sugars resulting in better availability of fermentable sugar for ethanol fermentation. The process has great commercial application specifically in sugar and molasses industries distilleries, alcohol-making industries. Specific enzymes such as alpha amylase, glucoamylase, alpha galactosidase, xylanase, amylase, cellulase are added to molasses during the process of fermentation *S. cerevisiae*. These enzymes, in specific amount and combination, break down the oligosaccharides to easily fermentable monosaccharides. Since xylose and other pentose sugars remain unutilized by *S. cerevisiae*, xylose isomerase along with the other carbohydrate depolymerizing enzymes has been used to isomerize the unfermented xylose to a readily fermentable xylulose. However, the isomerization step is a reversible process. Therefore, addition of the same enzymes (particularly xylose isomerase) is necessary during the fermentation process so that more xylose is isomerized as xylulose is fermented by the yeast. Addition of an antibiotic such as virginiamycin reduces contamination by opportunistic bacteria during molasses storage and fermentation.

Therefore, as a result of this enzymatic process, additional substrate in form of xylulose is available for yeast fermentation thereby utilizing non-fermentable sugar in molasses resulting in the enhanced alcohol production

Increase in other reducing sugars such as glucose and fructose was observed upon addition of enzymes to molasses as compared to the control due to the action of enzymes such as alpha amylase, glucoamylase and cellulase on complex carbohydrates such as starch and cellulose in the molasses. This increase also contributed to the increase in alcohol content. In case of the control set, the alcohol yield was found to be lower compared to the enzyme treated set. Addition of enzymes also resulted in higher biomass as compared to the control and this was due to availability of surplus carbon source in form of released reducing sugar due to enzymatic action Fermentation of molasses was carried out with the addition of carbohydrate depolymerizing enzymes but without the addition of xylose isomerase to study the individual effect of carbohydrate depolymerizing enzymes on increasing alcohol levels. Increase in alcohol was observed in this case as compared to the control without the addition of enzymes.

However, the alcohol yields were lower (76 g/L) as compared to the process where xylose isomerase was added along with the other enzymes (79 g/L). This establishes that although increase in alcohol can be obtained by addition of enzymes that specifically break down complex polysaccharides to reducing sugars, this is further enhanced by addition of xylose isomerase that isomerizes unfermentable xylose to fermentable xylulose and subsequent formation of alcohol.

13. Vinasses

Vinasse, a residual substance left after sugarcane alcohol distillation, represents a major environmental problem for the ethanol industry. No one has found a convenient and economical disposal solution for this black-reddish, solids ranging from 5-10% when it is obtained from sugarcane molasses, is produced at a rate of 10 to 15 times greater than the ethanol itself is a mixture of water and organic and inorganic compounds. These compounds remain after different steps involving the sugar cane production and processing. These hazardous substances cause the vinasse to have a very high Biological Oxygen Demand (BOD), ranging from 30-40000 and presenting a pH of 4-5 [36]. Research has demonstrated that vinasse disposal in river basins alone isn't a convenient disposal solution. Due its high B.O.D., this material can cause damage to aquatic life, especially when dumped in large volumes.

In Brazil, most of the vinasse that results from ethanol production is being used as fertilizer due to its high potassium content. The problem occurs when some soils don't respond positively to the application of this acid material. The disposal problem is aggravated by an economic drawback, because water must be evaporated for easy transportation at low costs. Some other applications, like methane gas production by anaerobic fermentation, have been investigated but the economics associated with technical problems acts as a limiting factor.

13.1. Utilization of Vinasses in Alcoholic Fermentation

Many considerable studies to decrease the volume of vinasses resulted from alcoholic fermentation for overcoming the disposal of the huge volumes of vinasses voiding its hazardous environmental effects, can be summarized in the following:

13.1.1. Recycling in Fermentation Process

Considerable efforts were accomplished to reduce the vinasse volume produced from alcoholic fermentation from 12-15L / liter alcohol to about 5-6Lvinasse/ L alcohol by recycling the produced vinasse in molasses dilution instead of water. Batch fermentation is more suitable than other fermentation methods. Selection salt tolerant yeast strains for this purpose is needed (benefit). Addition of enzymes which convert unfermentable sugars to fermentable sugars well makes molasses alcoholic fermentation more economic [38].

13.1.2. Yeast Strain

Yeast strains are limiting factor in fermentation process, in this concept high osmotolerat *S.cerevisiae* strains are obtained can ferment high concentrate molasses to produce till 23 % (v /v) ethanol and this mean saving water for molasses dilution and reduce vinasse production [38].

14. Distillery Yields

Just as in a sugar factory there are a number of measures of operational efficiency in a distillery. In the sugar industry ratios like extraction, boiling house recovery, and overall recovery are well defined and universally understood. There are four commonly used measures of yield include fermentation yield, fermentation efficiency, alcohol recovery, and overall conversion efficiency.

14.1. Fermentation Yield

Fermentation yield is measured in litres of absolute alcohol in beer per ton of sugars in molasses, and is calculated by the formula: ($Y_f = V_b \cdot a_b / \{M_m \cdot {}_fS_m\}$), Where, Y_f is fermented yield, V_b is volume of beer (litre), a_b is alcohol content of beer (v/v), M_m is mass of molasses (ton), and S_m is the fermentable sugars content of molasses (m/m).

14.2. Fermentation Efficiency

Fermentation efficiency is an expression of how much alcohol was actually produced in beer relative to the amount that could be theoretically produced, and is given by the following equation: ($E_f = Y_f \cdot 0.794 / 0.5111 \times \{100/1000\}$). The factor 0.794 corresponds to the specific gravity of absolute alcohol and the factor 0.5111 is best explained as follows: If one kilogram of sugar was completely fermented (using theoretical 100 % efficient yeast); 511.1 grams of alcohol and 1000 - 511.1 = 488.9 grams of carbon dioxide would result.

14.3. Alcohol Recovery

Alcohol recovery is a measure of how much alcohol was finally produced relative to the amount that was in the beer. It shows the amount of losses in the evaporation and distillation sections combined. Alcohol recovery is calculated as follows:($E_{de} = \{{}_{aa}V_p + {}_{ss}V_f\} / {}_{aa}V_b \cdot 100$), where: E_{de} is alcohol recovery or evaporation efficiency, ${}_{aa}V_p$ is the volume of potable alcohol as litres absolute alcohol, ${}_{ss}V_f$ is the volume of feints as litres absolute alcohol, and ${}_{aa}V_b$ is the volume of beer as litres absolute alcohol.

14.4. Overall Conversion Efficiency

Overall conversion efficiency is a measure of how much alcohol is finally produced relative to the amount that could be theoretically produced, and is given by the following equation: ($E_o = E_f \cdot E_{de} \cdot 100$), where, E_f is the fermentation efficiency; and E_{de} is alcohol recovery or evaporation efficiency.

14.5. Values of Yield

The following figure gives values of yield that one would expect in a well run distillery:

Parameter	Value
Alcohol Recovery	98.5%
Fermentation Yield	573
Fermentation Efficiency	89%
Overall Conversion Efficiency	87%

Figure 4. (No legend included/described with text).

Conclusion

Specific yeasts, product formulations and processes for their incorporation into ethanol plants - options for overall yeast strategies have been developed for the modern ethanol industry. The choice of an overall yeast strategy will depend mainly on the objectives of ethanol plant management. That choice, and selections from the possible operating approaches they can choose from, should lead to a unique yeast strategy dependent on the plant's design and operating constraints. The analysis leading to a winning yeast strategy for that plant is best undertaken in consultation with a knowledgeable and reputable yeast supplier. Yeast choices have a significant impact on each ethanol plant's profitability and long term viability.

Yeasts for ethanol production are subjected to new and different conditions and expectations with innovative process alterations in ethanol plant design attributed to microbiological and engineering advances. Driven by the explosive growth of the North American dry grind corn (maize) fuel ethanol market, modern yeast selection, production and formulation technologies have led to valuable specialty yeast products suited for today's ethanol production facilities. As a result, the ethanol producer has a number of viable commercial yeast product options to choose from, including a variety of yeast strains and formulations. Additionally, in-plant yeast management options abound. The optimal yeast choices - strain, formulation and management - for each facility depend on a number of interrelated factors. Decisions are getting more complicated - and more critical - as ethanol production technologies and markets mature.

Conflict of Interest

The author declares that there is no conflict of interest

References

[1] Webster N, Andrew P, Murray M, Christopher N (2004) Devirse microbial communities inhabit Antarctic sponges. *Env. Microbiol.* 6:288-300.

[2] Tessier J, Carla C (2004) Differentiation in sex investment by clones and populations of *Daphnia pulicaria*. *Ecology Lett.* 7:695-703.

[3] Ingledew WM (2001) *Alcohol production by Saccharomyces cerevisia*. 3:49-87.

[4] Lorenz M, Cutler N, Heitman J (2000) Characterization of alcohol-induced filamentous growth in *Saccharomyces cerevisiae*. *Mol. Biol. Cell* 11:183-199.

[5] Ceccato-Antonini S, Tosta C, Claudia da Silva A (2004) Determination of yeast killer activity in fermenting sugarcane juice using selected ethanol-making strains. *Braz. Arch. Biol. Technol.* 47:13-23.

[6] IFT (2003) http://ift.confex.com/ift/2003/techproqram/paper 16150.htm.

[7] Kelly J (2003) *Oakinwinemaking*. http.//www.wineanorak.com/brettanomvces.htm.

[8] Abboti J, Hyfiesr S, Ingledew W (2004) Growth rates of Dekkera / Brettanomyces yeasts hinder their ability to compete with *Saccharomyces cerevisiae* in batch corn mash fermentation. *Appl. Microbiol. Biotechnol.* 66:641-647.

[9] Aris (2002) http://www.thewinemerchantinc.corn/educational/veast.html.

[10] Deshpande GB (2002) Overview of continuous alcohol fermentation and multi pressure distillation technology. *Proc. SAfrSug. Technol.* 76:999-1007.

[11] Bhutto AW (2004) http://www.pakistaneconomist.corn/paae/issue38/i&e3.

[12] Lansing R (2003) *Recent Developments in Brett Management and Monitoring*. http://winebusiness.com/html/mQnthlv article.cfm.

[13] Connolly C (1997) Bacterial contaminants and their effects on alcohol production. In: Jacques KA, Lyons TP, Kelsall DR, (eds.) *The Alcohol TextBook*. 3rd. ed. Nottingham, UK: Nottingham University Press.

[14] Kelly AS, Timothy LD (2004) Bacterial contaminants of fuel ethanol production. *J. Ind. Microbiol. Biotechnol.* 31:401-408.

[15] Makanjuola D, Tymon A, Springham D (1992) Some effects of lactic acid bacteria on laboratory-scale yeast fermentations. *Enz. Microbiol. Tech.* 14:350-357.

[16] Narendranath N, Hynes S, Thomas K, Ingledew W (1997) Effects of *Lactobacilli* on yeast catalyzed ethanol fermentations. *Appl. Env. Microbiol.* 63:4158-4163.

[17] Day W, Serjak W, Stratton J, Stone L (1954) Antibiotics as contamination control agents in grain alcohol fermentations. *J. Agric. Food Chem.* 2:252-258.

[18] Narendranath N, Thomas K, Ingledew W (2000) Urea hydrogen peroxide reduces the numbers of *Lactobacilli*, nourishes yeast, and leaves no residues in the ethanol fermentation. *Appl. Environ. Microbiol.* 66:4187-4192.

[19] Chang I, Kim B, Shin P (1997) Use of sulfite and hydrogen peroxide to control bacterial contamination in ethanol fermentation. *Appl. Environ. Microbiol.* 63:1-9.

[20] Gibbons W, Westby C (1986) Use of potassium metabisulfite to control bacterial contaminants during fermentation of fodder beet cubes for fuel ethanol. *Biomass.* 11:9-13.

[21] Oliva-Neto P, Yokoya F (1998) Effect of 3,4,4-trichlorocarbanilide on growth of lactic acid bacteria contaminants in alcoholic fermentation. *Biores. Technol.* 63:17-23.

[22] Rückle L, Senn T (2006) Hop acids can efficiently replace antibiotics in ethanol production. *Int. Sugar J.* 108:139-147.

[23] Stroppa C, Andrietta M, Andrietta S, Steckelberg C, Serra G (2000) Use of penicillin and monensin to control bacterial contamination of Brazilian alcohol fermentations. *Int. Sugar J.* 102:78-82.

[24] Aquarone E (1960) Penicillin and tetracycline as contamination control agents in alcoholic fermentation of sugar cane molasses. *Appl. Microbiol.* 8:263-268.

[25] Bayrock DP, Thomas KG, Ingledew WM (2003) Control of *Lactobacillus* contaminants in continuous fuel ethanol fermentations by constant or pulsed addition of penicillin G. *Appl. Microbiol. Biotechnol.* 62:498-502.

[26] Hynes S, Kjarsgaard D, Thomas K, Ingledew W (1997) Use of virginiamycin to control the growth of lactic acid bacteria during alcohol fermentation. *J. Ind. Microbiol. Biotechnol.* 18:284-291.

[27] Zia M, Asghar M, Bhatti H (2011) Improving bio-ethanol yield using virginiamycin and sodium fluoride at a Pakistani distillery. *Afr. J. Biotecnol.* 53:11071-11074.

[28] NeelakantamV, Narendranath R (2004) Effect of yeast inoculation rate on the metabolism of contaminating *Lactobacilli* during fermentation of corn mash. *J. Ind. Microbiol. Biotechnol.* 31:581-584.

[29] Thomas K, Hynes S, Ingledew W (2001) Effect of *Lactobacilli* on yeast growth, viability and batch and semi-continuous alcoholic fermentation of corn mash. *J. Appl. Microbiol.* 90:819-828.

[30] Espinosa R, CojulunV, Marroquin F (1978) Alternatives for energy saving at plant level for the production of alcohol for use as automative fuel. *Biotecnol. Bioeng*. 8:69-74.

[31] Del Rosario EJ, Lee KJ, Rogers PL (1979) Kinetics of alcoholic fermentation at high yeast levels. *Biotechnol. Bioeng*. 27:1477-1482.

[32] Cysewski GR, Wilke CR (1977) Rapid ethanol fermentation using vacuum and cell recycle. *Biotechnol. Bioeng*. 19:1125-1143.

[33] Ghose TK, Tyagi RD (1979) Rapid ethanol fermentation of cellulose hydrolysate I. Batch versus continuous system. *Biotechnol. Bioeng*. 21:1387-1400.

[34] Aiba S, Shoda M, Nagatani M (1968) Kinetics of product inhibition in alcoholic fermentations. *Biotechnol. Bioeng*. 10:845-864.

[35] Abdel-Fattah WR, Fadil M, Nigam P, Banat IM (2000) Isolation of thermo tolerant ethanologenic yeasts and use of selected strains in Industrial scale fermentation in an Egyptian distillery. *Biotechnol. Bioeng*. 68:531-535.

[36] Polack JA, Day DF, Cho YK (1981) *Gasohol from sugar-cane- stillage, Disposition, Audubon,* Sugar Institute Louisina State University September., P4.

[37] Patent (2010) A process to enhance ethanol yield from molasses fermentation by addition of enzymes which convert unfermentable sugars into fermentable sugars. *WO* 2010086840 A2.

[38] Ingledew WM, Bellissimi E (2012) Active dry, High- alcohol yeast: The new yeast on the block. Ethanol producer magazine. *International fuel ethanol workshop and EXPO.* June, 4-7.

In: Microbes in Process
Editors: Neelam Garg and Abhinav Aeron

ISBN: 978-1-63117-127-7

Chapter 5

Neurospora crassa: A Promising Candidate for Bioethanol Production from Lignocellulosic Biomass

Ioannis Dogaris, Diomi Mamma and Dimitris Kekos*

Biotechnology Laboratory, School of Chemical Engineering,
National Technical University of Athens, Zografou Campus,
Athens, Greece

Abstract

Microbial production of ethanol might be a potential route to replace oil and chemical feedstocks. Bioethanol is by far the most common biofuel in use worldwide. Lignocellulosic biomass is the most promising feedstock for fuel bioethanol production. However, the main limiting factor is the higher degree of complexity inherent to the processing of this feedstock. Bioconversion of lignocellulosics to ethanol consists of four major unit operations: pretreatment, hydrolysis, fermentation and product separation/ distillation. Fungi are probably the most biotechnologically useful group of organisms and are used to synthesize a wide range of economically important compounds, enzymes, and secondary metabolites.

The mesophile filamentous fungus *Neurospora crassa* is the best known *Neurospora* species and has been used extensively for genetic, biochemical, and molecular studies as a model organism. Furthermore, it has been found to possess the ability of metabolizing a wide range of substrates, including cellulose and hemicelluloses, making it a promising candidate for bioethanol production from lignocellulosic biomass. The most limiting factor is the slow fermentation rate that needs further improvement.

* E-mail: kekos@chemeng.ntua.gr.

1. Introduction

With the inevitable depletion of the world's energy supply, there has been an increasing worldwide interest in alternative sources of energy, such as biofuels [2, 3, 4]. Biofuels are renewable fuels derived from biological feedstocks (biomass), and include both liquid forms such as bioethanol (gasoline-equivalent) or biodiesel (diesel-equivalent), and gaseous forms such as biogas (e.g. methane) or hydrogen [5, 6]. Bioethanol is by far the most common biofuel in use worldwide [5, 7]. Moreover, biomass energy can play an important role in reducing green house gas emissions. Ethanol production process only uses energy from renewable energy sources. Hence no net carbon dioxide is added to the atmosphere, making ethanol an environmentally beneficial energy source [2, 3].

A variety of natural sources exist that can provide the required sugars for bioethanol production and can be classified based on their composition: (a) sucrose-containing feedstocks (e.g. sugar beet, sweet sorghum and sugarcane), (b) starchy materials (such as wheat, corn and barley), and (c) lignocellulosic biomass (e.g. wood, straw, and grasses) [7, 8, 9].

Especially interesting are technologies for the production of second generation biofuels. The term second generation means that, contrary to biofuels of the first generation (bioethanol from sugar or starch containing plants), the raw materials are not used for the production of food products. These include mainly lignocellulosic raw materials such as energy plants, where the whole plant is processed, straw, wood and various agricultural and wood processing waste products, such as organic waste [10]. During the last two decades, technology for ethanol production from non-food-plant sources has been developed to the point at which large-scale industrial production will be a reality in the next few years [2, 3, 11].

The fungus *N. crassa* has the ability to synthesize and secrete high levels of all three enzyme types involved in lignocellulosics biomass degradation. In addition, *N. crassa* has been reported convert to ethanol various hexose and pentose sugars, cellulose polymers and agro-industrial residues. The literature demonstrating those interesting abilities of *N. crassa*, making it a promising candidate for ethanol production from lignocellulosics, is reviewed in the present work.

2. Overview of the Ethanol Production Process

Although there are similarities between ethanol processes for lignocellulosic biomass and the starchy or sucrose-containing feedstocks, the techno-economic challenges facing the former are large [12].

2.1. Structure of Lignocellulosic Biomass

Lignocellulosic materials are the most abundant biopolymers in the Earth. It is considered that lignocellulosic biomass comprises about 50% of world biomass, about 180-200 million dry tons per year [7, 13]. In general, prospective lignocellulosic materials for fuel ethanol production can be divided into six main groups: crop residues (cane bagasse, corn stover, wheat straw, rice straw, rice hulls, barley straw, sweet sorghum bagasse, olive stones and pulp), hardwood (aspen, poplar), softwood (pine, spruce), cellulose wastes (newsprint, waste office paper, recycled paper sludge), herbaceous biomass (alfalfa hay, switchgrass, reed canary grass, coastal Bermudagrass, thimothy grass), and municipal solid wastes [7, 13, 14, 15].

Numerous studies for developing large-scale production of ethanol from lignocellulosics have been carried out in the world. However, the main limiting factor is the higher degree of complexity inherent to the processing of this feedstock [7]. This is related to the nature and composition of lignocellulosic biomass. The main components of the biomass are cellulose, hemicellulose and lignin [8, 16, 17].

Cellulose is a linear homopolysaccharide composed of β-D-glucopyranose units, linked by β-(1→4)-glycosidic bonds. Cellulose fibers provide wood's strength. Cellobiose is the smallest repetitive unit of cellulose and can be converted into glucose residues. Hemicelluloses are heterogeneous polymers built up by pentoses (D-xylose, D-arabinose), hexoses (D-mannose, D-glucose, D-galactose) and sugar acids. Hemicelluloses in hardwood contain mainly xylans, while in softwood glucomannans are most common. Lignin is a highly branched, substituted, mononuclear aromatic polymer in the cell walls of certain biomass, especially woody species, and is often bound to adjacent cellulose fibers to form a lignocellulosic complex. Lignin is the most recalcitrant component of the plant cell wall, and the higher the proportion of lignin, the higher the resistance to chemical and enzymatic degradation.

2.2. Bioconversion of Lignocellulosics to Ethanol

Bioconversion of lignocellulosics to ethanol consists of four major unit operations: pretreatment, hydrolysis, fermentation and product separation/ distillation (Figure 1) [3, 7, 8, 12, 17, 18, 19]. The first step in the conversion of biomass to ethanol is pretreatment. The pretreatment aims to increase pore size and reduce cellulose crystallinity. In acid-catalyzed pretreatment, the hemicellulose layer is hydrolyzed, whereas in alkali-catalyzed pretreatment, mainly, a part of the lignin is removed and hemicellulose has to be hydrolyzed by the use of hemicellulases.

Hence, pretreatment is necessary to expose the cellulose fibers to the enzymes or to at least make the cellulose more accessible to the enzymes. This should be done with a minimum formation of compounds which inhibit the fermenting microorganisms [20]. The aim of the enzymatic hydrolysis is to depolymerize cellulose fibers and other oligo- and polysaccharides (e.g. from hemicellulose) released during pretreatment step [19, 20]. Therefore, the hemicellulose and cellulose polymers are hydrolyzed to release monomeric sugars.

The sugars from the pretreatment and enzymatic hydrolysis steps are fermented by bacteria, yeast or filamentous fungi. Contrary to sucrose- and starch-based ethanol production, lignocellulose-based production is a mixed-sugar fermentation in the presence of inhibiting compounds - low molecular weight organic acids, furan derivatives, phenolics and inorganic compounds - released and formed during pretreatment and/or hydrolysis of the raw material [12].

After final purification (by distillation and molecular sieves or other separation techniques), the ethanol is ready to be used as a fuel. A part of the lignin, the principal solid part of the biomass remaining, can be burnt to provide heat and electricity for the process, whereas the rest is retained as a valuable co-product [12, 18].

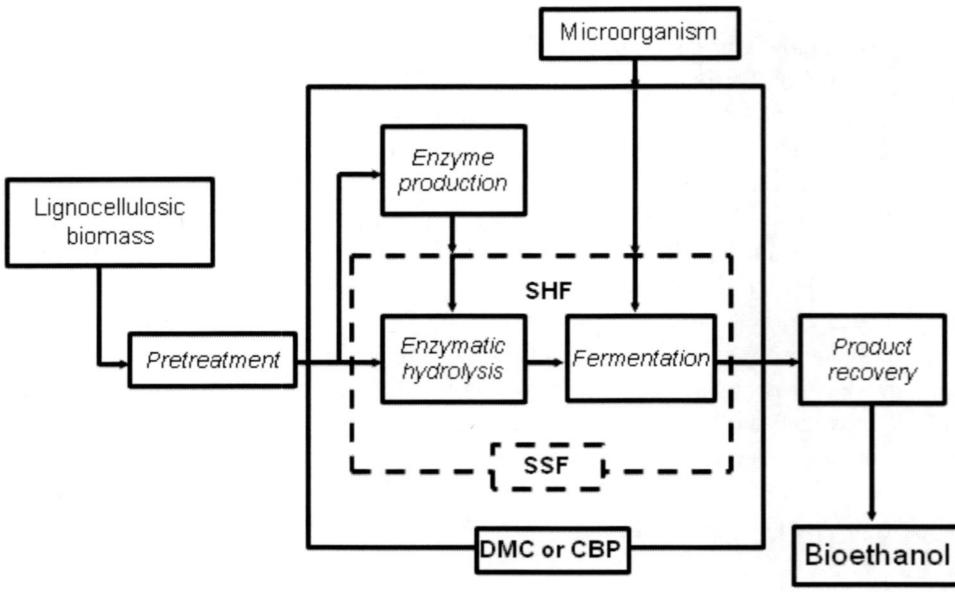

(Adapted and Modified from Cardona and Sánchez [1]).

Figure 1. General Process Scheme for Ethanol Production from Lignocellulosic Biomass, including Process Integration Options.

3. The Fungus *Neurospora crassa*

Fungi are probably the most biotechnologically useful group of organisms and are used to synthesize a wide range of economically important compounds, enzymes, and secondary metabolites, including antibiotics and other pharmaceuticals.

3.1. General

Neurospora is a mesophile filamentous fungus belonging to ascomycetes, commonly seen following agricultural burning in warm, moist climates [21]. The genus was first described in 1927 by Shear and Dodge [21] and thousands of papers have been published on *Neurospora* research since. Numerous natural populations of *Neurospora* have been identified and their characteristics extensively studied [22, 23, 24]. The genus includes the heterothallic species *N. crassa, N. discreta, N. intermedia* and *N. sitophila*, the pseudohomothallic *N. tetrasperma* and the homothallic species *N. africana, N. dodgei, N. galapagosensis, N. lineolata* and *N. terricola* [23]. Karyotypes, karyogamy, and meiotic chromosome behavior are similar for all the known *Neurospora* species, with seven chromosomes and a single terminal nucleolus organizer [24]. During their sexual reproduction cycle, the heterothallic species produce eight-spored asci (each ascospore *A* or *a*) and the pseudohomothallic *N. tetrasperma* four-spored acsi (each ascospore typically *A*+*a*). The homothallic species produce eight-spored asci with no mating type and the cultures originating from single-ascospores are self-fertile [23]. Only the nonhomothalic *Neurospora* species produce conidia (asexual reproduction), forming large orange or orange-yellow colonies. The existence of carotenoids in macroconidia and mycelia is the reason for their vivid color. The macroconidia are usually abundant, powdery and readily airborne. The homothallic species produce grey-brown mycelia which do not have any apparent carotenoids [21, 23].

N. crassa is the best known *Neurospora* species and has been used extensively for genetic, biochemical, and molecular studies as a model organism [25, 26]. It is a saprophyte microorganism, obtaining nutrition from substrates, and growing equally well in simple liquid or on solid media of known composition. It is nonpathogenic, although it is very closely phylogenetically related to pathogens. Both asexual development and sexual differentiation are highly influenced by environmental factors such as nutrient, light and temperature [27]. The genome sequence of *N. crassa* was recently reported [28, 29]. It possesses a large number of genes without homologues in the *Saccharomyces cerevisiae*. Some of the unshared genes encode for sugar transporters, transcription factors and environmental sensing pathways, as well as a diversified metabolic machinery [30].

3.2. Cellulases and Hemicellulases

Utilizing lignocellulosic biomass requires the action of specific hydrolytic enzymes, such as cellulases and hemicellulases, which decompose the biomass into simple fermentable sugars. Several microorganisms exist that naturally grow on lignocellulosic substrates, producing the required hydrolytic enzymes. Numerous bacteria as well as fungi, aerobic or anaerobic, mesophile or thermophile have been reported to produce cellulases and hemicellulases [8, 16, 20, 31]. However, filamentous fungi are still the main source for commercial cellulases and hemicellulases [16, 32].

3.2.1. Mode of Action

Eberhart et al. [33, 34] were the first to investigate cellulase production by *Neurospora*. The degradation of cellulose to glucose in *N. crassa* appears to be a sequential process involving the participation of three enzyme activities: endoglucanase, exoglucanase and β-glucosidase [33, 34, 35, 36, 37, 38]. These three types of enzymes act synergistically for the hydrolysis of cellulose and are common to all cellulolytic fungi [16, 19, 39]. Briefly, the endoglucanases (EC 3.2.1.4) attack regions of low crystallinity and cleave internal β-1,4-glucosidic bonds. The exoglucanases or cellobiohydrolases (EC 3.2.1.91) act on the reducing and non-reducing ends of cellulose chains to release short-chain cello-oligosaccharides such as cellobiose. Finally, β-glucosidases (EC 3.2.1.21) hydrolyze soluble cello-oligosaccharides, such as cellobiose, to glucose molecules.

Hemicellulose decomposition is also required for the efficient exploitation of lignocellulosic biomass. Xylan, the main hemicellulose polymer, is hydrolyzed to xylo-oligosaccharides by endo-β-1,4-xylanases (endoxylanases, EC 3.2.1.8), followed by the action of β-xylosidases (EC 3.2.1.37) on the produced xylo-oligosaccharides (mainly xylobiose) to release xylose molecules [16, 19, 39]. Studies on the mode of action of *N. crassa* β-xylosidase and xylanase revealed that xylanase has no action on lower xylo-oligosaccharides (X_3 and X_4), whereas β-xylosidase can hydrolyze them [40]. Thus, β-xylosidase complements the action of xylanase during the hydrolysis of xylan. The authors suggested that this may be attributed to the substantial transferase activity exhibited by the β-xylosidase of *N. crassa*. In addition, the differential kinetic parameters of the *N. crassa* xylanase towards the hydrolysis of xylan and CM-cellulose showed that the pure enzyme has two catalytic sites. The enzyme hydrolyzed both β-1,4-xylosidic and glucosidic linkages. Auxiliary enzymes that act on various hemicellulose side-chains include α-glucuronidases (EC 3.2.1.139), α-L-arabinofuranosidases (EC 3.2.1.55), acetyl esterases (EC 3.1.1.72) and ferulic acid esterases (FAEs) (EC 3.1.1.73) [16, 19, 39]. Other hemicellulose polymers, such as glucomannans are decomposed by mannanases and β-mannosidases [16].

A quantitative proteomics approach of the secretome of *N. crassa* during growth on microcrystalline cellulose led to the characterization of thirteen proteins, including 2 cellobiohydrolases (CBH-1 and GH6-2), an endoglucanase (GH5-1), and a β-glucosidase (GH3-4) [41]. The latter four enzymes were used to reconstitute a defined cellulase mixture in vitro and compared to the crude mixture of *N. crassa* secretome proteins. Interestingly, the authors reported that while these four proteins represented 65% of the total secretome by weight, they accounted for just 43% of the total activity on microcrystalline cellulose after 24 h of hydrolysis. Therefore, the authors suggested that proteins other than the standard fungal cellulases may also play an important role in cellulose degradation by fungi.

3.2.2. Induction and Regulation

Regulating enzyme synthesis is crucial for the economics of biofuel production from lignocellulosic biomass. Production of hydrolytic enzymes associated with plant cell wall degradation is induced to high levels only in the presence of plant cell wall biopolymers or their derivatives [42, 43]. The best inducers of hydrolytic enzyme expression by filamentous fungi are insoluble substrates that include cellulose, hemicellulose, or mixtures of plant polymers. Because these naturally inducing substances cannot enter fungal cells, it is believed that oligosaccharides released from polymers and their derivatives function as the actual molecules that trigger enzyme induction [44].

However, the use of insoluble substrates to induce enzyme secretion is not ideal for industrial processes [43]. The most problematic and expensive aspect of industrial cellulase production is providing the appropriate inducer for cellulases. Cellulase production on a commercial scale is induced by growing the fungus on solid cellulose or by culturing in the presence of a disaccharide inducer such as lactose. However, on an industrial scale, both methods of induction result in high costs [45]. *N. crassa* can use a wide variety of carbon sources. This requires an optimal adaption to the environment by synthesizing transporters and secreting enzymes, which are induced by specific substrates or metabolites. For *N. crassa* to grow on plant cell wall material, a low constitutive level of extracellular or cell wall-associated enzymes releases metabolites from plant biomass. These metabolites are transported into the cell and serve as inducers directly or are converted to signal molecules required for triggering transcription of genes encoding plant cell wall-degrading enzymes [46].

The simultaneous induction of all three cellulase enzymes on *N. crassa* was first reported by Eberhart et al. [33] using cellobiose as substrate. The induction properties of *β*-glucosidases of *N. crassa* have been reported by Eberhart and Beck [47], suggesting that aryl-*β*-glucosidase and cellobiase represent two fundamentally different classes of disaccharidiases. Aryl-*β*-glucosidase represents a broadly inducible (or derepressible) class of enzymes, while cellobiase represents enzymes with a highly specific induction requirement. Cellobiase was induced mainly by cellobiose and its activity was very low in all stages of the vegetative life cycle in the absence of di-*β*-glucoside inducer. Aryl-*β*-glucosidase was semiconstitutive at late stages of culture growth prior to conidiation. At early stages, aryl-*β*-glucosidase was induced by cellobiose, laminaribiose, and gentiobiose, and weakly induced by galactose, amino sugars, and aryl-*β*-glucosides. The induction of *β*-glucosidases was inhibited by glucose, 2-deoxy-D-glucose and sodium acetate. Sodium phosphate concentrations between 0.01 and 0.1 M stimulated induction of both enzymes, while concentrations above 0.1 M were inhibitory. The optimal pH for induction of both *β*-glucosidases was 6.0.

In later studies by Yazdi et al. [37], cellulase induction was also reported on other soluble substrates with *β*-1,4-glucan linkages, such as carboxymethyl cellulose. However, the best inducer for cellulase was cellulose, with absorbant cotton best for exoglucanase and *β*-glucosidase, and microcrystalline cellulose best for endoglucanase. Tian et al. [48] also reported a robust cellulolytic response of *N. crassa* to growth on plant cell walls and microcrystalline cellulose (Avicel), including induction and secretion of a large number of cellulases and hemicellulases. Xylanase and *β*-xylosidase production by *N. crassa* was induced on xylan as a carbon source [49].

Analysis of the *N. crassa* genome predicted genes for approximately 100 glycosyl hydrolases [30]. *N. crassa* genome is also predicted to contain twice as many cellulases as in the industrially important species *Trichoderma reesei*, as well as many hemicellulases and other enzymes involved in plant biomass degradation[48]. According to Myers and Eberhart [50] the expression of *β*-glucosidases in *N. crassa* is regulated by two types of genes. The *gluc-1* gene causes low aryl-*β*-glucosidase production, while the *cell-1* gene produces constitutively high levels of cellobiase and cellulase. Taleb and Radford [51] described the isolation, cloning and sequencing of the cellobiohydrolase 1 (EC 3.2.1.91)-encoding gene (*cbh-1*) of *N. crassa*. The *cbh-1* gene has high homology with cellulase genes from related fungi, such as *T. reesei*, *Humicola grisea* and *Phanerochaete chrysosporium*, with clear signal, catalytic, hinge and substrate-binding domains.

Cellobiohydrolase(I) (CBH(I)) of *N. crassa* is the most highly produced extracellular protein during growth on Avicel or *Miscanthus*, and deletion of this gene causes the most severe growth deficiencies on cellulosic substrates [48]. Deletion of cellobiohydrolase(II) also causes growth deficiencies on cellulosic substrates, but to a much lesser extent than CBH(I), suggesting that exoglucanase activity in *N. crassa* is predominantly from CBH(I) and that endoglucanases and other CBHs do not compensate for the loss of CBH(I). Tian et al. [48] also suggested that endoglucanase in *N. crassa* is not required for growth on crystalline cellulose and that different endoglucanases have functional redundancies.

In filamentous fungi, genes encoding hydrolytic enzymes involved in plant cell wall deconstruction are repressed during growth on easily metabolized carbon sources, such as glucose. Carbon catabolite repression (CCR) is an important mechanism to repress the production of plant cell wall degrading enzymes during growth on preferred carbon sources.

Some aspects of CCR are mediated by Mig1/CreA/CRE1, a zinc-finger transcription factor conserved in most fungal species [42]. Although CCR has been studied in filamentous fungi, only a limited number of genes/systems have been shown clearly to be subject to direct CreA/CRE1 repression. Sun and Glass [42] reported that deletion of *cre-1* caused sustained expression of cellulase genes in *N. crassa*, resulting in higher cellulolytic enzyme activity. The repression of cellulolytic genes during growth on Avicel was correlated with *cre-1* transcription levels. When switched from a favored carbon source such as sucrose to cellulose, *N. crassa* dramatically up-regulates expression and secretion of a wide variety of genes encoding lignocellulolytic enzymes [43]. However, the means by which *N. crassa* and other filamentous fungi sense the presence of cellulose in the environment remains unclear.

Znameroski et al. [43] suggested that the induction of lignocellulose-degrading enzymes in *N. crassa* may be accounted to cellodextrins. They examined the hypothesis that cellobiose functions as an inducer of cellulase gene expression and secretion when filamentous fungi, such as *N. crassa*, are exposed to cellulose, but that the action of extracellular and intracellular β-glucosidases and CCR mask this inducing activity. They reported that a strain of *N. crassa* carrying deletions for the three major β-glucosidase genes induces cellulases when exposed to cellobiose and this induction recapitulates, on both transcriptional and protein level, the wild-type response to crystalline cellulose (Avicel). Further deletion of *cre-1*, which encodes a CCR transcription factor, enables *N. crassa* to produce a higher level of secreted active cellulases when induced with cellobiose, as compared to enzyme levels observed during growth on Avicel.

The regulation of genes encoding xylanolytic enzymes in *N. crassa* is efficiently described in the work of Sun et al. [46]. A secretome and transcriptome analysis of *N. crassa* grown on beechwood xylan led to the identification of 34 secreted proteins and 353 genes with elevated transcription on xylan. A transcription factor XLR-1 (xylan degradation regulator 1) was identified to be essential for hemicellulose degradation that is an ortholog to XlnR/XYR1 in *Aspergillus* and *Trichoderma* species, respectively, a major transcriptional regulator of genes encoding both cellulases and hemicellulases. Deletion of *xlr-1* in *N. crassa* negated growth on xylan and xylose, but growth on cellulose and cellulolytic activity were only slightly affected. XLR-1 regulated only some predicted hemicellulase genes in *N. crassa* and was required for a full induction of several cellulase genes. Hemicellulase gene expression is induced by a combination of release from CCR and induction. Basal levels of secreted hemicellulases release small xylo-oligosaccharides and xylose from hemicellulose that trigger the transcription and activation of *xlr-1* and other unknown transcription factors (TFs), which sequentially activate or modulate the transcription of certain hemicellulase genes. Action of these TFs results in production and secretion of hydrolytic enzymes, which thereby deconstruct hemicellulose, produce metabolites and additional signaling molecules. Another unknown transcription factor(s) also coordinates with XLR-1 to modulate the regulation of genes associated with xylose metabolism.

Light represents an important environmental cue, which exerts considerable influence on the metabolism of fungi. It has been shown that light modulates transcription of cellulase genes in *T. reesei* [52]. Photoreceptors and their cognate signaling cascades have numerous regulatory targets and impact almost every aspect of physiology in *N. crassa* and other filamentous fungi [53, 54, 55, 56]. Recently, Schmoll et al. [57] reported that the *N. crassa* photoreceptor genes *wc-1*, *wc-2* and *vvd* are involved in regulation of cellulase gene expression, indicating that this phenomenon is conserved among filamentous fungi.

3.2.3. Localization

Plant cell-wall degrading enzymes exist either in complexed or non-complexed systems. The complexed systems are known as *cellulosomes* and tend to be present in anaerobic bacteria and fungi, whereas the aerobic microorganisms produce discrete enzymes secreted into the growth media [16, 39]. Eberhart and Beck [58] attempted to determine the location *β*-glucosidases within the cellular structure of *N. crassa*, using various wild-type strains from several genetic backgrounds. They found two *β*-glucosidases, an "aryl-*β*-glucosidase" and a "cellobiase", with distinct difference in specificity on *p*-NPG (*p*-nitrophenyl-*β*-D-glucopyranoside) and cellobiose, respectively. Aryl-*β*-glucosidase is primarily a detectable or cell wall bound enzyme during any stage of the vegetative life cycle of *Neurospora* in which it is induced. On the other hand, cellobiase is predominantly internal and cryptic at all stages, but its activity can be detected in intact-cell assays by specific physical or chemical treatments that alter the cell membrane. The ratio of external to cell-bound *β*-glucosidase was characteristic of each strain. Eberhart and Beck [58] also suggested that cellular location of enzymes is apparently not determined by molecular size alone, since aryl-*β*-glucosidase (168,000, molecular weight) is cell wall bound, whereas cellobiase (80,000, molecular weight) is essentially intracellular. Mishra et al. [49] reported secreted as well as intracellular hemicellulose activities by *N. crassa* 870. The intracellular xylanase activity was only 7% of the extracellular xylanase, but the *β*-xylosidase activity was almost equally distributed between the cell-free and cell-bound fractions.

3.2.4. Production

Commercial production of cellulases and hemicellulases has been tried in either solid state cultures (SSC) or submerged fermentations (SmF), including batch, fed batch and continuous flow processes [45]. Most manufacturers produce enzymes using submerged fermentation techniques [59]. However, during the last decades, an increased interest towards solid state conditions for the production of enzymes has been reported. Solid state fermentation should define any fermentation process occurring in the absence or near-absence of free water, employing a natural substrate or an inert substrate used as solid support [59, 60]. SSC processes have shown to be particularly suitable for the production of enzymes by filamentous fungi, since they reproduce the natural living conditions of such fungi [31, 59].

As stated earlier, *N. crassa* is able to synthesize and secrete all three enzyme types involved in cellulose degradation, as well as enzymes with hemicellulolytic activities. The fungus has been grown successfully and produced cellulases and hemicellulases on various substrates, in both submerged (Figure 2) and solid state cultures (Figure 3), at temperatures 25-30 °C and culture medium pH set at 5.0-7.0. Reported growth conditions and enzyme activities are discussed in more detail in the following paragraphs.

3.2.4.1. Cellulases

As discussed earlier, Eberhart et al. [33, 47, 58] reported the presence of cellulases and *β*-glucosidases of *N. crassa*. Later work by Eberhart at al. [34] described the production of extracellular endoglucanase and conditions that foster the release of this enzyme into the external media. Induction started immediately when already grown mycelia were added to 0.1 M phosphate buffer, pH 7.0, and a final concentration of 1.0 mM cellobiose and incubated at 25 °C for 6 h. *N. crassa* is able to produce significant levels of cellulase when grown in liquid medium with 1 or 2 % (w/v) microcrystalline cellulose (MC), for 8-11 days at 25 °C and pH set to 7.0 [36, 38, 61]. The cellulase enzymes were produced by growing the FGSC 4335 *cell-1* mutant strain (with elevated uninduced levels of both cellulase and cellobiase). Eight extracellular cellulase enzymes of *N. crassa*, three exoglucanases, four endoglucanases and one *β*-glucosidase [36] as well as two intracellular *β*-glucosidases [61] have been identified and characterized.

N. crassa strain	Substrate	Cultivation conditions	Cellulases (U/ml)			Hemicellulases (U/ml)		Reference
			EG[e]	EXG[f]	*β*-GLU[g]	XYL[h]	*β*-XYL[i]	
FGSC 4335 *cell-1*	Cellobiose	pH 7.0, 25°C, 11 days	23	1.4	0.7	n.d. *	n.d.	[37]
	Cellobiose + MC[a]		105	10	2.5	n.d.	n.d.	
	MC		110	9.8	0.4	n.d.	n.d.	
	CMC[b]		43	3.9	0.2	n.d.	n.d.	
	Glucose + MC		222	25	2.9	n.d.	n.d.	
	Absorbent cotton		80	12.5	1.1	n.d.	n.d.	
	Newspaper cellulose		52	4.1	0.9	n.d.	n.d.	
	Xylan		31	1.8	0.9	n.d.	n.d.	
FGSC 4335 *cell-1*	Wheat straw	pH 6.5, 30°C, 4-9 days	14.5 **	2.9 **	0.55 **	n.d.	n.d.	[65]
NCIM 870	Alkali treated CP[c]	pH 5.0, 28°C, 10 days	49.8	n.d.	0.4	n.d.	n.d.	[62]
	Untreated CP + WB[d]		95.8	n.d.	0.5	n.d.	n.d.	
NCIM 1021	Alkali-treated CP		49.6	n.d.	0.4	n.d.	n.d.	
	Untreated CP + WB		95.6	n.d.	n.d.	n.d.	n.d.	
NCIM 870	Xylan	pH 5.0, 28 °C, 4 days	1.2	n.d.	n.d.	14	0.025	[49]
	CP + WB		10	n.d.	n.d.	7	n.d.	

*not determined, ** estimated from graph

[a] microcrystalline cellulose, [b] Carboxymethyl cellulose, [c] cellulose powder, [d] wheat bran, [e] endoglucanase, [f] exoglucanase, [g] *β*-glucosidase, [h] xylanase, [i] *β*-xylosidase.

Figure 2. Production of Cellulases and Hemicellulases by *N. crassa* under Submerged Cultures (SmC).

The pH optima for all the enzymes were acidic, ranging between pH 4.0 and 7.0 and the highest activities were observed at temperatures between 45 and 65 °C. Endoglucanase, exoglucanase and *β*-glucosidase activities reached 222, 25 and 2.9 U/ml, respectively [37]. Addition of readily metabolized soluble carbon sources to the MC medium increased growth and improved cellulase production. Yazdi et al. [37] also reported that by increasing glucose concentration in the above medium, although growth might improve further, the lag time for cellulase production increased due to the inhibitory effect of glucose, with significant production only commencing after the glucose had been metabolized. Rao et al. [62] have identified two strains of *N. crassa*, NCIM 870 and NCIM 1021, which utilize cellulose and produce extracellular endoglucanase and *β*-glucosidase. The activities were detected as early as 48 h in the culture broth and maxed after 7 days. Higher activities, 96.5 U/ml endoglucanase and 0.5 U/ml *β*-glucosidase, were observed when both fungi were grown on alkali-treated cellulose powder (CP) compared to a mixture of untreated CP and wheat bran (WB).

N. crassa strain	Substrate	Cultivation conditions	Cellulases (U/g substrate)			Hemicellulases (U/g substrate)		Reference
			EG[f]	EXG[g]	*β*-GLU[h]	XYL[i]	*β*-XYL[j]	
40b	Untreated WS[a]	Initial pH 6.5, 30 °C, 5 days,	586.2	n.d.*	58.4	n.d.	n.d.	[63]
STA	Untreated WS	Initial pH 6.5, 30 °C, 5 days,	969.2	n.d.	169.4	n.d.	n.d.	[64]
DSM 1129	WS/BG[b] (1:1 ratio)	Initial pH 5.0, initial moisture 61.5% (w/w), 30 °C, 6 to 8 days	56	4.2	1.6	1073	n.d.	[66]
DSM 1129	Orange peels	Initial pH 5.0, initial moisture 80% (w/w), 30 °C	138.5	n.d.	7.9	56.8	n.d.	[67]
DSM 1129	CC[c]	Initial pH 5.0, initial moisture 70.5% (w/w), 30 °C, 6 to 8 days	22.00	0.51	2.26	12.23	0.03	[68]
	SB[d]		71.31	0.53	8.19	16.98	0.04	
	WB[e]		174.76	0.36	12.85	130.47	0.08	
	WS		277.60	1.72	23.40	169.71	0.17	
	WS/WB (5:1 ratio)		478.00	1.08	26.67	238.28	0.13	

* not determined
[a] wheat straw, [b] brewing spent grain, [c] corn cobs, [d] sorghum bagasse, [e] wheat bran, [f] endoglucanase, [g] exoglucanase, [h] *β*-glucosidase, [i] xylanase, [j] *β*-xylosidase.

Figure 3. Production of Cellulases and Hemicellulases by *N. crassa* under Solid State Cultures (SSC).

The production of the cellulase complex of *N. crassa* using low-cost lignocellulosic agricultural residues, such as wheat straw (WS) was first reported by Macris et al. [63, 64]. Two strains of *N. crassa* 40b [63] and STA [64] were grown on untreated WS in SSC at 30 °C for 5 days, and each strain produced 586.2 U and 969.2 U endoglucanase and 58.4 U and 169.4 U *β*-glucosidase per gram of WS, respectively. The authors also estimated the optimum enzyme activities of endoglucanase and *β*-glucosidase at pH 4.0 and 5.0 and temperature 50 and 60 °C, respectively [63].WS was also studied for the production of *N. crassa* cellulases by Romero et al. [65]. Highest activities for all cellulolytic enzymes were observed when the fungal strain 4335 *cell-1* was grown on liquid cultures of 5 % straw at initial pH value of 6.5 and a temperature of 30 °C. The authors also reported that changes in the initial pH, lowering or raising it, lowered the enzymes production, but favored fungal growth. The *β*-glucosidase and endoglucanase activities showed fluctuations with fermentation time that could be attenuated by adding a protease inhibitor.

Other residues from agri-food processes such as spent grain from brewing (BG) have also been used for cellulases production by *N. crassa* [66]. Large scale production by the strain DSM 1129 under SSC was conducted at 30 °C in a laboratory horizontal bioreactor using WS and BG in the ratio 1:1, initial moisture 61.5 %, initial pH 5.0 and nitrogen source content 5 % w/w. The reported activities of endoglucanase, exoglucanase and *β*-glucosidase were 56, 4.2 and 1.6 U per g carbon source respectively and were observed after 7 or 8 days growth.

Orange peels, the principal solid by-product of the citrus processing industry, was reported as a suitable substrate for cellulase production by *N. crassa* [67]. *N. crassa* DSM 1129 was grown on water unextractable orange peels (WUOP) under SSC. Endoglucanase and *β*-glucosidase activities reached 138.5 and 7.9 U per g substrate, respectively, under optimal conditions (30 °C, pH 5.0 and initial moisture 80 % w/w).

In our previous work [68] we demonstrated the ability of *N. crassa* DSM 1129 to produce a complete cellulolytic system under SSC on various agro-industrial byproducts, such as WS, WB, corn cobs (CC) and sorghum bagasse (SB). Highest endoglucanase, exoglucanase and *β*-glucosidase activities, 492.8, 1.08 and 26.7 U per g of carbon source, respectively, were reported when the fungus was grown on WS enriched with WB (5/1, w/w). Optimum conditions for cellulase production under SSC were at 30 °C, initial pH 5.0, initial moisture content 70.5 % (w/w) and ammonium sulfate as nitrogen source. The potential of the produced multi-enzyme system was tested successfully for the hydrolysis of SB into fermentable carbohydrates [68, 69].

3.2.4.2. Hemicellulases

The production of hemicellulases by *N. crassa* was first reported by Mishra et al. [49]. *N. crassa* NCIM 870 produced 14 and 0.025 U of extracellular xylanase and *β*-xylosidase per ml, respectively, in 4 days at 28 °C and pH 5.0 when commercial 2 % xylan was used as a carbon source. Two xylanases (I and II) were purified and had pI values of 4.8 and 4.5 and molecular weights of 33,000 and 30,000 D. The maximum degree of hydrolysis of xylan by the *N. crassa* xylanase was 63 %. Later, an extracellular *β*-xylosidase from *N. crassa* NCIM 870 was purified [40]. Its molecular weight was 83,000 D and the K_m on p-nitrophenyl-*β*-D-xyloside was 0.047 mM. The enzyme's optimum pH was in the range 4.5-5.0 with a pI value of 4.3 and showed maximum activity at 55 °C.

Other enzymes that take part in hemicellulose hydrolysis have also been identified in *N. crassa*, including *β*-galactosidases [70, 71, 72, 73, 74, 75], α-L-arabinofuranosidases [66, 76] and ferulic acid esterases (FAEs) [66, 77, 78]. Crepin et al. [77] reported the ability of *N. crassa* wildtype ST *A* (74 *A*) to produce multiple FAE activities in response to the availability of plant cell wall carbohydrates. FAE activity was detected following growth on liquid cultures with 2 % (w/v) sugar beet pulp (SBP) or WB during the first 3 days at 30 °C.

N. crassa strain DSM 1129 has been used to produce various hemicellulases from different lignocellulosic substrates under SSC, in 6 to 8 days at 30 °C and initial pH 5.0 [66, 67, 68]. Xiros et al. [66] reported the production of 1073, 3.1, 5.7 and 0.52 U per g carbon source of xylanase, α-L-arabinofuranosidase, acetyl esterase and FAE, respectively. The strain DSM 1129 was grown on WS/BG (1:1 ratio) in a laboratory horizontal bioreactor, with initial moisture 61.5 % and nitrogen source content 5 % (w/w). When orange peels were used as a carbon source (80 % w/w moisture content), 56.8 U per g substrate of xylanase were produced by the same strain [67]. Additionally, *N. crassa* DSM 1129 was able to produce extracellular xylanase and *β*-xylosidase when grown under SSC on other agro-industrial residues such as WS, WB, CC and SB, as described in our previous work [68]. The mixture 5/1 (w/w) WS-WB was found the best inducer for cellulases as well as hemicellulases production, leading to xylanase and *β*-xylosidase activities of 297.8 and 0.13 U per g of substrate. *N. crassa* was grown under SSC for 8 days at 30 °C, initial pH 5.0, initial moisture content 70.5 % (w/w) and ammonium sulfate as nitrogen source.

4. Ethanol Production from Lignocellulosic Biomass

Ethanol fermentation is a biological process in which organic material is converted by microorganisms to simpler compounds, such as sugars. These fermentable compounds are then fermented by microorganisms to produce ethanol and CO_2. Several reports and reviews have been published on production of ethanol fermentation by microorganisms, and several bacteria, yeasts, and fungi have been reportedly used for the production of ethanol [2]. The general requirements on an organism to be used in ethanol production is that it should give a high ethanol yield, a high productivity and be able to withstand high ethanol concentrations in order to keep distillation costs low. In addition to these general requirements, efficient de-polymerization of cellulose and hemicellulose to soluble sugars, ability to utilize multiple sugars (hexoses and pentoses), inhibitor and temperature tolerance are essential for lignocellulosic biomass fermentations [12, 20].

4.1. Process Integration

As discussed earlier, overall bioethanol production from lignocellulosic biomass includes four main steps: biomass pretreatment, enzyme hydrolysis, fermentation of hexoses and pentoses, and product recovery (Figure 3). Bioconversion can be performed either in separate enzyme hydrolysis and fermentation steps, or in a combined step.

The classic configuration employed for fermenting biomass hydrolyzates involves a sequential process where the hydrolysis of cellulose and the fermentation are carried out in different units [7]. This configuration is known as separate hydrolysis and fermentation (SHF). The solid fraction of pretreated lignocellulosic material undergoes hydrolysis (saccharification). This fraction contains the cellulose in an accessible to acids or enzymes form. Once hydrolysis is completed, the resulting cellulose hydrolyzate is fermented and converted into ethanol. One of the main features of the SHF process is that each step can be performed at its optimal operating conditions. The most important factors to be taken into account for saccharification step are reaction time, temperature, pH, enzyme dosage and substrate load. The disadvantage is that cellulolytic enzymes are end-product inhibited so that the rate of hydrolysis is progressively reduced when glucose and cellobiose accumulate [12].

The most important process improvement made for the enzymatic hydrolysis of biomass is the introduction of simultaneous saccharification and fermentation (SSF) [3]. The SSF process appears more attractive than the SHF, as higher ethanol yields can be obtained and less energetic consumption is required [7]. In this case, the enzymes and microorganisms are added to the same process unit allowing that the glucose formed during the enzymatic hydrolysis of cellulose be immediately consumed by the microbial cells converting it into ethanol. Thus, the inhibition effect caused by the sugars over the enzymes is neutralized. However, the need of employing more dilute media to reach suitable rheological properties will lead to low final product concentrations. In addition, this process operates at non-optimal conditions for hydrolysis and requires higher enzyme dosage, which positively influences on substrate conversion, but negatively on process costs.

An alternative technology for lignocellulosic biomass bioconversion has been proposed, known as Consolidated Bioprocessing (CBP) or Direct Microbial Conversion (DMC). In CBP ethanol and all required enzymes are produced by a single microorganism, in a single reactor [79]. CBP is gaining increasing recognition as a potential breakthrough for low-cost biomass processing. Application of CBP implies no capital or operating costs for dedicated enzyme production (or purchase), reduced diversion of substrate for enzyme production, and compatible enzyme and fermentation systems [79]. In addition, several factors support the possibility of realizing higher hydrolysis rates, and hence reduced reactor volume and capital investment, using CBP. However, there is no robust organism available that can produce cellulases or other cell wall degrading enzymes in conjunction with ethanol with high yields. Fungal candidates for CBP belong to *Neurospora*, *Fusarium*, *Monilia*, *Rhizopus*, *Aspergillus* and *Trichoderma* genus, which are able to produce the required hydrolytic enzymes as well as ferment to ethanol the produced sugars, but with low conversion rates.

4.2. Fermentation of Simple Sugars

Glycolysis is the process where sugars are metabolized to acetyl-CoA or pyruvate before oxidation during the citric acid cycle or fermentation to ethanol or lactate. Three metabolic pathways for the biosynthesis of ethanol from glucose have been described in various microorganisms, the Embden-Meyerhof-Parnas (EM), hexose monophosphate (HM), and Entner-Doudoroff (ED) pathways. The EM and HM pathways are prevalent in fungi, while evidence for the ED pathway has been demonstrated for only a few species [80]. Previous work had suggested that the predominant route for glycolysis in *Neurospora* is entry of glucose-6-phosphate into the EM pathway (80 to 90 %), with the remainder diverted to the HM pathway [25]. The analysis of the genome sequence is consistent with intact EM and HM glycolytic pathways in *Neurospora* [30].

The fermentation of glucose to ethanol in aerobic cultures of *N. crassa* has been documented more than 40 years ago [81, 82, 83, 84, 85, 86]. Weiss and Turian [87] observed that, in standing (anoxic) cultures of *N. crassa*, ethanol production was greater on media containing ammonia than on media containing nitrate as the nitrogen source. Howell et al. [88] also observed that, in anaerobic cultures of a facultative anaerobic mutant of *N. crassa* in media containing ammonia as nitrogen source, ethanol production was proportional to the initial glucose concentration. Colvin et al. [86] demonstrated that aerobic fermentation of glucose by wild-type *N. crassa* 74A8 can also occur and the ethanol yield is proportional to the initial glucose concentration. The authors reported that at high glucose concentrations, *N. crassa* cultures can accumulate up to 2 % ethanol, and the efficiency of conversion of glucose to ethanol can approach 100 %. In addition, the time for maximum ethanol accumulation depended on initial glucose concentration, and thereafter, in stationary phase the ethanol concentration declined to zero. Upon exhaustion of glucose from the media, the repression of the assimilatory isozyme of alcohol dehydrogenase [89] and the synthesis of glyoxylate cycle enzymes [90] are initiated. Thus, the stage may be set for the utilization of ethanol (via acetate) as a carbon source in gluconeogenesis. Comparison of the specific rates of glucose utilization and ethanol accumulation during the fermentation by *N. crassa* with 2 % initial glucose reveals a stoichiometric relationship on the basis of the Embden-Meyerhof glycolytic pathway, with 2 mol of ethanol being produced per mol of glucose utilized [86].

According to Xie et al. [91] pyruvate decarboxylase (PDC) is expressed at a higher level in the presence of glucose, and acetaldehyde produced in abundance. The acetaldehyde may be efficiently reduced to ethanol using NADH via alcohol dehydrogenase (ADH). In *S. cerevisiae* ADH1 generates ethanol and ADH2 is involved in its degradation. Two predicted *N. crassa* ADHs (NCU02476.1 and NCU01754.1) have strong identity (>50 %) to *S. cerevisiae* ADH1, ADH2 and ADH3. NCU02476.1 was not strongly regulated by glucose, while NCU01754.1 is strongly repressed by glucose. These two *N. crassa* proteins likely play roles in ethanol synthesis and degradation, respectively.

The fungus *T. reesei* primarily utilizes glucose via respiration and carries out a low rate of conversion of pyruvate to acetate [92]. In contrast, *N. crassa* utilizes glucose in a manner in which the rate of glycolysis exceeds the rate at which the products of glycolysis are utilized by the TCA cycle and respiration. Fermentation to ethanol rather than acetate is required to generate NAD^+ to support the high rate of glycolysis. *N. crassa* is well known for its rapid growth rate, and energy demands to support this high growth rate may require fermentation as well as respiration. In addition, rapid removal of glucose from a glucose rich environment and production of ethanol may represent a strategy that *N. crassa* utilizes for competition with other microbes [91].

Two glucose transport systems have been characterized in *N. crassa*, a constitutive low-affinity transport system and a glucose-repressible high-affinity transport system [93, 94]. Thirty-three hexose transporter homologs have been identified in the *N. crassa* genome sequence, representing one of the largest multigene families, and phylogenetic analyses indicate that the relationship between *S. cerevisiae* and *N. crassa* hexose transporter paralogs is poor and no clear orthologs for comparison of gene expression are present [28].

Although the pentose sugars D-xylose and L-arabinose represent the most abundant components of hemicellulose in this biomass [95, 96], the inability of established ethanol production organisms to effectively harness these pentose sugars represents a significant biotechnological bottleneck [19]. For example, the predominant ethanol producer *S. cerevisiae*, as well as the ethanologen bacterium *Zymomonas mobilis* are unable to utilize these hemicellulosic pentose sugars without the introduction of exogenous pathway genes [97, 98]. Enteric bacteria and some fungi and yeasts, are able to ferment pentoses, but with low ethanol yields. Furthermore, in the case of xylose fermenting yeasts (*Pachysolen tannophilus, Candida shehatae, and Pichia stipitis*), large-scale utilization is hampered by their sensitivity to high concentrations of ethanol ($\geq$40 g/l), the requirement for carefully monitored micro-aerophilic conditions, high sensitivity to inhibitors, and the inability to ferment xylose at low pH [99, 100].

In fungi, xylose is reduced to xylitol by NADH- or NADPH-dependent xylose reductase (XR) and thereafter is oxidized to xylulose by NAD^+-dependent xylitol dehydrogenase (XDH). The xylulose is phosphorylated by xylulose kinase (XK), channeled into the pentose phosphate pathway (PPP) [97, 101]. The enzymes XR and XDH of *N. crassa* have been identified and characterized [102, 103, 104], as well as a XR gene [105]. Rawat et al. [102] reported the induction of two XRs in *N. crassa*, one NADH-linked and one NADPH-linked, with the latter being the most abundant and stable. Maximum ethanol from xylose was obtained when high NADH-linked XR and NAD-linked XDH activities were observed. Two isoenzymes (EI and EII) with NADPH-linked xylose reductase activities were identified from *N. crassa*, one of which (EII) also had NADH-linked activity [106]. XR of most fungi, including most yeasts, prefers NADPH to NADH. Because of the cofactor preference of XR

(NADPH) and XDH (NAD^+), redox imbalance may occur under anaerobic fermentations of pentoses [107]. Rawat et al. [102] also suggested that the presence of NADH-linked XR in *N. crassa* probably prevents that imbalance, thus allowing the conversion of xylose to ethanol.

L-Arabinose and D-xylose can be catabolized through a common intermediate, D-xylulose 5-phosphate, via distinct pathways in bacteria [108] and fungi [109]. L-Arabinitol 4-dehydrogenase (LAD) catalyzes the second step of L-arabinose assimilation by oxidation of L-arabinitol to L-xylulose, with concomitant reduction of NAD^+ to NADH [110]. L-Xylulose reductase (LXR) produces xylitol in the successive step. An LXR identified from the genome sequence of *N. crassa* has been biochemically characterized [111]. LAD and LXR are the two unique enzymes required for L-arabinose utilization, while the other enzymes in the pathway are shared in the D-xylose pathway [112]. Due to their importance in cost-effective pentose sugar catabolism, several LADs from different fungal sources have been biochemically characterized, and *N. crassa* LAD was among the most active and stable LADs [113]. *N. crassa* LAD is predominantly NAD^+ dependent, as is the case with all LADs characterized to date, and is a member of the superfamily of medium-chain dehydrogenases/reductases [114] with homology to xylitol dehydrogenase and sorbitol dehydrogenase. The preferred substrate is L-arabinitol, with oxidation activity also present with the epimeric sugar alcohols xylitol and adonitol. In addition, *N. crassa* LAD shows low yet detectable activity with the $NADP^+$ cofactor, suggesting inherent relaxed cofactor specificity [113]. The crystal structure of LAD from *N. crassa* has been reported by Bae et al. [115].

Rao et al. [62] reported ethanol production from D-glucose by *N. crassa* strains NCIM 870 and 1021. The strains readily fermented D-glucose, producing 4-4.4 % (w/v) ethanol in quantitative conversion in 10 days. Ethanol production was carried out in two stages, an aerobic growth for 2 days and a non-aerated fermentation at 28 °C. Deshpande et al. [40] reported the fermentation of various simple sugars to ethanol by *N. crassa* NCIM 870. Ethanol production was carried out in two stages as described by Rao et al. [62]. D-Mannose was fermented rapidly giving 100 % conversion of sugar to ethanol in 2 days. D-Galactose gave 75 % conversion in 6 days. The pentose sugars, D-xylose and D-arabinose were also fermented but at a lower rate yielding 65 % and 44 % conversion to ethanol, respectively, in 6 days (based on the assumption that theoretically 1 mol D-xylose yields 1.7 mol ethanol). Zhang et al. [116] studied xylose metabolism of *N. crassa* AS3.1602 under oxygen-limited conditions. Ethanol production reached 6.7 g/l, leading to 66 % conversion yields, but productivity was low, 0.093 g/l·h.

4.3. Fermentation of Lignocellulosics

Yeasts cannot metabolize cellulose or hemicellulose, while other fungal systems which could grow on lignocellulosic substrates and directly ferment them to ethanol have been reported. These technologically very attractive systems may lead to alternative and more cost-efficient consolidated bioprocesses for bioethanol production, compared to establish technologies (SHF, SSF) with added hydrolytic enzyme production costs. The fungus *N. crassa* is one of the few microorganisms that can convert cellulose and hemicellulose to ethanol in two stages, by producing the required hydrolytic enzymes during aerobic growth stage and fermenting the produced simple sugars to ethanol when switched to anaerobic or micro-aerobic conditions (fermentation stage). However, the reported rates of alcohol fermentation by *N. crassa* are lower than yeasts. A summary of SSF by *N. crassa* of cellulosic, hemicellulosic and lignocellulosic substrates to ethanol is presented in Figure 4.

Rao et al. [62] screened a number of *Neurospora* cultures for cellulase activity and ethanol production. Two strains of *N. crassa*, NCIM 870 and NCIM 1021, were selected and the direct fermentation of alkali-treated CP and steam-treated bagasse to ethanol was reported with periodic addition of substrate. Both strains were grown aerobically at 28 °C in 100 ml culture medium (growth stage) and transferred to non-aerated flasks containing 1 g of substrate (fermentation stage). Additional substrate (1 g each in 5 ml H_2O) was added on the 5th and 7th day of fermentation and the fermentation continued up to 10 days. The alkali-treated CP conversion was 80-90 % at the end of the fermentation and the obtained ethanol concentrations were 18 and 11 g/l for NCIM 870 and NCIM 1021, respectively. Bagasse conversion to ethanol reached 60 % (based on cellulose and hemicellulose content) by both strains in 10 days. In addition, *N. crassa* NCIM 870 converted 86 % microcrystalline cellulose (Avicel), 66 % powdered cellulose (Solka-Floc), 60 % alkali-treated straw and >80 % alkali-treated bagasse to ethanol.

Substrate	Ethanol yield[b]	References
Steam-treated bagasse	60%	[62]
Microcrystalline cellulose (Avicel)	86%	
Powdered cellulose (Solka-Floc)	66%	
Alkali-treated straw	60%	
Alkali-treated bagasse	>80%	
Alkali-treated CP[a]	90% (18)[c]	
Alkali-treated CP	91% (9.9)	[40]
Microcrystalline cellulose (Avicel)	100% (5.5)	
Powdered cellulose (Solca floc)	66% (6.9)	
D-Xylan	58% (7.5)	
Gum arabic	70% (4)	
Alkali-treated bagasse	90% (6)	
Mesta wood (alkali)	100% (7)	[117]
Bagasse(ethanolic alkali)	72% (13)	
Straw (ethanolic alkali)	66 % (12)	
Alkali pretreated BG[d]	36% (5.5)	[66]
untreated SB[e]	12% (3.7)	[67]
Dilute-acid SB	85% (27.6)	[120]

[a] cellulose powder,
[b] based on maximum theoretical conversion of cellulose/hemicellulose sugars,
[c] ethanol concentration (g/l),
[d] brewing spent grain,
[e] sorghum bagasse

Figure 4. Bioconversion of Cellulose, Hemicellulose and Lignocellulosic Substrates to Ethanol by *N. crassa* SSF Processes.

Work on bioethanol production from lignocellulosic substrates by *N. crassa* continued by Rao et al. [117] and Deshpande et al. [40]. Rapidly growing tropical woods and other agricultural residues [117] as well as cellulose (Avicel, alkali-treated CP), xylan, gum arabic and treated bagasse [40] were utilized as substrates for *N. crassa* NCIM 870 two-stage bioconversion, as described above. The direct fermentation of pretreated wood (20 g/l) gave 90-100 % conversion of available hemicellulose/cellulose to ethanol in 5 days. Maximum ethanol concentration of 11 g/l was obtained in 5 days when the woods were pretreated with cold alkali. Increasing the substrate concentration to 50 g/l lowered the conversion to 40-60 %, yielding 12 g/l of ethanol. Fermentation of wood (50 g/l) pretreated with hot 1 M NaOH followed by neutralization with HCl gave only 6 g/l of ethanol. 12-13 g/l ethanol were obtained from ethanolic NaOH treated bagasse or straw. The optimization of fermentation conditions resulted in yields of 5.5 and 10 g/l of ethanol from 1 % Avicel and 2 % (w/v) alkali-treated CP, corresponding to 100 and 90 % conversion, respectively, in 4 days. The direct fermentation of D-xylan, gum arabic (a branched polymer of D-galactose, L-rhanmose, L-arabinose and D-glucuronic acid) and treated bagasse yielded 58, 70 and 90 % conversion to ethanol, respectively, in 4-6 days. Increasing the substrate concentration of treated bagasse (5 %) yielded increased concentration of ethanol (11 g/l) corresponding to 75 % conversion of available cellulose to ethanol in 6 days.

Xiros et al. [66] studied the ethanol production from brewing spent grain (BG), a brewing industry by-product, by *N. crassa* DSM 1129. Aerobic growth was carried out in SSC (BG/WS as carbon source, moisture 61.5 % w/w), while submerged culture was employed for the fermentation (production phase) of 75 g/l alkali treated or untreated BG, in a 2 l bioreactor under anaerobic and micro-aerobic (air flow 0.1 vvm) conditions, at 30 °C and pH maintained at 5.0. The authors reported that alkali pretreatment increased substantially the ethanol production from BG and the potential presence of inhibitors did not hamper the fermentation. Moreover, micro-aerobic conditions favored ethanol production by *N. crassa*, compared to anaerobic fermentation where the detected ethanol concentrations were almost half. Maximum ethanol yield was 74 g ethanol per kg dry BG. *N. crassa* has also been reported to convert into ethanol various food industry wastes [118]. Corn cobs, sugarcane bagasse, sugarcane molasses were fermented to ethanol with high yields, after incubation for 96 hours at a temperature of 37 °C [119].

Ethanol production by *N. crassa* from sorghum bagasse (SB), the solid residue obtained after extraction of sugars from the energy plant sweet sorghum, has been described in our previous works [68, 120]. The fungus was grown under SSC optimum conditions [68], followed by micro-aerobic fermentation of SB at 30 °C. The bioconversion of untreated SB to ethanol resulted in low ethanol production [68], thus a dilute-acid pretreatment of SB prior to SSF was later investigated.

We demonstrated an alternative cost-efficient bioconversion of SB to ethanol by the fungus *N. crassa* DSM 1129 at high solids loadings (16 % at pretreatment and 8 % at fermentation), low external cellulase loadings (6 FPU/g SB), and co-fermentation of hexoses and pentoses [120]. Ethanol production from SB by *N. crassa* (supplemented with cellulase at 6 FPU/g SB) reached 27.6 g/l or 84.7 % ethanol yield (based on SB cellulose and hemicellulose sugar content). In order to improve the sugar assimilation rate, mixed cultures of *N. crassa* DSM 1129 with *S. cerevisiae* 2541 were also applied with the same bioconversion setup. However, the bioconversion efficiency of *N. crassa* single cultures was found superior to *S. cerevisiae* as well as to their mixed cultures.

4.4. Factors Affecting the Fermentation

Various factors such as temperature, substrate concentration, aeration as well as pretreatment inhibitors have been studied that may significantly affect ethanol production from lignocellulosics by *N. crassa*.

4.4.1. Temperature

A small increase in the temperature of *N. crassa* fermentations has been reported to increase conversion yields as well as reduce the fermentation time. At 37 °C, an almost quantitative conversion of cellulose to ethanol in 4 days was observed [117], compared to 66-80 % conversion in 7 days at 28 °C [62]. The increased rates of alcohol production may be due to the faster sugar production from the cellulose by the increased cellulase activity of *N. crassa* at 37 °C [40].

Temperatures higher than 37 °C seem unfavorable for alcohol production but favorable for cellulase activity. The rate-limiting step in the overall conversion of cellulose to ethanol appears to be the inability of *N. crassa* to produce ethanol at higher temperatures [40]. Hence the development of thermotolerant mutants of *N. crassa* with higher cellulase activity might increase the rate of fermentation of cellulose to ethanol.

4.4.2. Substrate Concentration

Colvin et al. [86] reported that ethanol yield from glucose fermentation by wild-type *N. crassa* 74A8 is proportional to the initial glucose concentration. Initial glucose concentrations of up to 10 % were converted to ethanol without inhibition. However, the time for maximum ethanol accumulation depended on initial glucose concentration. Increasing the substrate concentration of treated bagasse from 2 to 5 % (w/v) resulted in increased ethanol concentration, but lowered the conversion yields from 90 to 75 % and prolonged fermentation by 2 days [40].

Rao et al. [117] also reported that increasing concentration of wood and straw from 2 to 5 % (w/v) lowered the percentage conversion although a higher concentration of ethanol was obtained. The decreased conversion of cellulose to ethanol at increased substrate concentrations may not be due to the product inhibition as *N. crassa* is able to ferment D-glucose and produce ethanol up to the level of 45 g/l [62]. In order to obtain increased bioconversion of cellulose to ethanol at higher substrate concentrations the amount of inoculum should be proportionately increased.

4.4.3. Aeration

The work by Zhang et al. [116] provided valuable information of xylose metabolism of *N. crassa* under oxygen-limited conditions. By lowering oxygen levels, xylose uptake and cell growth rates decreased, and the metabolic fluxes of ethanol production changed gradually. At oxygen transfer rate of 8.4 mmol/l·h, the productivity and final concentration of ethanol reached the highest values. Concentrations of most of the intracellular metabolites decreased with oxygen limitation. Intracellular enzyme activities of XR, XDH, and XK decreased with the increase in oxygen limitation, resulting in the decreased xylose uptake rate.

Oxygen limitation activated pyruvate kinase (PK) in the downstream of glycolysis, converting phosphoenolpyruvate (PEP) to pyruvate. PK is an important regulatable enzyme in glycolysis and is inhibited by adenosine triphosphate (ATP). The authors suggested that the decreased ATP production rate under oxygen-limited conditions could alleviate the inhibiting effect of ATP on PK. Oxygen limitation had little effect on transaldolase (TAL) and transketolase (TKL) activities (enzymes involved in Pentose Phosphate Pathway), which were maintained at low levels, showing control effects on xylose metabolism. The enzyme glucose-6-phosephate dehydrogenase (G6PDH) played a major role in NADPH regeneration, and its activity decreased remarkably with the increase in oxygen limitation.

In addition, change in oxygen levels during *N. crassa* xylose fermentations seem to affect pyruvate decarboxylase (PDC) and ADH, enzymes involved in ethanol production [116]. PDC decreased and ADH increased with the increase in oxygen limitation. At oxygen transfer rate of 8.4 mmol/l·h, the activities of both enzymes were high, and ethanol yield was the highest. The authors suggested that oxygen-limited rather than anaerobic conditions are ideal for bioconversion of xylose to ethanol by *N. crassa*, so that the accumulated reduced cofactor can be oxidized to reach redox balance. Therefore, a critical level of oxygen should exist for the highest ethanol yield and productivity.

The requirement for low oxygen conditions for *N. crassa* lignocellulosic fermentations was also demonstrated by Xiros et al. [66] on a hemicellulose-rich brewing by-product (BG). By increasing the aeration levels in the submerged bioreactor from zero to 0.1 vvm, maximum ethanol concentration increased from 54 mM to 120 mM.

4.4.4. Pretreatment Inhibitors

Depending on the pretreatment process of lignocellulosic biomass, some sugar degradation compounds such as furfural and 5-hydroxymethyl furfural (HMF) and aromatic lignin degradation compounds are detected, and may affect the microorganism metabolism in the fermentation step [7, 121]. HMF and furfural decrease the volumetric ethanol yield and productivity, inhibit growth or give rise to a longer lag phase [122]. More specifically, furfurals inhibit cellular enzymes such as ADH, aldehyde dehydrogenase (ALDH), pyruvate dehydrogenase [123], and two key glycolytic enzymes, hexokinase (HK) and glyceraldehyde-3-phosphate dehydrogenase (G3PDH) [124]. In general, microorganisms appear to have the ability to convert both HMF and furfural to less inhibitory compounds and inhibitory effects are therefore gradually reduced, as long as initial concentrations are not too high, but at the cost of cellular energy and co-factors thus resulting in an increased lag phase [122, 125, 126]. According to Olofsson et al. [20] detoxification should be avoided if possible, due to additional process cost as well as possible loss of fermentable sugars.

The effect of inhibitors formed during different pretreatment conditions of SB, on the fermentability by *N. crassa* DSM 1129 was evaluated in our previous work [120]. Monitoring the time course of the bioconversion by *N. crassa* of dilute-acid pretreated SB showed that the ethanol production phase starts after the substantial decrease of inhibitor concentrations (HMF and furfural) in the fermentation broth. This pattern indicates that metabolic activity of the fungus *N. crassa* halts until furfural levels drop enough, either by assimilating or transforming the inhibitors into less toxic compounds, such as furfuryl alchohol and furoic acid [127], after which sugar metabolism to ethanol can occur. A similar phenomenon has also been observed on fermentations by *S. cerevisiae* [122, 126]. In addition, the initial furfural concentrations seemed to account more for the observed lag phases than HMF or acetic acid [120]. No toxicity levels have been reported so far for *N. crassa*.

Conclusion

Microbial production of ethanol might be a potential route to partially replace oil and chemical feedstocks. Bioethanol production generally utilizes derivatives from food crops such as corn grain and sugarcane, but the limited supply of these crops can lead to competition between their use in bioethanol production and food provision.

Lignocellulosic biomass is the most promising feedstock considering its great availability and low cost, but the large-scale commercial production of fuel bioethanol from lignocellulosic materials has still not been implemented. Lignocellulosic biomass can be converted to bioethanol by hydrolysis and subsequent fermentation by microorganisms such as yeast or bacteria which have a limited range of substrate assimilation. Because such lignocellulose hydrolysate contains not only glucose, but also various monosaccharides (e.g. xylose, mannose, fructose, galactose, and arabinose) and oligosaccharides, microorganisms should be required to efficiently ferment these sugars for the successful industrial production of bioethanol.

Some fungal strains, particularly *N. crassa*, have been found to possess the ability of metabolizing a wide range of substrates, including cellulose and hemicelluloses making it a promising candidate for bioethanol production from lignocellulosic biomass. However, no available *Neurospora* strain is completely satisfactory for an industrial process of lignocellulosic biomass to ethanol. The most limiting factor is the slow fermentation rate that needs further improvement. Nevertheless, the abilities of *N. crassa* for single-step direct conversion of low-cost biopolymeric substrates and pentoses into ethanol as well as in situ cellulolytic and hemicellulolytic enzymes may have a significant impact on the production economics.

Conflict of Interest

The Authors declare that there is no conflict of interest.

References

[1] Cardona CA, Sánchez OJ (2007) Fuel ethanol production: Process design trends and integration opportunities. *Bioresource Technol*. 98(12):2415-2457.

[2] Lin Y, Tanaka S (2006) Ethanol fermentation from biomass resources: current state and prospects. *Appl. Microbiol. Biotechnol*. 69:627-642.

[3] Chandel AK, Chan ES, Rudravaram R, Narasu ML, Rao LV, Ravindra P (2007) Economics and environmental impact of bioethanol production technologies: an appraisal. *Biotechnol. Mol. Bio. Rev*. 2(1):14-32.

[4] Agarwal AK (2008) Biofuels (alcohols and biodiesel) applications as fuels for internal combustion engines. *Prog. Energ. Combust.* 33:233-271.
[5] Koh LP, Ghazoul J (2008) Biofuels, biodiversity, and people: Understanding the conflicts and finding opportunities. *Biol. Conserv.* 141:2450-2460.
[6] Demirbaş A (2009) Biofuels securing the planet's future energy needs. *Energ. Convers. Manag.* 50(9):2239-2249.
[7] Sánchez OJ, Cardona CA (2008) Trends in biotechnological production of fuel ethanol from different feedstocks. *Bioresource Technol.* 99(13):5270-5295.
[8] Balat M, Balat H, Öz C (2008) Progress in bioethanol processing. *Prog. Energ. Combust.* 34(5):551-573.
[9] Tomás-Pejó E, Oliva JM, Ballesteros M (2008) Realistic approach for full-scale bioethanol production from lignocellulose: A review. *J. Sci. Ind. Res.* 67(11):874-84.
[10] Festel GW (2008) Biofuels - Economic Aspects. *Chem. Eng. Technol.* 31(5):715-720.
[11] Wyman CE (2007) What is (and is not) vital to advancing cellulosic ethanol. *Trends Biotechnol.* 25:153-157.
[12] Hahn-Hägerdal B, Galbe M, Gorwa-Grauslund MF, Liden G and Zacchi G (2006) Bioethanol-the fuel of tomorrow from the residues of today. *Trends Biotechnol.* 24:549-556.
[13] Demain AL (2009) Biosolutions to the energy problem. *J. Ind. Microbiol. Biotechnol.* 36:319-332.
[14] Lee J-S, Parameswaran B, Lee J-P, Park S-C (2008) Recent developments of key technologies on cellulosic ethanol production. *J. Sci. Ind. Res.* 67:865-873.
[15] Saxena RC, Adhikari DK, Goyal HB (2009) Biomass-based energy fuel through biochemical routes: A review. *Renew Sust. Energ. Rev.* 13:167-178.
[16] Kumar R, Singh S, Singh OV (2008) Bioconversion of lignocellulosic biomass: Biochemical and molecular perspectives. *J. Ind. Microbiol. Biotechnol.* 35(5):377-391.
[17] Taherzadeh MJ, Karimi K (2008) Pretreatment of lignocellulosic wastes to improve ethanol and biogas production: A review. *Int. J. Mol. Sci.* 9:1621-1651.
[18] Wyman CE, Dale BE, Elander RT, Holtzapple M, Ladisch MR, Lee YY (2005) Coordinated development of leading biomass pretreatment technologies. *Bioresource Technol.* 96(18 SPEC. ISS.): 1959-1966.
[19] Margeot A, Hahn-Hägerdal B, Edlund M, Slade R, Monot F (2009) New improvements for lignocellulosic ethanol. *Curr. Opin. Biotechnol.* 20:372-380.
[20] Olofsson K, Bertilsson M, Lidén G (2008) A short review on SSF – an interesting process option for ethanol production from lignocellulosic feedstocks. *Biotechnology for Biofuels* 1:7, doi:10.1186/1754-6834-1-7 (http://www.biotechnologyforbiofuels.com/content/1/1/7).
[21] Perkins DD (1992) *Neurospora*: the organism behind the molecular revolution. *Genetics* 130:687-701.
[22] Perkins DD, Turner BC, Barry EG (1976) Strains of *Neurospora* collected from nature. *Evolution* 30:281-313.
[23] Perkins DD, Turner BC (1988) *Neurospora* from natural populations: Toward the population biology of a haploid eukaryote. *Exp. Mycol.* 12:91-131.
[24] Turner BC, Perkins DD, Fairfield A (2001) *Neurospora* from natural populations: a global study. *Fungal. Genet. Biol.* 32:67-92.

[25] Davis RH (2000) *Neurospora: Contributions of a Model Organism.* Oxford University Press, New York.

[26] Davis RH, Perkins DD (2002) *Neurospora*: a model of model microbes. *Nature Rev. Genet.* 3: 397-403.

[27] Dunlap JC, Borkovich KA, Henn MR, Turner GE, Sachs MS, Glass NL, McCluskey K, Plamann M, Galagan JE, Birren BW, Weiss RL, Townsend JP, Loros JJ, Nelson MA, Lambreghts R, Colot HV, Park G, Collopy P, Ringelberg C, Crew C, Litvinkova L, DeCaprio D, Hood HM, Curilla S, Shi M, Crawford M, Koerhsen M, Montgomery P, Larson L, Pearson M, Kasuga T, Tian C, Baştürkmen M, Altamirano L, Xu J (2007) Enabling a Community to Dissect an Organism: Overview of the *Neurospora* Functional Genomics Project. In: Dunlap JC (ed), *Advances in Genetics*, vol 57. Academic Press, pp 49-96.

[28] Galagan JE, Calvo SE, Borkovich KA, Selker EU, Read ND, Jaffe D, FitzHugh W, Ma LJ, Smirnov S, Purcell S, Rehman B, Elkins T, Engels R, Wang SG, Nielsen CB, Butler J, Endrizzi M, Qui DY, Ianakiev P, Pedersen DB, Nelson MA, Werner-Washburne M, Selitrennikoff CP, Kinsey JA, Braun EL, Zelter A, Schulte U, Kothe GO, Jedd G, Mewes W, Staben C, Marcotte E, Greenberg D, Roy A, Foley K, Naylor J, Stabge-Thomann N, Barrett R, Gnerre S, Kamal M, Kamvysselis M, Mauceli E, Bielke C, Rudd S, Frishman D, Krystofova S, Rasmussen C, Metzenberg RL, Perkins DD, Kroken S, Cogoni C, Macino G, Catcheside D, Li WX, Pratt RJ, Osmani SA, DeSouza CPC, Glass L, Orbach MJ, Berglund JA, Voelker R, Yarden O, Plamann M, Seller S, Dunlap J, Radford A, Aramayo R, Natvig DO, Alex LA, Mannhaupt G, Ebbole DJ, Freitag M, Paulsen I, Sachs MS, Lander ES, Nusbaum C, Birren B (2003) The genome sequence of the filamentous fungus *Neurospora crassa*. *Nature* 422:859-868.

[29] Mannhaupt G, Montrone C, Haase D, Mewes HW, Aign V, Hoheisel JD, Fartmann B, Nyakatura G, Kempken F, Maier J, Schulte U (2003) What's in the genome of a filamentous fungus? Analysis of the *Neurospora* genome sequence. *Nucleic Acids Res.* 31:1944-1954.

[30] Borkovich KA, Alex LA, Yarden O, Freitag M, Turner GE, Read ND, Seiler S, Bell-Pedersen D, Paietta J, Plesovsky N, Plamann M, Goodrich-Tanrikulu M, Schulte U, Mannhaupt G, Nargang FE, Radford A, Selitrennikoff CP, Galagan JE, Dunlap JC, Loros JJ, Catcheside D, Inoue H, Aramayo R, Polymenis M, Selker EU, Sachs MS, Marzluf GA, Paulsen I, Davis RH, Ebbole DJ, Zelter A, Kalkman ER, O Rourke R, Bowring F, Yeadon J, Ishii C, Suzuki K, Sakai W, Pratt RJ (2004) Lessons from the genome sequence of *Neurospora crassa*: tracing the path from genomic blueprint to multicellular organism. *Microbiol. Mol. Biol. Rev.* 68:1-108.

[31] Tergerdy RP, Szakacs G (2003) Bioconversion of lignocellulose in solid substrate fermentation. *Biochem. Eng. J.* 13:169-179.

[32] Gusakov AV, Salanovich TN, Antonov AI, Ustinov BB, Okunev ON, Burlingame R (2007) Design of highly efficient cellulase mixtures for enzymatic hydrolysis of cellulose. *Biotechnol. Bioeng.* 97:1028-1038.

[33] Eberhart BM, Cross DF, Chase LR (1964) β-Glucosidase system of *Neurospora crassa*. I. β-Glucosidase and cellulase activities of mutant and wildtype strains. *J. Bacteriol.* 87:761-770.

[34] Eberhart BM, Beck RS, Goolsby KM (1977) Cellulase of *Neurospora crassa*. *J. Bacteriol.* 130:181-186.

[35] Mahadevan PR, Eberhart BM (1964) The beta-glucosidase system of *Neurospora crassa*. II. Purification and characterization of aryl beta-glucosidase. *Arch. Biochem. Biophys*. 108:22-29.

[36] Yazdi MT, Radford A, Keen JN, Woodward JR (1990a) Cellulase production by *Neurospora crassa*: purification and characterization of cellulolytic enzymes. *Enzyme Microb. Technol.* 12(2):120-123.

[37] Yazdi MT, Woodward JR, Radford A (1990b) Cellulase production by *Neurospora crassa*: The enzymes of the complex and their regulation. *Enzyme Microb. Technol.* 12(2): 116-119.

[38] Yazdi MT, Woodward JR, Radford A (1990c) The cellulase complex of *Neurospora crassa*: activity, stability and release. *J. Gen. Microbiol.* 136(7):1313-1319.

[39] Gray KA, Zhao L, Emptage M (2006) Bioethanol. *Curr. Opin. Chem. Biol.* 10(2): 141-146.

[40] Deshpande V, Keskar S, Mishra C, Rao M (1986) Direct conversion of cellulose/hemicellulose to ethanol by *Neurospora crassa*. *Enzyme Microb. Technol.* 8(3):149-152.

[41] Phillips CM, Iavarone AT, Marletta MA (2012) A quantitative proteomic approach for cellulose degradationby *Neurospora crassa*. *J. Proteome. Res*. 10:4177-4185.

[42] Sun J, Glass L (2011) Identification of the CRE-1 Cellulolytic Regulon in *Neurospora crassa*. *PLoS ONE* 6(9):e25654.

[43] Znameroski EA, Coradetti ST, Roche CM, Tsai JC, Iavarone AT, Cate JHD, Glass NL (2012) Induction of lignocellulose-degrading enzymes in *Neurospora crassa* by cellodextrins. *Proc. Natl. Acad. Sci. USA* 109(16):6012-6017.

[44] Kubicek CP, Messner R, Gruber F, Mach RL, Kubicek-Pranz EM (1993) The *Trichoderma* cellulase regulatory puzzle: From the interior life of a secretory fungus. *Enzyme Microb. Technol*. 15:90-99.

[45] Sukumaran RK, Singhania RR, Pandey A (2005) Microbial cellulases - Production, applications and challenges. *J. Sci. Ind. Res*. 64:832-844.

[46] Sun J, Tian C, Diamond S, Glass NL (2012) Deciphering transcriptional regulatory mechanisms associated with hemicellulose degradation in *Neurospora crassa*. *Eukaryot. cell* 11(4):482-493.

[47] Eberhart BM, Beck RS (1973) Induction of β-glucosidase in *Neurospora crassa*. *J. Bacteriol.* 116:295-303.

[48] Tian C, Beeson WT, Iavarone AT, Sun J, Marletta MA, Cate JHD, Glass NL (2009) Systems analysis of plant cell wall degradation by the model filamentous fungus *Neurospora crassa*. *Proc. Natl. Acad. Sci. USA* 106:22157-22162.

[49] Mishra C, Keskar S, Rao M (1984) Production and properties of extracellular endoxylanase from *Neurospora crassa*. *Appl. Environ. Microbiol.* 48:224-228.

[50] Myers MG, Eberhart B (1966) Regulation of cellulase and cellobiase in *Neurospora crassa*. *Biochem. Biophys. Res. Commun.* 24(5):782-785.

[51] Taleb F, Radford A (1995) The cellulase complex of *Neurospora crassa*: *cbh-1* cloning, sequencing and homologies. *Gene* 161:137-138.

[52] Schmoll M, Franchi L, Kubicek CP (2005) Envoy, a PAS/LOV domain protein of *Hypocrea jecorina* (anamorph *Trichoderma reesei*), modulates cellulase gene transcription in response to light. *Eukaryot. Cell* 4(12):1998-2007.

[53] Chen CH, Ringelberg CS, Gross RH, Dunlap JC, Loros JJ (2009) Genome-wide analysis of light-inducible responses reveals hierarchical light signalling in *Neurospora. EMBO J.* 28(8):1029-1042.

[54] Smith KM, Sancar G, Dekhang R, Sullivan CM, Li S, Tag AG, Sancar C, Bredeweg EL, Priest HD, McCormick RF, Thomas TL, Carrington JC, Stajich JE, Bell-Pedersen D, Brunner M, Freitag M (2010) Transcription factors in light and circadian clock signaling networks revealed by genome-wide mapping of direct targets for *Neurospora* WHITE COLLAR COMPLEX. *Eukaryot. Cell* 9(10):1549-1556.

[55] Tisch D, Schmoll M (2010) Light regulation of metabolic pathways in fungi. *Appl. Microbiol. Biotechnol.* 85(5):1259-1277.

[56] Schmoll M (2011) Assessing the relevance of light for fungi: implications and insights into the network of signal transmission. *Adv. Appl. Microbiol.* 76:27-78.

[57] Schmoll M, Tian C, Sun J, Tisch D, Glass NL (2012) Unraveling the molecular basis for light modulated cellulase gene expression – the role of photoreceptors in *Neurospora crassa. BMC Genomics* 2012, 13:127, doi:10.1186/1471-2164-13-127 (http://www.biomedcentral.com/1471-2164/13/127).

[58] Eberhart BM, Beck RS (1970) Localization of the *β*-glucosidases in *Neurospora crassa. J. Bacteriol.* 101:408-417.

[59] Couto SR, Sanromán MA (2005) Application of solid-state fermentation to ligninolytic enzyme production. *Biochem. Eng. J.* 22:211-219.

[60] Pandey A, Soccol CR, Mitchell D (2000) New developments in solid state fermentation: I-bioprocesses and products. *Process Biochem.* 35:1153-1169.

[61] Yazdi MT, Khosravi AA, Nemati M, Motlagh NDV (2003) Purification and characterization of two intracellular *β*-glucosidases from the *Neurospora crassa* mutant *cell-1*. *World J. Microb. Biotechnol.* 19:79-84.

[62] Rao M, Deshpande V, Keskar S, Srinivasan MC (1983) Cellulase and ethanol production from cellulose by *Neurospora crassa Enzyme. Microb. Technol.* 5:133-136.

[63] Macris BJ, Kekos D, Evangelidou X, Galiotou-Panayotou M, Rodis P (1987) Solid state fermentation of straw with *Neurospora crassa* for CMCase and *β*-glucosidase production. *Biotechnol. Lett.* 9(9):661-664.

[64] Macris BJ, Kekos D, Evangelidou X (1989) A simple and inexpensive method for cellulase and beta-glucosidase production by *Neurospora crassa* (Short Contribution). *Appl. Microbiol. Biotechnol.* 31:150-151.

[65] Romero MD, Aguado J, González L, Ladero M (1999) Cellulase production by *Neurospora crassa* on wheat straw. *Enzyme Microb. Technol.* 25(3-5):244-250.

[66] Xiros C, Topakas E, Katapodis P, Christakopoulos P (2008) Hydrolysis and fermentation of brewer's spent grain by *Neurospora crassa. Bioresource Technol.* 99: 5427-5435.

[67] Mamma D, Kourtoglou E, Christakopoulos P (2008) Fungal multienzyme production on industrial by-products of the citrus-processing industry. *Bioresource Technol.* 99(7): 2373-2383.

[68] Dogaris I, Vakontios G, Kalogeris E, Mamma D, Kekos D (2009b) Induction of cellulases and hemicellulases from *Neurospora crassa* under solid-state cultivation for bioconversion of sorghum bagasse into ethanol. *Ind. Crop. Prod.* 29(2-3):404-411.

[69] Dogaris I, Karapati S, Mamma D, Kalogeris E, Kekos D (2009a) Hydrothermal processing and enzymatic hydrolysis of sorghum bagasse for fermentable carbohydrates production. *Bioresource Technol.* 100(24):6543-6549.

[70] Landman OE, Bonner DM (1952) *Neurospora* lactase. I: Properties of lactase preparations from a lactose utilizing and a lactose non-utilizing strain. *Arch. Biochem. Biophys.* 41(2):253-265.

[71] Landman OE (1954) *Neurospora* lactase. II. Enzyme formation in the standard strain. Arch Biochem Biophys 52(1):93-109

[72] Johnson HN, DeBusk A (1970a) The beta-galactosidase system of *Neurospora crassa*. I. Purification and properties of the pH 4.2 enzyme. *Arch. Biochem. Biophys.* 138(2): 408-11.

[73] Johnson HN, DeBusk A (1970b) The beta-galactosidase system of *Neurospora crassa*. II. Extracellular nature of the pH 4.2 enzyme. *Arch. Biochem. Biophys*. 138(2):412-7.

[74] Perry CB, Lester G (1973) The identification of a third b-galactosidase activity in *Neurospora crassa. Biochem. Biophys. Res. Commun.* 54(4):1476-1482.

[75] Stephens R, DeBusk A (1975) (75) β-Galactosidases from *Neurospora crassa*. In: Wood WA (ed) *Methods in Enzymology*, vol 42. Academic Press, pp 497-503.

[76] Takata R, Tokita K, Mori S, Shimoda R, Harada N, Ichinose H, Kaneko S, Igarashi K, Samejima M, Tsumuraya Y, Kotake T (2010) Degradation of carbohydrate moieties of arabinogalactan-proteins by glycoside hydrolases from *Neurospora crassa*. *Carbohydr. Res.* 345(17):2516-2522.

[77] Crepin VF, Faulds CB, Connerton IF (2003) A non-modular type B feruloyl esterase from *Neurospora crassa* exhibits concentration-dependent substrate inhibition. *Biochem. J.* 370:417-27.

[78] Crepin VF, Faulds CB, Connerton IF (2004) Identification of a type-D feruloyl esterase from *Neurospora crassa*. *Appl. Microbiol. Biotechnol.* 63:567-70.

[79] Lynd LR, Weimer PJ, van Zyl WH, McBride JE, Laser M (2005) Consolidated bioprocessing of cellulosic biomass: an update. *Curr. Opin. Biotechnol.* 16:577-583.

[80] Griffen DH (1994) *Fungal physiology*, 2nd ed. Wiley-Liss Inc., New York, N.Y.

[81] Perlman D (1950a) Observations on the production of ethanol by fungi and yeasts. *Amer. J. Bot.* 37:237-241.

[82] Perlman D (1950b) The metabolism of carbohydrates by *Neurospora*. *Bull. Torrey Bot. Club* 77:103-109.

[83] Tissieres A, Mitchell HK, Haskins FA (1953) Studies on the respiratory system of the poky strain of *Neurospora*. *J. Biol. Chem.* 205:423-433.

[84] Strauss BS, Pierog S (1954) Gene interactions. The mode of action of the suppressor of acetate-requiring mutants of *Neurospora crassa*. *J. Gen. Microbiol.* 10:221-235.

[85] Kobr MJ, Bianchi DE, Oulevey N, Turian G (1967) The effect of oxygen tension on growth, conidiation, and alcohol production of *Neurospora crassa*. *Can. J. Microbiol.* 13:805-817.

[86] Colvin HJ, Sauer BL, Munkres KD (1973) Glucose utilization and ethanolic fermentation by wild type and extrachromosomal mutants of *Neurospora crassa*. *J. Bacteriol.* 116(3):1322-1328.

[87] Weiss B, Turian G (1966) A study of conidiation in *Neurospora crassa*. *J. Gen. Microbiol.* 44:407-418.

[88] Howell N, Zuiches CA, Munkres KD (1971) Mitochondrial biogenesis in *Neurospora crassa*. I. An ultrastructural and biochemical investigation of the effects of anaerobiosis and chloramphenicol inhibition. *J. Cell Biol.* 50:721-736.

[89] Zink MW (1969) Regulation of NAD-specific alcohol dehydrogenase in *Neurospora crassa*. *Can. J. Microbiol.* 15:265-271.

[90] Flavell RB, Fincham JRS (1968) Acetate-nonutilizing mutants of *Neurospora crassa*. II. Biochemical deficiencies and the roles of certain enzymes. *J. Bacteriol.* 95:1063-1068.

[91] Xie X, Wilkinson HH, Correa A, Lewis ZA, Bell-Pedersen D, Ebbole DJ (2004) Transcriptional response to glucose starvation and functional analysis of a glucose transporter of *Neurospora crassa*. *Fungal. Genet. Biol.* 41:1104-1119.

[92] Chambergo FS, Bonaccorsi ED, Ferreira AJ, Junior JR, Abrahao-Neto J, Farah JP, El-Dorry H (2002) Elucidation of the metabolic fate of glucose in the filamentous fungus *Trichoderma reesei* using expressed sequence tag (EST) analysis and cDNA microarray. *J. Biol. Chem.* 277:13983-13988.

[93] Scarborough GA (1970) Sugar transport in *Neurospora crassa*. *J. Biol. Chem.* 245: 3985-3987.

[94] Schneider RP, Wiley WR (1971) Kinetic characteristics of the two glucose transport system in *Neurospora crassa*. *J. Bacteriol.* 106:479-486.

[95] Lee J (1997) Biological conversion of lignocellulosic biomass to ethanol. *J. Biotechnol.* 56:1-24.

[96] van Maris AJ, Abbott DA, Bellissimi E, van den Brink J, Kuyper M, Luttik MA, Wisselink HW, Scheffers WA, van Dijken JP, Pronk JT (2006) Alcoholic fermentation of carbon sources in biomass hydrolysates by *Saccharomyces cerevisiae*: current status. *Antonie Van Leeuwenhoek* 90:391-418.

[97] Zaldivar J, Nielsen J, Olsson L (2001) Fuel ethanol production from lignocellulose: a challenge for metabolic engineering and process integration. *Appl. Microbiol. Biotechnol.* 56(1-2):17-34.

[98] Matsushika A, Inoue H, Kodaki T, Sawayama S (2009). Ethanol production from xylose in engineered *Saccharomyces cerevisiae* strains: current state and perspectives. *Appl. Microbiol. Biotechnol.* 84:37-53.

[99] Hahn-Hägerdal B, Jeppsson H, Skoog K, Prior BA (1994) Biochemistry and physiology of xylose fermentation by yeasts. *Enzyme Microb. Technol.* 16:933-943.

[100] Chandrakant P, Bisaria VS (1998) Simultaneous bioconversion of cellulose and hemicellulose to ethanol. *Crit. Rev. Biotechnol.* 18:295-331.

[101] Skoog K, Hahn-Hägerdal B (1988) Xylose fermentation. *Enzyme. Microb. Technol.* 10: 66-80.

[102] Rawat U, Bodhe A, Deshpande V, Rao M (1993) D-xylose catabolizing enzymes in *Neurospora crassa* and their relationship to D-xylose fermentation. *Biotechnol. Lett.* 15(11):1173-1178.

[103] Rawat U, Rao M (1996) Purification, kinetic characterization and involvement of tryptophan residue at the NADPH binding site of xylose reductase from *Neurospora crassa*. *Biochim. Biophys. Acta* 1293:222-230.

[104] Phadtare SU, Rawat UB, Rao MB (1997) Purification and characterisation of xylitol dehydrogenase from *Neurospora crassa*. *FEMS Microbiol. Lett.* 146(1):79-83.

[105] Woodyer R, Simurdiak M, van der Donk WA, Zhao H (2005) Heterologous expression, purification, and characterization of a highly active xylose reductase from *Neurospora crassa. Appl. Environ. Microbiol.* 71(3):1642-1647.

[106] Zhao X, Gao P, Wang Z (1998) The production and properties of a new xylose reductase from fungus *Neurospora crassa. Appl. Biochem. Biotechnol.* 70-72:405-414.

[107] Vandeska E, Amartey S, Kuzmanova S, Jeffries T (1995) Effects on environmental conditions on production of xylitol by Candida boidinii. *World J. Microbiol. Biotechnol.* 11:213-218.

[108] Dien BS, Cotta MA, Jeffries TW (2003) Bacteria engineered for fuel ethanol production: current status. *Appl. Microbiol. Biotechnol.* 63:258-266.

[109] Jennings DH (1984) Polyol metabolism in fungi. *Adv. Microb. Physiol.* 25:149-193.

[110] Richard P, Londesborough J, Putkonen M, Kalkkinen N, Penttila M (2001) Cloning and expression of a fungal L-arabinitol 4-dehydrogenase gene. *J. Biol. Chem.* 276:40631-40637.

[111] Nair N, Zhao H (2007) Biochemical characterization of an L-xylulose reductase from *Neurospora crassa. Appl. Environ. Microb.* 73(6):2001-2004.

[112] Chiang C, Knight SG (1960) A new pathway of pentose metabolism. *Biochem. Biophys. Res. Commun.* 3:554-559.

[113] Sullivan R, Zhao H (2007) Cloning, characterization, and mutational analysis of a highly active and stable L-arabinitol 4-dehydrogenase from *Neurospora crassa. Appl. Microbiol. Biotechnol.* 77:845-852.

[114] Eklund H, Ramaswamy S (2008) Medium- and short-chain dehydrogenase/reductase gene and protein families: three-dimensional structures of MDR alcohol dehydrogenases. *Cell Mol. Life Sci.* 65:3907-3917.

[115] Bae B, Sullivan RP, Zhao H, Nair SK (2010) Structure and engineering of L-arabinitol 4-dehydrogenase from *Neurospora crassa. J. Mol. Biol.* 402:230-240.

[116] Zhang Z, Qu Y, Zhang X, Lin J (2008) Effects of oxygen limitation on xylose fermentation, intracellular metabolites, and key enzymes of *Neurospora crassa* AS3.1602. *Appl. Biochem. Biotechnol.* 145(1-3):39-51.

[117] Rao M, Mishra C, Keskar S, Srinivasan MC (1985) Production of ethanol from wood and agricultural residues by *Neurospora crassa. Enzyme Microb. Technol.* 7(12): 625-628.

[118] Okonko IO, Adeola OT, Aloysius FE, Damilola AO, Adewale OA (2009) Utilization of food wastes for sustainable development. *Electron. J. Environ. Agr. Food Chem.* 8(4): 263-286.

[119] Ashiru AW (2005) Production of ethanol from molasses and corncobs using yeasts and a mould. In: *the Book of Abstract of the 29th Annual Conference & General Meeting (Abeokuta 2005) on Microbes As Agents of Sustainable Development, organized by Nigerian Society for Microbiology* (NSM), University of Agriculture, Abeokuta, 6-10th November, p 22.

[120] Dogaris I, Gkounta O, Mamma D, Kekos D (2012) Bioconversion of dilute-acid pretreated sorghum bagasse to ethanol by *Neurospora crassa. Appl. Microbiol. Biotechnol.* 95(2):541-550.

[121] Alvira P, Tomás-Pejó E, Ballesteros M, Negro MJ (2010) Pretreatment technologies for an efficient bioethanol production process based on enzymatic hydrolysis: A review. *Bioresource Technol.* 101:4851-4861.

[122] Almeida JR, Modig T, Petersson A, Hahn-Hägerdal B, Lidén G, Gorwa-Grauslund MF (2007) Increased tolerance and conversion of inhibitors in lignocellulosic hydrolysates by *Saccharomyces cerevisiae*. *J. Chem. Technol. Biotechnol.* 82:340-9.

[123] Modig T, Lidén G, Taherzadeh MJ (2002) Inhibition effects of furfural on alcohol dehydrogenase, aldehyde dehydrogenase and pyruvate dehydrogenase. *Biochem. J.* 363:769-776.

[124] Banerjee N, Bhatnagar R, Viswanathan L (1981) Inhibition of glycolysis by furfural in *Saccharomyces cerevisiae*. *Appl. Microbiol. Biotechnol.* 11:226-8.

[125] Palmqvist E, Almeida JS, Hahn-Hägerdal B (1999) Influence of furfural on anaerobic glycolytic kinetics of *Saccharomyces cerevisiae* in batch culture. *Biotechnol. Bioeng.* 62:447-454.

[126] Almeida JRM, Bertilsson M, Gorwa-Grauslund MF, Gorsich S, Lidén G (2009) Metabolic effects of furaldehydes and impacts on biotechnological processes. *Appl. Microbiol. Biotechnol.* 82(4):625-638.

[127] Eilers FI, Sussman AS (1970) Conversion of furfural to furoic acid and furfuryl alcohol by *Neurospora* ascospores. *Planta* (Berl.) 94:253-264.

In: Microbes in Process
Editors: Neelam Garg and Abhinav Aeron
ISBN: 978-1-63117-127-7

Chapter 6

Microbial Chitinases: Manifestation and Prospective

Sneha Jha[1], Chaitanya Kumar Jha[*2,3], H. A. Modi[1] and Meenu Saraf[3]
[1]Department of Life Sciences, University School of Science, Gujarat University, Ahmedabad, Gujarat, India
[2]Department of Microbiology, Gujarat Arts and Science College, Ahmedabad, Gujarat, India
[3]Department of Microbiology, University School of Science, Gujarat University, Ahmedabad, Gujarat, India

Abstract

Relevant to important biophysiological functions and applications of chitinase, a considerable amount of research on fungal and bacterial chitinases has been carried out in recent years. Chitin is a common constituent of many organisms and is found in fungal cell walls, insect exoskeletons, and shells of crustaceans. All of these chitin containing organisms produce chitinase. Some other organisms, such as bacteria and higher plants, do not contain chitin but do produce chitinases. Chitinases are produced by several bacteria, actinomycetes, fungi and also by higher plants. The production of inexpensive chitinolytic enzymes not only solves environmental problems but also reduces the production cost of chito-oligosaccharides required in the manufacture of value added products such as sweeteners, several growth factors, and single cell protein. Furthermore, chitinase has many other applications such as in the biocontrol of plant pathogenic fungi and plant defense. Actinomycetes, Gram-positive mycelial bacteria, ubiquitous in soil, are well known producers of many extracellular enzymes with polymer-degrading properties, including chitinases. Therefore, the present review will focus on fungal as well as bacterial chitinases, containing their characterization and biotechnological potentials.

* jha_chaitanya@rediffmail.com.

Keywords: Chitin, Chitinase, Biological control, Plant defence, Microbial Chitinase

1. Introduction

Chitinase (EC 3.2.1.14) is a glycosyl hydrolase that catalyzes the degradation of chitin, an insoluble linear β-1, 4-linked polymer of *N*-acetylglucosamine. Chitinases are of great biotechnological interest. First, these enzymes may be used to convert chitin containing biomass into depolymerised components. Second, chitinases may be exploited for the control of fungal and insect pathogens of plants as they can inhibit growth of chitin-containing plant-pathogens and plague insects that need chitins for normal development [1, 2].

Characteristics	Family 18	Family 19
Catalytic Domain	(α/β)8-barrel domain, consisting of eight α-helices and eight β-strands	high α- helical content.
Distribution	Bacteria, Fungi, Plants, Insects, Mammals, and Viruses.	Plants and some *Streptomyces* strains.
Catalytic Mechanism	β-anomers are produced by hydrolysis of β-1,4- glycosidic linkages.	Similar to lysozyme and chitosanase
Accepted Model for Catalysis	Substrate assisted catalysis	General acid and-base mechanism
Sensitivity	Sensitive to allosamidin, a potent chitinase inhibitor	Insensitive
Hydrolysis	Hydrolyse GlcNAc-GlcNAc and GlcNAc-GlcN linkages	Hydrolyse GlcNAc- GlcNAc and GlcN-GlcNAc

Figure 1. Comparative Study of Chitinases Family 18 and Family 19.

Chitin was first discovered in 1811 by Prof. Henri Braconnott in the cell walls of mushrooms. Due to its strength, solubility and chelating properties, it was quickly realized that chitin held great biotechnological promise. For instance, the capacity of both chitin and chitosan (a deacetylated form of chitin) to bind heavy metals and other contaminants in water is currently exploited in a wide range of applications, including water purification and wastewater treatment. Furthermore, consumer products such as dietary supplements, cosmetics, internal sutures, burn dressings and even contact lenses may contain chitin or chitosan. The high carbon and nitrogen content of chitin, its resiliency to degradation and its large annual production make it a significant component of the organic matter in many environments. Thus, the mineralization of chitin, which is performed almost exclusively by microbes, is an essential component of nutrient cycling in most natural systems. The hypothesis drove the investigation of chitinases and GlcNAcidases in microbes from Mono Lake because of the high salinity, high pH and the overall unique environmental conditions there.

In general, chitinolytic enzymes can be divided into three categories (Fig 1): exochitinases, demonstrating activity only for the non-reducing end of the chitin chain; endochitinases, which hydrolyze internal β-1,4- glycoside; and β-*N*-acetylglucosaminidase, which cleaves GlcNAc units sequentially from the nonreducing end of the substrate [3]. Chitinases vary widely in their size, ranging from as low as 20 kDa to about 90 kDa. They are found to be active over a wide range of temperature and pH, depending on the source of bacteria from which they have been isolated.

Based on amino acid sequence similarity, chitinases are classified into families 18 and 19 of glycoside hydrolases. The members of the two different families differ in their amino acid sequences, three-dimensional structures, and molecular mechanisms of catalytic reactions. Chitinases belong to glycosyl hydrolase families 18 and 19 according to the classification made by Henrissat and Bairoch [4]. The chitinases are also classified according to N-terminal sequence, localization of the enzyme (Fig 2), isoelectric pH, signal peptide and the inducers into classes. There has been an all around effort to increase the production of chitinases by using genetically engineered organisms and by incorporating modern fermentation techniques. This review presents a brief outline on the potential biotechnological application of microbial chitinases.

Chitinases	**Source**
Class I	Plants
Class II	Plants, Fungi, Bacteria
Class III	Did not show any sequence similarity to enzymes of Class I or II.
Class IV	Similar characteristics to Class I chitinases including immunological properties, but they were significantly smaller than Class I chitinases
Class V	Has been detected in studies on symbiotic plant–microbe interactions

Figure 2. Classification of Chitinase on the Basis of Localization of the Enzyme.

2. Microbial Chitinase

Chitinases, which hydrolyze chitin, occur in a wide range of microorganisms including bacteria, fungi, and actinomycetes. The roles of chitinases in these organisms are diverse. In bacteria, chitinases play a role in nutrition and parasitism. Chitinases can be used for the production of chitooligosaccharides, which have been found to function as antibacterial agents, elicitors of lysozyme inducers. Chitinases can also be used in agriculture to control plant diseases. Some of the bacterial spp producing chitinases are *Serratia*, *Bacillus*, *Aeromonas*, *Vibrio, Pseudomonas, Alteromonas*, *Enterobacter* and *Clostridium*. Chitinase in fungi is thought to have autolytic, nutritional, and morphogenetic roles. In viruses, chitinases are involved in pathogenesis. Actinomycetes and particularly *Streptomyces* species are good sources of novel Chitinases and other enzymes, enzyme inhibitors, immunomodifiers, vitamins and antibiotics.

2.1. Bacterial Chitinase

Bacteria (*Bacillus, Aeromonas, Vibrio, Pseudomonas, Serratia, Enterobacter*) produce multiple chitinases to degrade the diverse configuration of chitins present in nature. Chitinase producing marine bacteria play an important role in the degradation on chitin in the oceans. Various species of *Bacillus* have been shown to secrete chitinase, including *Bacillus circulans, Bacillus licheniformis, Bacillus laterosporus, Bacillus amyloliquefaciens, Bacillus magaterium, Bacillus pabuli, Bacillus stearothermophilus, Bacillus subtilis, Bacillus thuringiensis* [5].

Chitinolytic enzymes which are grouped into families 18 (and 19) of glycosyl hydrolases based on their amino acid sequence similarity but family 18 is diverse and contains majority of chitinases from bacteria alongwith fungi, animals, and some plant chitinases. Bacterial chitinases are further grouped into three major subfamilies, A, B, and C, based on the amino acid sequence of individual catalytic domains [6]. Over the years, several bacterial species have been harvested for the production of chitinase for various applications [7]. Bacterial chitinases have a size range of ~20–60 kDa, which is similar to that of plant chitinases (~25–40 kDa), and are smaller than insect chitinases (~40–85 kDa).They are found to be active over a wide range of temperature, pH, and broad range of isoelectric points (pI~4.5–8.5) depending on the source of bacteria from which they have been isolated [8]. *Streptomyces violaceusniger* produces endochitinases and thermostable chitinase produced from *Streptomyces thermoviolaceus* OPC-520 have optimum temperatures of, respectively, 28^0C and ~80^0C [9, 10]. Also, the last enzyme has high pH optima in the range of 8.0 to 10.0 [10].

Chimeric chitinases were generated by swapping the chitin-binding domain between chitinases of *B. licheniformis* and *B. thuringiensis* resulting in increased affinity to colloidal chitin and insoluble chitinous substrates, and increased conformational stability [11]. Using chimeric constructs of chitinase, Huang et al. [12] showed an involvement of the chitin binding domain in the antifungal activity of a chitinase from *B. cereus*.

Many mechanisms of pathogen suppression by bacteria have been suggested: substrate competition and niche exclusion, production of siderophores and/or production of antibiotics and induced resistance. In addition, chitinases are present in a wide range of bacterial strains which catalyse the degradation of chitin. The prime application of chitinases is protection against fungal pathogens. Since chitin forms a major constituent of the fungal cell wall, they are readily susceptible to various chitinolytic enzymes. Also, chitinase has been identified as one of the pathogenesis related proteins (PR proteins) produced by plants. Thus, chitinase producing organisms are used in agriculture as an effective biocontrol agent against a number of phytopathogenic fungi [7]. *Serratia marcescens* controls *B.cinerea, Rhizoctonia solani,* and *F. oxysporum* [13]. *S. marcescens* controlled growth of *Sclerotinia minor*, the causal agent of basal drop diseases in lettuce [14].

2.2. Fungal Chitinase

The fungal cell wall is composed of chitin, 1, 3-β-glucan and 1, 6-β-glucan, mannan and proteins. Fungal chitinases belong to the family 18 of glycosyl hydrolase. Fungal chitinases share high amino acid homology with class III plant chitinases [15]. It consists of 5 regions in its structure:

- i Catalytic domain
- ii N-terminal signal peptide region
- iii Chitin binding domain
- iv Serine/threonine rich region
- v C- terminal extension region.

Although, in most of the fungal chitinase, chitin binding domain, serine/threonine rich region, C-terminal extension region are not present, yet they are enzymatically active. The opportunistic fungal pathogen *A. fumigatus* contains at least 11 conserved active-site region for chitinases of glycosyl hydrolase family 18. The chitinolytic activity in filamentous fungi, as reported for *Beauveria bassiana*, *Trichoderma harzianum*, *Aspergillus fumigatus*, *Lecanicillium lecanii*, *Lecanicillium fungicola*; *Metarhizium anisopliae*, *Fusarium* increases with time of incubation and depends on other factors such as pH, temperature or water activity [16]. The enzyme production starts approximately from 20 to 40 hours during the exponential phase of growth.

Many of the multiple chitinase-encoding genes of filamentous fungi may encode secreted enzymes with nutritional roles rather than cell wall hydrolases for growth and morphogenesis [17]. *Saccharomyces cerevisiae* (Yeast) possesses two chitinase encoding genes, the plant-type chitinase ScCTS1 and ScCTS2, with an apparent molecular weight of 130 kDa (including glycosylation) showing sequence similarity to hevamine; and which is more similar to bacterial chitinases like chitinase A and B from *Serratia marcescens* respectively [18]. A knockout of the cts1 gene in *S. cerevisiae* resulted in a ''clumping'' phenotype, where, after cell division, mother and daughter cells were unable to separate [19]. Treatment of *S. cerevisiae* with a derivative of allosamidin (a well-characterized chitinase inhibitor [20]) gave a remarkably similar phenotype [21]

The mycoparasitic activity is related with the production of cell wall degrading enzymes (CWDE), such as β-1, 3-glucanases, proteases and chitinases. The majority of known mycoparasite chitinase genes are from *Trichoderma harzianum*, *Trichoderma atroviride*, and *Trichoderma virens*. The mycoparasitic process involves chemothophic growth of

Trichoderma, recognition of the host by mycoparasite and its attachment, excretion of cell wall degrading enzymes (CWDE), hyphae penetration and lysis of the host [22]. The association of plant roots with mycorrhizal fungi provides the plant with a greater chance of encountering resources through rapid colonization by the extra-radical mycelium due to their physiological and morphological plasticity. *Hebeloma syrjense*, fungal symbionts excrete enzymes such as chitinases, phosphatases, phytases and proteases for conversion of organic residues to nutrients such as N and P. Fungal chitinases have shown a massive potential application in various fields. In recent times, the possibility for improving plant resistance using genetic manipulation techniques is promising. The chit42 gene of *T. harzianum*, encoding a powerful endochitinase with a much stronger antifungal activity to a wide range of phytopathogenic fungi, is expressed constitutively in tobacco, apple and potato, and these transgenic plants show a high level resistance against phytopathogenic fungi tested [23]. Chitinases from *T. harzianum* are used for the generation of fungal protoplasts [24] and in bioconversion of shellfish. The chitinases of *Myrothecium verrucaria* and *Saccharomyces cerevisiae* are utilized for the production of single-cell protein (SCP) from chitinous waste [25] and conversion of chitin to ethanol has been suggested [26]. Transgenic tobacco plants expressing the *S. cerevisiae* Cts1 chitinase can inhibit spore germination and hyphal growth of *Botrytis cinerea* [27] and are also used for insect control. The research intend to discovery of novel chitinases, especially from hostile environmental fungi, determining biological functions of chitinase in growth and development and mycoparasitism, signaling in chitinase expression, regulation mechanism between different chitinases for better biotechnological prospects.

2.3. Actinomycetes Chitinase

Actinomycetes are a group of Gram-positive filamentous bacteria with high G+C ratio. They play an important role in decomposition of organic materials, such as cellulose, lignocellulose, starch and chitin in soil. Furthermore, actinomycetes are a promising group of root-colonizing microbes where they may influence plant by promoting growth and protecting against disease. Actinomycetes and particularly *Streptomyces* species are important saprophytic soil microorganisms and good sources of novel antibiotics, enzymes, enzyme inhibitors, immunomodifiers, and vitamins [28]. Several chitinolytic enzymes have been identified in various *Streptomyces* sp., including, *Streptomyces plicatus, S. lividans, S. virdificans,S. halstedii, S. aureofaciens, S. diasitapiticus, S. thermoviolaceus* and *S. griceus.* The antifungal biocontrol agent *Streptomyces lydicus* WYEC108 was capable of not only destroying germinating oospores of *Pythium ultimum* but also of damaging the cell walls of fungal hyphae [29]. *Streptomyces lydicus* WYEC108 also produced high levels of chitinases when induced with fungal cell walls chitins in growth media. However, negligible levels of enzymes were detected when *S. lydicus* WYEC108 was grown in the absence of chitin. Chitinase production by *S. lydicus* WYEC108 was also induced by colloidal chitin, *N*-acetylglucosamine and chito-oligosaccharides. However, the synthesis was repressed by high (but not low) levels of glucose and carboxymethyl cellulose (CMC) [30].

Terayama et al. [31] used chitinase from *Streptomyces* sp. for large-scale preparation of *N, N'*-diacetylchitobiose by enzymatic degradation of chitin and its chemical modifications. Bacterial chitinase production on a large scale is still in its early stages possibly due to problems encountered when expressing enzymes with potential lysozyme activity. El-Sayed et al. [32] identified two chitinases, A and B, purified from a culture supernatant of *S. albocinaceus* S-22 that showed potential antifungal activity by cell wall lysis. Mitsutomi et al. [33] reported on the purification of two novel chitinases, C-1 and C-2, from *Streptomyces griseus* HUT 6037. Both enzymes had a molecular weight of 27 kDa and had an optimum pH of 55∘C. Antibiotics and protein inhibitors produuced by actinomycetes are formed during the late growth cycle, when familiar regulatory processes,such as transcriptional control, are ineffective [28]. Generally, low nitrogen conditions are conducive to the secretion of antifungal and antibacterial secondary metabolites by actinomycetes used as biocontrol agents.

3. Crystal Structures of Microbial Chitinolytic Enzymes

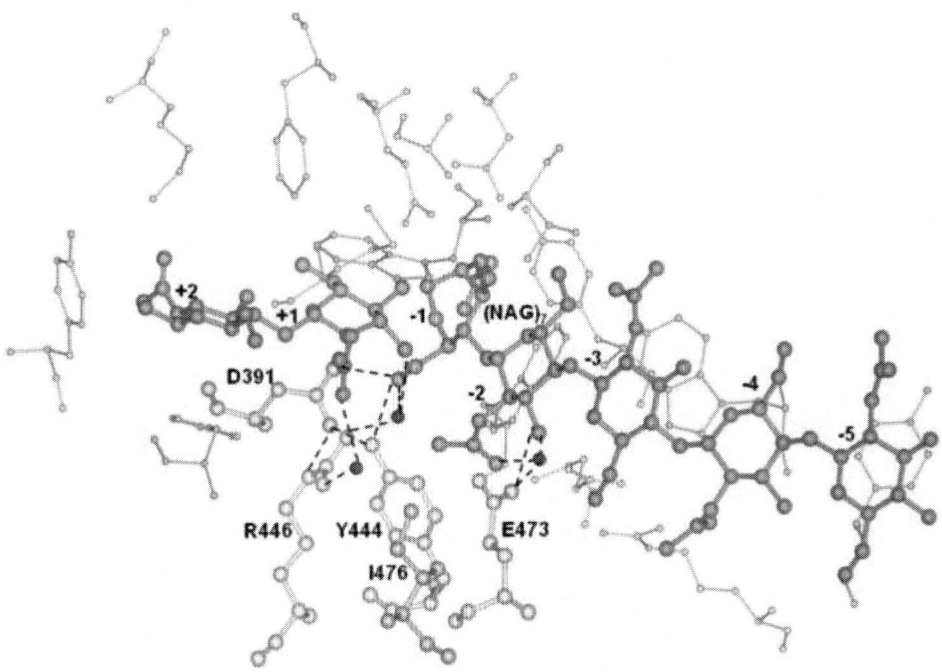

Adapted from Li H and Greene LH [56]

Figure 3. Residues on S. marcescens chiA Interact with 7-mer of NAG (N-acetylglucosamine) Substrate. Residues in Yellow on the CID (Chitinase Insertion Domain) of S. marcescens chiA form Hydrogen Bonds with the Substrate, although Some Interactions are Mediated by Asp391 and Water Molecules Colouredin Blue. Additional TIM Barrel Residues involving in Hydrogen Bonding and Hydrophobic Interactions are shown in Green and Brown, respectively.

Recently, crystal structures of chitinolytic enzymes have been successively reported, and the enzymes were found to have catalytic and substrate binding clefts, whose structure is different from that of the lysozyme. The crystal structure of chitinase from *Serratia marcescens* (Fig 3) reveals an $(\alpha/\beta)_8$- barrel structure. Since most of the chitinolytic enzymes have molecular weights of 15~80 kDa, X-ray crystallography is the main strategy for obtaining three-dimensional structures. Family 18 chitinases have catalytic domains of triosephosphateisomerase (TIM barrel) which catalyze the hydrolytic reaction by substrate-assisted mechanism whereas family 19 chitinases have high percentage of α-helices and adopt the single displacement catalytic mechanism. In family 18 chitinases, the leaving group is protonated by a conserved glutamic acid, the sugar at 21 subsite is distorted into a boat conformation, and an oxazolinium intermediate is stabilized by the sugar N-acetamido group and then hydrolyzed [34].

Barley chitinase, *Streptomyces* chitosanase, goose egg, white lysozyme, T4 phage lysozyme, and hen egg white lysozyme have a structurally invariant core consisting of two helices and a three-stranded β-sheet which form the substrate binding and catalytic cleft. The proteins share no significant sequence homologies, and belong to different enzyme families. Similarities in three-dimensional structure would not always originate from sequence homologies. Family 18 chitinases can be classified into three subfamilies A, B, and C, in terms of the amino acid sequence similarity [6]. The main structural difference between subfamilies A and B chitinasesis that a small α + β domain inserts into the TIM barrel catalytic domain in the subfamily A, while this insertion domain is absent in the subfamily B. *S. marcescens* chitinase C (chiC), belonging to the subfamily B, has a catalytic domain, a fibronectin type III-like domain, and a chitin-binding domain [35]. Therefore the presence or absence of the insertion domain appears to be subfamily specific. Examples of family 18 chitinases in the subfamily B are only limited to a few bacteria, such as *S. marcescens* and *Bacillus circulans* [36]. *S. marcescens* chitinase A (chiA) has an additional N-terminal domain while *S. marcescens* chitinase B (chiB) has an extra C-terminal domain which belongs to the carbohydrate binding domain superfamily.

4. Mechanism of Chitinase Action

Maximum parts of the chitin biosynthetic pathway (Leloir pathway) in fungi and other microbes would be similar or identical as in an insect which have been worked out extensively (Fig 4). The biosynthetic pathway of chitin can be thought of as consisting of three sub-reactions. The first set leads to the formation of the amino sugar, GlcNAc, the second to its activated form UDP-GlcNAc, and the last yields the polymeric chitin from the amino sugar.

The source of the sugar residues for chitin synthesis can be traced to fat body glycogen, which is acted upon by glycogen phosphorylase, Glucose-1-P produced by this reaction is converted to trehalose. Trehalose, the extracellular source of sugar is acted upon by a trehalase. The conversion of glucose-6-P to fructose-6-P needed for chitin synthesis involves two glycolytic enzymes present in the cytosol. These enzymes are hexokinase and glucose-6-P isomerase, which convert glucose-6-P to fructose-6-P. From the latter, the chitin biosynthetic pathway branches off, with the first enzyme catalyzing this branch being glutamine fructose-6-phosphate aminotransferase (GFAT, E. C. 2.6.1.16), which might be thought of as the first committed step in amino sugar biosynthesis. The conversion of fructose-6-P to GlcNAc phosphate involves amination, acetyl transfer, and an isomerization step, which moves the phosphate from C-6 to C-1 (catalyzed by a phospho-N-acetylglucosaminemutase). The conversion of this compound to the nucleotide sugar derivative follows the standard pathway and leads to the formation of an UDP-derivative of GlcNAc, which serves as the substrate for (chitin synthases) CHS. The entire chitin biosynthetic pathway is outlined in Figure 2. At this point, this possibility remains unproven. Similarly, the requirement for a primer to which the GlcNAc residues can be transferred also remains speculative.

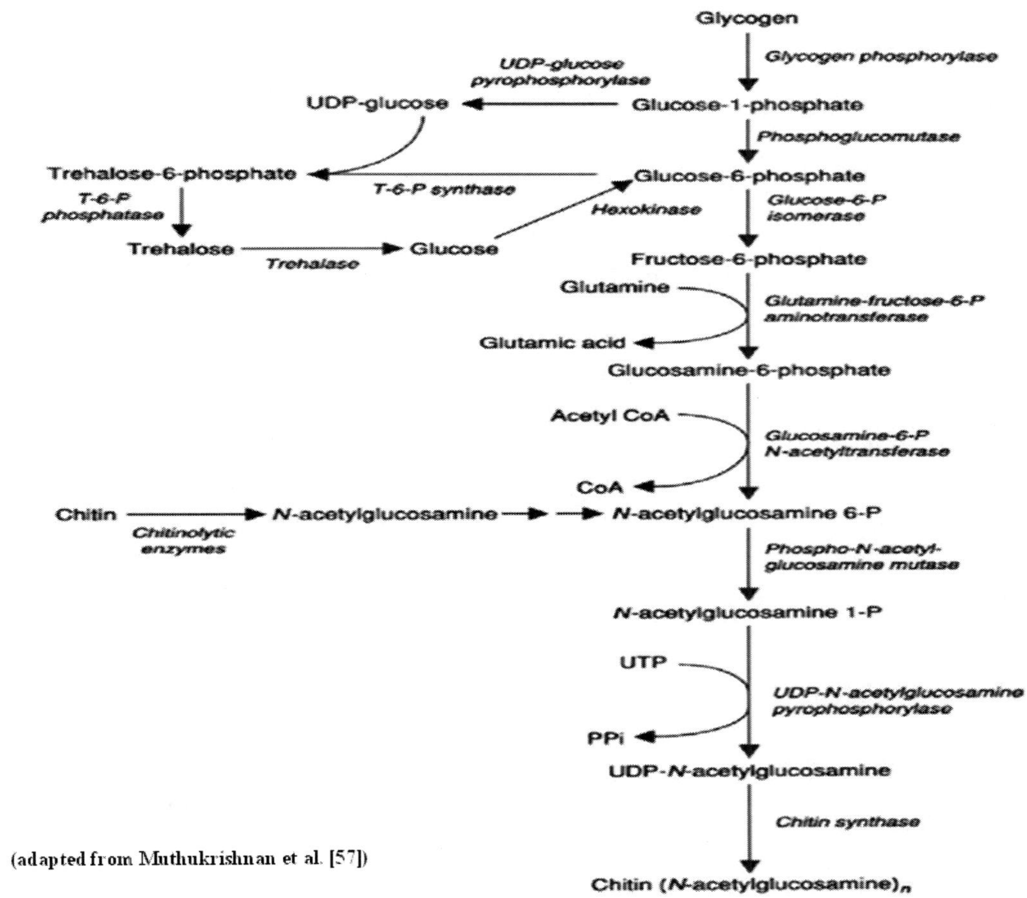

Figure 4. Chitin Biosynthetic Pathway.

5. Biotechnological Applications

Chitinases produced from the microbial source have many applications like working as a Biocontrol agent, Bioinsecticide, Pharmaceuticals and food industry. The use of chitinases for the biocontrol of plant pathogens, and for developing transgenic plants is one of the major applications. Chitinases can also be employed in human health care, such as making opthalmic preparations with chitinases and microbicides.

5.1. Biocontrol Agents

The chitin is a structural component of the cell wall of phytopathogenic and entomopathogenic microorganisms. Chitin has been considered as a target molecule to be attack by biological agents that might contain chitinases, chitin synthesis inhibitors or microorganisms as antagonists. For instance the chitinases from *Myrothecium verrucaria* that degrades the cuticle of the mosquito *Aedes aegypti*; and those from *B. bassiana* that attacks *Galleria mellonella* and *Trichoplusia ni.* Another example Skujins et al. [37] were first to observe *Streptomyces* chitinase that effectively lyse the hyphal wall of *Aspergillus oryzae* and *Fusarium solani.* Defence mechanism was observed when chitinase produced from micro-organisms was inserted into the rhizobacterial soil. In agriculture, chitinase-producing organisms act as an effective biocontrol agent against a number of phytopathogenic fungi [7].

5.2. Bioinsecticides

The insect molting enzyme, chitinase has been described from *Bombyx mori*, *Manduca sexta,* and several other insects. The pseudotrisaccharide, allosamidin, is a potent inhibitor of chitinases from most of the sources. The allosaminidin was found to be inhibitory after ingestion to the growth of mite, *Tetranychus urticae* and a larva of a housefly, *M. domestica* [38]. On the other hand chitinase inhibitors can be explored as potential biopesticides. Chitinases used in mosquito control, It has been seen that instar larvae of a mosquito, *A. aegypti*, can be killed within 48 h with the help of the crude preparation from *M. verrucaria* [39]. Though 100% mortality was observed with the crude enzyme (170-mg protein per liter) preparation within 48 h, with purified endo-chitinase lethal time (LT50) was 48 h and 120 h instar larvae, respectively. However, the time period was found to be decreased correspondingly to 24 h and 48 h when the purified chitinase was supplemented with lipolytic activity.

5.3. Plant Immunity

It is known that the plants have a complex defensive system of chitinases in combination of β-1, 3 glucanases to be protected by the attack of phytopathogen chitin-containing fungi by causing lysis of hyphal tips because the apex has the exposed nascent chitin chains that are accessible to hydrolysis. The cellular and secretory chitinases in plant presumably have different roles in defense reactions. Apoplastic chitinases are considered to be part of an early, induced response as these chitinases act directly by blocking the growth of the hyphae invading the intercellular space [40]. Further, it is indicated that apoplastic chitinases act possibly indirectly also by releasing fungal elicitors which in turn trigger downstream defense pathways. During the infection process, cellular chitinases are released when hyphae penetrate and affect cell integrity [41]. Chitinases are induced in *Brassica napus* after infection with *Sclerotinia sclerotiorum* [42]. The levels of chitinases in rice cultivars correlated with resistance to sheath blight pathogen *Rhizoctonia solani* [43].

5.4. Medical Application

It was found that human gastric juice contains chitin in it, but its function is still a mystry. Renkema et al. [44] described that patients affected by Gaucher disease had high levels of chitotriosidase in its plasma. Recent evidence shows that Chit was involved in innate immune responses and its concentration was elevated in diseases such as acute malaria, β-thalassemia and other hemoglobinopathies [45, 46]. Chitin is absent in mammals but sometimes chitin like structures may be present. These are T helper cells that induce type-1 response in mammals. A member of the mammalian chitinase family Chitinase 3-like-1 (CHI3L1), is induced particularly when inflammation such as inflammatory bowel disease, hepatitis and asthma occurs. The exact function of CHI3L1 in inflammation and cancer is not known but plays a crucial role in exacerbating the inflammatory processes, promoting angiogenesis and remodeling of the extracellular matrix. CHI3L1 has recently been proposed as a significant marker for indicating inflammatory activity High levels of CHI3L1 have been reported in inflammatory conditions such as IBD [Crohn's disease (CD) and ulcerative colitis (UC)], asthma and liver cirrhosis in the serum of patients [48, 49, 51].

5.5. Food Industry

Chitosan can be utilized as a tool for the purification of waste water because of its high sorption capacity. The capacity of chitin and chitosan to form complexes with metal ions has been exploited in Japan for water purification. The use of commercially available chitosan for potable water purification has been approved by the United States Environmental Protection Agency (USEPA) up to a maximum level of 10 mg/L [50]. Storage life of fruits and better control of decay of peaches could be increased for long period of time after coating with chitosan. Chen et al. [51] have observed that the packaging film prepared from methylcellulose, chitosan and preservatives possesses antimicrobial activity. Chitosan salts, which carry a strong positive charge, have been shown to be effective as dehazing agents; they may also be used to control acidity in fruit juices. Chitinase-producing organisms are also effectively used in the bioconversion process to treat shellfish waste and also to obtain value-added products from such wastes [52].

5.6. Oligosaccharides Production

Kobayashi et al. [53] however, reported use of *Bacillus* chitinase for the synthesis of chitobiose by combining GlcNAc and a sugar oxazoline derivative. The GlcNAc itself also is an anti-inflammatory drug. In the human body it is synthesized from glucose and is incorporated in to glycoproteins and glycosaminoglycans. Chito-oligosaccharides acts as elicitors of plant defense, involved in the signaling for root nodule formation, and are potentially useful in human medicine too. For example, chitohexaose and chitoheptaose showed anti-tumor activity. A chitinase from *Vibrio alginolyticus* was used to prepare chitopentaose and chitotriose from colloidal chitin [54]. *N*, *N'*-Diacetylchitobiose has been widely used as a starting material for synthesis of biologically active compounds.

5.7. Single Cell Protein

Single cell protein from chitin had been reported, however there is a low added value in the products of fermentation as it has a high content of nucleic acids, and sometimes limited nutritional and sensorial properties. There are four steps in production of single cell protein from shellfish waste. These are;

i Drying, size reduction and chemical purification of shellfish processing waste;
ii Extraction of chitinase enzyme from purified waste;
iii Chitin hydrolysis; and
iv Fermentation in submerged culture.

In this approach chitinases of *Serratia marcescens* were used to bioconvert chitin to aminosugars, mainly N-Acetylglucosamine that was used for growth of the yeast *Pichia kudriavzevii*, producing 45% of protein and 8 to 11% of nucleic acids [55].

Conclusion and Future Trends

Conventionally, Chitinases were produced by fermentation methods, but now it is improved by optimization process of culture conditions. In nature, there are many organisms that are a good source of chitinolytic enzymes, but strain improvement may play an important role in increasing the production of chitinase. Strain improvement is obtained by mutagenesis and selection. At present, possibility for industrial development of microorganisms and their products is shown by recombinant DNA technology. Usually, strain improvement is done to increase the production of chitinase enzyme which was obtained by mutagenesis and selection process. These days recombinant DNA technology is used for strain improvement. For the production of high amounts of pure chitinases having better binding properties and better activity can be achieved by combining the chitinase gene together with crop condition and protein engineering. Increasing the activity of strain boost the intensity of transcription of genes that encodes hydrolytic enzymes, these improved strains act as biocontrol agents or producers that are having superior chitinolytic properties that are essential for industrial use.

References

[1] Melchers LS, Stuiver MH (2000) Novel genes for Disease resistance breeding. *Curr Op Plant Biol* 3:147
[2] Spindler KD, Spindler-Barth M (1996) Chitin degradation and synthesis in arthropods. Giraud-Guille MM (ed) *Chitin in life sciences, J AndreÂ Publisher*, Paris, pp. 41-52
[3] Fukazimo T, Kramer KJ (1985) Mechanism of chitin hydrolysis by the binary chitinase system in insect moulting fluid. *Insect Biochem* 15:141-145
[4] Henrissat B, Bairoch A (1993) New families in the classification of glycosyl hydrolases based on amino acid sequence similarities. *Biochem J* 293(Pt 3):781-788

[5] Shanmugaiah N, Mathivanan N, Balasubramanian, Manoharan PT (2008) Optimization of Cultural conditions for production of Chitinase by *Bacillus laterosporous* MMI2270 isolated from rice rhizosphere soil. *Afr J Biotechnol* 7(15):2562-2568

[6] Watanabe T, Kobori K, Miyashita K, Fujii T, Sakai H, Uchida M, Tanaka H (1993) Identification of glutamic acid 204 and aspartic acid 200 in chitinase A1 of *Bacillus circulans*WL-12 as essential residues for chitinase activity. *J Biol Chem* 268:18567-18572

[7] Ajit NS, Verma R, Shanmugam V (2006) Extracellular chitinases of fluorescent pseudomonads antifungal to *Fusarium oxysporum* f. sp. *Dianthi* causing carnation wilt. *Curr Microbiol* 52:310-316

[8] Watanabe T, Oyanagi W, Suzuki K, Tanaka H (1990) Chitinase system of *Bacillus circulans* WL-12 and importance of chitinase A1 in chitin degradation. *J Bacteriol* 172:4017-4022

[9] Shekhar N, Bhattacharya D, Kumar D, Gupta RK (2006) Biocontrol of wood-rotting fungi using *Streptomyces violaceusniger* XL-2. *Can J Microbiol* 52:805-808

[10] Tsujibo H, Minoura K, Miyamoto K, Endo H, Moriwaki M, Inamori Y (1993) Purification and properties of a thermostable chitinase from *Streptomyces thermoviolaceus* OPC-520. *Appl Environ Microbiol* 59:620-622

[11] Neeraja C, Moerschbacher B, Podile AR (2009) Swapping of the chitin-binding domain in *Bacillus* chitinases improves the substrate binding affinity and conformational stability. *Mol Biosyst* DOI:10.1039/B923048C

[12] Huang CJ, Guo SH, Chung SC, Lin YJ, Chen CY (2009) Analysis of the involvement of chitin-binding domain of ChiCW in antifungal activity, and engineering a novel chimeric chitinase with high enzyme and antifungal activities. *J Microbiol Biotechnol* 19:1169-1175

[13] Someya N, Kataoka N, Komagata T, Hirayae K, Hibi T, Akutsu K (2000) Biological control of cyclamen soil borne diseases by *Serratia marcescens. Plant Dis* 84:334-340

[14] El-Tarabily KA, Soliman MH, Nassar AH, Al-Hassani HA, Sivasithamparam K, McKenna F, Hardy GESTJ (2000) Biological control of *Sclerotinia minor* using a chitinolytic bacterium and actinomycetes. *Plant Pathol* 49:573-583

[15] Hamel F, Boivin R, Temblay C, Bellemare G (1997) Structural and evolutionary relationships among chitinases of flowering plants. *J Mol Evol* 44:614-624

[16] Tronsmo A, Hjeljord L, Klemsdal SS, Varum KM, Nordtveit Hjerde R, Harman GE (1996) Chitinolytic enzymes from the biocontrol agent *Trichoderma harzianum.* In: *Chitin Enzymology*. R.A.A. Muzzarelli (ed). Atec Edizioni, Italia. 2:235-244

[17] Horsch M, Mayer C, Sennhauser U, Rast DM (1997) Beta-N-acetylhexosaminidase: a target for the design of antifungal agents. *Pharmacol Ther* 76(13):187-218

[18] Brurberg MB, Nes IF, Eijsink VGH (1996) Comparative studies of chitinases A and B from *Serratia marcescens*. *Microbiol* 142:1581-1589

[19] Kuranda MJ, Robbins PW (1991) Chitinase is required for cell-separation during growth of *Saccharomyces cerevisiae*. *J Biol Chem* 266:19758-19767

[20] Sakuda S, Isogai A, Matsumoto S, Suzuki A, Koseki K (1986) The structure of allosamidin, a novel insect chitinase inhibitor produced by *Streptomyces* sp. *Tetrahed Lett* 27:2475-2478

[21] Sakuda S, Nmoto Y, Ohi M, Watanabe M, Takayama S, Isogai A, Yamada Y (1990) Effects of demethylallosamidin, a potent yeast chitinase inhibitor, on the cell-division of yeast. *Agric Biol Chem* 54:1333-1335

[22] Gokul B, Lee JH, Song KB, Rhee SK, Kim CH, Panda T (2000) Characterization and applications of chitinases from *Trichoderma harzianum*-A Review. *Bioprocess Eng.* 23(6):691-694

[23] Lorito M, Woo SL, Fernandez IG, Colucci G, Harman GE, Pintor-Toro JA, Filippone E, Muccifora S, Lawrence CB, Zoina A, Tuzun S, Scala F (1998) Genes from mycoparasitic fungi as a source for improving plant resistance to fungal pathogens. *Proc Natl Acad Sci USA* 95:7860-7865

[24] Kitomoto Y, Mori N, Yamamoto M, Ohiwa T, Ichiwaka Y (1988) A simple method of protoplast formation and protoplast regeneration from various fungi using an enzyme from *Trichoderma harzianum*. *Appl Microbiol Biotechnol* 28:445-450

[25] Vyas PR, Deshpande MV (1991) Enzymatic hydrolysis of chitin by *Myrothecium verrucaria* chitinase complex and its utilization to produce SCP. *J Gen Appl Microbiol* 37:267-275

[26] Cody RM, Davis ND, Lin J, Shaw D (1990) Screening microorganisms for chitin hydrolysis and production of ethanol from amino sugars. *Biomass* 21:285-295

[27] Carstens M, Vivier MA, Pretorius IS (2003) The *Saccharomyces cerevisiae* chitinase, encoded by the CTS1-2 gene, confers antifungal activity against *Botrytis cinerea* to transgenic tobacco. *Trans Res* 12:497-508

[28] Kumar D, Gupta RK (2006) Biocontrol of wood-rotting fungi. *Ind J Biotechnol* 5:20-25

[29] Yuan WM, Crawford DL (1995) Characterization of *Streptomyces lydicus*WYEC108 as a potential biocontrol agent against fungal root and seed rots. *Appl Env Microbiol* 61:3119- 3128

[30] Mahadevan B, Crawford DL (1997) Properties of the chitinase of the antifungal biocontrol agent *Streptomyces lydicus*WYEC108. *Enz Microb Technol* 20:489-493

[31] Terayama H, Takahashi S, Kuzuhara H (2005) Large scale preparation of *N, N'*-diacetylchitobiose by enzymic degradation of chitin and its chemical modifications. *J Carbohydr Chem* 12:81-93

[32] El-Sayed ESA, Ezzat SM, Ghaly MF, Mansour M, El-Bohey MA (2000) Purification and characterization of two chitinases from *Streptomyces albovinaceus* S-22. *World J Microbiol Biotechnol* 16:87-89

[33] Mitsutomi M, Hata T, Kuwahara T (1995) Purification and characterization of novel chitinases from *Streptomyces griseus* HUT 6037. *J Ferment Bioeng* 80:153-158

[34] Van Aalten DM, Komander D, Synstad B, Ga°seidnes S, Peter MG, Eijsink VJ, Weirenga RK (2001) Structural insights into the catalytic mechanism of a family 18 exo-chitinase. *Proc Natl Acad Sci USA* 98:8979-8984

[35] Suzuki K, Sugawara N, Suzuki M, Uchiyama T, Katouno F, Nikaidou N, Watanabe T (2002) Chitinases A, B, and C1 of *Serratia marcescens* 2170 produced by recombinant Escherichia coli: enzymatic properties and synergism on chitin degradation. *Biosci Biotechnol Biochem* 66:1075-1083

[36] Suzuki K, Taiyoji M, Sugawara N, Nikaidou N, Henrissat B, Watanabe T (1999) The third chitinase gene (chiC) of *Serratia marcescens* 2170 and the relationship of its product to other bacterial chitinases. *Biochem J* 343:587-596

[37] Skujins JJ, Potgieter HJ, Alexander M (1965) Dissolution of fungal cell walls by *Streptomyces* chitinase and β-1, 3-glucanase. *Arch Biochem Biophys* 111:358-364

[38] Sakuda S (1996) Studies on the chitinase inhibitors, allosamidins. In: Muzzarelli RAA, editor. *Chitin Enzymology, Vol. 2*. Grottammare: Eur. Chitin Soc. p. 203-12

[39] Mendonsa ES, Vartak PH, Rao JU, Deshpande MV (1996) An enzyme from *Myrothecium verrucaria* that degrades insect cuticles for biocontrol of *Aedes aegypti* mosquito. *Biotechnol Lett* 18:373-376

[40] Gerhardt LB, Sachetto-Martins G, Contarini MG, Sandroni M, de Lima VM, Cordeiro MC, de Oliveira DE, Margis-Pinheiro M (1997) *Arabidopsis thaliana* class IV chitinase is early induced during the interaction with *Xanthomonas campestris*. *FEBS Lett* 419:69-75

[41] Collinge DB, Kragh KM, Mikkelsen JD, Nielsen KK, Rasmussen U, Vad K (1993) Plant chitinases. *Plant J* 3:31-40

[42] Zhao J, Wang J, An L, Doerge RW, Chen ZJ, Grau CR, Meng J, Osborn TC (2007) Analysis of gene expression profiles in response to *Sclerotinia sclerotiorum* in *Brassica napus*. *Planta* 227:13-24

[43] Shrestha CL, O□na I, Muthukrishnan S, Mew TW (2007) Chitinase levels in rice cultivars correlate with resistance to the sheath blight pathogen *Rhizoctonia solani*. *Eur J Plant Pathol* 120:69-77

[44] Renkema GH, Boot RG, Au FL, Donker- Koopman WE, Strijland A, Muijsers AO, Hrebicek M, Aerts JM (1998) Chitotriosidase, a chitinase, and the 39-kDa human cartilage glycoprotein, a chitin-binding lectin, are homologues of family 18 glycosyl hydrolases secreted by human macrophages. *Eur J Biochem* 251:504-509

[45] Barone R, Simporé J, Malaguarnera L, Pignatelli S, Musumeci S (2003) Plasma chitotriosidase activity in acute *Plasmodium falciparum* malaria. *Clin Chim Acta* 331: 79-85

[46] Barone R, Bertrand G, Simporé J, Malaguarnera M, Musumeci S (2001) Plasma chitotriosidase activity in beta-thalassemia major: a comparative study between Sicilian and Sardinian patients. *Clin Chim Acta* 306:91-96

[47] Vind I, Johansen JS, Price PA, Munkholm P (2003) Serum YKL-40, a potential new marker of disease activity in patients with inflammatory bowel disease. *Scand J Gastroenterol* 38:599-605

[48] Ober C, Tan Z, Sun Y, Possick JD, Pan L, Nicolae R, Radford S, Parry RR, Heinzmann A, Deichmann KA, Lester LA, Gern JE, Lemanske RF Jr, Nicolae DL, Elias JA, Chupp GL (2008) Effect of variation in CHI3L1 on serum YKL-40 level, risk of asthma, and lung function. *N Engl J Med* 358:1682-1691

[49] Johansen JS, Moller S, Price PA, Bendtsen F, Junge J, Garbarsch C, Henriksen JH (1997) Plasma YKL-40: a new potential marker of fibrosis in patients with alcoholic cirrhosis. *Scand J Gastroenterol* 32:582-590

[50] Knorr D (1984) *`Use of Chitinous Polymers in Food-A Chal- lenge for Food Research and Development' in Food Technol* 38:85-97

[51] Chen M, Yeh GH, Chiang B (1996) `Antimicrobial and Physicochemical Properties of Methylcellulose and Chitosan Films Containing a Preservative'. *J Food Proc Preserv* 20:379-390

[52] Carroad PA, Tom RA (1978) Bioconversion of shellfish waste process conception and selection of microorganisms. *J Food Sci* 43:1158-1161

[53] Kobayashi S, Kiyosada T, Shoda S-I (1997) A novel method for synthesis of chitobiose via enzymatic glycosylation using a sugar oxazoline as glycosyl donor. *Tetrahedron Lett* 38:2111-2112

[54] Murao S, Kawada T, Itoh H, Oyama H, Shin T (1992) Purification and characterization of a novel type of chitinase from *Vibrio alginolyticus*. *Biosci Biotech Biochem* 56:368-369

[55] Revah-Moiseev S, Carroad PA (1981) Conversion of enzymatic hydrolysate of shellfish waste chitin to SCP. *Biotechnol Bioeng* 23:1067-1078

[56] Li H, Greene LH (2010) Sequence and Structural Analysis of the Chitinase Insertion Domain Reveals Two Conserved Motifs Involved in Chitin-Binding. *PLoS ONE* 5(1): e8654. doi:10.1371/journal.pone.0008654

[57] Muthukrishnan S, Morzendorfer H, Arakane Y, Kramer KJ (2012) Chitin metabolism in insects. *In*: *Insect molecular biology and biochemistry*, *Ed*: Gilbert L.I. Academic press; Elsevier B.V. UK. pp 193

In: Microbes in Process
Editors: Neelam Garg and Abhinav Aeron
ISBN: 978-1-63117-127-7

Chapter 7

Fungal Laccases: A Promising Eco-Friendly Approach for the Removal of Pharmaceutical Compounds

Susana Rodríguez-Couto [*1,2]
[1]CEIT, Unit of Environmental Engineering, Paseo Manuel de Lardizabal, San Sebastian, Spain
[2]IKERBASQUE, Basque Foundation for Science, Bilbao, Spain

Abstract

In the last decade, advances in analytical chemistry have made possible the detection of very low concentrations of chemicals in environmental samples thus detecting the presence of an ever-increasing number of pharmaceutical compounds. This causes serious concern due to the potential adverse effects of such compounds on the ecosystem and human health. The existing technologies to treat wastewater are either inefficient or costly for the treatment of pharmaceutical compounds which has impelled the search for new technological approaches. In this sense, fungal laccases have attracted recent interest due to both their broad substrate specificity and their simple catalysis requirements. This chapter reviews the latest published literature dealing with the removal of pharmaceutical compounds by laccase-producing fungi or their laccases.

1. Introduction

Pharmaceutical compounds have been receiving increasing attention as potential bioactive chemicals in the environment in recent years [1], particularly in the industrialised countries. Such compounds belong to the so-called emerging pollutants because they still remain unregulated or are under a regularisation process [2, 3]. These pollutants and their bioactive metabolites do not need to be persistent in the environment to cause adverse effects

* srodriguez@ceit.es, Tel.: +34 943 212800 (ext. 2239).

since their high transformation/removal rates can be counteracted by their continuous entry into the aquatic environment [4], where they are detected at trace concentrations (usually in the range of μg/L or ng/L).

The use of pharmaceutical compounds has been increased in the last decade, especially those that do not need medical prescription. After their intake, the non-metabolised active substances of these compounds are excreted in urine and faeces and, thus, entering into the water cycle. In addition to pharmaceutical production sites, hospitals are also important sources of pharmaceutical residues. Also, pharmaceutical residues can be found in wastewater treatment plant (WWTP) effluents due to their inefficient removal by the conventional systems [5, 6, 7, 8]. For example, in Italy, France, Greece and Sweden, carbamazepine, clofibrate, phenazone, aminopyrine, clofibric acid, diclofenac, fenofibrate, fenoprofen, flurbiprofen, gemfibrozil, ibuprofen, ketoprofen and naproxen were found in the effluents of sewage treatment plants [9]. These effluents containing such bioactive molecules are discharged into receiving surface waters, mainly rivers [9]. Several reviews have reported the occurrence and fate of pharmaceutical compounds in the environment, where they are ubiquitous in soil and water [1, 2, 10, 11, 12]. The presence of such compounds in the environment is being cause of concern in recent times due to their potential adverse effects on the ecosystem and human health. Therefore, pharmaceutical compounds must be removed prior to their discharge into the environment.

Figure 1. The White-rot-fungi *Trametes pubescens* (left) and *Trametes versicolor* (right) as they Grow in Nature (photographs reprinted from http://commons.wikimedia.org/).

The removal of pharmaceutical compounds is a challenge task due to: (i) the non-existence of limiting regulations for water and wastewater treatment, (ii) some of these compounds have been classified as endocrine disrupter compounds (EDCs) and (iii) their measurement and detection is difficult, since they are discharged into the environment at very low concentrations. The physicochemical processes such as coagulation-flocculation, flotation, sedimentation and filtration have shown to be inefficient in the removal of pharmaceutical compounds [13]. Advanced Oxidation Processes (AOPs), such as UV/H_2O_2, Fenton's reagent (H_2O_2+ Fe^{2+}), O_3/UV, O_3/H_2O_2 and photo-catalysis, have shown to be effective but costly. This has impelled the search for alternative approaches to remove such compounds. In this regard fungal laccases have attracted recent interest due to their broad substrate specificity and their simple catalysis requirements, since they use molecular oxygen (easily available in the environment) to oxidise the substrate and produce water as the only by-product. This latter feature renders laccases as eco-friendly catalysts.

Laccases (p-diphenol:oxygen oxidoreductase, EC 1.10.3.2) belong to a group of enzymes called blue multi-copper oxidases. They are often found in white-rot fungi (Figure 1), which are the only organisms able to degrade the whole wood components [14]. Thus, one of the roles of laccases in nature is to degrade the bulky recalcitrant polymer lignin (Figure 2) in order to gain access to the other carbohydrates in wood (cellulose and hemicellulose). Their low substrate specificity allows laccases degrading compounds with a structure similar to lignin, such as polyaromatic hydrocarbons (PAHs), textile dyes and other xenobiotic compounds [16], and hence their industrial interest.

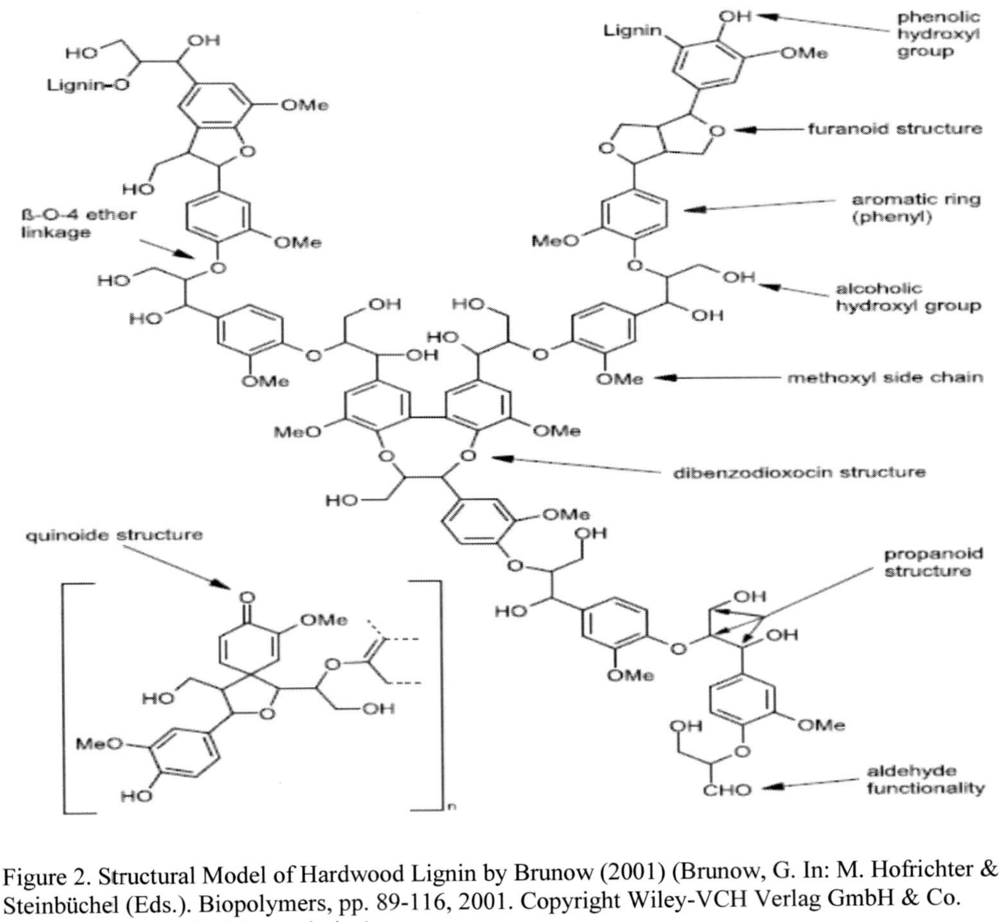

Figure 2. Structural Model of Hardwood Lignin by Brunow (2001) (Brunow, G. In: M. Hofrichter & A. Steinbüchel (Eds.). Biopolymers, pp. 89-116, 2001. Copyright Wiley-VCH Verlag GmbH & Co. KGaA. Reproduced with permission).

According to my knowledge there are no review articles dealing with the removal of pharmaceutical compounds by fungal laccases. This chapter reviews the published literature up today dealing with this topic in order to highlight the great potential of laccases in the treatment of these important emerging micro-pollutants.

2. Fungal Laccases

Laccase (benzenediol:oxygen oxidoreductase; p-diphenol oxidase EC 1.10.3.2) was first discovered by Yoshida [17] in the exudates of the Japanese lacquer tree *Rhus vernicifera*, from which the name laccase was taken, and characterised as a metal containing oxidase by Bertrand [18]. Fungal laccases were also discovered during the 19th century [19, 20]. Among fungi basidiomycetes, ascomycetes and deuteromycetes can produce laccase, the white-rot fungi (Figure 1) being the most efficient laccase producers [21].

2.1. Mechanism of Catalysis

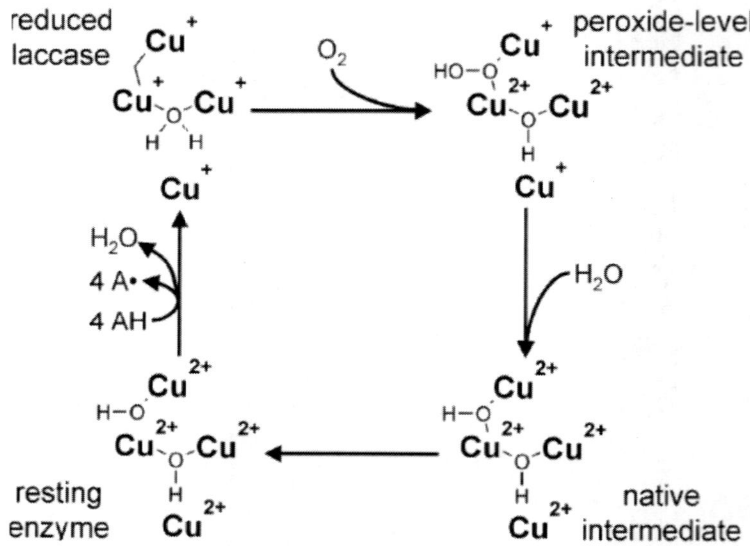

Figure 3. Catalytic Cycle of a Four-Copper Laccase. Reprinted from Biotechnology Advances 22, Wesenberg D, Kyriakides I, Agathos SN. Autoreduction and aggregation of fungal laccase in solution phase: possible correlation with a resting form of laccase, 161-187. Copyright (2003), with permission from Elsevier Ltd., UK.

Laccases contain four copper atoms in different sites, which are classified into three categories according to their spectroscopic characteristics [22]: one Type 1 (T1) copper showing an intense band at 600 nm, which is responsible for the blue colour of the enzyme, and is EPR (electronic paramagnetic resonance) detectable, one Type 2 (T2) copper without observable bands in the absorption spectra but it is EPR detectable and two Type 3 (T3) coppers, which show an absorption spectra at 330 nm but they are no EPR detectable [23]. T2 and T3 coppers form a tri-nuclear cluster. Figure 3 shows the catalytic cycle of laccase enzyme [24]. Laccase catalysis occurs due to the reduction of one oxygen molecule to water together with the oxidation of one electron of a wide range of aromatic compounds such as polyphenols [25], methoxy-substituted monophenols and aromatic amines [26]. This occurs in three steps: (1) T1 copper reduction by substrate; (2) electron transfer from the T1 copper to the T2 and T3 copper tri-nuclear cluster; (3) reduction of oxygen to water at the tri-nuclear cluster [27].

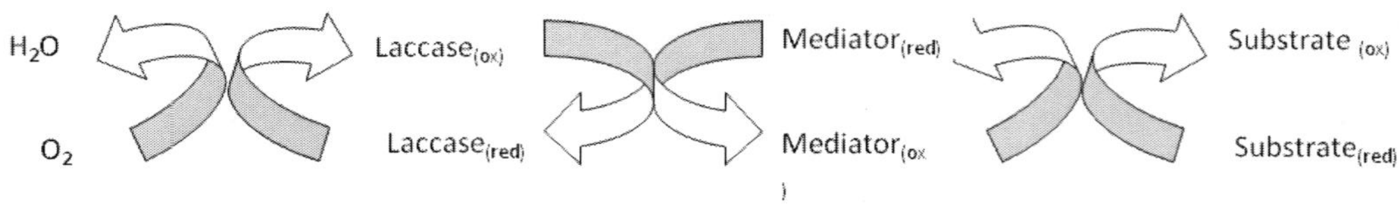

Figure 4. Schematic Representation of Laccase Mediator System (LMS) for Substrate Oxidation (adapted from Riva 2006).

synthetic

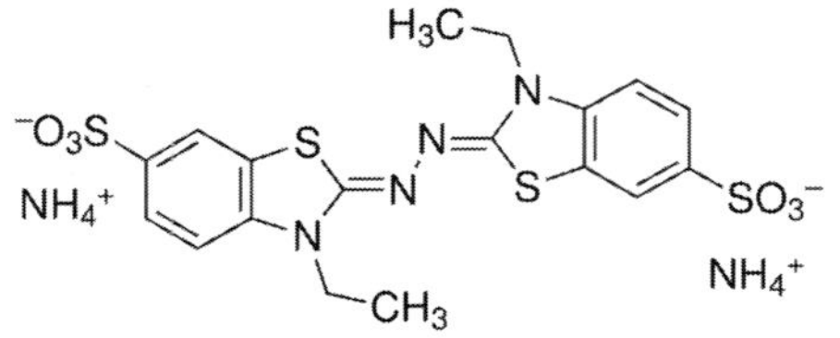

2,2'- azine-bis (3-ethylbenzothiazoline-6-sulfonic acid (ABTS)

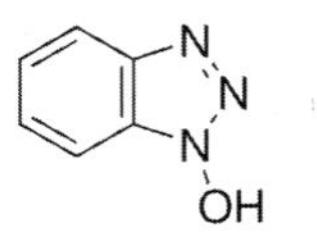

1-hydroxybenzotriazole (HBT)

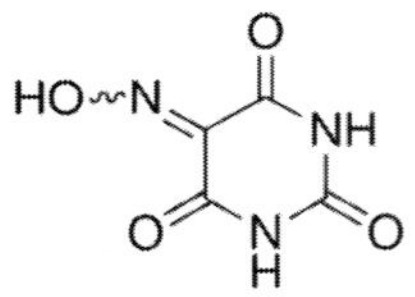

violuric acid (2,4,5,6(1*H*,3*H*)-pyrimidinetetrone 5-oxime ; 5-isonitrosobarbituric acid; alloxan-5-oxime)

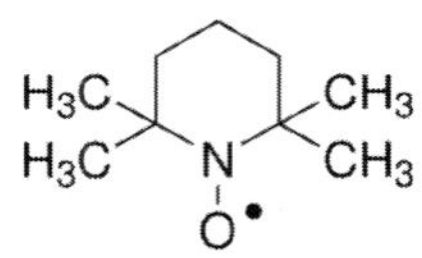

TEMPO (2,2,6,6-tetramethyl-1-piperidinyloxy;2,2,6,6-tetramethylpiperidine 1-oxyl)

natural

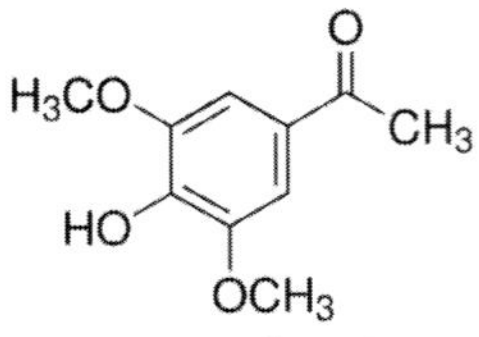

acetosyringone (4'-hydroxy-3',5'-dimethoxyacetophenone)

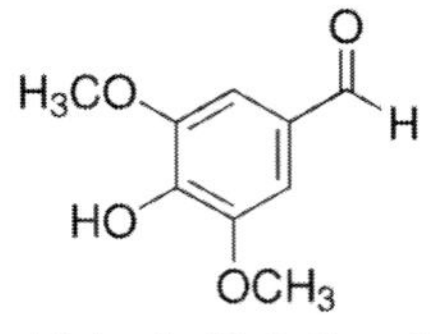

syryngaldehyde (3,5-dimethoxy-4-hydroxybenzaldehyde; 4-Hydroxy-3,5-dimethoxybenzaldehyde)

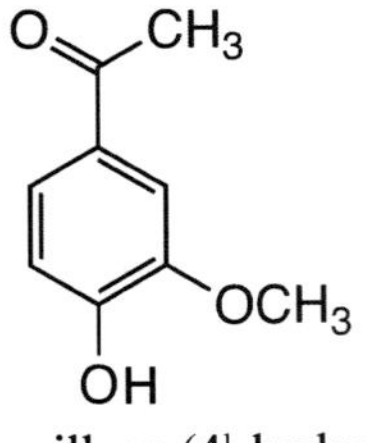

acetovanillone (4'-hydroxy-3'-methoxyacetophenone)

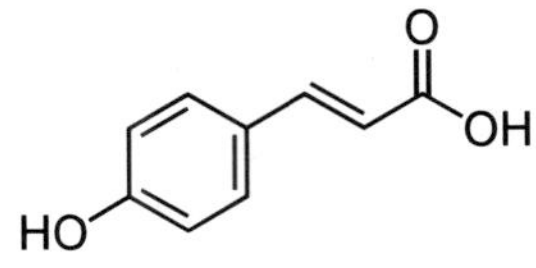

p- coumaric acid (*trans*-4-hydroxycinnamic acid)

Figure 5. Chemical Structures of Several Synthetic and Natural Redox Mediators.

The laccase-mediated catalysis can be extended to non-phenolic substrates by the inclusion of the so-called redox mediators. These redox mediators are a group of low-molecular weight organic compounds that are oxidised by laccase forming highly active radical cations capable of oxidising non-phenolic compounds that laccase alone cannot oxidise (Figure 4). More than 100 redox mediators have been described but the most commonly used are 2, 2'- azine-bis (3-ethylbenzothiazoline-6-sulfonic acid (ABTS) and 1-hydroxybenzotriazole (HBT). However, these synthetic mediators are toxic and costly which has impelled the search for natural ones. Thus, Camarero et al. [29] reported that several lignin-derived phenols (such as syringaldehyde and acetosyringone) represented eco-friendly alternatives to synthetic mediators for the degradation of different types of dyes and other recalcitrant compounds by laccase in terms of both efficiency and rate of oxidation. Figure 5 shows the chemical structures of several synthetic and natural redox mediators.

2.2. Properties

The biochemical and catalytic properties of fungal laccases have been comprehensively reviewed by Baldrian [30]. Typical fungal laccases have a molecular mass of 60-70 kDa. Most fungal laccases are monomeric proteins but some exhibit homodimeric, heterodimeric or oligomeric structures. Several laccase isoenzymes have been detected in many fungal species which vary with both fungal species and environmental conditions. Like most extracellular fungal enzymes laccases are glycoproteins [30]. Glycosylation of fungal laccase usually constitutes between 15 and 25% of the molecular mass and may contribute to the high stability of the enzyme [31].

The isoelectric point of laccases is around 4 and the optimal pH depends on the substrate used to test the enzymatic activity. Usually, fungal laccases exhibit pH optima in the range of 3-5, when the substrate is a hydrogen atom donor compound (e.g. ABTS). When the substrate is a phenolic compound (e.g. syringaldazine), the optimal pH is shifted to 6-7. This shift in pH is a result of the balance of redox potentials between the substrate and the inhibition of the T2/T3 copper site by binding of an OH^- ion. The optimal temperature differs with the source of laccase but usually varies from 50 to 70°C [30].

3. Occurrence and Fate of Pharmaceutical Compounds

In recent years the occurrence and fate of pharmaceutical active compounds in the environment is becoming a matter of serious concern due to their potential adverse effects on living organisms, including humans. The main classes of human pharmaceuticals are shown in Figure 6 [32].

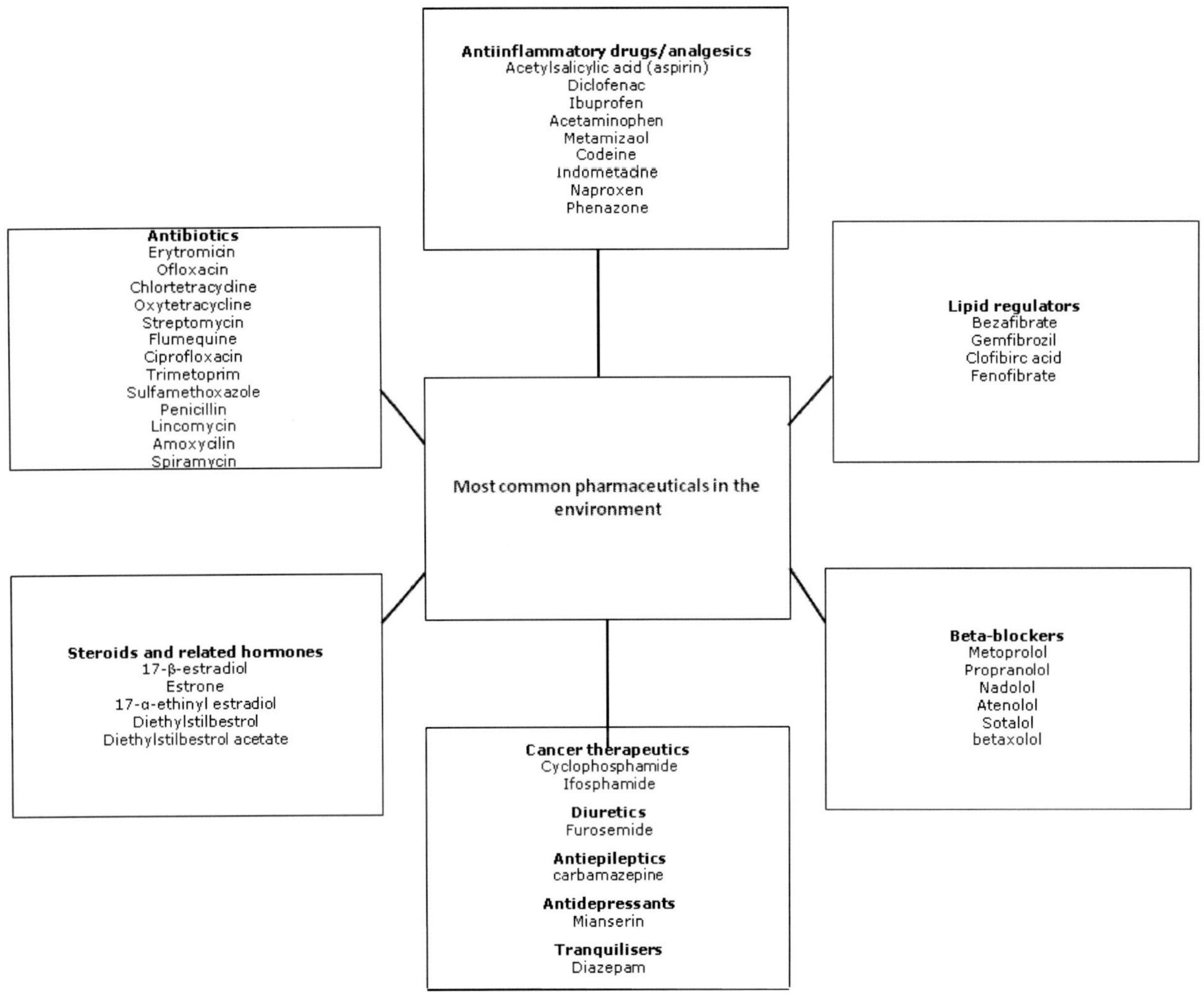

Figure 6. The Most Common Pharmaceutical Compounds in the Environment (extracted from Nikolaou et al. [32]).

Among them, diclofenac, clofibric acid, acetaminophen, ibuprofen, aspirin, carbamazepine, artorvasatin, gemfibrozil, fluoxetine and 17β-ethynylestradiol have become ubiquitous in surface waters and wastewater. Figure 7 shows the chemical structures of several commonly used pharmaceutical compounds. Pharmaceutical active compounds are continuously discharged into the environment as a consequence of human and veterinary medicine. Thus, a wide variety of such compounds has been detected at low concentrations world-wide in samples from sewage-treatment plant effluents, surface water, groundwater [33, 34], drinking water [35, 36, 37], sediment and soil [38]. Figure 8 shows possible sources and pathways for the occurrence of pharmaceutical residues in the aquatic environment [34]. Figure 9 collects the concentrations of different pharmaceutical compounds detected in the influent and effluent of several WWTPs [39].

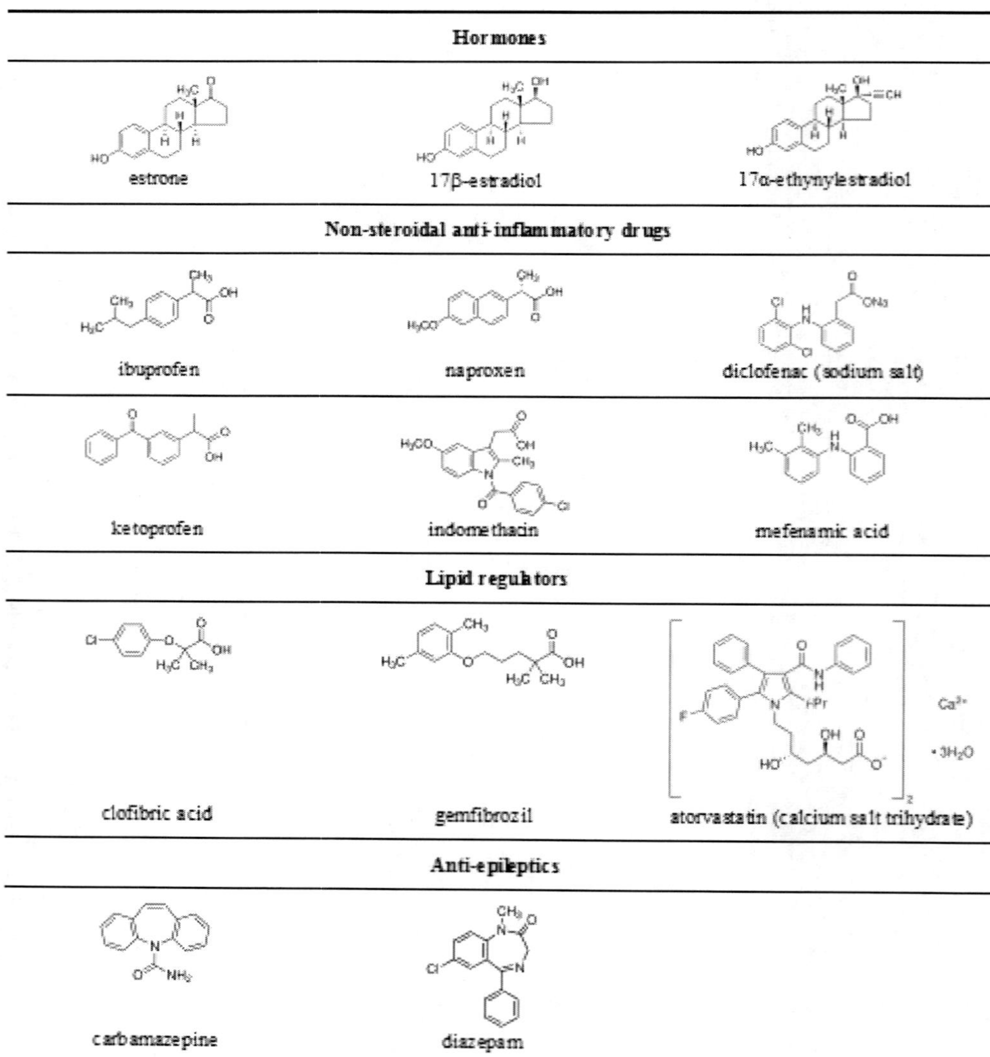

Figure 7. (Continued).

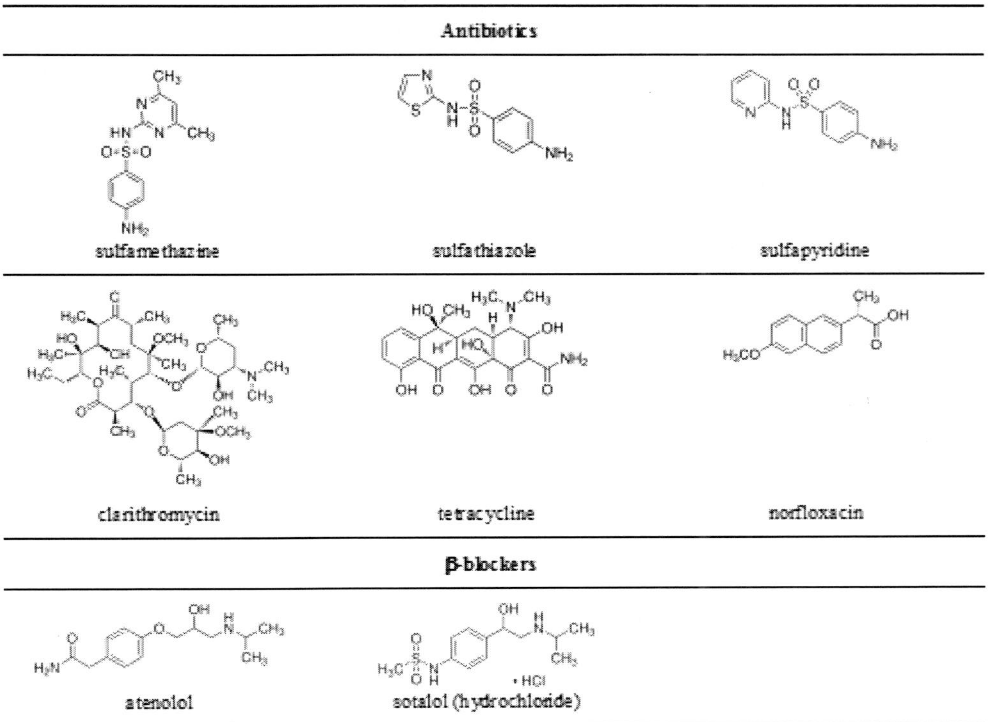

Figure 7. Chemical Structures of Different Pharmaceutical Compounds.

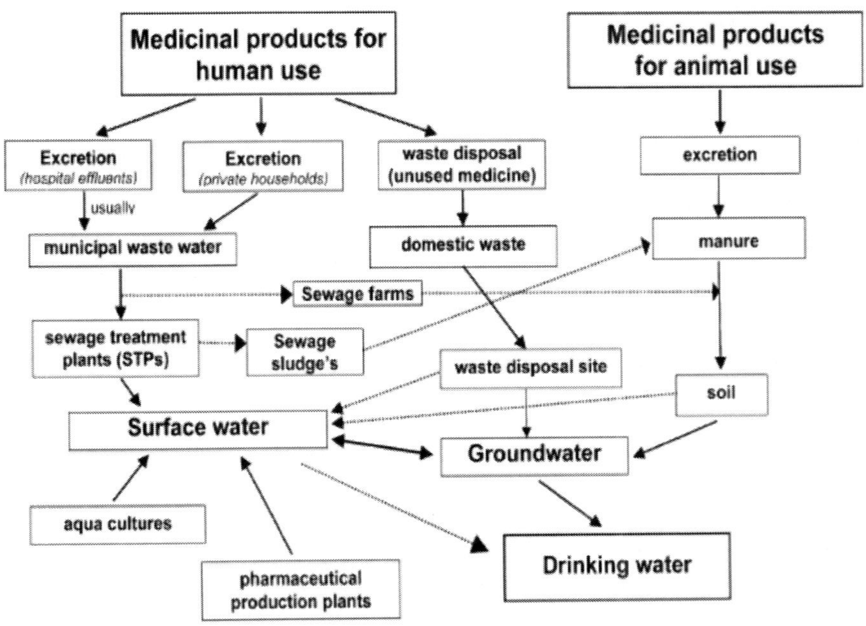

Figure 8. Scheme Showing the Possible Routes and Pathways for the Occurrence of Pharmaceutical Compounds in the Aquatic Environment. Reprinted from Toxicology Letters, 131, Heberer, T. Occurrence, fate, and removal of pharmaceutical residues in the aquatic environment: a review of recent research data, 5-17. Copyright (2002), with permission from Elsevier Ltd., UK.

	Pharmaceutical	Influent (μg/L)*	Effluent (μg/L)*
Antibiotics	clarithromycin	0.344	0.15
	ciprofloxacin	0.62	0.234
	doxycyclin	0.65	0.420
	erithromycin	0.58	0.297
	erythromycin-H_2O	2.025	0.59
	methronidazole	0.09	0.055
	norfloxacin	0.115	0.0526
	ofloxacin	0.482	0.171
	roxithromycin	0.78	0.472
	sulfamethoxazole	0.32	0.264
	sulfapyridin	0.492	0.081
	tetracyclin	48	2.375
	trimethoprim	0.43	0.424
Anti-epileptics	carbamazepine	0.732	0.774
	4-aminoantipyrine	1.517	0.676
	antipyrin	0.04	0.027
	codein	2.8605	1.93
Analgesics and anti-inflammatories	diclofenac	1.039	0.679
	ibuprofen	13.482	3.480
	indomethacine	0.136	0.166
	ketoprofen	0.483	0.333
	ketorolac	0.407	0.228
	naproxen	5.077	0.934
Lipid regulators	clofibric acid	0.215	0.131
	fenofibric acid	0.079	0.196
	bezafibrate	1.948	0.763
	gemfibrozil	1.562	0.757
Antidepressants	fluoxetin	5.85	0.112

*The values are referred to mean values

Figure 9. Concentrations of Some Emerging Pollutants found in Influents and Effluents of Different WWTPs (extracted from Deblonde et al. [39]).

4. Degradation of Pharmaceutical Compounds by Laccase-Producing Fungi and Their Laccases

One of the latest applications of laccase-producing fungi and their laccases is the degradation of pharmaceutical compounds. Figure 10 reports the degradation of different classes of pharmaceutical compounds by different laccase-producing fungi or their laccases.

Fungus	Pharmaceutical	Removal	Reference
		Hormones	
commercial laccase from *Trametes sp* (Daiwa Kasei)	17α-ethynylestradiol (2 mol/g sea sand)	90% in 48 h (test tube); 80% in 8 h (rotating reactor)	(40)
commercial laccase from *P. coccineus* (Daiwa Kasei)	17α-ethynylestradiol (2 mol/g sea sand)	80% in 48 h	(40)
partially-purified laccase from *Trametes versicolor* IFO-6482	17β-estradiol (10^{-5} M), ethinylestradiol (10^{-5} M)	≅100% in 1 h (HBT)	(41)
commercial laccase from *T. versicolor* (Sigma-Aldrich)	estrone, 17β-estradiol, estriol, and 17α-ethynylestradiol (0.4 nM each)	100% in 1 h	(42)
commercial laccase from *T. versicolor* (Sigma-Aldrich)	estrone, 17β-estradiol, estriol, and 17α-ethynylestradiol (0.4 nM each)	100% in 1 h	(43)
T. versicolor ATCC 45230	17β-estradiol (18.8 mg/L)	97 % in 24 h (batch)	(44)
	17α –ethynylestradiol (7.3 mg/L)	100% (continuous, HRT 120 h)	
Irpex lacteus 617/93, *Pleurotus ostreatus* 3004 CCBAS 278, *Pycnoporus cinnabarinus* CCBAS 595	17α-ethynylestradiol (10 mg/L)	below the detection limit in 3 days	(45)
Bjerkandera adusta 606/93	17α-ethynylestradiol (10 mg/L)	≅25% in 3 days	(45)
Phanerochaete chrysosporium ME 446	17α-ethynylestradiol (10 mg/L)	≅20% in 3 days	(45)
Phanerochaete magnoliae CCBAS 134/I	17α-ethynylestradiol (10 mg/L)	≅40% in 3 days	(45)
T. versicolor 167/93	17α-ethynylestradiol (10 mg/L)	≅45% in 3 days	(45)

Figure 10. (Continued).

Dichomitus squalens CCBAS 750	17α-ethynylestradiol (10 mg/L)	≅65% in 3 days	(45)
commercial laccase from *Trametes* sp Ha 1 (Daiwa Kasei)	estrone	---	(46)
	17β-estradiol	---	
	17α-ethynylestradiol	---	
commercial laccase from *Myceliophthora thermophila* (Novozymes)	estrone (5 mg/L)	65% in 24 h (laccase); below the detection limits in 24 h (laccase + violuric acid); 88% in 24 h (laccase + HBT); 74% in 24 h (laccase + syringaldehyde); 0% in 24 h (laccase + vanillin); 46 % in 24 h (laccase + p-coumaric acid); 37% in 24 h (laccase + ferulic acid)	(47)
	17β-estradiol (5 mg/L)	below the detection limits in 24 h (laccase; laccase + HBT; laccase + syringaldehyde; laccase + violuric acid; laccase + vanillin; laccase + p-coumaric acid; laccase + ferulic acid)	
	17α-ethynylestradiol (5 mg/L)	below detection limits in 24 h (laccase; laccase + HBT; laccase + syringaldehyde; laccase + violuric acid; laccase + vanillin; laccase + p-coumaric acid; laccase + ferulic acid)	
commercial laccase from *T. versicolor* (Fluka)	17β-estradiol (20 μM)	50% in 5 min	(48)
Lentinus tigrinus	17α-ethynylestradiol	---	(49)
purified laccase from *Lentinus tigrinus*	17α-ethynylestradiol	---	(49)
commercial laccase from *M. thermophila* (Novozymes)	estrone (5 mg/L)	94.1% (fed-batch)	(50)
	estrone (4 mg/L)	95% (continuous)	
	estradiol (5 mg/L)	95.5% (fed-batch)	
	estradiol (4 mg/L)	≅100% (continuous)	

Figure 10. (Continued).

	Non-steroidal anti-inflammatory drugs		
T. versicolor ATCC 45230	ibuprofen (10 mg/L)	no detectable levels in 7 days	(51)
Ganoderma lucidum (Leysser) karsten FP-58537-Sp	ibuprofen (10 mg/L)	no detectable levels in 7 days	(51)
I. lacteus AX1	ibuprofen (10 mg/L)	no detectable levels in 7 days	(51)
P. chrysosporium ME-446	ibuprofen (10 mg/L)	70% in 7 days (MEM); 88% in 7 days (LNM)	(51)
commercial laccase from *M. thermophila* (Novozymes)	diclofenac (5 mg/L)	83% in 24 h (laccase); below the detection limits in 24h (laccase + HBT; laccase + violuric acid; laccase + syringaldehyde; laccase + vanillin); 92% in 24 h (laccase + p-coumeric acid); 83% in 24 h (ferulic acid)	(47)
	naproxen (5 mg/L)	0% in 24 h (laccase; laccase + syringaldehyde; laccase + vanillin; laccase + p-coumaric acid; laccase + ferulic acid); 68% in 24 h (laccase + HBT); 36% in 24 h (laccase + violuric acid)	
T. versicolor ATCC 45230	naproxen (10 mg/L)	100% in 6 h	(52)
	naproxen (55 μg/L)	95% in 5 h	
commercial purified laccase from *T. versicolor* (Fluka)	naproxen (20 mg/L)	< 10% (laccase); ≅100% (laccase + HBT)	(52)
T. versicolor ATCC 45230	diclofenac (10 mg/L)	94% in 1 h; 100% in 4 h	(53)
	diclofenac (45 μg/L)	100% in 0.5 h	
commercial purified laccase from *T. versicolor* (Fluka)	diclofenac (40 mg/L)	>95% in 4.5 h	(53)
T. versicolor ATCC 45230	ketoprofen (10 mg/L)	no detectable levels in 24 h	(54)
	ketoprofen (40 μg/L)	95% in 5 h	
commercial purified laccase from *T. versicolor* (Fluka)	ketoprofen (10 mg/L)	0% (laccase; laccase+ABTS)	(54)

Figure 10. (Continued).

T. versicolor ATCC 45230	naproxen (0.067 mg/g dry solid)	47% in 24 h (bioslurry); 100% in 72 h (solid-phase plus 38% of wheat straw)	(55)
T. versicolor ATCC 45230	diclofenac (10 µg/L)	100% in 48 h	(56)
	naproxen (10 µg/L)	100% in 48 h	
	indomethacin (10 µg/L)	100% in 48 h	
	ibuprofen (10 µg/L)	100% in 48 h	
	fenoprofen (10 µg/L)	100% in 48 h	
	ketoprofen (10 µg/L)	100% in 48 h	
	propyphenazone (10 µg/L)	68% in 48 h	
crude laccase from *T. versicolor* ATCC 45230	diclofenac (10 µg/L)	100% in 12 h	(56)
	naproxen (10 µg/L)	100% in 12 h	
	indomethacin (10 µg/L)	100% in 12 h	
	ibuprofen (10 µg/L)	≅15% in 12 h	
	fenoprofen (10 µg/L)	≅25% in 12 h	
	ketoprofen (10 µg/L)	≅12% in 12 h	
	propyphenazone (10 µg/L)	≅5% in 12 h	
commercial laccase from *T. versicolor* (Sigma-Aldrich)	diclofenac (10 µg/L)	100% in 12 h (laccase; laccase + ABTS; laccase + HBT)	(56)
	naproxen (10 µg/L)	≅37% in 12 h (laccase); 100% in 12 h (laccase + ABTS; laccase + HBT)	
	indomethacin (10 µg/L)	≅20% in 12 h (laccase); 100% in 12 h (laccase + ABTS; laccase + HBT)	
	ibuprofen (10 µg/L)	≅15% in 12 h (laccase); 27% in 12 h (laccase + ABTS); 18% in 12 h (laccase + HBT)	
	fenoprofen (10 µg/L)	≅25% in 12 h (laccase); 50% in 12 h (laccase + ABTS); 37% in 12 h (laccase + HBT)	
	ketoprofen (10 µg/L)	≅26% in 12 h (laccase); 55% in 12 h (laccase + ABTS); 40% in 12 h (laccase + HBT)	
	propyphenazone (10 µg/L)	≅3% in 12 h (laccase); 12% in 12 h (laccase + ABTS); 19% in 12 h (laccase + HBT)	

Figure 10. (Continued).

T. versicolor ATCC 45230	naproxen (0.1 mg/g)	50% in 24 h (sterilised soil amended with wheat straw); 20-30% in 24 h (non-sterilised soil amended with wheat straw)	(57)
T. versicolor ATCC 45230	ibuprofen (85.9 ng/g)	75% in 42 days	(58)
	diclofenac (60.3 ng/g)	64% in 42 days	
	mefenamic acid (17.9 ng/g)	72% in 42 days	
	phenazone (9.6 ng/g)	100% in 42 days	
T. versicolor	diclofenac (20 mg/L)	100% in 5 h	(59)
	mefenamic acid (20 mg/L)	100% in 5 h	
T. versicolor ATCC 45230	ketoprofen (42.4 ng/g)	49.8% in 26 days	(60)
	naproxen (6.17 ng/g)	45.3% in 26 days	
	ibuprofen (135 ng/g)	91.9% in 26 days	
	indomethacine (9.60 ng/g)	61.9% in 26 days	
	diclofenac (209 ng/g)	59.8 in 26 days	
	mefenamic acid (14.2 ng/g)	41.1 in 26 days	
laccase from *Pleurotus florida* NCIM 1243 (immobilised on PLGA nanofiber)	diclofenac (50 mg/L)	100% for 3 cycles (laccase); 100% for 6 cycles (laccase + syringaldehyde)	(61)
Lipid regulators			
T. versicolor ATCC 45230	clofibric acid (10 mg/L)	91% in 7 days	(51)
G. lucidum (Leysser) Karsten FP-58537-Sp	clofibric acid (10 mg/L)	≅45% in 7 days	(51)
I. lacteus AX1	clofibric acid (10 mg/L)	≅20% in 7 days	(51)
P. chrysosporium ME-446	clofibric acid (10 mg/L)	≅25% in 7 days (LNM); 0% in 7 days (MEM)	(51)
T. versicolor ATCC 45230	clofibric acid (10 μg/L)	80% in 48 h (without ABTS); 75% in 48 h (with ABTS)	(56)
	gemfibrozil (10 μg/L)	80% in 48 h (without ABTS); 68% in 48 h (with ABTS)	
crude laccase from *T. versicolor* ATCC 45230	clofibric acid (10 μg/L)	5% in 12 h	(56)
	gemfibrozil (10 μg/L)	4% in 12 h	
commercial laccase from *T.*	clofibric acid (10 μg/L)	≅5% in 12 h (laccase); 27% in 12 h (laccase + ABTS);	(56)

Figure 10. (Continued).

versicolor(Sigma-Aldrich)		24% in 12 h (laccase + HBT)	
	gemfibrozil (10 μg/L)	≅10% in 12 h (laccase); 37% in 12 h (laccase + ABTS); 30% in 12 h (laccase + HBT)	
T. versicolor ATCC 45230	bezafibrate (4.5 ng/g)	100% in 42 days	(58)
	fenofibrate (4.2 ng/g)	100% in 42 days	
	atorvastatin (37.8 ng/g)	80% in 42 days	
T. versicolor ATCC 45230	bezafibrate (11.8 ng/g)	45.3% in 26 days	(60)
	gemfibrozil (14.2 ng/g)	41.1% in 26 days	
	atorvastatin (35 ng/g)	49.8% in 26 days	
	Anti-epileptic drugs		
T. versicolor ATCC 45230	carbamazepine (10 mg/L)	58% in 7 days	(51)
G. lucidum (Leysser) Karsten FP-58537-Sp	carbamazepine (10 mg/L)	47% in 7 days	(51)
I. lacteus AX1	carbamazepine (10 mg/L)	≅5% in 7 days	(51)
P. chrysosporium ME-446	carbamazepine (10 mg/L)	0% in 7 days	(51)
T. versicolor ATCC 45230	carbamazepine (10 mg/L)	22% in 24 h (single treatment); 60% in 48 h (repeated treatments)	(62)
T. versicolor ATCC 45230	carbamazepine (0.067 mg/g dry solid)	57% in 24 h (bioslurry); 48% in 72 h (solid-phase)	(55)
T. versicolor ATCC 45230	carbamazepine (10 μg/L)	≅75% in 48 h (with and without ABTS)	(56)
crude laccase from *T. versicolor* ATCC 45230	carbamazepine (10 μg/L)	≅4% in 12 h	(56)
commercial laccase from *T. versicolor* (Sigma-Aldrich)	carbamazepine (10 μg/L)	≅10% in 12 h (laccase); 10% in 12 h (laccase + ABTS); 18% in 12 h (laccase + HBT)	(56)
T. versicolor ATCC 45230	diazepam (19.3 ng/g)	43% in 42 days	(58)
	carbamazepine (25.6 ng/g)	43% in 42 days	
T. versicolor ATCC 45230	carbamazepine (9 mg/L)	94% in 6 days	(63)
	carbamazepine (50 mg/L)	61% in 7 days	
	carbamazepine (200 μg/L)	96% in 2 days (in batch); 90% in 25 days (in	

Figure 10. (Continued).

		continuous mode)	
T. versicolor ATCC 45230	diazepam (7.71 ng/g)	26.1% in 26 days	(60)
	carbamazepine (29.2 ng/g)	30.7% in 26 days	
	Antibiotics		
commercial laccase from *T. versicolor* (Fluka)	sulphanilamide (10^{-3} M)	10% in 15 days	(64)
	sulfadimethoxine (10^{-3} M)	75.1% in 15 days	
	sulfapyridine (10^{-3} M)	95.6% in 15 days	
T. versicolor ATCC 45230	sulphametazine (20 mg/L)	>95% in 20 h	(65)
commercial purified laccase from *T. versicolor* (Sigma-Aldrich)	sulphametazine (20 mg/L)	22% in 48 h; 93-100% in 48 h (LMS)	(65)
T. versicolor ATCC 45230	sulfamethazine, sulfapyridine and sulfathiazole (from 19.1 to 71.1 ng/g)	100% (sewage sludge sterilised systems)	(65)
T. versicolor ATCC 45230	norfloxacin (2 mg/L)	>90% in 7 days	(66)
	ciprofloxacin (2 mg/L)	>90% in 7 days	
commercial purified laccase from *T. versicolor* (Fluka)	norfloxacin (10 mg/L)	0%; 33.7% (ABTS 1 mM)	(66)
	ciprofloxacin (10mg/L)	16.3% in 20 h; 97.7% (ABTS 1 mM)	
T. versicolor ATCC 45230	clarithromycin (21 ng/g)	100% in 42 days	(58)
	sulfamethazine (20.4 ng/g)	100% in 42 days	
T. versicolor ATCC 45230	sulfamethazine (6.1 ng/g)	91.0% in 26 days	(60)
	sulfapyridine (21.4 ng/g)	100% in 26 days	
	sulfathiazole (143 ng/g)	85.9% in 26 days	
T. versicolor ATCC 45230	sulfapyridine (10 mg/L)	100% in 48 h	(67)
	sulfathiazole (10 mg/L)	≅80% in 72 h	
	sulfapyridine, sulfathiazole and sulfamethazine (at 5 mg/L each)	≅94% in 216 h (continuous mode)	
commercial purified laccase from *T. versicolor* (Sigma-Aldrich)	sulfapyridine (20 mg/L)	0% in 50 h; 75-98% in 50 h (LMS)	(67)
	sulfathiazole (16 mg/L)	≅15% in 50 h; 82-100% in 50 h (LMS)	
partially-purified laccase	tetracycline (10^{-4} M)	100% (1 h)	(68)

Figure 10. (Continued).

from *T. versicolor* IFO-6482	chlortetracycline (10^{-4} M)	100% (15 min)	
	doxycycline (10^{-4} M)	100% (15 min)	
	oxytetracycline (10^{-4} M)	100% (1 h)	
Laccase from *Perenniporia* strain TRFI 707	sulfadimethoxine (50 mg/L)	≅8% (laccase) in 330 min; 100% in 30 min (laccase+ABTS; laccase + violuric acid); 90% in 330 min (laccase + syringaldehyde); 67% in 330 min (laccase + hydroxybencyl alcohol); ≅10% in 330 min (laccase + hydroxyacetophenone; laccase + cyanophenol)	(69)
	sulfamonomethoxine (50 mg/L)	≅9% (laccase) in 330 min; 100% in 30 min (laccase+ABTS; laccase + violuric acid); 83% in 330 min (laccase + syringaldehyde); 70% (laccase + hydroxybencyl alcohol); 32% in 330 min (laccase + hydroxyacetophenone; 19% in 330 min (laccase + cyanophenol)	
	***β*-blockers**		
T. versicolor ATCC 45230	atenolol (13.6 ng/g)	100% in 42 days	(58)
T. versicolor ATCC 45230	atenolol (1.70 ng/g)	100% in 26 days	(67)
	sotalol (4.88 ng/g)	100% in 26 days	

Figure 10. Removal of Different Pharmaceutical Compounds by Laccase-Producing Fungi or their Laccases.

4.1. Hormones

Various natural and synthetic steroid hormones are used in the treatment of different types of medical situations such as menopausal symptoms, growth hormone deficiency, hypothyroidism and some types of cancer [70] and as oral contraceptives (estrogens). The latter are an important group of EDCs, which cause negative effects particularly on aquatic organisms. Thus, several studies showed that these compounds mimic hormones or interfere with the action of endogenous hormones [71, 72]. They are frequently found in surface waters and in wastewater in higher quantities than those of their predicted no-effect concentrations [73] and are reported to be responsible for the majority of the estrogenic activity of municipal wastewater [74, 75].

Tanaka et al. [40] studied the remediation of soil contaminated with the synthetic estrogen 17α-ethynylestradiol (2 mol/g sea sand) by commercial laccases (800 U/L) from

Trametes sp and *P. coccineus* in test tubes and found that it was removed 90% by the former and of 80% by the latter in 48 h. Also, they tested the commercial laccase (800 U/L) from *Trametes* sp in a rotating reactor at 10 rpm and found that the 80% of 17α-ethynylestradiol was degraded in 8 h.

Suzuki et al. [41] reported the nearly complete removal of the steroidal hormones 17β-estradiol (10^{-5} M) and 17α-ethynylestradiol (10^{-5} M) by a partially-purified laccase (600 U/L) from *Trametes versicolor* in the presence of the redox mediator HBT (0.2 mM). In addition, they found that the estrogenic activity of the treated solution, using the yeast two-hybrid assay system, was removed in a percentage higher than 80%. Auriol et al. [42] investigated the removal of estrone, 17β-estradiol, estriol (natural estrogen) and 17α-ethynylestradiol (synthetic estrogen) from synthetic and municipal wastewater by a commercial laccase from *T. versicolor*. These compounds (0.4 nM) were totally removed by laccase (20,000 U/L) from both types of wastewater, thus, indicating that the municipal wastewater constituents did not inhibit the laccase action. They also found that the addition of the redox mediator HBT (100 μM) reduced the laccase dosage which would reduce the costs, however it is necessary to take into account the toxic effects exerted by this synthetic mediator. Later [43], they studied the residual estrogenic activity, using the recombinant yeast assay, after laccase treatment of both synthetic and municipal wastewater containing estrone, 17β-estradiol, estriol and 17α-ethynylestradiol (0.4 nM each) and found that 1 h of laccase treatment totally removed the estrogenic activity in the former and by 98% in the latter.

Blánquez and Gieβe [44] showed the technical feasibility of 17β-estradiol and 17α-ethynylestradiol removal by the white-rot fungus *T. versicolor* at reactor scale (working volume 100 mL) operating in continuous mode. Thus, the bioreactor operated for 26 days at a hydraulic retention time (HRT) of 120 h and 17β-estradiol and 17α-ethynylestradiol were completely removed (volumetric removal rates of 0.16 mg/L h and 0.09 mg/L h, respectively) when fed at 18.8 mg/L and 7.3 mg/L, respectively. Cajthaml et al. [45] studied the degradation of 17α-ethynylestradiol (10 mg/L) by the eight white-rot fungi *Irpex lacteus, Bjerkandera adusta, Phanerochaete chrysosporium, Phanerochaete magnoliae, Pleurotus ostreatus, T. versicolor, Pycnoporus cinnabarinus* and *Dichomitus squalens*. *I. lacteus* and *P. ostreatus* were found to be the most efficient degraders, thus, degrading 17β-ethynylestradiol below the detection limits in 3 cultivation days. In general 17α-ethynylestradiol degradation, assayed with a recombinant yeast test, was accompanied by a decrease in the estrogenic activity of the solution except in the case of *P. magnoliae,* indicating that a highly estrogen-active metabolite was formed in this case.

Tanaka et al. [46] investigated the degradation of estrone, 17β-estradiol, and 17α-ethynyletstradiol by a commercial laccase from *Trametes* sp (800 U/L). They suggested that the estrogens were oligomerised. Also, they showed that estrogenic activities, using the medaka (*Oryzias latipes*) vitellogenin assay, were removed. Lloret et al. [47] assessed the degradation of estrone, 17β-estradiol and 17α-ethynylestradiol by a commercial laccase from *Myceliophthora thermophila*. The two latter were totally removed by laccase in 24 h whereas the former was only removed by 35%. Further, they tested the effect of different natural (vanillin, syringaldehyde, p-coumaric acid and ferulic acid) and synthetic (HBT and violuric acid) mediators at 1 mM on the degradation of the above-mentioned compounds by laccase. The synthetic mediators and the natural mediator syringaldehyde led to the highest degradation percentages (Figure 10). Later [50], they studied the degradation of estrone and

17β-estradiol by the same commercial laccase in fed-batch (250 mL) and enzymatic reactors (370 mL). They found a degradation of 94.1% for estrone and 95.5% for 17β-estradiol in fed-batch reactors (initial concentration of each estrogen was 5 mg/L), under the best conditions tested, and 95% and nearly 100%, respectively, in membrane enzymatic reactors operating in continuous mode (estrogen concentrations in the influent were 4 mg/L each). In addition, the residual estrogenic activity, measured by LYES (lyticase yeast estrogen screen), of the effluent was reduced up to 97%.

Cardinal-Watkins and Nicell [48] reported the conversion of estradiol by a commercial laccase from *T. versicolor* immobilised by covalent bonding onto silica beads in a packed-bed reactor (porosity 0.68) operating in continuous mode. A residence time of only about 5 min was required to obtain a conversion of estradiol of 50% for an initial concentration of 20 μM. Covino et al. [49] studied the degradation of 17α-ethynyletstradiol by whole fungal cultures of the white-rot fungus *Lentinus tigrinus* and by a purified laccase from this fungus. They found that both treatments degraded 17α-ethynyletstradiol significantly accompanied by the complete removal of the estrogenic activity.

4.2. Non-Steroidal Anti-Inflammatory Drugs

Non-steroidal anti-inflammatory drugs (NSAIDs) comprise one of the main classes of pharmaceutical compounds commonly consumed in both prescription and non-prescription drugs. They have analgesic, antipyretic and anti-inflammatory activities. Marco-Urrea et al. [51] studied the degradation of 10 mg/L ibuprofen by the white-rot fungi *T. versicolor, I. lacteus, Ganoderma lucidum* and *P. chrysosporium* after 7 days of incubation. They found that ibuprofen was extensively degraded by all the fungi tested. Thus, *T. versicolor, I. lacteus* and *G. lucidum* degraded ibuprofen to non-detectable levels and *P. chrysosporium* around 70% (malt extract medium) and 88% (low nitrogen medium).

Lloret et al. [47] assessed the degradation of diclofenac (5 mg/L) and naproxen (5 mg/L) by a commercial laccase (2000 U/L) from *M. thermophila*. The former was totally removed by laccase in 24 h, whereas the latter was not oxidised. Further, they tested the effect of different natural (vanillin, syringaldehyde, p-coumaric acid and ferulic acid) and synthetic (HBT and violuric acid) mediators at 1 mM on the degradation of the above-mentioned compounds by laccase. The synthetic mediators and the natural mediator syringaldehyde led to the highest degradation percentages. Thus, diclofenac was totally degraded in the presence of violuric acid, syringaldehyde and vanillin and above 80% in the presence of p-coumaric acid and ferulic acid in 24 h. Naproxen was only partially degraded in the presence of HBT and violuric acid, reaching degradation percentages of 68% and 36%, respectively, after 24h.

Marco-Urrea et al. [52] investigated the degradation of naproxen by the white-rot fungus *T. versicolor* and found that it was totally degraded in 6 h for a concentration of 10 mg/L and by 95% in 5 h for a concentration of 55 μg/L. A commercial purified laccase (1000 U/L) from *T. versicolor* alone and plus the redox mediator HBT (1 mM) showed less than 10% and almost complete naproxen (20 mg/L) degradation, respectively. They also found that the cytochrome P450 system could play a role in naproxen degradation by *T. versicolor*. Two degradation intermediates were detected, which together with the parent compound disappeared after 6 h. The non-toxicity of the treated medium was confirmed by the Microtox test.

Marco-Urrea et al. [53] studied the degradation of diclofenac by the white-rot fungus *T. versicolor* and found a degradation percentage of 94% in 1 h and of 100% in 4 h for a diclofenac concentration of 10 mg/L and 100% in 0.5 h for a concentration of 45 μg/L. They identified two hydroxylated metabolites, 4-hydroxydiclofenac and 5-hydroxydiclofenac, which together with the parent compound disappeared after 24 h coinciding with a decrease in ecotoxicity assessed by the Microtox test. They suggested that cytochrome P450 system played a key role in the first step of diclofenac degradation. A commercial purified laccase from *T. versicolor* (2000 U/L) was also able to degrade diclofenac (40 mg/L), leading to a percentage higher than 95% after 4.5 h. One degradation intermediate (4-(2,6-dichlorophenylamino)-1,3-benzenedimethanol) was identified, which, however, was not detected in diclofenac degradation by *T. versicolor*. They [54] also studied the degradation of ketoprofen by the same fungal strain and found that the fungus removed ketoprofen to non-detectable levels in 24 h for an initial concentration of 10 mg/L and 95% in 5 h for an initial concentration of 40 μg/L. They suggested that the first oxidation step was mediated by cytochrome P 450 but laccase did not play a role in ketoprofen degradation as confirmed experiments with a commercial purified laccase (10,000 U/L) from *T. versicolor* and the redox mediator ABTS (0.8 mM).

Rodriguez-Rodriguez et al. [55] reported the degradation of naproxen (0.067 mg/g dry solid) on sewage sludge by the white-rot fungus *T. versicolor*. Bioslurry (25% of sludge) cultures showed around 47% naproxen removal in 24 h, whereas solid-phase cultures amended with 38% of wheat straw showed complete removal of naproxene in 72 h. Tran et al. [56] studied the degradation of ibuprofen, fenoprofen, ketoprofen, naproxen, diclofenac, indomethacin and propyphenazone at concentrations of 10 μg/L by whole fungal cultures of *T. versicolor* in the presence and absence of ABTS (1 mM), crude laccase (1500 U/L) from *T. versicolor* and a commercial laccase (2000-6000 U/L) preparation from *T. versicolor* in the presence and absence of ABTS or HBT (1 mM). They obtained complete removal of ibuprofen, fenoprofen, ketoprofen, naproxen, diclofenac and indomethacin and partial degradation of propyphenazone (around 68%) after 48 h of incubation with the 7-day-old liquid fungal culture both in the presence and absence of the redox mediator ABTS and suggested that degradation may be caused by the presence of laccase, MnP (manganese-dependent peroxidase) and cytochrome P450 oxygenase in the fungal culture. Crude laccase removed completely naproxen, diclofenac and indomethacin in 12 h. Since laccase was the dominant enzyme in the cell-free extract, they suggested that the complete removal of naproxen, diclofenac and indomethacin may involve laccase alone or in combination with natural mediators contained in the extract. In addition, naproxen, diclofenac and indomethacin were totally transformed by the commercial laccase in the presence of the redox mediators ABTS or HBT (Figure 9).

Borrás et al. [57] studied the degradation of naproxen (0.1 mg/g) in soil by the white-rot fungus *T. versicolor* and found that *T. versicolor* was able to degrade significant amounts of naproxen after 24 h in sterile and non-sterile soil cultures, especially when using a material based on wheat straw as a carrier (50% and 20-30%, respectively).

Rodriguez-Rodriguez et al. [58] reported the degradation of ibuprofen, diclofenac, mefenamic acid and phenazone at environmentally relevant concentrations in sewage sludge by the fungus *T. versicolor*. Thus, 75% of ibuprofen, 64% of diclofenac, 72% of mefenamic acid and 100% of phenazone were removed after the fungal treatment. Toxicological analyses

(*Daphnia magna*, *Vibrio fischeri* and seed germination) showed an important reduction in sludge toxicity after fungal treatment.

Margot et al. [59] determined the optimal conditions for the degradation of diclofenac and mefenamic acid by laccase from *T. versicolor*. They found that optimal conditions were: pH values between 4.5 and 6, temperatures from 30°C to 40°C and a laccase activity of 725 U/L. Thus, operating under such conditions, diclofenac (20 mg/L) and mefenamic acid (20 mg/L) were completely removed after 5 h.

Rodriguez-Rodriguez et al. [60] studied the removal of ibuprofen, indomethacine, diclofenac, ketoprofen, naproxen and mefenamic acid from the sewage sludge of a wastewater treatment plant located in El Prat de Llobregat (Barcelona, Spain) by the white-rot fungus *T. versicolor* after 26 days. Ibuprofen showed the highest removal percentage (91.9%) followed by indomethacine and diclofenac (around 60%) and ketoprofen, naproxen and mefenamic acid (about 40-50%).

Sathiskumar et al. [61] studied the removal of diclofenac (50 mg/L) by a laccase from *Pleurotus florida* immobilised on poly(lactic-co-glycolic acid) (PLGA) nanofiber. The immobilised laccase (4000 U/L) totally transformed diclofenac in three successive cycles, which was extended to 6 cycles in the presence of syringaldehyde (250 μM).

4.3. Lipid Regulators

Lipid regulators are substances used to lower the levels of triglycerides and low-density lipoproteins (LDL) and increase the levels of high-density lipoproteins (HDL) in the blood [76]. These substances are used among people at risk of heart attack. There are three kinds of lipid regulators: fibric acid derivatives (or fibrates), statins [or 3-hydroxy-3-methylglutarylcoenzyme A (HMG-CoA) reductase inhibitors and niacin (or nicotinic acid).

Marco-Urrea et al. [51] studied the degradation of 10 mg/L clofibric acid by the white-rot fungi *T. versicolor, I. lacteus, G. lucidum* and *P. chrysosporium* after 7 days of incubation. They found that *T. versicolor* was the only strain able to degrade the clofibric acid (91%). *In vitro* experiments using MnP and LMS showed that extracellular fungal enzyme systems did not play a role in the first step of degradation. However, *in vivo* studies using cytochrome P450 inhibitors suggested that the intracellular cytochrome P450 system may be involved in the first step of clofibric acid degradation by *T. versicolor*.

Tran et al. [56] studied the degradation of clofibric acid and gemfibrozil at concentrations of 10 μg/L by whole fungal cultures of *T. versicolor* in the presence and absence of ABTS (1 mM), crude laccase (1500 U/L) from *T. versicolor* and a commercial laccase (2000-6000 U/L) preparation from *T. versicolor* in the presence and absence of ABTS or HBT (1 mM). They obtained a partial degradation after 48 h of incubation with the 7-day-old liquid fungal culture both in the presence (around 80%) and absence of the redox mediator ABTS (around 75% for clofibric acid and about 68% for gemfibrozil). Clofibric acid and gemfibrozil degradation by crude and commercial laccase led to lower removal percentages (Figure 9).

Rodriguez-Rodriguez et al. [58] reported the complete removal of bezafibrate and fenofibrate and 80% removal of atorvastatin at environmentally relevant concentrations in sewage sludge by the fungus *T. versicolor*. None of these pharmaceutical compounds had shown to be degradable by white-rot fungi before this study. In addition, toxicological analyses (*Daphnia magna*, *Vibrio fischeri* and seed germination) showed an important reduction in sludge toxicity after fungal treatment.

Rodriguez-Rodriguez et al. [60] studied the removal of the lipid regulators bezafibrate, gemfibrozil and atorvastatin from the sewage sludge of a wastewater treatment plant located in El Prat de Llobregat (Barcelona, Spain) by the white-rot fungus *T. versicolor* and found a removal of 45.3, 41.1 and 49.8%., respectively However, toxicological analyses (*Daphnia magna*, *Vibrio fischeri*) showed an acute increase in toxicity after fungal treatment.

4.4. Anti-Epileptics

Anti-epileptic drugs are used in the treatment of epilepsy. They act by inhibiting the propensity of seizures. Some anti-epipeltics are also useful to treat some types of psychiatric disorders such as anxiety, depression and mood disorder.

Marco-Urrea et al. [51] studied the degradation of 10 mg/L carbamazepine by the white-rot fungi *T. versicolor, I. lacteus, G. lucidum* and *P. chrysosporium* after 7 days of incubation. They found that *T. versicolor* and *G. lucidum* were the only fungi able to remove carbamazepine. *In vitro* experiments using MnP and LMS showed that extracellular fungal enzyme systems did not play a role in the first step of degradation. However, *in vivo* studies using cytochrome P450 inhibitors suggested that the intracellular cytochrome P450 system may be involved in the first step of carbamazepine degradation by *T. versicolor*. Hata et al. [62] reported that a single treatment with laccase (600 U/L) from *T. versicolo*r and the redox mediator HBT (0.22 mM) removed about 22% of carbamazepine (2×10^{-5} M) in 24 h and additions of laccase and HBT to the reaction mixture every 8 h removed about 60% in 48 h. They identified 2 degradation products of carbamazepine: 10,11-dihydro-10,11-epoxycarbamazepine and 9(10H)-acridone. The former is believed to be as pharmaceutically active as the parent drug [77, 78].

Rodriguez-Rodriguez et al. [55] investigated the degradation of carbamazepine (0.067 mg/g dry solid) on sewage sludge by the white-rot fungus *T. versicolor*. Bioslurry (25% of sludge) cultures showed around 57% removal in 24 h, whereas sludge solid cultures amended with 38% of wheat straw led to a degradation of about 48% in 72 h. Tran et al. [56] studied the degradation of carbamazepine (10 μg/L) by whole fungal cultures of *T. versicolor* in the presence and absence of ABTS (1 mM), crude laccase (1500 U/L) from *T. versicolor* and a commercial laccase (2000-6000 U/L) preparation from *T. versicolor* in the presence and absence of ABTS or HBT (1 mM). They obtained a degradation percentage of around 75% after 48 h of incubation with the 7-day-old liquid fungal culture both in the presence and absence of the redox mediator ABTS. Crude laccase and commercial laccase led to lower percentages of carbamazepine degradation (Figure 9). Rodriguez-Rodriguez et al. [58] showed the degradation of diazepam and carbamazepine at environmentally relevant concentrations in sewage sludge by the fungus *T. versicolor*. Both compounds were removed by 43% after the fungal treatment. Furthermore, toxicological analyses (*Daphnia magna*, *Vibrio fischeri* and seed germination) showed an important reduction in sludge toxicity after fungal treatment.

Jelic et al. [63] studied the degradation of carbamazepine by the white-rot fungus *T. versicolor* at both flask and reactor scale. They found that at flask scale 94% was degraded in 6 days for an initial concentration of 9 mg/L and 61% in 7 days for an initial concentration of

50 mg/L and acridone, acridine, 10,11-dihydro-10,11-dihydroxy-carbamazepine and 10, 11-epoxy-carbamazepine were identified as main metabolites. Then, they conducted the degradation in an air-pulsed fluidised-bed bioreactor (1500 mL; air flow 12 L/h) and found that operating in batch mode 96% of carbamazepine (200 μg/L) was degraded in 2 days, 10, 11-dihydro-10,11-epoxycarbamazepine being the unique metabolite found. Operating the bioreactor in continuous mode with an HRT of 3 days, they found that 54% of carbamazepine (200 μg/L) was removed in 25 days (steady state) and no metabolite was detected. Toxicity tests (Microtox) indicated that the final solutions obtained at both batch and continuous mode were not toxic. Rodriguez-Rodriguez et al. [60] reported the removal of carbamazepine and diazepam from the sewage sludge of a wastewater treatment plant located in El Prat de Llobregat (Barcelona, Spain) by the white-rot fungus *T. versicolor* and found a degradation percentage of 30.7 and 26.1%, respectively, in 26 days.

4.5. Antibiotics

Antibiotics are a group of pharmaceutical compounds used for the treatment of both human and animals with bacterial and fungal infections. Their accumulation in the environment causes especially concern because such compounds can cause resistance in bacteria.

Schwarz et al. [64] investigated the degradation of the sulfonamides sulfanilamide, sulfadimethoxine and sulfapyridine at concentrations of 10^{-3} M by a commercial laccase (48,000 U/L) from *T. versicolor*. They found a degradation percentage, after a reaction time of 15 days, of 10.0% for sulfanilamide, 75.1% for sulfadimethoxine and 95.6% for sulfapyridine. These differences in the degradation obtained were attributed to the different molecular substituents of each compound. They also found that some of the metabolites formed were similar to those reported for sulfonamide photodegradation and degradation in soil.

García-Galán et al. [65] studied the degradation of sulfamethazine by the white-rot fungus *T. versicolor*. Removal to nearly undetectable levels after 20 h in liquid medium was achieved for an initial concentration of 9 mg/L. In addition, sulfamethazine (20 mg/L) degradation by a commercial purified laccase (343 U/L) from *T. versicolor* in the absence and presence of different redox mediators (1 mM HBT, 0.8 mM DMHAP (3, 5-dimethoxy-4-hydroxyacetophenon) or 0.8 mM ABTS) was tested. They found that laccase alone degraded 22% of sulfamethazine, while degradation by LMS ranged from 93% to 100% in 48 h. Further, they studied the degradation capacity of *T. versicolor* in sewage sludge sterilised systems and obtained a degradation percentage of 100% for sulfamethazine, sulfapyridine and sulfathiazole (concentrations in the raw sludge ranged from 19.1 to 71.1 ng/g).

Prieto et al. [66] reported a degradation percentage higher than 90% for ciprofloxacin and norfloxacin at initial concentrations of 2 mg/L by liquid cultures of *T. versicolor* after 7 days of incubation. They also found that a commercial purified laccase (60,000 U/L) from *T. versicolor* removed 16.3% of ciprofloxacin (10 mg/L) after 20 h; however norfloxacin (10 mg/L) was not degraded. The addition of the laccase mediator ABTS (1 mM) led to 97.7% and 33.7% degradation of ciprofloxacin and norfloxacin, respectively. They suggested that both laccase and the cytochrome P450 of *T. versicolor* were involved in ciprofloxacin and norfloxacin degradation.

Rodriguez-Rodriguez et al. [58] showed the total degradation of the antibiotics clarithromycin and sulfamethazine at environmentally relevant concentrations in sewage sludge by the fungus *T. versicolor*. In addition, toxicological analyses (*Daphnia magna*, *Vibrio fischeri* and seed germination) showed an important reduction in sludge toxicity after fungal treatment.

Rodriguez-Rodriguez et al. [60] found high removal of the sulfonamide antibiotics sulfapyridine (100%), sulfamethazine (91%) and sulfathiazole (86%) from the sewage sludge of a wastewater treatment plant located in El Prat de Llobregat (Barcelona, Spain) by the white-rot fungus *T. versicolor* in a bioslurry reactor (working volume 6 L) after 26 days. However, toxicological analyses (*Daphnia magna*, *Vibrio fischeri*) showed an acute increase in toxicity after fungal treatment.

Rodriguez-Rodriguez et al. [67] evaluated the degradation of the sulfonamides sulfapyridine and sulfathiazole by the white-rot fungus *T. versicolor*. at initial concentrations of 10 mg/L. They found that the former was totally removed in 48 h and the latter was degraded by 80% in 72 h. Then, they assessed the degradation of sulfapyridine (20 mg/L) and sulfathiazole (16 mg/L) by a commercial purified laccase from *T. versicolor* (50.4 U/L for sulfapyridine and 55.4 U/L for sulfathiazole) and found minimal degradation for both antibiotics in 50 h (0% and 15%, respectively). The addition of redox mediators (VA, DMHAP or ABTS at 0.8 mM) led to a removal from 75 to 98% for sulfapyridine and from 82 to 100% for sulfathiazole in 50 h. Eight degradation intermediates for sulfapyridine and five for sulfathiazole were detected for both fungal and laccase experiments. Also, they found that a fluidised-bed reactor (working volume 1500 mL) with *T. versicolor* pellets degraded a mixture of sulfonamides (sulfapyridine, sulfathiazole and sulfamethazine, 5 mg/L each) by greater than 94% each after 3 HRTs (HRT 72 h).

Suda et al. [68] studied the degradation of tetracycline (10^{-4} M), chlortetracycline (10^{-4} M), doxycycline (10^{-4} M) and oxytetracycline (10^{-4} M) by a partially-purified laccase (600 U/L) from *T. versicolor* in the absence and presence of the redox mediator HBT (0.2 mM). They found that laccase showed a limited ability to remove these tetracycline antibiotics Thus, tetracycline, chlortetracycline, doxycycline and oxytetracycline decreased by about 16%, 48%, 34% and 14%, repectively, after 4 h of treatment with laccase alone. The addition of HBT had an acute effect on the removal of the four tetracycline antibiotics; chlortetracycline and doxicycline were totally removed in 15 min, whereas tetracycline and oxytetracycline were removed after 1 h. In addition, the treated solutions showed no ecotoxicity against the bacteria *Escherichia coli* and *Bacillus subtilis* and the green alga *Pseudokirchneriella subcapitata*.

Weng et al. [69] investigated the implication of the redox mediators ABTS, violuric acid, syringaldehyde, 4-hydroxybenzyl alcohol, 4-hydroxyacetophenone and 4-cyanophenol at 1 mM in the degradation of sulfonamide antibiotics (50 mg/L) by laccase (6000 U/L) from the white-rot fungus *Perenniporia*. They found that laccase alone had limited effect on sulfadimethoxine and sulfamonomethoxine removal (92 and 91% of their initial concentrations, respectively, were retained after 330 min). The redox mediators ABTS and violuric acid provided the fastest transformations of sulfonamide antibiotics, thus, both antibiotics were degraded to non-detectable levels in 30 min. The addition of syringaldehyde and 4-hydroxybenzyl alcohol led to a partial degradation of both antibiotics (90% and 67%, in 330 min, respectively for sulfadimethoxine and 83% and 70%, in 330 min, for sulfamonomethoxine). The addition of 4-hydroxyacetophenone and 4-cyanophenol led to low degradation percentages (Figure 10). In addition, they found that de-aniline and oxidative coupling were the two detectable pathways for sulfonamide transformation by LMS.

4.6. β-Blockers

β-blockers are a type of drugs that act on blood vessels preventing vasodilatation and reducing the speed and force of heart contractions. They are used in the treatment of hypertension, ischemic heart disease, congestive heart failure and certain arrhythmias and also to prevent migraines.

Rodriguez-Rodriguez et al. [58] reported the total degradation of atenolol at environmentally relevant concentrations in sewage sludge by the fungus *T. versicolor*. In addition, toxicological analyses (*Daphnia magna*, *Vibrio fischeri* and seed germination) showed an important reduction in sludge toxicity after fungal treatment.

Rodriguez-Rodriguez et al. [67] reported the total removal of the β-blockers atenolol y sotalol from the sewage sludge of a wastewater treatment plant located in El Prat de Llobregat (Barcelona, Spain) by the white-rot fungus *T. versicolor* in 26 days. Toxicological analyses (*Daphnia magna*, *Vibrio fischeri*) showed however an acute increase in toxicity after fungal treatment.

Conclusion

The presence of pharmaceutical compounds in the environment is causing serious concern in recent times. One of the major sources of introduction of pharmaceutical compounds into the environment is through WWTPs effluents, since WWTPs are not designed to treat this type of compounds. There is no clear evidence of adverse effects of pharmaceutical compounds at trace concentrations in human health. However, there are several classes of pharmaceuticals with toxic and estrogenic properties such as antibiotics and natural and synthetic hormones, which have shown to produce adverse effects in aquatic organisms. Furthermore, the potential adverse effects on human health of a mixture of pharmaceutical active compounds are still unknown. Therefore, the removal of pharmaceutical compounds before entering into the environment is highly desirable. The physicochemical processes have shown to be inefficient in the removal of pharmaceutical compounds. AOPs have shown to be effective but economically expensive. The data reported by different authors on the removal of pharmaceutical compounds by laccase-producing fungi or their laccases are very promising. However, much research is still needed before processes based on laccase-producing fungi or their laccases can be implemented in real cases.

Acknowledgments

This research was financed by the Spanish Ministry of Science and Innovation (Project CTM2008-02453/TECNO).

References

[1] Kümmerer K (2009) The presence of pharmaceuticals in the environment due to human use – present knowledge and future challenges. *J Environ Manag* 90:2354-2366

[2] Daughton CG, Ternes TA (1999) Pharmaceuticals and *personal care products in the environment: agents of subtle change? Environ* Health Perspect 107:907-938

[3] Esplugas S, Bila D, Krause LGT, Dezoti M (2007) Ozonation and advanced oxidation technologies to remove endocrine disrupting chemicals (EDCs) and pharmaceuticals and personal care products (PPCPs) in water effluents. *J Hazard Mater*149:631-642

[4] Petrovic M, Gonzalez S, Barceló D (2003) Analysis and removal of emerging contaminants in wastewater and drinking water. *Trends Anal Chem* 22:685-696

[5] Kümmerer K (2001) Drugs in the environment: emission of drugs, diagnostic aids and disinfectants into wastewater by hospitals in relation to other sources—a review. *Chemosphere* 45:957-969

[6] Carballa M, Omil F, Lema JM, Llompart M, García-Jares C, Rodríguez I, Gómez M, Ternes TA (2004) Behaviour of pharmaceuticals, cosmetics and hormones in a sewage treatment plant. *Water Res* 38:2918-2926

[7] Onesios KM, Yu JT, Bouwer EJ (2009) Biodegradation and removal of pharmaceuticals and personal care products in treatment systems: a review. *Biodegradation* 20:441-466

[8] Andreozzi R, Caprio V, Marotta R, Vogna D (2003) Paracetamol Oxidation from Aqueous Solutions by Means of Ozonation and H2O2/UV System. *Water Res* 37:993-1004

[9] Klavarioti M, Mantzavinos D, Kassinos D (2009) Removal of residual pharmaceuticals from aqueous systems by advanced oxidation processes. *Env Int* 35:402-417

[10] Khetan SK, Collins TJ (2007) Human pharmaceuticals in the aquatic environment: a challenge to green chemistry. *Chem Rev* 107:2319-2364

[11] Pauwels B, Verstraete W (2006) The treatment of hospital wastewater: an appraisal. *J Water Health* 4:405-416

[12] Rahman MF, Yanful EK, Jasim SY (2009) Occurrences of endocrine disrupting compounds and pharmaceuticals in the aquatic environment and their removal from drinking water: challenges in the context of the developing world. *Desalination* 248:578-585

[13] Carballa M, Omil F, Lema JM (2005) Removal of cosmetic ingredients and pharmaceuticals in sewage primary treatment. *Water Res* 39:4790-4796.

[14] Kirk TK, Fenn, P (1982) Formation and action of the ligninolytic system in basidiomycetes in: Franklin JC, Hedges JN, Swift MJ (eds) *Decomposer Basidiomycetes, British Mycological Society Simposium 4*, Cambridge University Press, pp 67-90

[15] Brunow, G (2001) Methods to reveal the structure of lignin in: Hofrichter M, Steinbüchel A (eds) *Biopolymers, vol. 1*, Wiley-VCH, Weinheim, pp. 89-116

[16] Mayer AM, Staples RC (2002) Laccase: new functions for an old enzyme. *Phytochemistry* 60:551-565

[17] Yoshida H (1883) Chemistry of lacquer (Urushi). *J Chem Soc* 43:472-486

[18] Bertrand G (1895) Sur la laccase et sur le pouvoir oxydant de cette diastase. *CR Acad Sci (Paris)* 120:266-269

[19] Bertrand G (1896) Sur la presence simultanee de la laccase et de la tyrosinase dans le suc de quelques champignons. *CR Hebd Seances Acad Sci* 123:463-465

[20] Laborde J (1896) Sur la casse des vins. *CR Hebd Seances Acad Sci* 123:1074-1075

[21] Arora DA, Sharma RK (2010) Ligninolytic Fungal Laccases and Their Biotechnological Applications. *Appl Biochem Biotechnol* 160:1760-1788

[22] Malmström BG (1982) Enzymology of oxygen. *Annu Rev Biochem* 51:21-59

[23] Thurston CF (1994) The structure and function of fungal laccases. *Microbiol* 140:19-26

[24] Wesenberg D, Kyriakides I, Agathos SN (2003) White-rot fungi and their enzymes for the treatment of industrial dye effluents. *Biotechnol Adv* 22:161-187

[25] Bourbonnais R, Paice MG (1990) Oxidation of non-phenolic substrates. An expanded role for laccase in lignin biodegradation. *FEBS Lett* 267:99-102

[26] Bourbonnais R, Paice MG, Reid ID, Lanthier P, Yaguchi M (1995) Lignin oxidation by laccase isozymes from *Trametes versicolor* and role of the mediator 2,2'-azinobis(3-ethylbenzthiazoline-6-sulfonate) in kraft lignin depolymerization. *Appl Environ Microbiol* 61:1876-1880

[27] Gianfreda L, Xu F, Bollag J-M (1999) Laccases: a useful group of oxidoreductive enzymes. *Bioremediation* J 3:1-25

[28] Riva S (2006) Laccases: blue enzymes for green chemistry. *Trends Biotechnol* 24:219-226

[29] Camarero S, Ibarra D, Martínez MJ, Martínez AT (2005) Lignin-derived compounds as efficient laccase mediators for decolorization of different types of recalcitrant dyes. *Appl Environ Microbiol* 71:1775-1784

[30] Baldrian P (2006) Fungal laccases-occurrence and properties. *FEMS Microbiol Rev* 30:215-242

[31] Claus H (2003) Laccases and their occurrence in prokaryotes. *Arch Microbiol* 179:145-150

[32] Nikolaou A, Meric S, Fatta D (2007) Occurrence patterns of pharmaceuticals in water and wastewater environments. *Anal Biol Chem* 387:1225-1234

[33] Halling-Sørensen B, Nors Nielsen S, Lanzky PF, Ingerslev F, Holten Lutzhoft HC, Jørgensen SE (1998) Occurrence, fate and effects of pharmaceutical substances in the environment – a review. *Chemosphere* 36:357-393

[34] Heberer T (2002) Tracking persistent pharmaceutical residues in the aquatic environment: a review of recent research data. *Toxicol Lett* 131:5-17

[35] Webb SF, Ternes T, Gibert G, Olejnicak K (2003) Indirect human exposure to pharmaceuticals via drinking water. *Toxicol Lett* 142:157-167

[36] Stackelberg PE, Furlong ET, Meyer MT, Zaugg SD, Henderson AK, Reissman DB (2004) Persistence of pharmaceutical compounds and other organic wastewater contaminants in a conventional drinking-water-treatment plant. *Sci Total Env* 329:99-113

[37] Jones OAH, Voulvoulis N, Lester JN (2005) Human pharmaceuticals in wastewater treatment processes. *Crit Rev Environ Sci Technol* 35:401-427

[38] Zuccato E, Calamari D, Natangelo M, Fanelli R (2000) Presence of therapeutic drugs in the environment. *Lancet* 355:1789-1790

[39] Deblonde T, Cossu-Leguille C, Hartemann P (2011) Emerging pollutants in wastewater: A review of the literature. *International J Hyg Environ Health* 214:442-448

[40] Tanaka T, Tonosaki T, Nose M, Tomidokoro N, Kadomura N, Fujii T, Taniguchi M (2001) Treatment of model soils contaminated with phenolic endocrinedisrupting chemicals with laccase from *Trametes* sp. in a rotating reactor. *J Biosci Bioeng* 92:312-316

[41] Suzuki K, Hiraia H, Muratab H, Nishida T (2003) Removal of estrogenic activities of 17ßβ-estradiol and ethinylestradiol by ligninolytic enzymes from white rot fungi. *Water Res* 37:1972-1975

[42] Auriol M, Filali-Meknassi Y, Tyagi RD, Adams CD (2007) Laccase-catalyzed conversion of natural and synthetic hormones from municipal wastewater. *Water Res* 41:3281-3288

[43] Auriol M, Filali-Meknassi Y, Adams CD, Tyagi RD, Noguerol T-N, Piña B (2008) Removal of estrogenic activity of natural and synthetic hormones from a municipal wastewater: efficiency of horseradish peroxidase and laccase from *Trametes versicolor*. *Chemosphere* 70:445-452

[44] Blánquez P, Guieysse B (2008) Continuous biodegradation of 17β-estradiol and 17α-ethynylestradiol by *Trametes versicolor*. *J Hazard Mater* 150:459-462

[45] Cajthaml T, Kresinová Z, Svobodová K, Sigler K, Rezanka T (2009) Microbial transformation of synthetic estrogen 17β-ethinylestradiol. *Environ Pollut* 157:3325-3335

[46] Tanaka T, Tamura T, Ishizaki Y, Kawasaki A, Kawase T, Teraguchi M, Taniguchi M (2009) Enzymatic treatment of estrogens and estrogen glucuronide. *J Environ Sci* 21:731-735

[47] Lloret L, Eibes G, Lú-Chau TA, MoreiraMT, Feijoo G, Lema JM (2010) Laccase catalyzed degradation of anti-inflammatories and estrogens. Biochem Eng J 51:124-131

[48] Cardinal-Watkins C, Nicell JA (2011) Enzyme-Catalyzed Oxidation of 17*β*-Estradiol Using Immobilized Laccase from *Trametes versicolor*. *Enzym Res* doi:10.4061/2011/725172

[49] Covino S, Stella T, Kresinova Z, D´Annibale A, Petruccioli M, Cajthaml (2012) Degradation of endocrine disrupting chemicals and removal of estrogenic activities by laccase from *Lentinus tigrinus* and its extracellular enzymes. *Environ Eng Manag J* 11:S44

[50] Lloret L, Eibes G, Feijoo G, Moreira MT, Lema JM (2012) Degradation of estrogens by laccase from *Myceliophthora thermophila* in fed-batch and enzymatic membrane reactors. *J Hazard Mater* 213-214:175-183

[51] Marco-Urrea E, Perez-Trujillo M, Vicent T, Caminal G (2009) Ability of white-rot fungi to remove selected pharmaceuticals and identification of degradation products of ibuprofen by *Trametes versicolor*. *Chemosphere* 74:765-772

[52] Marco-Urrea E, Perez-Trujillo M, Blanquez P, Vicent T, Caminal G (2010a) Biodegradation of the analgesic naproxen by *Trametes versicolor* and identification of intermediates using HPLC-DAD-MS and NMR. *Bioresour Technol* 101:2159-2166

[53] Marco-Urrea E, Perez-Trujillo M, Cruz-Morato C, Caminal G, Vicent T (2010b) Degradation of the drug sodium diclofenac by *Trametes versicolor* pellets and identification of some intermediates by NMR. *J Hazard Mater* 176:836-842

[54] Marco-Urrea E, Perez-Trujillo M, Cruz-Morato C, Caminal G, Vicent T (2010c) White-rot fungus-mediated degradation of the analgesic ketoprofen and identification of intermediates by HPLC-DAD-MS and NMR. *Chemosphere* 78:474-481

[55] Rodríguez-Rodríguez CE, Marco-Urrea E, Caminal G (2010a) Degradation of naproxen and carbamazepine in spiked sludge by slurry and solid-phase *Trametes versicolor* systems. *Bioresour Technol* 101:2259-2266

[56] Tran NH, Urase T, Kusakabe O (2010) Biodegradation characteristics of pharmaceutical substances by whole fungal culture *Trametes versicolo*r and its Laccase. *J Water Environ Technol* 8:125-140

[57] Borràs E, Llorens-Blanch G, Rodríguez-Rodríguez CE, Sarrà M, Caminal G (2011) Soil colonization by *Trametes versicolor* grown on lignocellulosic materials: Substrate selection and naproxen degradation. *Int Biodeterior Biodegrad* 65:846-852

[58] Rodriguez-Rodriguez CE, Jelic A, Llorca M, Farre M, Caminal G, Petrovic M, Barcelo D, Vicent T (2011) Solid-phase treatment with the fungus *Trametes versicolor* substantially reduces pharmaceutical concentrations and toxicity from sewage sludge. *Bioresour Technol* 102:5602-5608

[59] Margot J, Maillard J, Rossi L, Barry DA, Holliger C (2012) Determination of optimal conditions for the degradation of micropollutants by laccase from *Trametes versicolor. Environ Eng Manag* J 11:S125

[60] Rodriguez-Rodriguez CE, Baron E, Gago-Ferrero P, Jelic A, Llorca M, Farre M, Diaz-Cruz MS, Eljarrat E, Petrovic M, Caminal G, Barcelo D, Vicent T (2012a) Removal of pharmaceuticals, polybrominated flame retardants and UV-filters from sludge by the fungus *Trametes versicolor* in bioslurry reactor. *J Hazard Mater* doi:10.1016/j.jhazmat.2012.07.024

[61] Sathishkumar P, Chae JC, Unnithan AR, Palvannan T, Kim HY, Lee KJ, Cho M, Kamala-KannanS, Oh BT (2012) Laccase-poly(lactic-co-glycolic acid) (PLGA) nanofiber: Highly stable, reusable, and efficacious for the transformation of diclofenac. *Enzym Microb Technol* 51:113-118

[62] Hata T, Shintate H, Kawai S, Okamura H, Nishida T (2010) Elimination of carabamazepine by repeated treatment with laccase in the presence of 1-hydroxybenzotriazole. *J Hazard Mater* 181:1175-1178

[63] Jelic A, Cruz-Morató C, Marco-Urrea E, Sarrà M, Perez S. Vicent T, Petrovic M, Barcelo D (2012) Degradation of carbamazepine by Trametes versicolor in an air pulsed fluidized bed bioreactor and identification of intermediates. *Water Res* 46:955-964

[64] Schwarz J, Aust M-O, Thiele-Bruhn S (2010) Metabolites from fungal laccase catalysed transformation of sulfonamides. *Chemosphere* 81:1469-1476

[65] García-Galán MJ, Rodríguez-Rodríguez CE, Vicent T, Caminal G, Díaz-Cruz MS, Barceló D (2011) Biodegradation of sulfamethazine by *Trametes versicolor*: removal from sewage sludge and identification of intermediate products by UPLC-QqTOF-MS. *Sci Total Env* 409:5505-5512

[66] Prieto A, Möder M, Rodil R, Adrian L, Marco-Urrea E (2011) Degradation of the antibiotics norfloxacin and ciprofloxacin by a white-rot fungus and identification of degradation products. *Bioresour Technol* 102:10987-10995

[67] Rodríguez-Rodríguez C, García-Galán M, Blánquez P, Díaz-Cruz S, Barceló D, Caminal G, Vicent T (2012b) Continuous degradation of a mixture of sulfonamides by

Trametes versicolor and identification of metabolites from sulfapyridine and sulfathiazole. *J Hazard Mater* 213-214:347-354

[68] Suda T, Hata T, Kawai S, Okamura H, Nishida T (2012) Treatment of tetracycline antibiotics by laccase in the presence of 1-hydroxybenzotriazole. *Bioresour Technol* 103:498-501

[69] Weng SS, Ku KL, Lai HT (2012) The implication of mediators for enhancement of laccase oxidation of sulfonamide antibiotics. *Bioresour Technol* 113:259-264

[70] Arcand-Hoy LD, Nimrod AC, Benson WH (1998) Endocrine-Modulating Substances in the Environment Estrogenic Effects of Pharmaceutical Products. *Int J Toxicol* 17:139-158

[71] White R, Jobling S, Hoare SA, Sumpter JP, Parker MG (1994) Environmentally persistent alkylphenolic compounds are estrogenic. *Endocrinology* 135:175-182

[72] Ishibashi H, Matsumura N, Hirano M, Matsuoka M, Shiratsuchi H, Ishibashi Y, Takao Y, Arizono K (2004) Effects of triclosan on the early life stages and reproduction of medaka *Oryzias latipes* and induction of hepatic vitellogenin. *Aquat Toxicol* 67:167-179

[73] Khan U, Nicell JA (2007) Horseradish peroxidase-catalysed oxidation of aqueous natural and synthetic oestrogens, *J ChemTechnol Biotechnol* 82:818-830

[74] Eggen RIL, Bengtsson BE, Bowmer CT, Gerritsen AAM, Gibert, M, Hylland K, Johnson, AC, Leonard P, Nakari T, Norrgren L, Sumpter JP, Suter MJF, Svenson A, Pickering AD (2003) Search for the evidence of endocrine disruption in the aquatic environment: lessons to be learned from joint biological and chemical monitoring in the European Project COMPREHEND. *Pure Appl Chem* 75:2445-2450

[75] Vethaak AD, Lahr J, Schrap SM, Belfroid AC, Rijs GBJ, Gerritsen A, de Boer J, Bulder AS, grinwins GCM, Kuiper RV, Legler J, Murk TAJ, Peijnenburg W, Verhaar HJM, de Voogt P (2005) An integrated assessment of estrogenic contamination and biological effects in the aquatic environment of the Netherlands. *Chemosphere* 59:511-524

[76] Mahley RW, Bersot TP (2001) Drug therapy for hypercholesterolemia and dyslipidemia in: Hardman JG, Limbird LE (eds) *Goodman & Gilman's - The Pharmacological Basis of Therapeutics 10th edn,* McGraw-Hill, New York, USA, pp 971-1002

[77] Zhang Y, Geißen SU, Gal C (2008) Carbamazepine and diclofenac: removal in wastewater treatment plants and occurrence in water bodies. *Chemosphere* 73:1151-1161

[78] Cunningham VL, Perino C, D'Aco VJ, Hartmann A, Bechter R (2010) Human health risk assessment of carbamazepine in surface waters of North America and Europe. *Regul Toxicol Pharmacol* 56:343-351

In: Microbes in Process
Editors: Neelam Garg and Abhinav Aeron

ISBN: 978-1-63117-127-7

Chapter 8

Inulinases

Camelia Neagu**, *Oana Constantin and Gabriela Bahrim
Dunarea de Jos University, Faculty of Food Science and Engineering,
Department of Food Science, Food Engineering and Food Biotechnology,
Galati, Romania

Abstract

Inulinases are versatile enzymes that act on β-2, 1 linkage of fructopolymers, like inulin. Inulinases can be produced by plants and many microorganisms, and the latter are the best sources for inulinases production, as they are easy to be cultivated and produce high yields of enzymes. Crude inulinases obtained through fermentation are usually purified in order to remove all contaminants that can interfere with their intended use and to increase their specific activity. The inulinases characterized until now in the literature show considerable variability with respect to biophysical and biochemical characteristics. Inulinases have a molecular weight over 50.0 kDa, an optimal pH and temperature between 4.5-6.0, respectively between 30°C and 60°C, depending on the producing microorganisms. Purified enzyme activity can be promoted by some ions (e.g. Ca^{+2}, K^{+}, Na^{+}), while other metallic ions have an inhibitory effect on the enzyme (e.g. Mg^{+2}, Ag^{+}). Inulinases are used in a wide range of industries, ranging from food and bioethanol production to pharmacology and chemical industry. In this chapter, the obtaining, characterisation and potential applications of microbial inulinases are reviewed.

1. Introduction

Inulin rich feedstock has lately gained researchers attention as sustainable raw material for fructose and fructooligosaccharides production, important ingredients in the food and pharmaceutical industries. Also, inulin represents the start point for bioethanol and other chemicals production (like alcohols, lactic acid, single cell proteins, single cells oils etc.) [1, 2, 3]. Inulin is a mixture of linear β-(1-2) - linked fructose molecules with a terminal glucose

* camelia.neagu@ugal.ro.

residue at the reducing end. There can be also a small number of branched inulin chain [4]. It can be found in a wide range of plants, such are Jerusalem artichoke, chicory and dahlia, and in onion and garlic in smaller amounts respectively [1, 5, 6, 7].

Inulin can be hydrolyzed by acids, but low pH determines the degradation of fructose and formation of some undesirable compounds. Thus, the use of inulinases has been proposed as the most promising approach to obtain pure fructose syrups from inulin and fructooligosaccharides [5]. Inulinases are hydrolases that act on the β-2, 1-linkage from inulin molecule [8]. They are divided in exo-inulinases that act on the terminal fructose residues of inulin and endo-inulinases that hydrolyze the internal linkages of inulin, resulting inulooligosaccharides [8]. Most inulinases are, in fact, β-fructofuranosidases [9]. Inulinases can be produced by plants and microorganisms. Due to their easy cultivation and high yields of enzyme biosynthesis, microorganisms seem to be the best sources for these enzymes [10]. The commercial inulinases used today in food and industrial processes are obtained from microbial fermentation. All classes of microorganisms (bacteria, moulds and yeasts) are sources of commercial enzymes. Strains of *Aspergillus niger* and *Kluyveromyces marxianus* were the most used microorganisms for inulinase production. But other moulds, yeasts and bacteria have been also widely researched as inulinase producers, like *Aspergillus* spp., *Geotrichum candidum*, *Rhizoctonia* spp., [11-20] *Streptomyces* spp., *Bacillus* spp. [21, 22, 23], *Pichia* spp., *Cryptococcus* spp. [24, 10, 25] etc.

Obtained inulinases have been purified using different specific methods, from protein fractionation by the addition of salts, organic solvents or polymers, to the use of most powerful tools, like liquid chromatography, affinity chromatography and electrophoresis [9, 26]. The purified enzymes were characterized for their molecular weight, optimum pH and temperatures and their stability. Furthermore, the effect of metal ions and protein inhibitors upon catalytic activity was studied. Microbial inulinases have the molecular weight higher than 50 kDa [10] and different optimum pH and temperature according to their provenience. Some metals added to the reaction medium promote enzyme activity, while others are inhibitory. Also, some enzymes inhibitors like EDTA and pepstatin were studied [27, 28, 29]. The most important developments in the inulinase production will be presented in this chapter.

2. Definition and Classification

Inulinase (2, 1-β-D-fructan fructanohydrolase, EC3.2.1.7) is a versatile enzyme included in the glycoside hydrolase family (fructanohydrolases), which targets the β-2, 1 linkage of fructopolymers like inulin [22]. Inulinases are classified as endo- or exo-inulinases, depending on their mode of action. Endoinulinases (2,1-β-D-fructan fructanohydrolase; EC 3.2.1.7) hydrolyse inulin by breaking the bonds between fructose units that are located away from the ends of the polymer chain, to produce oligosaccharides [30]. Exo-inulinases (β-D-fructohydrolase; EC 3.2.1.80) split the terminal fructose units in sucrose, raffinose and inulin to liberate fructose [31, 32, 33]. Unfortunately, only a few inulinases required for industrial applications have an optimum temperature of 50°C or higher, the minimum temperature for inulin solubility in water [31, 34, 35, 36].

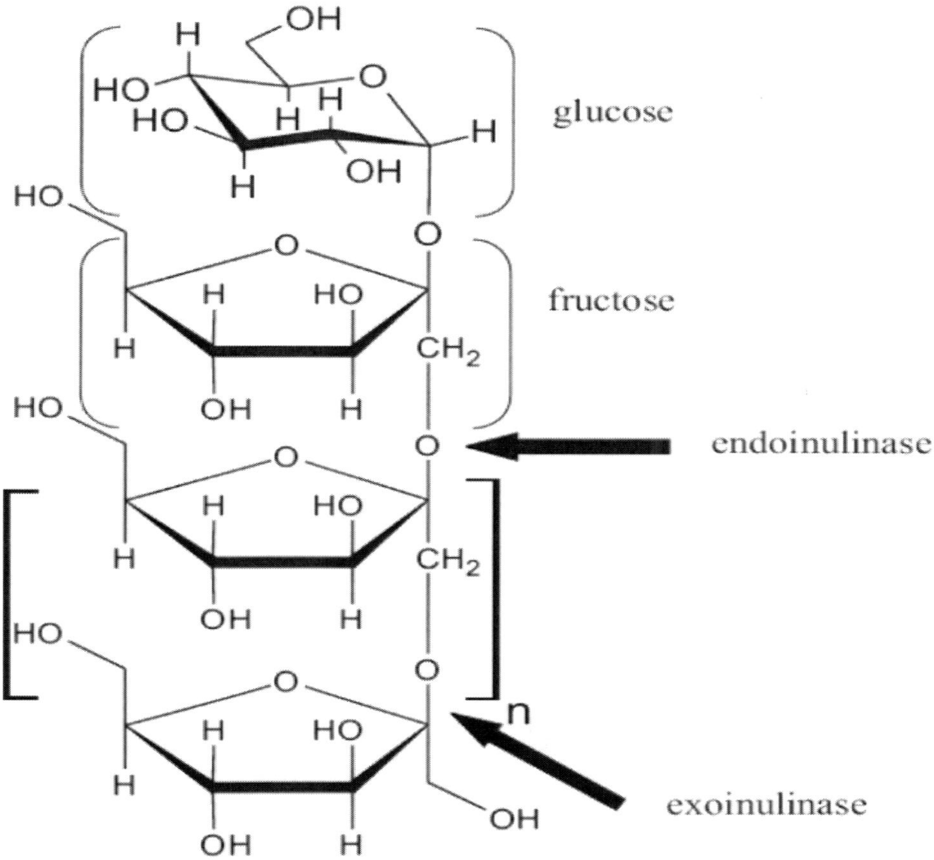

Figure 1. Action Pattern of Inulinases on Inulin [40].

Inulinases can be produced by many microorganisms using an inulin-based medium [37]. Inulin, a polyfructan, occurs as a reserve carbohydrate in numerous plant families and it represents more than 30 000 species [38, 39]. It consists of linear chains of β-2,1-linked fructosyl units terminating at the reducing end with a glucose residue attached through a sucrose-type linkage as illustrated in Figure 1 [40]. Inulin has received also, a great interest as a renewable raw material both for the production of high fructose syrup [41] and fructooligosaccharides [42].

3. Substrates Used for Inulinase Production

Inulin and inulin rich feedstock are the main substrates used for inulinase production. Inulin is a polydisperse carbohydrate consisting of β-(2,1) fructosyl-fructose linkages (F_m). A starting β-D-glucose residue can be present, but is not necessary (GF_n). Until recently inulin was considered a linear molecule, but it have been demonstrated that even native inulin has a small degree of branching ($GF_ç$), as the optimized permethylation analysis showed (Figure 2). Inulin exists also in cyclic form [43].

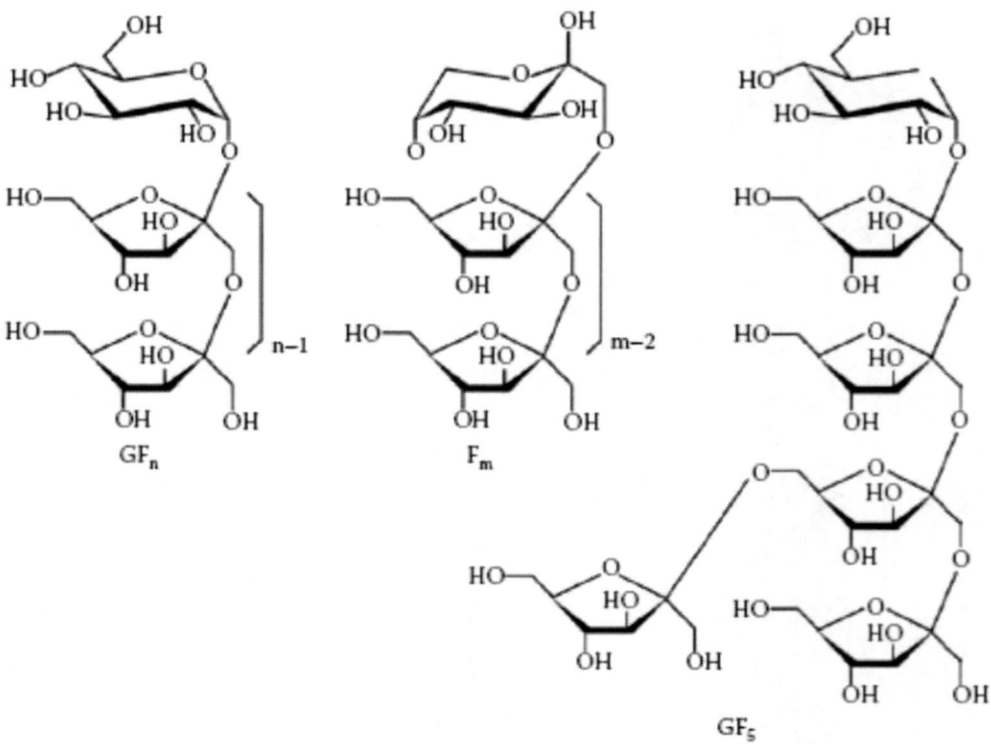

Figure 2. Inulin Structure; GFn – Inulin with a Terminal Glucose Unit, Fm – Inulin with a Terminal Fructose Unit, GFς – branched Inulin [4].

The degree of polymerization (DP) of inulin and the degree of branching are the most important properties that influence its functionality. The DP of inulin varies from 3 to 200 for plant inulins and up to 100 000 for bacterial inulin. Also, the branching degree varies for 1-2% (for chicory inulin) to 15% (bacterial inulin) [43]. Inulin is present in significant quantities in some plants, algae and microorganisms (Figure 3). Naturally inulin rich feedstock was preferred as substrate for inulinase production and lately, agroindustrial residues attracted researchers' attention.

Pure substrates were widely used for inulinase production. Among them, inulin, sucrose, fructose and glucose were the best substrates used to obtain inulinases with different microorganisms, like *A. niger*, *A. oryzae*, *Kluyveromyces marxianus*, *Streptomyces* spp. and *Pichia guilliermondii* [12, 25, 44, 35, 45, 46]. Naturally occurring inulin rich materials, like Jerusalem artichoke, chicory, dandelion and buckwheat are the preferred substrates used for inulinase obtaining as these contain large amounts of inulin and because of their availability [1, 47-50]. Jerusalem artichoke or topinambour (*Helianthus tuberosus* L) is a perennial highly competitive plant rich in inulin. Jerusalem artichoke produces large amounts of biomass, grows fast, even on marginal land, grows in poor conditions, does not need pesticides, fertilizer and large quantities of water, it can be stored *in situ* (in the cultivating place) or in cold deposits for 12 months [4].

Occurrence of inulins	Inulin content, g/100g
Plants:	
- Oat (*Avena sativa*)	NA*
- Barley (*Hordeum vulgare*)	0.5-1.5
- Rye (*Secale sativa*)	0.5-1
- Wheat (*Triticum durum, Triticum aestivum*)	1-4
- Leek (*Allium ampeloprasum*)	3-10
- Onion (*Allium cepa*)	1-7.5
- Garlic (*Allium sativum*)	9-16
- Asparagus (*Asparagus officinalis, A. racemosus*)	15
- Palm lily (*Cordyline terminalis*)	NA
- Chicory (*Cichorium intybus*)	15-20
- Dandelion (*Taraxacum officinale*)	12-15
- Jerusalem artichoke (*Helianthus tuberosus*)	17-20.5
- Salsify (*Tragopogon porrifolius*)	4-11
- Yacon (*Polymnia sonchifolia*)	3-19
- Banana (*Musa cavendishii*)	0.7-1
- Corn (*Zea mays*)	NA
- Buckwheat (*Fagopyrum* spp.)	NA
- Dahlia (*Dahlia pinnata*)	10-12
- Globe artichoke (*Cynara scolymus* L)	3
Algae, fungi and bacteria	
- *Acetabularia mediterranea*	
- *Cladophora* spp.	
- *Rhizoclonium* spp.	NA
- *Aspergillus sydowi*	
- *Streptococcus mutans*	
- *Lactobacillus reuteri*	

NA* - data not available.

Figure 3. Occurrence of Inulins in Plants, Algae and Microorganisms [104, 9, 43, 4, 7, 52].

Sharma et al. [21] used garlic powder for inulinase production using *Streptomyces* spp. strain and a maximum inulinase activity of 524 IU/l was obtained. The garlic powder was chosen from various substrates studied, like rye, barley, bananas, inulin, wheat, chicory, onion, garlic and dahlia. Also, Neagu Bonciu et al. [47] studied few carbon sources (onion, garlic, dahlia, Jerusalem artichoke, rye, buckwheat, oat, malt, lyophilized acid whey, pure inulin and rice) as substrates for inulinase production using *Rhizoctonia* spp. strain and buckwheat flour was the variable with the highest impact on inulinase production. Kango [13] studied the use of tap root of dandelion as substrate for inulinase production with an *A. niger* strain. He obtained a maximum inulinase activity of 55 IU/mL in medium with 40% dandelion tap root extract. *Asparagus* sp. was successfully used as substrate for inulinase production by Singh et al. [51] and Singh and Bhermi [52] as it contains up to 15% inulin and the annual production of this plant is higher than chicory and Jerusalem artichoke.

Lately, agroindustrial residues, like cassava flour, corncob, oat meal, rice straw, sugar cane bagasse, wheat bran and pressmud attracted researchers' attention, as these raw materials are cheap and easy to find [15, 53-59]. Sguarezi et al. [57] optimized the conditions for inulinase production from different agroindustrial residues (sugarcane molasses and corn steep liquor) and obtained a maximum inulinase activity of 1139 U/mL. Also, Dilipkumar et al. [53] used pressmud as substrate for *Streptomyces* spp. for inulinase production. Maximum inulinase production of 89 U/gds was obtained, at pressmud concentration of 12 g. Complex nitrogenous sources, like yeast extract, peptone, corn steep liquor, urea or inorganic sources, like ammonium sulphate, ammonium phosphate or ammonium chloride have been used for inulinase production. It have been established some nitrogen sources preferences of the strains used for inulinase production.

Inulinases production is also influenced by presence of metals ions, pH of the medium and environmental conditions (temperature, aeration etc.) [9], factors that usually enhance the inulinase activity or in some cases have inhibitory effects on the enzyme.

4. Microorganisms Producing Inulinases

Microorganisms can be considered the greatest sources for commercial production of inulinases because of their easy and short time cultivation and high yields of the enzyme production. So far, it has been found that the microorganisms which can produce high level of inulinases include species of *Aspergillus* spp., *Penicillium* spp., *Sporotrichum* spp., *Kluyveromyces* spp., *Cryptococcus* spp., *Pichia* spp., *Candida* spp., *Bacillus* spp., *Clostridium* spp., *Pseudomonas* spp., *Arthrobacter* spp., *Staphylococcus* spp. and *Xanthomonas* spp., [24, 60]. Inulinases are both extra- and intracellular, and several microorganisms have the capacity to express this enzyme in both ways [48]. Similarly, inulinases are both exo- and endo- acting but many microorganisms are known to produce mixed forms [1, 61-66]. The localization of an enzyme, yield and its mode of action depend upon the kind of microorganism and the substrate used during fermentation [41].

4.1. Yeasts

Although inulin-hydrolyzing activity has been reported from various microbial strains [48], yeasts (*Kluyveromyces* spp.), and the *Aspergillus* spp. have proven to be the most resourceful source of inulinases. Goa et al. [60] reported the production of inulinase of four marine yeasts: *Pichia guilliermondii, Cryptococcus aureus, Yarwonia lipolytica and Debaromyces hansenii.* In 2007, Sheng et al. [67], informed about the inulinase production by *Cryptococcus aureus* and in 2008, Sheng et al. [27], optimized the process parameters for high inulinase production by the marine yeast strain *Cryptococcus aureus* G7a.*C. aureus* G7a isolated from sediment of South China Sea was found to produce a large amount of inulinase into the medium. Under the optimal conditions, within 42 h of fermentation at a shake flask level, over 85.0±1.1 U/mL of inulinase activity is produced [67]. The marine-derived yeast *P. guilliermondii* strain 1, isolated from the surface of marine algae, was also found to secrete a large amount of inulinase using a submerged fermentation [68].Terrestrial yeasts, which are reported to produce inulinases, include strains of *Candida* spp., *Sporotrichum* spp., *Pichia* spp., and *Kluyveromyces* spp. So far, it has been shown that the inulinase activity produced by the terrestrial yeasts is less than 60 U/mL [51, 52, 69].

Kluyveromyces spp. is definitely, the widest used yeast for inulinase production. Two strains of *K. marxianus* (A1 and A2) isolated from "aguamiel" (agave sap) and one strain of *Kluyveromyces lactis* var. *lactis* (P7) isolated from "pulque" (product of fermented sap) show good inulinase production (less than 32 U/mL) in the liquid medium containing different concentrations of inulin, and they have low susceptibility to catabolic repression [70]. Maximum inulinase production (55.4 U/mL) by a newly isolated strain of *K. marxianus* YS-1 was obtained using an agitation rate of 200 rpm and aeration of 0.75 v/v per minute in a stirred tank reactor (1.5 l) with a fermentation time of 60 h [51]. In Singh and Bhermi [52] study, inulinase activity of 50.2 U/mL was obtained from the same yeast strain under agitation (200 rpm) and aeration (0.75 v/v per minute) condition, after 60 h of fermentation at 30°C, using a 1.5 l of bioreactor. Raw inulin (4.0%) was used as the production medium, from *Asparagus officinalis* root tubers. Inulinase yield in the bioreactor was almost six times higher, than the basal medium used initially in a shake flask. When *K. marxianus* NRRL Y-7571 was grown in the culture containing sugarcane bagasse as principal substrate and carbon source and corn steep liquor as nitrogen supplement, the extracellular inulinase concentration reaches 391.9 U/g of dry fermented bagasse [55]. The use of soybean bran decreased the time for reaching the maximum activity from 96 to 24 h, and the maximum productivity achieved *K. marxianus* NRRL Y-7571 is 8.87 U/gds per hour [71]. The average inulinase activity produced by a newly isolated strain *Kluyveromyces* S120 in SSF is 409.8 U/g initial dry substrate in the optimal medium [72].

Kluyveromyces marxianus was used in many optimization studies for inulinase production, when different yields of inulinase activity were obtained. Mazutti et al. [54] obtained a maximum inulinase activity of 250 U/gds by solid-state fermentation of sugar cane bagasse and 47.2 U/mL using a submerged liquid fermentation. Treichel et al. [73] obtained 1317 U/mL using agro-industrial residues as a substrate, whereas Sguarezi et al. [57] obtained 1139 U/mL by using similar substrates.

4.2. Moulds

Inulinases from filamentous fungi are in general exo- acting enzymes [48], but some fungal strains like *Aspergillus ficuum* [64], *Chrysosporium pannorum* [74] and *Penicillium rugulosum* [75] produce both exo- and endo-inulinases. Inulinases have been purified from acidophilic fungus, *Scytalidium acidophilum* [76]. The maximum activity of the enzyme from *S. acidophilum* was observed between pH=3.0–3.5, and it retained about 95 and 85 % of activities at 60 and 65°C, respectively, after 6h of incubation [76]. In several studies carried out on the *A. ficuum* inulinase activity the presence of multiple isoforms of exoinulinases and endoinulinases was revealed [64, 77]. In another interesting case, one invertase, five exoinulinases and three endoinulinases were isolated from a commercial preparation derived from *Aspergillus ficuum* (61). The strains of *A. niger* have been most extensively investigated for inulinase production of the *Aspergillus* spp. species. Most of the strains of *Aspergillus niger* have been reported to produce only endoinulinases extracellularly [11, 65, 13], while surprisingly *A. niger* strain 12 has been reported to produce both exo- and endoinulinases, intracellular [63]. Inulinase activity has also been determined from other *Aspergillus* spp., which include *A. fumigatus* [78], *A. awamori* [6], *A. ochraceus* [15], and *A. parasiticus* [16].

In the research of Mughal et al. [17], eight strains of *Geotrichum candidum* were isolated from soil samples and tested for inulinase activity. The activities of these strains ranged from 0.12 to 1.38 IU/mL. The strains provided a 50-fold improved inulinase activity (45.65 IU/mL) by inducing mutagenesis, using methyl methane sulphonate, ethyl methane sulphonate and UV exposure. Two forms of exo-acting extracellular inulinases have been reported from *Penicillium* spp. [79]. The inulin hydrolytic activity produced by different *Penicillium* spp. also illustrates the variability regarding various enzyme characteristics.From *Penicillium* spp. different inulinases, like *P. aculeatum*, *P. digitatum,* and *P. cyclopium*, the inulinase from *P. aculeatum* was more thermoresistant, whereas the inulinase from *P. cyclopium* was the most labile, particularly at 65°C [80]. Recently, *Rhizopus* spp. [20, 32, 47] and *Alternaria alternata* [81] have also been studied for inulinase production. The inulinase activity reported from both strains was produced extracellular.

4.3. Bacteria

Inulinases concentrations in bacteria are not equivalent to those of yeast and fungi. However, due to the ability of several bacteria to tolerate high temperatures, attempts have been made to isolate bacterial strains which can produce high quantities of thermally stable inulinases [51]. Although it has been reported that yeasts can produce more inulinase than bacteria and filamentous fungi, strain *Streptomyces* spp. GNDU 1 was still used to produce high levels of extracellular inulinase (0.552 U/mL) after 24 h at pH 7.5, temperature 46°C, in the presence of 1.0% inulin by submerged cultivation in the liquid medium [82]. Gao et al. [31], after screening over 50 bacteria strains, selected only one thermophilic strain (*Bacillus smithii* T7) from Tianjin (China), which was found to produce a large amount of extracellular inulinase. In this case the inulin was the only carbon source in the growing medium.

5. Purification of Inulinases

After obtaining through fermentation, crude inulinases can be purified in order to remove all contaminants that can interfere with their intended use and to increase their specific activity. The purification process is different for intracellular enzymes than for extracellular ones [26]. For intracellular inulinases the cell wall destruction step should be included as a step of purification [9]. Most reported protocols used for enzyme purification at laboratory scale are rarely applicable in production as the yields of recovery and productivity are very low. The basic methods for protein purification consist on the reduction in solubility by the addition of salts, organic solvents or polymers. Due to their low selectivity, they are used mostly as preliminary concentration step that can facilitate subsequent more selective operations [26]. Water-miscible organic solvents, like ethanol and acetone, have been also used since the beginnings of enzyme purification. This method is more selective and strictly based on the physicochemical properties of the enzyme [26]. Golunski et al. [83] used ethanol 55% (v/v) followed by ultrafiltration step for the purification of inulinases produced by a *Kluyveromyces marxianus* strain. The enzyme could be purified 5.5 fold with 86.1% yield proving that this simple purification method can be more efficient than some chromatographic techniques [83].

Liquid chromatography, next to size exclusion and ion exchange chromatography are powerful tools used for enzyme purification by protein fractionation, especially at laboratory scale. Affinity chromatography seems to be the most powerful purification technique in terms of resolution, as it is based mostly on the functionality of proteins; the enzymes can be specifically purified based on their catalytic and immunogenic properties, bringing the enzyme to the final desired level of purity [26]. Inulinases from *F. oxysporum* were purified by column chromatography, with yields of 65-75%. Also, inulinases from *Penicillium* spp. were purified by ion exchange chromatography after a preliminary step of ammonium sulphate precipitation [9, 84). Souza-Motta et al. [85] used molecular exclusion chromatography to purify inulinases from *A. niveus*, after enzyme fractionation with different concentrations of ammonium sulphate. The enzyme was purified 34.65 fold, and the purification yield was over 53% [85]. Fawzi [30] used Sephadex G-100 column to purify the exo-inulinases produced by *Thielavia terrestris* and *Aspergillus foetidus*, after the preliminary step of ammonium sulphate treatment while ethanol precipitation followed by chromatography on Sephadex G200 DEAE cellulose and carboxymethyl cellulose helped for the *K. fragilis* inulinase purification [9].

The inulinases produced by the filamentous fungus, *Rhizopus* spp., isolated from rhizosphere soil samples, were purified from the culture filtrate by DEAE-Cellulofine A-500 and Sephacryl S-200 HP chromatography, after ultrafiltration. The protocol conducted to 12 fold enzyme purification from culture filtrate, with a yield of 0.57% [32]. The inulinases produced by *Rhizoctonia solani* mould were purified 21.2 fold, with a yield of 11.67% by dialysis, concentration in silica gel 60 and chromatography on Sephadex G-150 and DEAE cellulose [86]. The extracellular inulinase from *A. niger* 20 OSM was purified by ion exchange chromatography on DEAE Sepharose, hydrophobic interaction on phenyl Sepharose, chromatofocusing on PBE-94 and size exclusion chromatography on Sephadex G-200 and yields over 58%, and purification factor about 22.5 were obtained [87].

The inulinase obtained from *Bacillus smithii* was precipitated with ammonium sulphate and the precipitate was loaded on anion-ion-exchange column of DEAE Sepharose, which resulted in 30 fold purification as compared to the crude extract, yielding 27.3% [31]. Kalil et al. [88] investigated the main parameters that could affect inulinase purification using an ion exchange fixed bed column. In the optimum conditions, the enzyme recovery was 67.5% and the purification factor was 6.6 fold.

6. Characteristics of Inulinases

The inulinases characterized until now in the literature show considerable variability with respect to biophysical and biochemical characteristics. Some proprieties of microbial inulinases are presented in Figure 4.

6.1. Molecular Weight of Inulinases

It has been reported that most of the inulinases from fungi have more than 50.0 kDa of molecular weight [48]. The molecular mass of the purified inulinase from the supernatant of the cell culture of the marine yeast strain *C. aureus* G7a was estimated to be 60.0 kDa [27]. The molecular mass of the purified inulinase from *P. guilliermondii* strain 1 was estimated to be 50.0 kDa [28]. The molecular mass of the purified exoinulinase from bacteria was estimated at approximately 54.0 kDa [89-90]. These studies suggest that molecular weight of the exoinulinases from bacteria is approximately as the exoinulinases from yeasts.

Source	Molecular weight, kDa	Optimum pH	Optimum temperature, °C	pH Stability Range	Temperature Stability, °C	Reference
K. marxianus	-	3.5	60	-	-	[54]
K. marxianus	-	4.4	50	-	-	[56]
K. marxianus	-	4.75	55	-	40	[94]
K. fragilis	250	-	55	-	-	[48]
Y. lipolitica	-	4.5	50	3-7	50	[3]
P. guilliermondii	50	6.0	60	6-7	60	[28]
C. aureus	60	5.0	50	4.0-6.5	65	[27]
Arthrobacter spp.	75	7.5	50	5-10.5	30...40	[29]
Bacillus spp.	-	7.0	-	6-8	25...40	[23]
Streptomyces spp.	-	6.0	60	-	60...70	[21]
A. niger	-	4.4	-	-	-	[21]
A. niger	70	5.0	40	-	-	[48]
P. janczewskii	-	4.8-5.0	35...45	-	-	[21]
A. ochraceus	79	4.5	60	-	60	[15]
F. oxysporum	300	5.8-6.2	30...37	-	-	[48]

Figure 4. Properties of Some Microbial Inulinases (adapted from Neagu Bonciu and Bahrim, [19]).

6.2. Optimum pH and Temperature for Production and Action of Inulinases

It is well known that the initial pH of the fermentation medium and incubation temperature are very important factors for production in high yield of a microbial metabolite [91]. The optimal pH activity of the purified inulinases from fungi and yeasts are in the range of 4.5–6.0 [27, 28, 48, 69, 51]. Quite the opposite, the exoinulinase activity produced by *Bacillus polymyxa* MGL21 occurs at optimal pH 7 [90]. The endoinulinase activity from *Penicillium* TN-88 is the highest at pH 5.2 [63]. In contrast, the endoinulinase from *Arthrobacter* sp S37 hydrolyzes inulin optimally at pH 7.5 [29]. Studies on *A. tamarii* [13] and *A. niger* [92] showed that during inulinase production, pH values decrease from 5.5 to 3.5 and 6.0 to 4.5 after 72 h and 96 h, respectively. This may be due to the deamination of some amino acids or the formation of organic acids [85]. The optimum temperature of inulinases activity is an extremely important factor for the application of these enzymes for commercial production of fructose or fructooligosaccharides from inulin, since high temperatures (60°C or higher) ensure proper solubility of inulin and also prevent microbial contamination [93]. Usually, the purified inulinases have optimal activity at 50-60°C.

Kluyveromyces spp. strains produce inulinases with maximum activity at 50-55°C [56, 48, 94], while inulinases obtained from *Pichia guilliermondii* [10, 28], *A. ochraceus* [15] and *Streptomyces* spp. [35] have optimum activity at 60°C, being stable at this temperature. *F. oxysporum* [48], *Penicillium janczewskii* [21] and *A. niger* [48] produce inulinases optimal at 30-40°C, indicating that the optimal temperatures of inulinases from different species of microorganisms can be significantly different. Lower temperatures have also been reported for other microorganisms, for example 45°C for *A. fumigatus* [95], 37°C for *A. niveus Blochwitz* 4128 URM [85], and 40°C for *A. tamarii* [92].

6.3. Effect of Metal Ions and Protein Inhibitors

Some metal ions available in the reaction medium can promote the inulinase activity, while others have inhibitory effect on the enzyme. Thereby, *Kluyveromyces marxianus* inulinase is activated by Co^{+2}, Mn^{+2}, Mg^{+2} and low concentrations of SDS (0.001%) and is inactivated by Cu^{+2}, Fe^{+3}, Zn^{+2}, Tween 20, Tween 80 and Brij-35 [52] and also by Ca^{+2}, Ba^{+2}, Zn^{+2}, Na^{+} [94]. Inulinase obtained from *Cryptococcus aureus* isolated from marine environment is activated by low concentrations of Ca^{+2}, K^{+}, Na^{+}, Zn^{+2} and Cu^{+2} and inhibited by Mg^{+2}, Hg^{+2} and Ag^{+} [27]. Inulinase acquired from *Pichia guilliermondii* isolated from marine environment also had a similar reaction of metal ions according to Gong et al. [28]. *Arthrobacter* spp. [29], *Streptomyces* spp. [35] and *Aspergillus ochraceus* [15] inulinases are inhibited by the presence of Hg^{+2} and Ag^{+} into the reaction medium.

The inulinase activity for *Aspergillus ficuum* JNSP5-06 was completely inhibited by Ag^{+}, strongly inhibited by Fe^{2+} and Al^{3+}, moderately inhibited by Cu^{2+} and slightly inhibited by Mn^{2+}, Zn^{2-} and SDS. K^{+}, Ca^{2+}, Li^{2+}, EDTA and urea had no significant influence on the inulinase activity, while Mg^{2+} enhanced the inulinase activity [66]. Furthermore, pepstatin, EDTA, 1,10-phenenthroline have inhibitory effects on some inulinase activity [27, 28, 29], demonstrating that the characterized enzymes were metalloenzymes.

7. Applications of Inulinases

Inulinases found their applications in a wide range of industries, ranging from food and bioethanol production to pharmacology and chemical industry.

7.1. High Fructose Syrup Obtaining

Production of fructose syrup from inulin rich feedstock is the major application of inulinases [9]. In this case, fructose is formed using a single step reaction compared to the conventional enzymatic method that converts starch to fructose using three enzymes (α-amylase, glucoamylase and glucose isomerase) [5, 9]. Compared to fructose obtaining from starch, which yields 45%, production of fructose from inulin with inulinases can yield over 95% D-fructose [5]. Free fructose can be obtained also by acid hydrolysis, although in this case, fructose is easily degraded and coloured inhibitory by-products like difructose anhydrides can be produced [5, 84]. Fructose is a sweetener used in food and beverages industry, having a sweetening power with 70% higher than sucrose [96]. Also, fructose is more soluble than sucrose, it is less viscous and can be metabolized without insulin when present in low concentrations. Plus, sucrose can cause problems related to obesity, carcinogenicity and atherosclerosis [5]. Fructose production from chicory inulin enzymatic hydrolysis was studied by Ricca et al. [97]. The possible complete reaction scheme was presented for inulin hydrolysis and the kinetic study for the hydrolysis of inulin by a commercial enzyme was established.

7.2. Inulo-Oligosaccharide Production

Inulooligosaccharides (IOS) are functional molecules with prebiotic activity that positively influence the gut microbiota, increasing the bifidobacteria population and improving mineral absorption [8, 13, 84, 87]. Also, IOS are used as fibre source in foods [5, 31]. Besides, IOS have been suggested to have growth inhibition effect on tumors [31]. Inulinases were widely used for IOS production from inulin, which have been suggested to have similar physiological functions to those of fructooligosaccharides [8, 31, 87]. IOS obtaining from inulin using *Pseudomonas* spp. had a maximum yield of 75.6% and IOS with DP2 to DP7 were produced [8]. Zhengyu et al. [98] studied the production of inulooligosaccharides from inulin, Jerusalem artichoke powder and juice, using endoinulinases from *Aspergillus ficuum*. The best yield was obtained from Jerusalem artichoke juice (79.8%), followed by inulin (70.37%) and IOS with DP2 to DP8 were obtained [98].

7.3. Bioethanol Production

Another application of inulinases is the production of fuel ethanol from inulin rich feedstock [13, 87]. Alcohol production from Jerusalem artichoke was studied since the beginning of 19th century [99]. Since then, different studies concerning the bioethanol obtaining from inulin rich raw materials were made, as ethanol is the most important liquid biofuel used either as fuel or as gasoline enhancer. Zhang et al. [100] obtained around 15 mL of ethanol in 100 mL of medium and over 98.9% of the total sugars utilization by inulin hydrolysis with inulinases produced by *Pichia pastoris* and sugar fermentation by *Saccharomyces* sp. yeast. High ethanol concentrations (88.1 g/l) were obtained also by Thuesombat et al. [101] by batch ethanol fermentation of the acid hydrolyzed inulins from

Jerusalem artichoke juice using *S. cerevisiae* yeast strain. To shorten the fermentation cycle and enhance the conversion efficiency, Ge and Zhang [102] used *A. niger* exoinulinase hyper producing strain and *S. cerevisiae* for rapid and complete saccharification and fermentation of Jerusalem artichoke flour. 19.6% ethanol was produced after 48 hours, and a 90% of the theoretical yield was obtained.

7.4. Chemicals Production

Inulin rich feedstock have recently received attention as a renewable raw material for the production of other chemicals also, like citric acid, lactic acid, 2,3-butanediol, mannitol, etc. Liu et al. [3] obtained a 12% of inulin conversion by the inulinases produced by surface-engineered *Yarrowia lipolytica*, and 77.9 g/l of citric acid after fermentation of the released sugars. Mannitol, a naturally occurring polyol used in food, pharmaceutical, medicine and chemical industries, was obtained from inulin by *Lactobacillus intermedius* using simultaneous saccharification and fermentation approach and over 220 g/l mannitol was obtained [2]. Choi et al. [103] performed lactic acid fermentation of Jerusalem artichoke tuber extract using *Lactobacillus paracasei* without prior hydrolysis of inulin and obtained a conversion efficiency of inulin to lactic acid of 98% of the theoretical yield.

Conclusion

From the information presented above, it is very obvious that many advances in microbial inulinases production and characterization were made. A part of the research work has been also focused on practical applications as inulin-oligosaccharides, bioethanol, and fructose production from inulin and inulin-containing materials. Various inulin-containing materials could be economically hydrolyzed with inulinases into fructose and inulooligosaccharides, which can be used for many industrial areas. Fungal inulinases have also many therapeutic and industrial applications and beneficial environmental impact as a result of biovalorisation of agrowastes.

Conflict of Interest

The authors declare that there are no conflicts of interest.

References

[1] Chi ZM, Zhang T, Cao TS, Liu XY, Cui W, Zhao CH (2011) Biotechnological potential of inulin for bioprocesses. *Bioresource Technol* 102:4295-4303

[2] Saha BC (2006) Production of mannitol from inulin by simultaneous enzymatic saccharification and fermentation with *Lactobacillus intermedius* NRRL B-3693. *Enzyme and Microb Tech* 39:991-995

[3] Liu XY, Chi Z, Liu GL, Wang F, Madzak C, Chi ZM (2010) Inulin hydrolysis and citric acid production from inulin using the surface-engineered *Yarrowia lipolytica* displaying inulinase. *Metab Eng* 12:469-476

[4] Kays SJ, Nottingham SF (2007) *Biology and chemistry of Jerusalem artichoke* (*Helianthus tuberosus L.*). CRC Press USA

[5] Singh P, Gill PK (2006) Production of inulinases: recent advances. *Food Technol Biotech* 44(2):151-162

[6] Nagem RAP, Rojas AL, Golubev AM, Korneeva OS, Eneyskaya EV, Kulminskaya KN, Neustroev KN, Polikarpov I (2004) Crystal structure of exo-inulinase from *Aspergillus awamori*: the enzyme fold and structural determinants of substrate recognition. J Mol Biol 344:471-480

[7] Molina DL, Martinez MDN, Melgarejo FR, Hiner ANP, Chazarra S, Lopez JNR (2005) Molecular properties and prebiotic effect of inulin obtained from artichoke (*Cynara scolymus* L.). *Phytochemistry* 66:1476-1484

[8] Chi Z, Chi Z, Zhang T, Liu G, Yue L (2009b) Inulinase-expressing microorganisms and applications of inulinases. *Appl Microbiol Biotechnol* 82:211-220

[9] Pandey A (2004*) Concise Encyclopedia of Bioresource Technology, Food Products* Press, Binghamtom, NY, USA

[10] Chi Z, Chi Z, Zhang T, Liu G, Li J, Wang X (2009a) Production characterization and gene cloning of the extracellular enzymes from the marine-derived yeasts and their potential applications. *Biotechnol Adv* 27:236-255

[11] Gern RMM, Furlan SA, Ninow JL, Jonas R (2001) Screening for microorganisms that produce only endo-inulinase. *Appl Microbiol Biotechnol* 55:632-635

[12] Ge XY, Zhang WG (2005a) Effects of octadecanoylsucrose derivatives on the production of inulinase by *Aspergillus niger* SL-09. *World J Microb Biot 21*:1633-1638

[13] Kango N (2008) Production of inulinase using tap roots of dandelion (*Taraxacum officinale*) by *Aspergillus niger*. *J Food Eng* 85:473-478

[14] Gill PK, Manhas RK, Singh P (2006) Comparative analysis of thermostability of extracellular inulinase activity from *Aspergillus fumigates* with commercially available (Novozyme) inulinase. *Bioresource Technol* 97:355-358

[15] Guimaraes LHS, Terenzi HF, Polizeli ML, Jorge JA (2007) Production and characterization of a thermostable extracellular β-D-fructofuranosidase produced by *Aspergillus ochraceus* with agroindustrial residues as carbon sources. *Enzyme Microb Tech* 42:52-57

[16] Ertan F, Ekinci F, Aktac T (2003b) Production of inulinases from *Penicillium spinulosum Aspergillus parasiticus* NRRL 2999 and *Trichoderma viride*. Pak J Biol Sci 6(15) 1332-1335

[17] Mughal MS, Ali S, Ashiq M, Talish AS (2009) Kinetics of an extracellular exo-inulinase production from a 5-flourocytosine resistant mutant of *Geotrichum candidum* using two-factorial design. *Bioresource Technol* 100:3657-3662

[18] Ertan F, Aktac T, Kaboglu C, Ekinci F, Bakar E (2003a) Determination of optimum cultivation conditions on the production of inulinase from *Rhizoctonia solani*. *Pak J Biol Sci* 6(16):1386-1388

[19] Neagu (Bonciu) C, Bahrim G (2011) Inulinases - a versatile tool for biotechnology. *Innov Rom Food Biotechnol* 9(2):1-11

[20] Neagu Bonciu C, Constantin O, Bahrim G (2012) Increase in extracellular inulinase production for a new *Rhizoctonia* ssp. strain by using buckwheat (*Fagopyrum esculentum*) flour as a single carbon source. *Lett Appl Microbiol* 55 (3):195-201

[21] Sharma AD, Kainth S, Gill PK (2006) Inulinase production using garlic (*Allium sativum*) powder as a potential substrate in *Streptomyces* sp.. *J Food Eng* 77:486-491

[22] Dilipkumar M, Rajasimman M, Rajamohan N (2011a) Response surface methodology for the optimization of inulinase production by *K. marxianus* var. *marxianus*. *JASES* 6(1):85-95

[23] Zherebtsov NA, Shelamova SA, Abramova IN (2002) Biosynthesis of inulinases by *Bacillus* bacteria. *Appl Biochem Microb* 38(6):634-638

[24] Gao L, Chi Z, Sheng J, Wang L, Li J, Gong F (2007b) Inulinase-producing marine yeasts: evaluation of their diversity and inulin hydrolysis by their crude enzymes. *Microbial Ecol* 54:722-729

[25] Yu X, Guo N, Chi Z, Gong F, Sheng J, Chi Z (2009) Inulinase overproduction by a mutant of the marine yeast *Pichia guilliermondii* using surface response methodology and inulin hydrolysis. *Biochem Eng J* 43:266-271

[26] Illanes A (2008) *Enzyme Biocatalysis – Principles and Applications*, Eds. Andres Illanes, Springer

[27] Sheng J, Chi Z, Yan K, Wang X, Gong FJL (2008) Use of response surface methodology for optimizing process parameters for high inulinase production by the marine yeast *Cryptococcus aureus* G7a in solid-sate fermentation and hydrolysis of inulin. *Bioprocess Biosys Eng* 32:333-339

[28] Gong F, Zhang T, Chi Z, Sheng J, Li J, Wang X (2008) Purification and characterization of extracellular inulinase from a marine yeast *Pichia guilliermondii* and inulin hydrolysis by the purified inulinase. *Biotechnol Bioproc E* 13:533-539

[29] Kang SI, Chang YJ, Oh SJ, Kim SI (1998) Purification and properties of an endo-inulinase from an *Arthrobacter* sp. *Biotechnol Lett* 20 (10):983-986

[30] Fawzi EM (2011) Comparative study of two purified inulinases from thermophile *Thielavia terrestris* NRRL 8126 and mesophile *Aspergillus foetidus* NRRL 337 grown on *Cichorum intybus*. L. *Braz J Microbiol* 42:633-649

[31] Gao W, Bao Y, Liu Y, Zhang X, Wang J, An L (2009) Characterization of thermo-stable endoinulinase from a new strain *Bacillus smithii* T7. *Appl Biochem Biotech* 157:498-506

[32] Ohta K, Suetsugu N, Nakamura T (2002) Purification and properties of an extracellular inulinase from *Rhizopus* sp. strain TN-96. *J Biosci Bioeng* 94(1):78-80

[33] Pessoni RA, Freshour G, Figueiredo-Ribeiro RCL, Hahn MG, Braga MR (2002) Woronin Bodies in *Penicillium janczewskii* Zaleski. *Braz J Microbiol* 33:127-130

[34] Gill P, Manhas R, Singh J, Singh P (2006) Purification and properties of a heat-stable exoinulinase isoform from *Aspergillus fumigatus*. *J Food Eng* 76:369-375

[35] Sharma AD, Gill PK (2006) Purification and characterization of heat-stable exo-inulinase from *Streptomyces* sp.. *J Food Eng* 79(4):1172-1178

[36] Kochhar A, Gupta AK, Kaur N (1999) Purification and immobilization of inulinase from *Aspergillus candidus* for producing fructose. *J Sci Food Agric* 79:549-554

[37] Vijayaraghavan K, Yamini D, Ambika V, Sowdamini NS (2009) Trends in inulinase production – A review. *Crit Rev Biotechnol* 29:67-77

[38] Kaur N, Gupta AK (2002) Applications of inulin and oligofructose in health and nutrition. *J Biosci* 27:703-714

[39] Franck A (2006) Inulin. In: *Food Polysaccharides and Their Applications* AM Stephen GO Phillips PA Williams (Eds.), CRC Press Boca Raton, FL, USA

[40] Singh RS, Singh RP (2010) Fructooligosaccharides from inulin as prebiotics. *Food Technol Biotech* 48 (4):435-450

[41] Ricca E, Calabrò V, Curcio S, Lorio G (2007) The state of the art in the production of fructose from inulin enzymatic hydrolysis. *Crit Rev Biotechnol* 27:129-145

[42] Sangeetha PT, Ramesh MN, Prapulla SG (2005) Recent trends in microbial production analysis and applications of fructooligosaccharides. *Trends Food Sci Technol* 16:442-457

[43] Roberfroid M (2005) *Inulin - type fructans – functional food ingredients* CRC Press LLC USA

[44] Guerrero AEC, Olvera JL, Garibay MG, Ruiz LG (2006) Inulinase hyperproducing strains of *Kluyveromyces* sp. isolated from aguamiel (*Agave* sap) and pulque. *World J Microbiol Biotechnol* 22:115-117

[45] Kalil SJ, Suzan R, Maugeri F, Rodrigues MI (2001) Optimization of inulinase production by *Kluyveromyces marxianus* using factorial design. *Appl Biochem Biotech* 94:257-264

[46] Santisteban BOYS, Converti A, Filho FM (2009) Effects of carbon and nitrogen sources and oxygenation on the production of inulinase by *Kluyveromyces marxianus*. *Appl Biochem Biotechnol* 152:249-261

[47] Neagu (Bonciu) C, Constantin OE, Bahrim G (2011) Screening of biotechnological parameters for fructofuranosidases production by newly isolated fungal strain by using Plackett-Burman design. *Not Bot Hort Agrobot Cluj* 39(2):271-275

[48] Pandey A, Soccol CR, Selvakumar P, Soccol VT, Krieger N, Fontana JD (1999) Recent developments in microbial inulinases. *Appl Biochem Biotech* 81:35-52

[49] Danilcenko H, Jariene E, Aleknaviciene P (2008) Quality of Jerusalem artichoke tubers in relation to storage conditions. *Not Bot Hort Agrobot Cluj* 36(2):23-27

[50] Bekers M, Grube M (2008) *Inulin syrup from dried Jerusalem artichoke*. LLU Raksti. 21(315):116-121

[51] Singh RS, Dhaliwal R, Puri M (2006) Production of inulinase from *Kluyveromyces marxianus* YS-1 using root extract of *Asparagus racemosus*. *Process Biochem* 41:1703-1707

[52] Singh RS, Bhermi HK (2008) Production of extracellular exoinulinase from *Kluyveromyces marxianus* YS-1 using root tubers of *Asparagus officinalis*. *Bioresource Technol* 99:7418-7423

[53] Dilipkumar M, Rajasimman M, Rajamohan N (2011b) Application of statistical design for the production of inulinase by *Streptomyces* sp. using pressmud. *Front Chem Sci Eng* 5(4):463-470

[54] Mazutti M, Ceni G, Luccio M, Treichel H (2007) Production of inulinase by solid-state fermentation: effect of process parameters on production and preliminary characterization of enzyme preparations. *Bioprocess Biosyst Eng* 30:297-304

[55] Bender JP, Mazutti MA, Oliveira D, Luccio M, Treichel H (2006) Inulinase production by *Kluyveromyces marxianus* NRRL Y-7571 using solid state fermentation. *Appl Biochem Biotech* 129-132:951-958

[56] Mazutti MA, Skrowonski A, Boni G, Zabot GL, Silva MF, Oliveira D, Luccio M, Filho FM, Rodrigues MI, Treichel H (2010) Partial characterization of inulinases obtained by submerged and solid-state fermentation using agroindustrial residues as substrates: a comparative study. *Appl Biochem Biotechnol* 160:682-693

[57] Sguarezi C, Longo C, Ceni G, Boni G, Silva MF, Luccio M, Mazutti MA, Maugeri F, Rodrigues MI, Treichel H (2009) Inulinase production by agro-industrial residues: optimization of pretreatment of substrates and production medium. *Food Bioprocess Tech* 2:409-414

[58] Treichel H, Mazutti MA, Filho FM, Rodrigues MI (2009a) Technical viability of the production partial purification and characterization of inulinase using pre-treated agroindustrial residues. *Bioprocess Biosyst Eng* 32:425-433

[59] Chen HQ, Chen XM, Chen TX, Xu XM, Jin ZY, (2011) Extraction optimization of inulinase obtained by solid state fermentation of *Aspergillus ficuum* JNSP5-06. *Carbohyd Polym* 85:446-451

[60] Gao LM, Chi ZM, Sheng J, Ni XM, Wang L (2007a) Single-cell protein production from Jerusalem artichoke extract by a recently isolated marine yeast *Cryptococcus aureus* G7a and its nutritive analysis. *Appl Microbiol Biotechnol* 77:825-832

[61] Ettalibi M, Baratti JC (1987) Purification properties and comparison of invertase exoinulinases and endoinulinases of *Aspergillus ficuum*. *Appl Microbiol Biotechnol* 26:13-20

[62] Wei W, Yu X, Dai Y, Zheng J, Xie Z (1997) *Purification and properties of inulinase from Kluyveromyces sp.* Y-85. Wei Sheng Wu Xue Bao 37(6):443-448

[63] Nakamura T, Kuramori K, Zaita N, Akimoto H, Ohta K (2001) Purification and properties of intracellular exo- and endo-inulinases from *Aspergillus niger* strain 12. *Bull Fac Agric* Miyazaki University 4849-4858

[64] Jing W, Zhengyu J, Bo J, Xueming X (2003) Separation and identification of exo- and endoinulinase from *Aspergillus ficuum*. *Curr Microbiol* 47:109-112

[65] Skowronek M, Fiedurek J (2003) Selection of biochemical mutants of *Aspergillus niger* resistant to some abiotic stresses with increased inulinase production. *J Appl Microbiol* 95:686-692

[66] Chen HQ, Chen XM, Li Y, Wang J, Jin ZY, Xuet XM (2009) Purification and characterization of exo- and endoinulinase from *Aspergillus ficuum* JNSP5-06. *Food Chem* 115:1206-1212

[67] Sheng J, Chi Z, Li J, Gao L, Gong F (2007) Inulinase production by the marine yeast *Cryptococcus aureus* G7a and inulin hydrolysis by the crude inulinase. *Process Biochem* 42(5):805-811

[68] Gong F, Sheng J, Chi ZM, Li J (2007) Inulinase production by a marine yeast *Pichia guilliermondii* and inulin hydrolysis by the crude inulinase. *J Ind Microbiol Biotechnol* 34:179-185

[69] Zhang L, Zhao C, Wang J, Ohta Y, Wang Y (2005) Inhibition of glucose on an exoinulinase from *Kluyveromyces marxianus* expressed in *Pichia pastoris*. *Process Biochem* 40:1541-1545

[70] Cruz-Guerrero AE, Olvera JL, Garcia-Garibay M, Gomez-Ruiz L (2006) Inulinase-hyperproducing strains of *Kluyveromyces* sp. isolated from aguamiel (*Agave sap*) and pulque. *World J Microbiol Biotechnol* 22:115-117

[71] Mazutti M, Bender JP, Treichel H, Luccio MD (2006) Optimization of inulinase production by solid-state fermentation using sugarcane bagasse as substrate. *Enzyme Microb Technol* 39:56-59

[72] Chen X, Wang JH, Li DS (2007) Optimization of solid-state medium for the production of inulinase by *Kluyveromyce*s S120 using response surface methodology. *Biochem Eng J* 34:179-184

[73] Treichel H, Mazutti MA, Maugeri F, Rodrigues MI (2009b) Use of a sequential strategy of experimental design to optimize the inulinase production in a batch bioreactor. *J Ind Microbiol Biot* 36:895-900

[74] Xiao R, Tanida *M, Takao S (1988) Inulinase from Chrysosporium pannorum. J Ferment* Technol 66(5):553-558

[75] Barthomeuf C, Regerat F, Pourrat H (1991) Production of inulinase by a new mould of *Penicillium rugulosum*. *J Ferment Bioeng* 72:491-494

[76] Kim MK, Kim YH, Kim HR, Kim BI, Byun SM, Uhm TB (1994) Thermal stability of an acidic inulinase from *Scytalidium acidophilum*. *Biotechnol Lett* 16:965-966

[77] Ettalibi M, Baratti JC (1990) Molecular and kinetic properties of *Aspergillus ficuum* inulinases. *Agr Biol Chem* 5461-5468

[78] Gill PK, Manhas RK, Singh P (2006) Hydrolysis of inulin by immobilized thermostable extracellular exoinulinase from *Aspergillus fumigates*. *J Food Eng* 76:369-375

[79] Wenling W, Zhaong X, Qian Z, Xinsheng Z (1997) Purification and properties of *Penicillium* extracellular inulinase. *Ind Microbiol* 27:8-12

[80] Hazaa MM (1999) Purification and characterization of inulinase enzymes from *Penicillium* species. *J Union Arabian Biol* 8:467-482

[81] Hamdy HS (2002) Purification and some important characters of extracellular inulinase of *Alternaria alternata* (Fr.) Keissler. *Indian J Exp Biol* 40:1393-1398

[82] Gill P, Sharma A, Harchand R, Singh P (2003) Effect of media supplements and culture conditions on inulinase production by an actinomycete strain. *Bioresource Technol* 87(3):359-362

[83] Golunski SM, Silva MF, Luccio M, Treichel H (2011) Ethanol precipitation and ultrafiltration of inulinases from *Kluyveromyces marxianus*. *Sep Pur Tech* 78(3):261-265

[84] Pessoni RAB, Figueiredo-Ribeiro RCL, Braga MR (1999) Extracellular inulinases from *Penicillium janczewskii* a fungus isolated from the rhizosphere of *Veronia herbacea* (Asteraceae). *J Appl Microbiol* 87:141-147

[85] Souza-Motta CM, Queiroz Cavalcanti MA, Figueirido Porto AL, Moreira KA, Lima Filho JL (2005) *Aspergillus niveus* Blochwitz 4128URM: new source for inulinase production. *Braz Arch Biol Techn* 48(3):343-350

[86] Ertan F, Sanal FE, Kaboglu A, Aktac T, Bakar E (2005) Some properties of inulinase from *Rhizoctonia solani*. *J Biol Sci* 5(3):330-334

[87] Skowronek M, Fiedurek J (2006) Purification and properties of extracellular endoinulinase from *Aspergillus niger* 20 OSM. *Food Technol Biotech* 44(1):53-58

[88] Kalil SJ, Silveira ST, Maugeri-Filho F, Rodrigues MI (2010) Evaluation of different parameters for the purification of inulinase using an ion exchange fixed bed. *Biotechnol Bioproc E* 15:676-679

[89] Tsujimoto Y, Watanabe A, Nakano K, Watanabe K, Matsui H, Tsuji K, Tsukihara T, Suzuki Y (2003) Gene cloning, expression, and crystallization of a thermostable exo-

inulinase from Geobacillus stearothermophilus KP1289. *Appl Microbiol Biotechnol* 62:180-185

[90] Kwon HJ, Jeon SJ, You DJ, Kim KH, Jeong YK, Kim YH, Kim YM, Kim BW (2003) Cloning and characterization of an exoinulinase from *Bacillus polymyxa*. *Biotechnol Lett* 25:155-159

[91] Erdal S, Canli O, Algur OF (2011) Inulinase production by *Geotrichum candidum* using Jerusalem artichoke as sole carbon source. *Rom Biotech Lett* 16(4):6378-6383

[92] Saber W, El-Naggar N (2009) Optimization of fermentation conditions for the biosynthesis of inulinase by the new source *Aspergillus tamarii* and hydrolysis of some inulin containing agro-wastes. *Biotechnology* 8(4):425-433

[93] Vandamme EJ, Derycke DG (1983) Microbial inulinases: Fermentation process properties and applications. *Adv Appl Microbiol* 29:139-176

[94] Kushi RT, Monti R, Contiero J (2000) Production, purification and characterization of an extracellular inulinase from *Kluyveromyces marxianus* var. *bulgaricus*. *J Ind Microbiol Biotechnol* 25:63-69

[95] Gouda M (2002) Some properties of inulinase from *Aspergillus fumigatus*. *Pak J Biol Sci* 5(5):589-593

[96] Pessoni RAB, Braga MR, Figueiredo-Ribeiro RCL (2007) Purification and properties of exo-inulinases from *Penicillium janczewskii* growing on distinct carbon sources. *Mycologia* 99(4):493-503

[97] Ricca E, Calabro V, Curcio S, Iorio G (2009) Fructose production by chicory inulin enzymatic hydrolysis: a kinetic study and reaction mechanism. *Process Biochem* 44:466-470

[98] Zhengyu J, Jing W, Bo J, Xueming X (2005) Production of inulooligosaccharides by endoinulinases from *Aspergillus ficuum*. *Food Res Int* 38:301-308

[99] Matias J, Gonzalez J, Royano L, Barrena RA (2011) Analysis of sugars by liquid chromatography-mass spectrometry in Jerusalem artichoke tubers for bioethanol production optimization. *Biomass Bioenerg* 35:2006-2012

[100] Zhang T, Chi Z, Zhao CH, Chi ZM, Gong F (2010) Bioethanol production from hydrolysates of inulin and the tuber meal of Jerusalem artichoke by *Saccharomyces* sp. W0. *Bioresource Technol* 101:8166-8170

[101] Thuesombat P, Thanonkeo P, Laopaiboon L, Laopaiboon P, Yunchalard S, Kaewkannetra P, Thanonkeo S (2007) The batch ethanol fermentation of Jerusalem artichoke using *Saccharomyces cerevisiae*. KMITL *Sci Tech J* 7(S2):93-96

[102] Ge XY, Zhang WG (2005b) A shortcut to the production of high ethanol concentration from Jerusalem artichoke tubers. *Food Technol Biotech* 43(3):241-246

[103] Choi HY, Ryu HK, Park KM, Lee EG, Lee H, Kim SW, Choi ES (2012) Direct lactic acid fermentation of Jerusalem artichoke tuber extract using *Lactobacillus paracasei* without acidic or enzymatic inulin hydrolysis. *Bioresource Technol* 114:745-747

[104] Vandamme E, De Baets S, Steinbuechel A (editors) (2004) *Biopolymers vol. 5-6 –* Polysaccharides I and II Inulin, Wiley VCH, Germany

In: Microbes in Process
Editors: Neelam Garg and Abhinav Aeron

ISBN: 978-1-63117-127-7

Chapter 9

Functional Foods: Health Benefits and Future Challenges

Shadia M. Abdel-Aziz[1*] and Foukia E. Mouafi[2]
[1]Microbial Chemistry Dept.,
[2]Microbial Biotechnology Dept, Genetic Engineering and Biotechnology Division,
National Research Center, Dokki, Cairo, Egypt

Abstract

The term "functional foods" comprises safely bacterial strains and some products of plant and animal origin which physiologically containing active compounds beneficial for human health and reducing the risk of chronic diseases. Functional foods represent one of the most intensively investigated and widely promoted areas in the food and nutrition sciences today. Among the best known functional compounds, probiotics, prebiotics, and natural antioxidants are good examples. Functional foods can considered to be those whole, fortified, enriched or enhanced foods that provide health benefits beyond the provision of essential nutrients (e.g., vitamins and minerals). Scientific progress in understanding the relationship of diet to disease, along with increasing health-care costs and consumers' desires to make healthy lifestyle improvements, provides a significant impetus for the development of novel foods with health benefits (functional foods). These functional food products result from: technological innovation at the processing level, such as cholesterol lowering spreads, xylitol-sweetened chewing gum, and dairy products fermented with specific lactic acid bacteria and technological innovation.

1. Introduction

Consumer interest in healthy eating is shifting towards the potential health benefits of specific foods and food ingredients. Moreover, scientific evidence supports the idea that some of these might have positive effects on our health and well-being, beyond the provision of

* Email: *abdelaziz.sm@gmail.com, nannarose58@yahoo.com.*

basic nutritional requirements. There is a growing interest in the subject of *functional foods*. Over 60 percent of the risk for chronic diseases such as heart disease, stroke, colon cancer, and type II diabetes is potentially preventable by life-style modifications, including changes in diet. This scientific progress in understanding the relationship between diet and disease along with the consumer's desire to initiate healthy eating and lifestyle habits, led to the development of food products that meet these demands [1]. In relation to heart health, increasing consumption of plant-based foods is considered a key life style recommendation for whose heart disease is a leading cause of death. Food products enriched for soy protein, plant sterols and stanols, omega-3 fatty acids as well as antioxidants and fiber are being formulated of functional foods to the consumer. Functional food products are also being promoted for the improvement and health of joints, muscles, and bones due to the actions of such ingredients as glucosamine, calcium, anti-inflammatory and anti-oxidant nutrients.

Health benefits of functional foods in various areas such as obesity, diabetes, allergy, osteoporosis, mental health, cancer and cardiovascular disease, are well documented. All foods have a function when consumed in proper balance as part of an overall healthy diet. Functional foods may include whole foods, such as fruits and vegetables, which represent the simplest example. Those foods that have been fortified, enriched, or enhanced with nutrients, phytochemicals, or botanicals, as well as dietary supplements, also fall within the realm of functional foods. The functional attributes of many traditional foods are now being discovered. Examples include phytoestrogens in soy foods, omega-3 fatty acids, and a variety of antioxidants in fruits and vegetables, such as lycopene in tomatoes. Still, new food products are being developed with beneficial components, with a focus on wellness and the reduced risk of chronic disease (i.e., foods and beverages containing pre-and probiotics to maintain gastrointestinal health, calcium-fortified beverages to maintain bone health, and dressings which may decrease the risk of heart disease).

New promising technologies such as nutrigenomics, imaging techniques, converging technologies are increasingly being used in nutrition research. Their huge potential will be apparent in the short and medium term; this will further enable the development of foods for targeted population groups with defined risk factors or diseases such as allergy, diabetes, obesity and cardiovascular disease. Even more innovative is the possibility of merging information about the physiological responses to food with individual genetic information to design personalized food and diets. The ingenuity of food technology might also contribute to further advances in the development of food products that can support optimum health.

The functional food industry, consisting of food, beverage and supplement sectors, is one of the several areas of the food industry that is experiencing fast growth in recent years [2]. It is estimated by BCC Research that the global market of functional food industry will reach 176.7 billion in 2013 with a compound annual growth rate of 7.4%. This kind of growth is fueled not only by industrial innovation and development of new products that satisfy the demand of health consumers but also by health claims covering a wide range of health issues [3]. The industry suggests the establishment of a health claim regulating agency, which may increase consumer confidence. Strict examination of some of the functional food claims may discourage some companies from launching their products [3].

2. History

Over two thousand years ago Hippocrates (Greek physician regarded as the father of medicine) said, "Let food be thy medicine and medicine be thy food" [4]. Although the concept of functional foods is not entirely new, it has evolved considerably over the years. In the early 1900s food manufacturers in the United States began adding iodine to salt in an effort to prevent goiter, representing one of the first attempts at creating a functional food through fortification. The term Functional food was first used in Japan in the 1980s where there is a government approval process for functional foods called Foods for Specified Health Use (FOSHU). Other twentieth-century examples include vitamin A and D fortification of milk and niacin and folic acid fortification of grains. These early fortification examples, however, focused on reducing the risk of diseases of deficiency. In the latter part of the twentieth century, consumers began to focus on wellness and the reduction of chronic disease. Research now focuses frequently on the promotion of health through many lifestyle factors, including the consumption of an optimal diet. As of 2002, researchers have identified hundreds of food components with functional qualities, and they continue to make new discoveries surrounding the complex benefits of phytochemicals in foods. As in 1998 and 2000, the vast majority of consumers believe that they have a "great amount" of control over their own health. Also, in comparing the effects of nutrition, exercise, and family health history on health, consumers believe that nutrition plays the greatest role (71 percent versus 63 percent and 41 percent, respectively). Therefore, it is no surprise that 93 percent of Americans believe that some foods have health benefits that go beyond basic nutrition and that 85 percent are interested in learning more about such foods [5]. These levels of interest have been consistently strong since 1998.

3. Definition

Functional food is a whole food or a food where a new ingredient(s) has been added and the new product has a new function (often related to health-promotion or disease prevention). The general category of functional foods includes processed food or foods fortified with health-promoting additives, like "vitamin-enriched" products. Products considered functional generally do not include products where fortification has been done to meet government regulations and the change is not recorded on the label as a significant addition ("invisible fortification"). An example of this type of fortification would be the historic addition of iodine to table salt, or Vitamin D to milk, done to resolve public health problems such as rickets [5].While soy, which contains soy protein associated with a reduction in cardiovascular disease, is a functional food as well as oat bran (for the fiber content). Functional foods are part of the continuum of products that individuals may consume to increase their health and/or contribute to reducing their disease burden. Functional Food is a Natural or processed food that contains known biologically-active compounds which when in defined quantitative and qualitative amounts provides a clinically proven and documented health benefit, and thus, an important source in the prevention, management and treatment of chronic diseases of the modern age. Many definitions exist worldwide for functional foods, but there is no official or commonly accepted definition. One view is that any food is indeed functional because it provides nutrients and has a physiological effect [5]. So, functional food should be considered a marketing term for a food whose attraction lies in its health claims and the way the product is perceived. Some even believe that, any food, if marketed appropriately, particularly with an accompanying health claim, is a functional food. Some foods considered to be functional are actually natural whole foods where new scientific information about their health qualities can be used to proclaim benefits.

The EC Concerted Action on Functional Food Science in Europe (FUFOSE) proposed a working definition of functional food: a food that beneficially affects one or more target functions in the body beyond adequate nutritional effects in a way that is relevant to either an improved state of health and well-being and/or reduction of risk of disease. It is consumed as part of a normal food pattern. It is not a pill, a capsule or any form of dietary supplement. While there is increased awareness of the links between diet and disease such as certain fats and cardiovascular disease, calcium and osteoporosis, fibre and gastrointestinal health, it is important, for a functional food, to identify the specific food constituents that could promote health and well-being as well as the exact conditions where they can have this beneficial effect. Functional food must be a natural food such as fruit or grain which may or may not be modified by plant breeding or other technologies (e.g. lycopene-enhanced tomatoes, vitamin E-enriched vegetable oils, vitamin A-enriched rice). Functional food must also be a food from which a component has been removed or reduced (e.g. a yogurt with reduced fat); or a food in which one, or several components, have been modified, replaced or enhanced to improve its health properties (e.g. a juice drink with enhanced antioxidant content, a yogurt with added prebiotic or probiotic). Many, if not most, fruits, vegetables, grains, fish, and dairy and meat products contain several natural components that deliver benefits beyond basic nutrition. Examples include lycopene in tomatoes, omega n-3 fatty acids in salmon or saponins in soy. Even tea and chocolate have been noted in some studies as possessing functional attributes, i.e. attributes beyond the provision of traditional nutrients. Others think that only fortified, enriched or enhanced foods with a component having a health benefit beyond basic nutrition should be considered as functional. Most definitions also suggest that a functional food should be, or look like, a traditional food and must be part of our normal diet. A functional food can be targeted at the whole population or at particular groups, which may be defined, for example, by age or genetic constitution.

Functional foods have been developed due to the growing interest in the relationship between diet, specific food ingredients and health. Healthy eating can make a key contribution to health and well-being, but busy consumers may not have the time to access their optimal diet. Functional foods can provide health enhancing ingredients in a convenient form. Increase of chronic diseases may be attributed to the diet problems and changes in lifestyle. We now have a real challenge to balance energy intake and expenditure as more of us are becoming overweight. In 2008, across the 27 countries of the EU, 59 % of adult men and 48 % of adult women were either overweight or obese. For much of its history, nutrition science has focused on the role of essential nutrients in preventing deficiencies. Further, the prevalence of chronic non-communicable diseases such as cardiovascular disease, high blood pressure and type 2-diabetes has increased, leading to concerns about escalating costs of healthcare and reduced quality of life. Modern lifestyles and longer life are also linked to various mental health problems such as depression, poor concentration and loss of memory.

3.1. Health Benefits of Functional Foods

Mother's diet and nutrition during pregnancy and lactation can influence the short and long-term development of the child. It might even be linked to the development of conditions such as high blood pressure, heart disease and diabetes in later life. More specifically, the course of pregnancy and childbirth and the composition of breast milk can be influenced by energy and protein intake as well as nutrients such as n-3 and n-6 polyunsaturated fatty acids, amino acids and micronutrients including folic acid, iron, zinc and iodine. Such nutrients may be useful as ingredients in functional foods. Calcium, vitamins D and K, fluoride as well as prebiotic fructans could offer possibilities as ingredients in functional foods. Health benefits of some marketed functional foods are represented in Figure 1 [6, 7]. Gastrointestinal growth during the early years may be influenced by food ingredients such as probiotics and also prebiotics, such as oligosaccharides and inulin [8].

Ingredient	Example product	Health benefits
Probiotic bacteria	Yoghurt and milk products	Improve the gut functions and enhance immunity
Omega-3 and Omega-6 fatty acids	Fish, tuna, shrimp, fruits, nut, eggs, cereals, some vegetables	Immune defense, heart protection, eyesight development
Isoflavones (phyto-estrogens) Flavonoides	Soy products Green tea, mint, chocolate, lettuce, fresh vegetables.	Reduce osteoporosis, prevent cardiovascular disease, improve cardiovascular health
Plant phytosterols and phytosanols	Yoghurt, margarine, barley,	Lower LDL cholesterol
Conjugated linoleic acid	Yoghurt, dairy products, meat, food supplements	Weight loss, cancer prevention
Herbs and herbal extracts	Green tea, drinks, cereals	Energy enhancement, memory improvement, weight loss
Non-starch polysaccharides and other fibers (Oats*)	Cereals, soy-been, fresh fruits	Improve mineral absorption, lower cholesterol, weight loss, improve gut health, potential anticancer effect
Antioxidants, vitamins, and minerals	Fresh fruits and vegetables, cereals (folic acid)	Reduce oxidative stress, prevent cells oxidation

Figure 1. Health Benefits of Some Marketed Functional Foods.

3.1.1. Immune Function

Immune function throughout life can be influenced by nutrition. Possible ingredients for the development of functional foods that could contribute to optimal immune response include the antioxidant vitamins, trace elements (e.g. zinc, copper and manganese), n-3 and n-6 polyunsaturated fatty acids, l-arginine, nucleotides and nucleosides, probiotics, prebiotics and synbiotics [9].

3.1.2. Gastrointestinal Health

The gastrointestinal health tract is an obvious target for the development of functional foods because it acts as an interface between the diet and all other metabolic functions. Gastrointestinal (GI) function depends on an appropriate balance of healthy bacteria to prevent the invasion of harmful bacteria. One of the most promising areas for the development of functional foods lies in the use of ingredients to modify the composition and metabolic activity of the gut microflora using foods contain probiotics; prebiotics; and synbiotics. Potential health benefits of probiotics include: reduced incidence or severity of GI infections; alleviation of lactose intolerance; and overall improvement in gut function, including reduction in constipation as well as diarrhea [8].

Prebiotics, non-digestible food components that can stimulate growth and/or modify the metabolic activity of specific bacterial species already present in the gut, might also improve gut health and have the potential to reduce risk of colon cancer. Their ability to increase production of short chain fatty acids in the colon is associated with increased absorption of minerals such as calcium and magnesium. Dairy products, table spreads, baked goods and breads, breakfast cereals and bars, salad dressings, meat products and some confectionery items are all foods which do or might contain prebiotics.\

3.1.3. Cardiovascular Disease

Optimal heart health can be promoted by several dietary components, of which dietary fats are the best studied. Functional foods low in saturated fatty acids could help to promote optimal low density lipoprotein and cholesterol levels. Mono-unsaturates, such as olive oil, and poly-unsaturates (e.g. linoleic and alpha-linolenic acid) can reduce plasma (low density lipoprotein) concentrations. Olive oil has been the subject of many research studies for its antioxidant and anti-inflammatory properties in reducing cardiovascular disease risk. The long chain n-3 fatty acids found in fish oils can reduce plasma tri-acylglycerols, counteract blood clotting and promote blood vessel integrity. Functional foods enriched in these unsaturated fatty acids could reduce cardiovascular disease risk. Soluble fiber can influence cholesterol and lipoprotein metabolism. In particular, it can reduce lipoprotein concentrations. Sources of soluble fiber currently used in functional foods include dietary fructans (e.g. inulin, oligofructose) as well as phytosterols (plant sterols and stanols) which are naturally present in many fruits, vegetables, nuts, seeds, legumes, cereals, vegetable oils and other plant sources. Diets high in plant foods and rich in polyphenols have been associated with reduced risk for cardiovascular disease and other chronic diseases. Possible mechanisms include their anti-inflammatory, vasodilatory, anti-platelet and antioxidant effects [10]. Flavonoids represent a diverse range of polyphenolic compounds, including flavonols, flavones, flavanones, flavan-3-ols, isoflavones, anthocyanins and proanthocyanidins. These compounds, found naturally in plant foods (e.g., fruits, vegetables, grains, and herbs), could offer potential as functional food ingredients.

3.1.4. Obesity

As already mentioned, obesity has become a global public health issue. In Europe and America, the prevalence of obesity has tripled during the past two decades. More than half of all adults and one in five children are overweight or obese. The nutritional approach to weight management involves reducing energy intake, which may be achieved by reducing the energy density of the diet, reducing appetite and/or enhancing satiety or reducing fat absorption [11]. Fat and sugar replacers are potential candidates for functional foods to reduce energy density, while foods with low glycaemic index or dietary fiber may enhance satiety. A number of proposed functional food ingredients have been shown to act before the food is absorbed in the gastrointestinal tract. These include chitosan, conjugated linoleic acid, diglycerides, medium-chain triglycerides, green tea, caffeine, calcium and capsaicin.

3.1.5. Bone Health

Bone health is dependent on a variety of nutrients, including calcium, magnesium, vitamin D, vitamin K and vitamin C as well as trace minerals such as manganese, copper and zinc. Vitamin D is attracting attention for its role in bone mineral density and bone growth. A significant proportion of the European population has poor vitamin D status, due to lack of sunlight exposure [8]. High fruit and vegetable intake is also associated with bone health and plant bioactive ingredients may emerge as useful ingredients for bone protection. Osteoporosis and bone fracture is a cause of considerable morbidity and mortality in older Europeans, making the search for bone protective ingredients of economic importance to the food industry and the well-being of the European population.

3.1.6. Musculoskeletal Disease

Osteoarthritis is one of the most prevalent and debilitating chronic conditions affecting older people. Pharmacological interventions such as non-steroidal anti-inflammatory drugs are associated with severe adverse effects; this creates a need for safe and alternative dietary therapies. Substances such as glucosamine and chondroitin are used as food supplements but could find application in functional foods. Evidence is also emerging for collagen hydrolysate, methyl-sulfonylmethane, S-adenosyl methionine and soybean un-saponifiables, all of which could be explored as functional food ingredients.

3.1.7. Diabetes Mellitus

Overweight and lack of physical activity have been consistently associated with increased risk of type-2 diabetes. However, diet also appears to be important. Evidence supports the use of whole grain foods, vegetables, fruits, foods low in saturated fat and also starchy foods with a low glycaemic index. Soluble fibres with a low glycaemic index, such as psyllium and inulin, can have potential beneficial effects on glucose metabolism and insulin sensitivity. Since compliance with dietary recommendations in diabetes is often poor, functional foods may be valuable in both treatment and prevention [12]. Oral amino acids included in snacks have also been studied to explore positive benefits in blood glucose control and insulin sensitivity. Spices such as cinnamon, coriander, garlic, and turmeric may also be beneficial anti-diabetic food ingredients. Chromium may be effective in optimizing insulin metabolism and lowering plasma cholesterol levels. Genetics appear to have an influence, and nutrigenomic studies may help to shed light on the individuals who could benefit from additional chromium.

3.1.8. Mental Health

Some functional foods could potentially promote optimal mental state and mental performance and influence behavior. They may influence cognitive performance; mood and vitality; reaction to stress; short-term memory; vigilance and attention; changes in memory and other mental processes during ageing. Glucose may have a beneficial influence on aspects of mental performance, including memory and decision time. Sucrose may reduce pain perception. Caffeine can lead to improvement in cognitive performance with effects on reaction time, vigilance, memory and psychomotor performance. Cognitive performance and maintenance of mental health in older people may be improved with B vitamins [13]. People often find that meals high in carbohydrate are associated with sleepiness and calmness. The amino acid tryptophan can reduce the time taken to fall asleep, while tyrosine and tryptophan may help recovery from jet lag. Several ingredients, such as n-3 fatty acids, S-adenosylmethione and folic acid have attracted attention as potential functional ingredients for improving depression.

3.1.9. Health and Well-being in Ageing

Now we are living longer, there is an increased prevalence of the chronic conditions of ageing (e.g. cardiovascular disease, cancer, cataract, age-related macular degeneration, Parkinson's disease, Alzheimer's disease, osteoarthritis). Such conditions, and indeed ageing itself, may put increased oxidative stress on the body. So, if we can reduce this, we can potentially delay or prevent some of these diseases.Our bodies have several defenses against oxidative stress, including antioxidant enzymes; minerals and trace elements, such as selenium, manganese and copper, which act as co-factors for the antioxidant enzymes; vitamins, such as vitamins C and E, and carotenoids, all of which can act as free radical quenchers with an antioxidant effect; glutathione, a cysteine-containing peptide with a thiol group which acts as a reducing agent that can be reversibly oxidised and reduced. Antioxidants naturally present in foods (e.g. vitamins C and E, carotenoids, flavonoids and other polyphenols) are potentially useful candidates for functional ingredients. Plant foods which contain these substances such as berries (e.g. cranberry, blueberry, goji, acai), mangosteen, pomegranate, tomato and grapes are now being explored by the functional food industry as potential antioxidants.Various other ingredients are potential functional ingredients for the prevention of age-related mental changes. The n-3 fatty acids may have a role in reducing the risk of depression and the prevention of age-related dementia.

3.2. Type of Functional Foods

There is a difference between the Western and Eastern perspective on functional foods. In the West, functional foods are considered revolutionary and represent a rapidly growing segment of the food industry. Food and pharmaceutical companies alike are competing to bring functional foods into the mass market. On the other hand, functional foods have been a part of Eastern cultures for centuries. Foods were used for medicinal purposes in traditional Chinese medicine as early as 1000 B.C.E [14, 15]. From ancient times, the Chinese have used foods for both preventive and therapeutic health effects, a view that is now being increasingly recognized around the world. Clearly, most foods are functional in some way. What makes a "functional food," however, is its potential ability to positively affect health. Functional foods range from fruits, vegetables, and whole grains, which are naturally high in phytochemicals, to products in which a specific ingredient is added, removed, increased, or decreased [14, 15]. Examples of functional foods include soy, oats, flaxseed, grape juice, broccoli and other cruciferous vegetables, phytosterol/stanol-enriched margarine, eggs enhanced with omega-3 fatty acids, and foods fortified with herbal preparations.

3.2.1. Functional Foods from Plant Sources

All *in vivo*, *in vitro*, and clinical trial data indicates that a plant-based diet can reduce the risk of chronic disease, particularly cancer. In 1992, a review of 200 epidemiological studies [16] showed that cancer risk in people consuming diets high in fruits and vegetables was only one-half that in those consuming few of these foods. It is now clear that there are components in a plant-based diet other than traditional nutrients that can reduce cancer risk. Steinmetz and Potter [17] identified more than a dozen classes of these biologically active plant chemicals, now known as, phytochemicals. *Phytochemicals* are naturally occurring chemicals in plants that provide flavor, color, texture, and smell. Phytochemicals have potential health effects, as they may boost enzyme production or activity, which may, in turn, block carcinogens, suppress malignant cells, or interfere with processes that can cause heart disease and stroke. Phytochemical-rich foods include *cruciferous* vegetables (e.g., broccoli, Brussels sprouts, cauliflower, cabbage), *umbelliferous* vegetables (e.g., carrots, celery, parsley, parsnips), *allium* vegetables (e.g., garlic, onions, leek), berries, citrus fruits, whole grains, and legumes (e.g., soybeans, beans, lentils, peanuts). In the early twenty-first century, identification of the role of phytochemicals in health is an emerging area of science, and the global health community does not recommend supplementation with any specific phytochemicals. Examples include:

3.2.1.1. Oats

Oat products are a widely studied dietary source of the cholesterol-lowering soluble fiber b-glucan. There is now significant scientific agreement that consumption of this particular plant food can reduce total and low density lipoprotein (LDL) -cholesterol, thereby reducing the risk of coronary heart disease. For this, the Food and Drug Administration awarded the first food-specific health claim in January 1997. In its health claim petition, the Quaker Oats Company summarized 37 human clinical intervention trials conducted between 1980 and 1995. The majority of these studies revealed statistically significant reductions in total and LDL-cholesterol in hypercholesterolemic subjects consuming either a typical American diet or a low fat diet [18]. A food bearing the health claim must contain 13 g of oat bran or 20 g oat meal, and provide, without fortification, at least 1.0 g of b-glucan per serving.

3.2.1.2. Soy

Soy has been in the spotlight during the 1990s. Not only to consider that soy is a high quality protein, it is thought to play preventive and therapeutic roles in cardiovascular disease (CVD), cancer, osteoporosis, and the alleviation of menopausal symptoms. The cholesterol-lowering effect of soy is the most well-document physiological effect. A 1995 meta-analysis of 38 separate studies (involving 743 subjects) found that the consumption of soy protein resulted in significant reductions in total cholesterol (9.3%), LDL cholesterol (12.9%), and triglycerides (10.5%), with a small but insignificant increase (2.4%) in high density lipoprotein (HDL) cholesterol [18]. Regarding the specific component responsible for the cholesterol-lowering effect of soy, attention has focused on the isoflavones. Several classes of anticarcinogens have been identified in soybeans, including protease inhibitors, phytosterols, saponins, phenolic acids, phytic acid, and isoflavones [19]. Of these, isoflavones are particularly noteworthy because soybeans are the only significant dietary source of these compounds. Isoflavones are heterocyclic phenols structurally similar to the estrogenic steroids. Because they are weak estrogens, isoflavones may act as antiestrogens by competing with the more potent, naturally-occurring endogenous estrogens for binding to the estrogen receptor. This may explain why populations that consume significant amounts of soy (e.g., Southeast Asia) have reduced risk of estrogen-dependent cancer.

3.2.1.3. Flaxseed

Flaxseed oil contains the most of the omega-3 fatty acid, a linolenic acid (57%), among the major seed oils. Recent research, however, has focused more specifically on fiber-associated compounds known as lignans. The two primary mammalian lignans, enterodiol and its oxidation product, enterolactone, are formed in the intestinal tract by bacterial action on plant lignan precursors. Flaxseed is the richest source of mammalian lignan precursors. Enterodiol and enterolactone may play a role in the prevention of estrogen-dependent cancers. In rodents, flaxseed has been shown to decrease tumors of the colon and mammary gland [20] as well as of the lung [21]. Fewer studies have evaluated the effects of flaxseed feeding on risk markers for cancer in humans. Phipps et al. [22] demonstrated that the ingestion of 10 g of flaxseed per day elicited several hormonal changes associated with reduced breast cancer risk. Consumption of flaxseed has also been shown to reduce total and LDL cholesterol [18].

3.2.1.4. Tomatoes

Tomatoes have received significant attention because of interest in lycopene, the primary carotenoid found in this fruit. Lycopene is associated with reduced cancer risk of the prostate, breast, digestive tract, cervix, bladder, and skin. Proposed mechanisms by which lycopene could influence cancer risk are related to its anti-oxidant function [18]. Proposed mechanisms by which lycopene could influence cancer risk are related to its antioxidant function.

3.2.1.5. Garlic

Garlic is likely the herb most widely quoted in the literature for medicinal properties. The health benefits of garlic are numerous, including cancer chemopreventive, antibiotic, anti-hypertensive, and cholesterol-lowering properties. The sulfur-containing compounds in garlic have been studied for their ability to reduce cancer of the gastrointestinal tract. An average of 900 mg garlic/day could decrease total serum cholesterol levels by approximately 9%. Moreover, 800 mg garlic/day reduced total cholesterol levels by 12% [23].

3.2.1.6. Broccoli and other Cruciferous Vegetables

Epidemiological evidence has associated the frequent consumption of cruciferous vegetables with decreased cancer risk. In a review of 87 case-control studies it was demonstrated that, an inverse association between consumption of total brassica vegetables and cancer risk was found. The percentages of case-control studies showing an inverse association between consumption of cabbage, broccoli, cauliflower, and Brussels sprouts and cancer risk were 70, 56, 67, and 29%, respectively [24]. The anticarcinogenic properties of cruciferous vegetables were attributed to their relatively high content of glucosinolates. Glucosinolates are a group of glycosides stored within cell vacuoles of all cruciferous vegetables. Myrosinase, an enzyme found in plant cells, catalyzes these compounds to a variety of hydrolysis products, including isothiocyanates and indoles. Attention has been focused on a particular isothiocyanate isolated from broccoli, known as sulforaphane. Sulforaphane has been shown to be the principal inducer of a particular type of enzyme, quinone reductase. A 3-day-old broccoli sprouts contained 10-100 times higher levels of glucoraphanin (the glucosinolate of sulforaphane) than did corresponding mature plants.

3.2.1.7. Tea

Tea is second only to water as the most widely consumed beverage in the world. A great deal of attention has been directed to the polyphenolic constituents of tea, particularly green. Polyphenols comprise up to 30% of the total dry weight of fresh tea leaves. Catechins are the predominant and most significant of all tea polyphenols. The four major green tea catechins are epigallocatechin-3-gallate, epigallocatechin, epicatechin-3-gallate, and epicatechin [25]. There has been a great deal of interest in pharmacological effects of tea. By far, most research on health benefits of tea has focused on its cancer chemopreventive effects. The consumption of five or more cups of green tea per day was associated with decreased recurrence of stage I and II breast cancer in Japanese women [26]. There is some evidence that tea consumption may also reduce the risk of heart diseases. Tea consumption is the major source of flavonoids.

3.2.2. Functional Foods from Animal Sources

Although the vast number of naturally occurring health-enhancing substances are of plant origin, there are a number of physiologically-active components in animal products that deserve attention for their potential role in optimal health [18]. Functional foods from animal sources include omega-3 fatty acids, dairy products, and beef.

3.2.2.1. Omega-3 Fatty Acids

Omega-3 fatty acids are essential polyunsaturated fatty acids found primarily in fish. They have been shown to lower LDL-cholesterol and reduce risk of cardiovascular mortality (discussed below).

3.2.2.2. Dairy Products

There is no doubt that dairy products are functional foods. They are one of the best sources of calcium, an essential nutrient which can prevent osteoporosis and possibly colon cancer. In view of the former, the National Academy of Sciences recently increased recommendations for this nutrient for most age groups. In addition to calcium, however, recent research has focused specifically on other components in dairy products, particularly fermented dairy products known as probiotics. Probiotics are defined as live microbial feed supplements which beneficially affect the host animal by improving its intestinal microbial balance [27]. It is estimated that over 400 species of bacteria, separated into two broad categories, inhabit the human gastrointestinal tract. The categories are: those considered to be beneficial (e.g., *Bifidobacterium* and *Lactobacillus* spp.) and those considered detrimental (e.g., *Enterobacteriaceae* and *Clostridium* spp.). Of the beneficial microorganisms traditionally used in food fermentation, lactic acid bacteria have attracted the most attention [18]. In addition to probiotics, prebiotics (defined as nondigestible food ingredients) play also an important role that beneficially affect the host by selectively stimulating the growth and/or activity of a number of bacteria in the colon and thus improves host health [28].

3.2.2.3. Beef

An anticarcinogenic fatty acid known as conjugated linoleic acid (CLA) was first isolated from grilled beef in 1987 [29]. CLA refers to a mixture of positional and geometric isomers of linoleic acid in which the double bonds are conjugated instead of existing in the typical methylene interrupted configuration. Nine different isomers of CLA have been reported as occurring naturally in food. CLA is unique in that it is found in highest concentrations in fat from ruminant animals (e.g., beef, dairy, and lamb). Beef fat contains 3.1 to 8.5 mg CLA/g fat contributing 57- 85% of the total CLA [30]. Interestingly, CLA increases in foods that are cooked and/or otherwise processed. This is significant in view of the fact that many mutagens and carcinogens have been identified in cooked meats. Over the past decade, CLA has been shown to be effective in suppressing fore-stomach tumors in mice, aberrant colonic crypt foci in rats, and mammary carcinogenesis in rats [31]. In the mammary tumor model, CLA is an effective anti-carcinogen in the range of 0.1-1% in the diet. These results are not due to displacement of linoleic acid in cells, suggesting that there may be unique mechanism(s) by which CLA modulates tumor development. Thus, there has been research designed to increase the CLA content in dairy cow milk through dietary modification. More recently, CLA has been investigated for its ability to change body composition, suggesting a role as a weight-reduction agent. Mice fed CLA-supplemented diets (0.5%) exhibited 60% lower body fat and 14% increased lean body mass relative to controls [32], possibly by reducing fat deposition and increasing lipolysis in adipocytes.

3.3. Research Challenges for Functional Foods

Key research challenges for functional foods are the need to identify new functional food ingredients and to gain consumer acceptance of such products. In particular, research needs to define the bioavailability of functional food ingredients, identify potential functional ingredients that could provide human health benefits, and develop appropriate biomarkers for a wider range of functional endpoints. Research needs also to identify individual biological responses to functional foods and their potential role, ensure stability of functional food ingredients during manufacturing, and ensure passage through the GI tract to reach the target organ intact.

4. Emerging Functional Foods

Functional foods are an emerging field in food science due to their increasing popularity with health-conscious consumers and the ability of marketers to create new interest in existing products. Examples for functional foods are:

4.1. Dietary Supplements

Dietary supplements are defined as any product that is intended to supplement the diet and contains one or more of the following: a vitamin, mineral, herb or other botanical; an amino acid or metabolite; an extract; or any combination of the previously mentioned items. According to U.S. Food and Drug Administration (FDA) regulations YEAR, a dietary supplement may be marketed in food form if it is not "represented" as a conventional food and is clearly labeled as a dietary supplement.

4.2. Nutraceuticals

Nutraceutical, a portmanteau of the words "nutrition" and "pharmaceutical", is a food or food product that reportedly provides health and medical benefits, including the prevention and treatment of disease. Health Canada defines the term as "*a product isolated or purified from foods that is generally sold in medicinal forms not usually associated with food. A nutraceutical is demonstrated to have a physiological benefit or provide protection against chronic disease.* The term, Nutraceutical, was coined in the 1990s by Dr. Stephen DeFelice, who defined it as any substance that is a food or a part of a food and provides medical or health benefits, including the prevention and treatment of disease [33]. Such products may range from isolated nutrients, dietary supplements and specific diets to genetically engineered designer foods, herbal products and processed foods such as cereals, soups and beverages. Since the term was coined, its meaning has been modified. The term also has been defined as a product isolated or purified from foods and generally sold in medicinal forms not usually associated with food and demonstrated to have a physiological benefit or provide protection against chronic disease. A nutraceutical may be a naturally nutrient-rich or medicinally active food, such as garlic or soybeans, or it may be a specific component of a food, such as the omega-3 fish oil that can be derived from salmon and other cold-water fish. Other nutraceutical such as Beta-carotene, lycopene, resveratrol, and vitamins are thought to play significant roles in treating or preventing disease. Resveratrol, for example, is found in red grape products and is believed to help prevent cancer and benefit cardiovascular health. There are many methods for adding nutraceuticals to foods, but they all have one common characteristic as being non-stable due to non-rigid structures. However, a method was developed for creating crystalline-like fibers to envelop nutraceuticals and protect them from external influences and revent degradation. Once the nutraceutical is enveloped, it is thermally protected.

4.3. Fortified Foods

Fortified foods are enriched with vitamins and minerals, usually at a range up to 100 percent of the Dietary Reference Intake (DRI, formally called the Recommended Daily Allowance or RDA) for that nutrient. Often, these foods are mandated by law to be fortified to a level that replaces nutrients lost during processing, as in adding B vitamins to many baked goods [59]. Breakfast cereals are a food category that has been fortified since the 1940s.

substances is plant material. Garlic, broccoli, green tea, soybean, tomato, carrot, Brussels sprouts, kale, cabbage, onions, cauliflower, red beets, cranberries, cocoa, blackberry, blueberry, red grapes, prunes, and citrus fruits are mentioned as the richest sources of antioxidants [35]. The chemical classification of antioxidants is very complicated. According to the mode of action, two main groups of antioxidants can be distinguished. The first comprises the chemical substances which interrupt the propagation of the free radical chain by hydrogen donation to radicals or stabilization of relocated radical electrons. The second group is characterized by a synergistic mode of action. It includes oxygen scavengers and chelators which bind ions involved in free radical formation. Their activity consists in hydrogen delivery to phenoxyradicals that leads to the reconstitution of the primary function of antioxidants. This role is played by substances binding metal ions, e.g. citric acid, and by secondary antioxidants, such as amino acids, flavonoids, β-caroten, selenium and many others.

Free radicals are a major cause of many degenerative diseases, such as atherosclerosis, cancer, cardiovascular diseases, inflammatory bowel diseases, skin aging, old age dementia and arthritis.

Clinical trials provide indications that antioxidants play a fundamental role in the prevention of cancer and cardiovascular diseases [36]. They act as scavengers of reactive oxygen species and metal chelators that protect human cells and reduce oxidative damages. The development of existing tumors is also significantly inhibited by antioxidants. It has been demonstrated that antioxidants, such as α-tocopherol and β-carotene increase the production of cytotoxic cells, activate them for the production of large quantities of cytokines and facilitate their migration to cancer cells.

It should be strongly emphasized that the action of antioxidants depends on the dose taken [37]. At too high doses, their role changes and instead of acting as protective substances, they themselves become pro-oxidants and result in very dangerous disease symptoms. The concentration of antioxidants in raw materials depends, to a considerable extent, on the variety of crop plants and conditions of their cultivation. In addition, also chemical changes and quantitative losses of selected groups of antioxidants and their biological activities in the entire processing chain starting with the raw material, through technological treatment, storage and culinary preparation until their absorption in the alimentary tract, all play an important role in the assessment of the antioxidant activity. Figure 3, represents the most important types of functional foods.

5. Industrial Functional Foods

Functional foods have been the topic of considerable interest in the food and nutrition industry for years. These functional food products result from: technological innovation [38] at the processing level, such as cholesterol lowering spreads, xylitol-sweetened chewing gum, and dairy products fermented with specific lactic acid bacteria [39]; technological innovation at the production level such as genetically modified crop seeds (e.g., Golden rice fortified with vitamin A) and planting materials derived through conventional breeding (e.g., orange fleshed sweet potato, fortified with vitamin A); or crops that naturally contain components that have a physiological function (e.g., soy bean and oats, which are good for cardiovascular health).

Functional food	Potential health benefit
Whole foods	
Oats	Reduces cholesterol and constipation, reduces risk of heart disease
Soy	Reduces cholesterol, reduces risk of osteoporosis, certain cancers, and heart disease
Fruits and vegetables	Reduces risk of certain cancers and heart disease; reduces hypertension
Fish	Reduces cholesterol and triglycerides
Garlic	Reduces risk of heart disease and certain cancers, reduces cholesterol
Flaxseed	Reduces risk of heart disease and certain cancers; reduces triglycerides; increases blood-glucose control
Nuts	Reduces risk of heart disease
Enriched foods	
Grains	Reduces risk of certain cancers, heart disease, and nutrient deficiencies
Fortified foods	
Juices with calcium	Reduces risk of osteoporosis, reduces hypertension
Grains with folic acid	Reduces risk of heart disease and neural tube birth defects
Infant formulas with iron	Reduces risk of iron deficiency
Grains with added fiber	Reduces risk of certain cancers and heart disease; reduces cholesterol and constipation; increases blood-glucose control
Milk with vitamin D	Reduces risk of osteomalacia and osteoporosis
Juices with added fiber	Reduces risk of certain cancers and heart disease; reduces cholesterol, hypertension, and constipation
Enhanced foods	
Dairy products with probiotics	Reduces risk of colon cancer and candidal vaginitis; controls inflammation; treatment of respiratory allergies, diarrheal disorders, and eczema
Beverages and salad dressings with antioxidants	May support overall health
Eggs with omega-3 fatty acids	Reduces risk of heart disease

Figure 3. The Most Important Types of Functional Foods.

The functional food industry, consisting of food, beverage and supplement sectors, is one of the several areas of the food industry that is experiencing fast growth in recent years. It is estimated by BCC Research that the global market of functional food industry will reach 176.7 billion in 2013 with a compound annual growth rate (CAGR) of 7.4%. Specifically, the functional food sector will experience 6.9% CAGR, the supplement sector will rise by 3.8% and the functional beverage sector will be the fastest growing segment with 10.8% CAGR [2]. This kind of growth is fueled not only by industrial innovation and development of new products that satisfy the demand of health conceious consumers but also by health claims covering a wide range of health issues. Yet, consumer skepticism persists mainly due to the fact that benefits associated with consuming the products may be difficult to be detected. The industry suggests the establishment of a health claim regulating agency, which may increase consumer confidence. Strict examination of some of the functional food claims may discourage some companies from launching their products [3]. The US functional foods market is estimated to be the largest in the world, representing between 35 and 50 percent of global sales. Asia-Pacific is the next biggest market. Together, the US and Asia-Pacific are estimated to account for approximately three-quarters of the current global market for functional foods. Food and beverage companies are the primary force and the market is largely consolidated, with US companies playing a major role. Nutraceutical extraction can be relatively costly; research and development is often significant, and there can be specialized technology needs as well.

5.1. Yoghurt as Functional Food

Fermented dairy products, having the tradition as healthy foods, are a natural choice for their make-over as functional foods. The word 'yogurt' was derived from the Turkish word 'Jugurt'. A vast array of yogurts is now available in the market to suit all palates and meal occasions. Yogurts are available in a variety of textures (e.g. liquid, set, smooth), fat contents (luxury, low-fat, virtually fat-free) and flavors (natural, fruit, cereal). The low-fat varieties of yogurt provide an array of important nutrients in significant amounts in relation to their energy and fat content, therefore making them a nutrient-dense food. The healthy image of yogurt is further endorsed by the addition of various fruit preparations in yogurt to include the health benefits of fruits such as providing fibre and antioxidants [40]. In recent years soymilk, corn milk, and peanut milk yogurts are being developed as a vegetarian alternate to bovine milk yogurt that can also overcome the problem of milk protein allergenicity. Further, inclusion of plant extracts to enhance yogurt functionality, such as tea catechins for antioxidative and antimicrobial properties, is also being considered.

Our digestive system normally has what we would call "good" bacteria and "bad" bacteria. Maintaining the correct balance between the "good" bacteria and the "bad" bacteria is necessary for optimal health. Things like medications, diet, diseases, and your environment can upset that balance. Keeping your intestinal bacteria in their optimal balance gives you your best shot at a lifestyle full of the things you want to do all year round. After all, who has time to be sick? Plus, your GI tract produces about 75% of your immune cells. So keeping your digestive tract robust only makes sense.

5.1.1. Probiotics

Probiotic organisms are live microorganisms that are thought to be beneficial to the host organism. Probiotics are live microorganisms which are intended to colonize the large intestine and confer physiological health benefits to the host. Historically, the original observation of the positive role played by certain bacteria was first introduced by Russian scientist and Nobel laureate Élie Metchnikoff, who in the beginning of the 20th century suggested that it would be possible to modify the gut flora and to replace harmful microbes with useful microbes [41]. Metchnikoff, at that time a professor at the Pasteur Institute in Paris, proposed the hypothesis that the aging process results from the activity of putrefactive (proteolytic) microbes producing toxic substances in the large bowel. Proteolytic bacteria such as clostridia, which are part of the normal gut flora, produce toxic substances including phenols, indols and ammonia from the digestion of proteins. According to Metchnikoff these compounds were responsible for what he called "intestinal auto-intoxication". It was at that time known that milk fermented with lactic-acid bacteria inhibits the growth of proteolytic bacteria because of the low pH produced by the fermentation of lactose. Metchnikoff had also observed that certain rural populations in Europe, for example in Bulgaria and the Russian steppes whose lived largely on milk fermented by lactic-acid bacteria were exceptionally long lived. Based on these facts, Metchnikoff proposed that consumption of fermented milk would "seed" the intestine with harmless lactic-acid bacteria and decrease the intestinal pH and that this would suppress the growth of proteolytic bacteria. Metchnikoff himself introduced in his diet sour milk fermented with the bacteria he called "*Bulgarian Bacillus*" and found his health benefited. Friends in Paris soon followed his example and physicians began prescribing the sour milk diet for their patients [42]. Lactic acid bacteria and bifidobacteria are the most common types of microbes used as probiotics; but certain yeasts and bacilli may also be used. Probiotics are commonly consumed as part of fermented foods with specially added active live cultures, such as in yogurt, soy yogurt, or as dietary supplements. Probiotics are also delivered in fecal transplants, in which stool from a healthy donor is delivered like a suppository to an infected patient.

Bifidobacteria were first isolated from a breast-fed infant by Henry Tissier who also worked at the Pasteur Institute, Paris, France. The isolated bacterium named *Bacillus bifidus communis* [43] was later renamed to the genus *Bifidobacterium*. Tissier found that *bifidobacteria* are dominant in the gut flora of breast-fed babies and he observed clinical benefits from treating diarrhea in infants with bifidobacteria. The claimed effect was bifidobacterial displacement of proteolytic bacteria causing the disease. In 1920, Rettger demonstrated that Metchnikoff's "Bulgarian Bacillus", later called *Lactobacillus delbrueckii subsp. bulgaricus,* could not live in the human intestine [44], and the fermented food phenomenon petered out. Metchnikoff's theory was disputable (at this stage), and people doubted his theory of longevity. After Metchnikoff's death in 1916, the centre of activity moved to the United States. It was reasoned that bacteria originating from the gut were more likely to produce the desired effect in the gut, and in 1935 certain strains of *Lactobacillus acidophilus* were found to be very active when implanted in the human digestive tract [45]. Trials were carried out using this organism, and encouraging results were obtained especially in the relief of chronic constipation. The term "probiotics" was first introduced in 1953 by Werner Kollath [46]. Contrasting antibiotics, probiotics were defined as microbially derived factors that stimulate the growth of other microorganisms. In 1989, Roy Fuller suggested a definition of probiotics that has been widely used: "*A live microbial feed supplement which beneficially affects the host animal by improving its intestinal microbial balance*". Fuller's definition emphasizes the requirement of viability for probiotics and introduces the aspect of a beneficial effect on the host [27]. In the following decades, intestinal lactic acid bacterial species with alleged health beneficial properties have been introduced as probiotics, including *Lactobacillus rhamnosus, Lactobacillus casei*, and *Lactobacillus johnsonii* [47].

5.1.1.1. Selection Criteria for Probiotics

In order to be able to exert its beneficial effects, a successful potential probiotic strain is expected to have a number of desirable properties. A potential probiotic strains does not need to fulfill all such selection criteria [48]. For organisms to achieve probiotic status, they must fulfill a number of criteria such as being non-pathogenic and isolated from the same species as its intended host; have a demonstrable beneficial effect on the human health; and can be able to colonize the human intestine. Probiotics must exhibit good growth characteristics, phage resistance and genetic stability, and be able to survive transit through the GIT by its resistance to acid and bile. Persist probiotics must also be able to attach to human epithelial cells and can produce an antimicrobial substance as well as to survive during manufacture processing and prolonged periods of storage.

5.1.1.2. Therapeutic Role of Probiotics

The efficacy of probiotics to exert a positive influence on host health or physiology was not confirmed for long, as evidence for their efficacy was low and information on the stability of the strains in the products and their survival in the GIT was often lacking. However, a pharmacological approach has now been used to assess the effects and it is now known that that the pharmacokinetics of probiotics varies between strains and that many are indeed effective. The most interesting studies are related to the immune-modulatory effects of probiotics with regards to treatment of allergic disease or inflammatory bowel diseases [49]. Although a variety of health benefits have been attributed to probiotics, their anticarcinogenic, hypocholesterolemic and antagonistic actions against enteric pathogens and other intestinal organisms have received the most attention. More evidence supports the role of probiotics in cancer risk reduction, particularly colon cancer [50]. This observation may be due to the fact that lactic acid cultures can alter the activity of fecal enzymes (e.g., b-glucuronidase, azoreductase, nitroreductase) that are thought to play a role in the development of colon cancer. In addition to probiotics, there is growing interest in fermentable carbohydrates (prebiotics) that feed the good microflora of the gut.

5.1.1.3. Effects of Probiotics on Health

There are lots of studies on searching the health benefits of fermented foods and probiotics. However, in most of these studies researchers did not use sufficient test subjects or they use microorganisms were not identified definitely. So, while a number of reported effects have been only partially established, some can be regarded as well-established and clinically well documented for specific strains. These health-related effects (Wikipedia, Probiotics) can be considered as:

Managing lactose intolerance.
Improving immune system.
Prevention of colon cancer.
Reduction of cholesterol and triacylglycerol plasma concentrations (weak evidence).
Lowering blood pressure.- Reducing inflammation.
Reduction of allergic symptoms.
Beneficial effects on mineral metabolism, particularly bone density and stability.
Reduction of Helicobacter pylori infection.
Suppression of pathogenic microorganisms (antimicrobial effect).
Prevention of osteoporosis.
Prevention of urogenital infections.

Treat diarrhea, especially following treatment with certain antibiotics

Prevent and treat vaginal yeast infections and urinary tract infections
Treat irritable bowel syndrome
Reduce bladder cancer recurrence
Speed treatment of certain intestinal infections
Prevent and treat eczema in children
Prevent or reduce the severity of colds and flu

5.1.2. Prebiotic

Prebiotics were first identified and named by Marcel Roberfroid [28]. As a functional food component, prebiotics, like probiotics, are conceptually intermediate between foods and drugs. Using prebiotics and probiotics in combination is often described as symbiotic. Prebiotics and probiotics can restore the balance of bacteria in your digestive tractn [51]. Probiotic bacteria colonize the gut and help in proper digestion, they also avoid the adhering of pathogens to the intestinal tract. In prebiotic a fiber such as fructose oligosaccharide, galactose oligosaccharide etc. are consumed that is intended as a food for the microbes in the large intestine that promote and stimulate the growth of intestinal beneficial microflora responsible for health benefits. Prebiotics are 'non-digestible carbohydrates (oligosaccharides) that act as food for probiotics and beneficially affect the host by selectively stimulating the growth and/or activity of the colon microflora' such as bifidobacteria, which confers benefits upon host well-being and health. These oligosaccharides are not or incompletely absorbed in the small intestine.Trans-galactooligosaccharides, inulin, fructo-oligosaccharides as well as lactulose and soy-bean, and flaxseeds are classified as prebiotics. Several reports are also focused on the distinction between short-chain, long-chain, and full-spectrum prebiotics. *Short-chain prebiotics*, e.g. oligofructose, contain 2-8 links per saccharide molecule and are typically fermented more quickly in the right-side of the colon providing nourishment to the bacteria in that area [52]. *Longer-chain prebiotics*, e.g. inulin, contain 9-64 links per saccharide molecule, and tend to be fermented more slowly, nourishing bacteria predominantly in the left-side colon [53]. *Full-spectrum prebiotics* provide the full range of molecular link-lengths from 2-64 links per molecule, and nourish bacteria throughout the colon, e.g. oligofructose-enriched inulin. Traditional dietary sources of prebiotics include soybeans, inulin sources (such as Jerusalem artichoke, jicama, and chicory root), raw oats, unrefined wheat, unrefined barley, and yacon. Some of the oligosaccharides that naturally occur in breast milk are believed to play an important role in the development of a healthy immune system in infants. It is becoming more common to properly distinguish between prebiotic substances and the food that contains them. No plant or food is a prebiotic; wheat, honey and many other foods contain prebiotics to a greater or lesser extent, ranging from fairly large portions (chicory root, Jerusalem artichoke) to only trace quantities (thousands of other plant-based foods). Figure 4, represents top 10 foods containing prebiotics.

Preliminary reports have demonstrated potential effects of prebiotics on calcium and other mineral absorption [54], immune system effectiveness [55], bowel pH, reduction of colorectal cancer risk [56], inflammatory bowel disorders [57], hypertension, and intestinal regularity. Recent human trials have provided further evidence for the potential role of prebiotics in lowering risk of colon cancer [58]. It has been argued that many of these health effects emanate from increased production of short-chain fatty acids (SCFA) by the stimulated beneficial bacteria. Thus food supplements specifically enhancing the growth of SCFA producing intestinal bacteria (such as clostridia and bacteroides species) are widely recognized to have such potential. Production of SCFA and fermentation quality is reduced during long-term diets of low fiber intake [59]. Until bacterial flora are gradually established to habilitate or restore intestinal tone, nutrient absorption will be impaired and colonic transit time temporarily increased with an immediate addition of higher prebiotic intake [60]. While there is no broad consensus on an ideal daily serving of prebiotics, recommendations typically range from 4 to 8 grams (0.14–0.28 oz) for general digestive health support, to 15 grams

(0.53 oz) or more for those with active digestive disorders. Figure 5, represents the amounts of prebiotic foods required to achieve a daily serving of prebiotic fiber (Wikipedia).

Food	Prebiotic fiber content by weight
Raw Chicory Root	64.6%
Raw Jerusalem Artichoke	31.5%
Raw Dandelion Greens	24.3%
Raw Garlic	17.5%
Raw Leek	11.7%
Raw Onion	8.6%
Cooked Onion	5%
Raw Wheat bran	5%
Whole Wheat flour	4.8%
Raw Banana	1%

Figure 4. Top 10 Foods Containing Prebiotics.

Food	Amount of food to achieve 6g serving of prebiotics
Raw Chicory Root	9.3 g (0.33 oz)
Raw Jerusalem Artichoke	19 g (0.67 oz)
	24.7 g (0.87 oz)
Raw Dandelion Greens	34.3 g (1.21 oz)
Raw Garlic	51.3 g (1.81 oz)
Raw Leek	69.8 g (2.46 oz)
Raw Onion	120 g (4.2 oz)
Cooked Onion	120 g (4.2 oz)
Raw Wheat bran	125 g (4.4 oz)
Whole Wheat flour	600 g (1.3 lb)
Raw Banana	

Figure 5. The Amounts of Prebiotic Foods Required to Achieve a Daily Serving of Prebiotic Fiber.

5.1.3. Synbiotic

Synbiotics are mixtures of probiotics and prebiotics in a form of synergism that beneficially affect the host by improving the survival and implantation of selected live microbial strains in the gastrointestinal tract. The rationale behind synbiotic treatment is that the desired probiotic and prebiotic would together exert a greater beneficial effect than when administered individually. Probiotics are found in foods such as yogurt, while prebiotics are found in whole grains, bananas, onions, garlic, honey and artichokes. In addition, probiotics and prebiotics are added to some foods and available as dietary supplements. Side effects are rare, and most healthy adults can safely add foods that contain prebiotics and probiotics to

4.4. Medical Foods

Medical foods refer to a food formulated to be consumed or administered internally while under the supervision of a physician. The food product is intended for the specific dietary management of a disease or condition for which distinctive nutritional requirements are established by medical evaluation. Medical foods may be used to treat diabetes, obesity or heart disease, for example, and may carry specific claims, but in a strict sense are sold through physicians and not through conventional probiotic retail outlets. Fermented foods with live cultures are considered functional foods with benefits [34]. Functional foods are part of the continuum of products that individuals may consume to increase their health and/or contribute to reducing their disease burden. Functional food may be a natural or processed food that contains known biologically-active compounds which when in defined quantitative and qualitative amounts provides a clinically proven and documented health benefit, and thus, an important source in the prevention, management and treatment of chronic diseases of the modern age. Figure 2, represents a comparison of functional foods with medical foods and drugs.

Difference	Functional foods	Medical foods	Prescription drugs
Uses	Energy enhancement; weight management; bolster gut, bone or heart health; disease risk reduction; memory improvement	Dietary management of a disease or condition with distinctive nutritional requirements (e.g. difficulty swallowing, loss of appetite, nutrition repletion post-surgery)	Treatment of disease, symptom, or condition
Method of obtainment	No prescription or supervision needed; consumer selects	Used with medical supervision	Prescribed by health provider
Distribution channels	Supermarkets, drugstores, online, major retailers	Hospitals, pharmacies, drugstores, online	Pharmacies, hospitals
Regulatory body	No specific body, but is considered food and is therefore subject to FDA regulation (FDA regulates any specific health claims that might be made)	No additional FDA review/approval needed, but must abide by regulations concerning foods, e.g., labeling (FDA regulates any specific health claims that might be made)	FDA approval needed, a multiyear, multistage review process
Amount consumed	As desired	As needed	As prescribed

Sources: Company reports, US Food and Drug Administration, Institute of Food Technoplogists.

Figure 2. Comparing Functional Foods with Medical Foods and Drugs.

4.5. Plant Antioxidants

Plant antioxidants constitute one of the most active food compounds, such as phenolic antioxidants in green leaves of tea (226g/Kg of plant dry matter). The main source of these their diet. Fortified yoghurt (probiotic) with inulin fibers (prebiotic), fruits, and beneficial yeast is widely marketed. Probiotics are live bacteria which are intended to colonize the large intestine and confer physiological health benefits to the host. A prebiotic is a food or dietary supplement product that confers a health benefit on the host associated with modulating the microbiota. Prebiotics are not drugs, not functioning because of absorption of the component, not due to the component acting directly on the host, and are due to changes to the resident bacteria – either changing the proportions of the resident bacteria or the activities thereof. Measurable changes to the microbiota in the absence of a desirable physiological consequence in the host, does not qualify as a prebiotic. A prebiotic may be a fiber, but a fiber is not necessarily a prebiotic. Using prebiotics and probiotics in combination is often described as synbiotic, but the United Nations Food & Agriculture Organization (FAO) recommends that the term "synbiotic" be used only if the net health benefit is synergistic. A further restriction is to require that the prebiotic be shown to increase the population and/or function of the probiotic it is paired with. Probiotic bacteria colonize the gut with all good microorganisms that help in proper digestion, they also avoid the adhering of pathogens to the intestinal tract.In pre biotic a fiber such as fructose oligosaccharide, galactose oligosaccharide etc. are consumed that is intended as a food for the microbes in the large intestine that promote the growth of intestinal microflora responsible for health benefits (Wikipedia, Probiotics).

5.2. Functional Beverage

A functional beverage can be defined as a drink product that is non-alcoholic, ready to drink and includes in its formulation non-traditional ingredients. This includes herbs, vitamins, minerals, amino acids or additional raw fruit or vegetable ingredients, so as to provide specific health benefits that go beyond general nutrition. Sports and performance drinks, energy drinks, ready to drink teas, enhanced fruit drinks, soy beverages and enhanced water, among others, are some of the product segments rolled out as functional beverages in the market space [2]. Functional beverages have become popular due to its appeal to consumers who are seeking specific health benefits in their foods and beverages with their 'healthiness-on-the-go' idea. Both convenience and health have been identified as important factors when consumers make decisions about purchasing foods and beverages. Functional drinks are promoted with benefits such as heart health, improved immunity and digestion, joint health, satiety, and energy-boosting.

5.2.1. Kombucha Tea

Kombucha tea is an effervescent fermentation of sweetened tea that is used as a functional food. Sometimes referred to as a mushroom, the kombucha culture is actually a symbiotic colony of bacteria and yeast comprising Acetobacter (a genus of acetic acid bacteria) and one or more yeasts. A kombucha culture may contain one or more of the yeasts *Saccharomyces cerevisiae, Brettanomyces bruxellensis, Candida stellata, Schizosaccharomyces pombe,* and *Torulaspora delbrueckii.* Alcohol production by the yeast(s) contributes to the production of acetic acid by the bacteria. Alcohol concentration also plays a role in triggering cellulose production by the bacterial symbionts [61]. Although the bacterial component of a kombucha culture comprises several species, it almost always includes *Gluconacetobacter xylinus* (formerly Acetobacter xylinum), which ferments the alcohols produced by the yeast(s) into acetic acid. This increases the acidity while limiting the alcoholic content of kombucha. *G. xylinum* is responsible for most or all of the physical structure of a kombucha mother, and has been shown to produce microbial cellulose [62]. This is likely due to artificial selection by brewers over time, selecting for firmer and more robust cultures. The acidity and mild alcoholic element of kombucha resists contamination by most airborne molds or bacterial spores. As a result, kombucha is relatively easy to maintain as a culture outside of sterile conditions. The bacteria and yeasts in kombucha may also produce antimicrobial defense molecules [61].

Some promotional kombucha sources suggest the history of this tea-based beverage originated in ancient China or Japan, though no written records support these assumptions. Kombucha tea is a fermented drink made with tea, sugar, bacteria and yeast. Although it's sometimes referred to as kombucha mushroom tea, kombucha is not a mushroom — it's a colony of bacteria and yeast. Kombucha tea is made by adding the colony to sugar and tea, and allowing the mix to ferment. The resulting liquid contains vinegar, B vitamins and a number of other chemical compounds. Kombucha contains multiple species of yeast and bacteria along with the organic acids, active enzymes, amino acids, and polyphenols produced by these microbes [63]. Finished kombucha may contain any of the following components: Acetic acid, Butyric acid, B-vitamins, Ethanol, as well as gluconic, lactic, oxalic, and malic acids.Normally, kombucha contains less than 0.5% ethanol, which classifies it as a nonalcoholic beverage. Older, more acidic, kombucha might contain 1.0% or 1.5% alcohol, depending on more anaerobic brewing time and higher proportions of sugar and yeast.

5.2.1.1. Health Benefits of Kombucha Tea

Health benefits attributed to kombucha tea include stimulating the immune system, preventing cancer, and improving digestion and liver function. However, there's no scientific evidence to support these health claims. There have, however, been reports of adverse effects such as stomach upset, infections and allergic reactions in kombucha tea drinkers. Kombucha tea is often brewed in homes under non-sterile conditions, making contamination likely. If ceramic pots are used for brewing, lead poisoning might be a concern.

5.2.1.2. Health Claims of Kombucha Tea

Kombucha producers often claim that kombucha 'detoxifies the body and energizes the mind', although little research of its health benefits has been published. While no randomized, case-controlled studies have been published in relation to its effect on humans, there has been suspicion in isolated incidents of its effect on the central nervous system, liver, metabolic acidosis, and toxicity in general [64, 65], though no specific links have been established. Acute conditions, such as lactic acidosis, caused by drinking of kombucha, are more likely to occur in persons with pre-existing medical conditions [65]. Other reports suggest care should be taken when taking medical drugs or hormone replacement therapy while regularly drinking kombucha [66]. It may also cause allergic reactions [67]. In short, there isn't good evidence that kombucha tea delivers on its health claims. At the same time, several cases of harm have been reported. Therefore, the prudent approach is to avoid kombucha tea until more definitive information is available.

5.2.2. Cranberry Juice Cocktail

Cranberry juice has been recognized as efficacious in the treatment of urinary tract infections since 1914 [68]. This benzoic acid-rich fruit caused acidification of the urine and recent investigations have focused on the ability of cranberry juice to inhibit the adherence of *Escherichia coli* to uro-epithelial cells. This phenomenon has been attributed to two compounds: fructose and a non-dialyzable polymeric compound. The latter compound, subsequently isolated from cranberry and blueberry juices [69], was found to inhibit adhesins present on the pili of the surface of certain pathogenic *E. coli.* Both cranberry and blueberry juices contain oligomeric proanthocyanidin compounds which inhibit the adherence of *E. coli* to cell walls, and thus help protect against urinary tract infections. Cranberry juice cocktail is now marketed as a functional food with a structure/function claim regarding support of urinary tract health. Recent research has shown that cranberry may also inhibit plaque-forming bacteria in the mouth by the same basic mechanism [70] and marketers are looking for new applications of cranberry in products such as toothpaste, dental floss, mouthwash and rinses.

5.3. Omega-3-Diet

The highly unsaturated fatty acids present in marine oils are responsible for many beneficial effects. Docosahexaenoic acid (DHA; C22:6n-3) is indeed a major fatty acid constituent of the grey matter of the brain and the retina of the eye, among others [71]. The organs most influenced by long-chain omega-3 fatty acids are those with electrical activity such as the brain, the heart and the eye. **Omega-3 fatty acids are essential** polyunsaturated fatty acids found primarily in fish. They have been shown to lower LDL-cholesterol. Eating 35 grams or more of fish daily reduces the risk of death from non-sudden myocardial infarction. As little as one fish meal per week has been associated with a significantly reduced risk of total cardiovascular mortality. Three months of supplementation with **Omega-3** DHA may boost eyesight in older people. Increased levels of omega-3 fatty acids may reduce the risk of atrial fibrillation, the most common irregular heart beat (chronic arrhythmia) in adults, by about 30%. High intake of the omega-3 fatty acids docosahexaenoic acid DHA could help protect the aging brain. Marine species are rich in omega-3 fatty acids and the corresponding oils are obtained from the flesh of fatty fish such as herring and mackerel, liver of white lean fish. Marine oils contain a generally high level of long-chain omega-3. This leads to high sensitivity of the oil to oxidative deterioration and off-flavor formation. Thus the oils need to be subjected to refining, bleaching, and deodorization (RBD). The final RBD oil needs to be stabilized by appropriate antioxidants in order to eliminate the deleterious effects of oxidized compounds and to take advantage of the beneficial health effects of their omega-3 components without changes in their sensory quality [72]. Products in which marine oils may be incorporated include bread, cereal products, spread, milk, mayonnaise and salad dressing, crackers and bars, as well as infant formulas. The oils may be microencapsulated and then used in different applications. However, the solid microcapsule particles must be washed properly with a non-polar solvent to remove any unencapsulated oil that may otherwise be oxidized. However, application of the oil is most successful in products that are usually consumed within a short period of time. Our experience shows that bread and milk products, commonly used within 2 weeks from their production date, may well serve as the best carriers for omega-3 fatty acids.

6. Health Claims

Health claims on food labels are claims by manufacturers of food products that their food will reduce the risk of developing a disease or condition. For example, it is claimed by the manufacturers of oat cereals that oat bran can reduce cholesterol, which will lower the chances of developing serious heart conditions. Disease-risk reduction claims, typically called health claims, do require FDA approval before they can be used on products and must reflect scientific consensus. For example, the health claim for soy protein and its relation to cardiovascular disease reads: "Diets low in saturated fat and cholesterol that include 25 grams of soy protein, a day, may reduce the risk of heart disease. Health claims describe the relationship of diet to a disease, examples are: 1) healthful diets with adequate folic acid may reduce a woman's risk of having a child with a brain or spinal cord defect; 2) Ca^{2+} is necessary for the normal structure of bones and maintain their strengthen as well as helps prevent the risk of osteoporosis; 3) **structure/Function claims** are statements of health-promoting or nutritional benefit allowed on dietary supplement labels. They are not allowed to mention disease conditions; they must describe the support or maintenance of the normal functioning of the body. "Cranberry supports the health of the urinary tract," is an example of a model structure/function claim [73]. Health claims are among the various types of claims allowed in food labeling. They show a relationship between a nutrient (or other substances in a food) and a disease or health-related condition. They differ from the more common claims that highlight a food's nutritional content, such as "low fat," "high fiber" and "low calorie. Health claims are also different from structure/ function claims, which also may appear on conventional food or dietary supplement labels. Manufacturers may make statements about a food substance's effect on the structure or function of the body, for example, calcium builds strong bones. Unlike health claims, structure/function claims do not deal with disease-risk reduction. Also, the FDA does not authorize structure/function claims. Rather, when the manufacturer uses a structure/function claim, the company is responsible for making sure the claim is truthful and not misleading.

7. Scientific Evidenc for Functional Foods

It is possible to identify and measure the biomarkers of health and well-being. Biomarkers could reflect an improved state of health and well-being or reduction in disease risk. Biomarkers could be chosen to reflect: 1) a key target biological function: e.g. Bacterial populations in the gut can be measured to demonstrate that a probiotic has successfully passed through the stomach and could potentially have a beneficial effect in the lower gastrointestinal tract; 2) a key stage in the development of a disease: e.g. Bone mineral density can be used as a marker in the study of a functional food evaluating potential benefit in reducing the risk of osteoporosis.

8. Future Challenges

The future of functional foods will undoubtedly involve a continuation of the labeling and safety debates. As consumers become more health conscious, the demand and market value for health-promoting foods and food components is expected to grow. Before the full market potential can be realized, however, consumers need to be assured of the safety and efficacy of functional foods [74]. Future research will focus on mechanisms by which food components such as phytochemicals positively affect health, and whether these components work independently or synergistically. According to the American Dietetic Association, dietetics professionals will be increasingly called upon to develop preventive meal plans, to recommend changes in food intake, to enhance phytochemical and functional food intake, and to evaluate the appropriateness of functional foods and dietary supplements to meet preventive (and therapeutic) intake levels for both healthy persons and those diagnosed with disease. The challenges related to functional foods are to develop effective synergies between science and food product development for the benefit of the consumer. Aspects of health linked with some of the most developed aspects of nutrition science include better early development and growth; health maintenance (e.g. immune function, gastrointestinal health, mental health, health in aging, physical performance); and reduced risk of obesity. Moreover, reduced risk of chronic diet-related diseases (e.g. cardiovascular disease, type-2 diabetes and metabolic disease, musculoskeletal disease).

A number of challenges exist for the vitality of the functional foods market in the years ahead. First, there are a number of scientific challenges for those attempting to unlock the food/health "equation." How do these food components act to modify disease processes and for whom are these benefits most pronounced? The food/health equation includes a complex set of interactions that must be better understood so that new food product efficacy can be appropriately characterized. The factors contributing to disparities in individual health responses to foods need to be clarified. These include genetic variation, gender differences, individual differences in metabolism, and how bioactive phytochemicals alter the routine functions of traditional nutrients. This challenge is underscored by the emergence of the field of nutrigenomics, in which investigators worldwide are studying how food constituents acting as "dietary signals" can modulate multiple gene-expression patterns in our cells and tissues, and ultimately reduce disease risk [75]. Another scientific challenge involves understanding our complex behaviors associated with food consumption. For example, the sheer magnitude of sensory dynamics is illustrated by the fact that we may consume over 10,000 different food chemicals in our diet which can interact with thousands of chemosensor receptors around our bodies [76]. How is this information coded in the human brain and how does it impact eating behaviors? Related to this issue, and perhaps one of the greatest challenges confronting those attempting to unlock the food/health equation, is gaining a meaningful understanding of the contributors to the international prevalence of obesity.

The focus of food science and technology has shifted from previous goals of improving food safety and enhancing food taste toward providing healthy and functional foods. Today's consumers desire foods that go beyond basic nutrition, foods capable of promoting better health, or even playing a disease-prevention role. To meet this need for innovation, academic research must be combined with the development and commercialization strategies of industry. **Innovation in Healthy and Functional Foods** brings together this knowledge, with contributions from experts in biological science, food science, engineering, marketing, regulation, law, finance, sustainability, and management.

Additional Knowledge

Glossary

Amino acid	One of the building blocks of protein
Antioxidant	A substance that blocks or inhibits the actions of free radicals, molecules that speed up the aging process and contribute to illness
Botanical	A plant-based product
Carotenoids	One of the most widespread groups of naturally occurring pigments which are responsible for the red, yellow and orange color of fruits and vegetables, and are found in many dark green vegetables.
Chelation	A process that binds the minerals in amino acids, it uses an agent, such as the chemical compound EDTA, to remove heavy metals from the body
Enzyme	A complex protein produced by cells that acts as a catalyst in specific biochemical reactions
Vitamin	An organic substance essential in small quantities to normal metabolism
Nutrient	Any substance that can be metabolized by an organism to give energy and build tissue
Mineral	A naturally occurring inorganic substance with a definite and predictable chemical composition and physical properties
Herb	A plant lacking a permanent woody stem
Isoavones	A class of organic compounds and biomolecules act as phytoestrogens, which may be useful in treating cancer.
Phytoestrogens	Compounds that occur naturally in plants (phyto-) and under certain circumstances can have actions like human estrogen
Sulfides	Refers to several types of chemical compounds containing sulfur
Thiols	Compounds containing the functional group composed of a sulfur atom and a hydrogen atom
Flavonoids	A class of water-soluble plant pigments
Genetically engineered food	A food substance that has foreign genes inserted into its genetic code. Genetic engineering can be done with plants, animals or microorganisms.
Holistic medicine	
Complementary and alternative medicine	

(Continued)

	An approach to medical care that emphasizes the study of all aspects of a person's health, including physical, psychological, social, economic and cultural factors A group of diverse medical and health care systems, practices and products that presently are not considered part of conventional medicine. Complementary medicine is used together with conventional medicine. Alternative medicine is used in place of conventional medicine

Conflict of Interest

The author declares that there is no conflict of interest.

References

[1] Willett WC (2002) Dietry fat plays a major role in obesity. *Obes Rev* 3:59-68.

[2] Roberts W (2009) *Benefiting Beverages*. Prepared Foods August 2009.

[3] Scholan I (2007) Functional Beverages- where next? Innovation in functional beverages market is set to continue. International Food Ingredients December 2007.

[4] Walter P (2007) History and importance of functional foods (Germany). Ther Umsch 64:131-134.

[5] Danik MM (*Ed.*) (2011) *Functional Foods and Chronic Diseases: Science and Practice.* Food Science Publisher.

[6] Katan M, Roos N (2005) Promises and problems of functional foods. *Crit Rev Food Sci Nutr* 44:369-377.

[7] Arvanitoyannis I, Houwelingen M (2005) Functional foods: A survey of health claims, pros and cons, and current legislation. Critical Rev Food Sci Nutr 45:385-404.

[8] De Froidmont-Görtz I (2010) Functional Foods. European Commission (e-mail: RTD-FOOD-NEWS@ec.europa.eu).

[9] Lopez-Varela S, Gonzalez-Gross M, Marcos A (2002) Functional Foods and the immune system: a review. *Eur J Clin Nutr* 56:S29-S33.

[10] Hasler C, Kundrat S, Wool D (2000) Functional Foods and cardiovascular disease. *Curr Atheroscler Rep* 2:467-475.

[11] Masahito S, Masaya K, Takuji T, Hisataka M (2012) Nutraceutical approach for preventing obesity related colorectal and liver carcinogenesis. *Int J Mol Sci* 13:579-595.

[12] Rudkowska I (2009) Functional Foods for health: focus on diabetes. *Maturitus* 62:263-269.

[13] Shimazaki Y, Soh I, Saito T, Yamashita Y, Koga T, Miyazaki H, Takehara T (2001) Influence of dentition status on physical disability, mental impairment, and mortality in instituationalized elderely people. *J Dental Res* 80:340-345.

[14] Mazza G (*Ed.*) (1998) *Functional Foods: Biochemical and Processing Aspects.* Lancaster, PA: Technomic Publishing.

[15] Wildman R (*Ed.*) (2001) *Handbook of Nutraceuticals and Functional Foods.* Boca Raton, FL: CRC Press.

[16] Block G, Patterson B, Subar A (1992) Fruit, vegetables, and cancer prevention: A review of the epidemiological evidence. *Nutr Cancer* 18:1-29.

[17] Steinmetz K, Potter J (1991) Vegetables, fruit and cancer II. Mechanisms. *Cancer Causes Control* 2:427-442.

[18] Hasler CM (2001) Functional Foods: Their Role in Disease Prevention and Health Promotion. (Institute of Food Technologists).

[19] Messina M, Barnes S (1991) The role of soy products in reducing risk of cancer. *J Natl Cancer Inst* 83:541-546.

[20] Thompson L (1995) Flaxseed, lignans, and cancer. In: *"Flaxseed in Human Nutrition,"* ed. S. Cunnane and L.U. Thompson, pp. 219-236.

[21] Yan L, Yee J, Li D, McGuire M, Thompson L (1998) Dietary flaxseed supplementation and experimental metastasis of melanoma cells in mice. *Cancer Lett* 124:181-186.

[22] Phipps W, Martini M, Lampe J, Slavin J, Kurzer M (1993) Effect of flax seed ingestion on the menstrual cycle. *J Clin Endocrin Metab* 77:1215-1219.

[23] Silagy C, Neil A (1994) Garlic as a lipid-lowering agent - a meta- analysis. *J Royal Coll Physicians Lond* 28:39-45.

[24] Verhoeven D, Verhagen H, Goldbohm R, Van den Brandet P, Van Poppel G (1997) A review of mechanisms underlying anticarcinogenicity by brassica vegetables. *Chem Bio Interactions* 103:79-129.

[25] Graham HN (1992) Green tea composition, consumption and polyphenol chemistry. *Prev Med* 21:334-350.

[26] Nakachi K, Suemasu K, Suga K (1998) Influence of drinking green tea on breast cancer malignancy among Japanese patients. Jap J Cancer Res 89:254-261.

[27] Fuller R (1989) Probiotics in man and animals. *J Appl Bacteriol* 66:365-78.

[28] Gibson G, Roberfroid M (1995) Dietary modulation of the human colonic microbiota: introducing the concept of prebiotics. *J Nutr* 125:1401-1412.

[29] Ha Y, Grimm N, Pariza M (1987) Anticarcinogens from fried ground beef: Health-altered derivatives of linoleic acid. *Carcinogen* 8:1881-1887.

[30] Decker EA (1995) The role of phenolics, conjugated linoleic acid, carnosine, and pyrroloquinoline quinone as nonessential dietary antioxidants. Nutr Rev 53:49-58.

[31] Ip C, Scimeca J (1997) Conjugated linoleic acid and linoleic acid are distinctive modulators of mammary carcinogenesis. *Nutr Cancer* 27:131-135.

[32] Park Y, Albrigh K, Il W (1997) Effect of conjugated linoleic acid on body composition in mice. *Lipids* 32:853-858.

[33] Kalra EK (2003) *Nutraceutical-definition and introduction. AAPS pharm.* Sci 5 (3):E25. doi:10.1208/ ps050325.

[34] Shibamoto T, Kanazawa K, Shahidi F, Ho C (*Eds*) (2008) *Functional Food and Health.* ACS Symposium series 993. p. 993.

[35] King A, Young G (1999) Characteristics and occurrance of phenolic phytochemicals. *J Am Diet Assoc* 99:213-219.

[36] Shklar G (1998) Mechanisms of cancer inhibition by anti-oxidant nutrients. *Oral Oncol* 34:24-29.

[37] Stahl W, van den Berg H, Arthur J, Bast A, Dainty J, Faulks R, Christine G (2002) Bioavailability and metabolism. *Mol Aspects Med* 23:39-100.

[38] Bio-Trade Peru (2004). PROMPEX. Export promotion strategy for natural ingredients for food, cosmetics and pharmaceuticals from Peru. (Available at:http:// www. biotrade.org).

[39] Niba L (2003) The relevance of biotechnology in the development of functional foods for improved nutritional and health quality in developing countries. *Afr J Biotechnol* 2:631-635.

[40] Panesar P (2011) Fermented dairy products: starter cultures and potential nutritional benefits. *Food Nutr Sci* 2:47-51.

[41] Metchnikoff E (1907) Essais optimistes. Paris. The prolongation of life. Optimistic studies. Translated and edited by P. Chalmers Mitchell. London: Heinemann, 1907.

[42] Vaughan RB (1965) The romantic rationalist: A study of Elie Metchnikoff". *Medical History* 9, 201-215.

[43] Tissier H (1900) Recherchers sur la flora intestinale normale et pathologique du nourisson. Thesis, University of Paris, Paris, France.

[44] Cheplin H, Rettger L (1920) Studies on the Transformation of the Intestinal Flora, with Special Reference to the Implantation of Bacillus Acidophilus: II. Feeding Experiments on Man. *Proc Nat Acad Sci USA* 6:704-705.

[45] Rettger L, Levy W, Weinstein L, Weiss J (1935) *Lactobacillus acidophilus* and its therapeutic application. Yale University Press, New Haven.

[46] Hamilton-Miller JM (2003) The role of probiotics in the treatment and prevention of *Helicobacter pylori* infection. *Intr J Antimicrobial Agents.*

[47] Tannock GW (2003) Probiotics: time for a dose of realism. *Intestinal Microbiol* 4:33-42.

[48] Chemlal-Kherraz D, Sahnouni F, Matallah A, Boutiba Z (2012) The probiotic potential of *lactobacilli* isolated from Nile tilapia (*Oreochromis niloticus*)'s intestine. *Afr J Biotech* 11:13220-13227.

[49] Marteau P, Shanahan F (2003) Basic aspects and pharmacology of probiotics: an overview of pharmacokinetics, mechanics of action and side-effects. *Best Pract Res Clin Gastroenterol* 17725:740-740.

[50] Mital B, Garg S (1995) Anticarcinogenic, hypocholesterolemic and antagonistic activities of *Lactobacilus acidophilus*. *Crit Rev Microbiol* 21:175-214.

[51] Jegtvig SH (2012) About.com Health's Disease and Condition (Medical Review Board)..

[52] Bouhnik Y, Vahedi K, Achour L (1999) Short-chain fructo-oligosaccharide administration dose-dependently increases fecal *bifidobacteria* in healthy humans. *J Nutr* 129:113-116.

[53] Kleessen B, Hartmann L, Blaut M (2001) Oligofructose and long-chain inulin: influence on the gut microbial ecology of rats associated with a human faecal flora. *Brit J Nutr* 86:291-300.

[54] Scholz-Ahrens K, Schrezenmeir J (2007) Inulin and oligofructose and mineral metabolism: the evidence from animal trials. *J Nutr* 137:2513S–2523S.

[55] Lomax A, Calder P (2009) Prebiotics, immune function, infection and inflammation: a review of the evidence. *Brit J Nutr* 101:633-658.

[56] Geier M, Butler R, Howarth G (2006) Probiotics, prebiotics and synbiotics: a role in chemoprevention for colorectal cancer?. *Cancer Biol Ther* 5:1265-1269.
[57] Hedin C, Whelan K, Lindsay J (2007) Evidence for the use of probiotics and prebiotics in inflammatory bowel disease: a review of clinical trials. *Proc Nutr Soc* 66:307-315.
[58] Munjal U, Glei M, Pool-Zobel B, Scharlau D (2009) Fermentation products of inulin-type fructans reduce proliferation and induce apoptosis in human colon tumour cells of different stages of carcinogenesis. *Brit J Nutr* 102:663-671.
[59] Roberfroid M (2000) Prebiotics and probiotics: are they functional foods? *Amr J Clin Nutr* 71:1682S-7S.
[60] Macfarlane S, Macfarlane G, Cummings J (2006) Review article: prebiotics in the gastrointestinal tract. *Aliment Pharmacol Ther* 24:701-714.
[61] Sreeramulu G, Zhu Y, Knol W (2000) Kombucha Fermentation and Its Antimicrobial Activity. *J Agr Food Chem* 48:2589-2594.
[62] Nguyen T, Flanagan B, Gidley M, Dykes G (2008) Characterization of Cellulose Production by a *Gluconacetobacter xylinus* strain from Kombucha. *Curr Microbiol* 57:449-453.
[63] Velicanski A, Cvetkovic D, Markov S, Tumbas V, Savatovic S (2007) Antimicrobial and antioxidant activity of lemon balm Kombucha. Acta Period Technol 38:165-72.
[64] Ernst E (2003) Kombucha: A Systematic Review of the Clinical Evidence. Forschende Komplementärmedizin und Klassische Naturheilkunde 10(2):doi:10.1159/000071667.
[65] Sunghee K, Jones H, Christensen R, Gladstein J (2009) A Case of Kombucha Tea Toxicity. *J Intensive Care Med* 24:205-207.
[66] Srinivasan R, Susan S, David G (1997) Probable gastrointestinal toxicity of kombucha tea. *J Gen Internal Med* 12:643-644.
[67] Perron A, Patterson J, Yanofsky N (1995) Kombucha "mushroom" hepatotoxicity. *Ann Emergency Med* 26:660–661.
[68] Blatherwick NR (1914) The specific role of foods in relation to the composition of the urine. *Arch Int Med* 14:409-450.
[69] Ofek I, Goldhar J, Zafriri D (1991) Anti-Escherichia coli adhesin activity of cranberry and blueberry juices. *New Eng J Med* 324:1599.
[70] Weiss E, Lev-Dor R, Kashamn Y, Houri-Haddad Y, Greenbaum E (1998) Inhibiting interspecies co-aggregation of plaque bacteria with cranberry juice constituent. *J Amr Dent Assoc* 129:1719-1723.
[71] Shahidi F, Finely J (2001) Omega-3 fatty acids: Chemistry, nutrition and health effects. ACS Symposium Series 788, Washington, D.C., American Chemical Society.
[72] Shahidi F, Kim S (2002) Quality management of marine nutraceuticals. In: Ho C-T, Zheng QY, (Eds). Quality management of nutraceuticals. ACS Symposium Series 803. Washington, D.C., American Chemical Society, pp 76-87.
[73] Avorn J, Monane M, Gurwitz J, Glynn R, Shoodnovskiy I, Lipsitz L (1994) Reduction of bacteriuria and pyuria after ingestion of cranberry juice. *JAMA* 271:751-754.
[74] Keith W, Michelle A (2004) Functional Foods: Consumer Issues and Future Challenges. Department of Agricultural and Consumer Economics, University of Illinois, Urbana, IL.
[75] Kaput J, Rodriguez R (2004) Nutritional Genomics: The Next Frontier in the Post-genomic Era. *Physiol Genom* 16:166-177.
[76] Le Coutre J (2003) Taste: The Metabolic Sense. *Food Technol* 57:34-39.

In: Microbes in Process
Editors: Neelam Garg and Abhinav Aeron
ISBN: 978-1-63117-127-7

Chapter 10

Probiotics' Potentialities in Dairy and Non-Dairy Products

Carlos Ricardo Soccol*[*], *Adenise Lorenci Woiciechowski, Luciana Porto de Souza Vandenberghe, Adriane Bianchi Pedroni Medeiros, Julio Cesar de Carvalho, Vanete Thomaz Soccol, Michele Rigon Spier, Luiz Alberto Junior Letti, Caroline Tiemi Yamaguishi and Maria Rosa Prado

Federal University of Parana, Bioprocess Engineering and Biotechnology Department
Curitiba, Paraná, Brazil

Abstract

Significant information about the consumption of probiotic products have been published emphasizing their health benefits when consumed regularly. Antimicrobial activity, improvement in lactose metabolism, reduction in serum cholesterol, antimutagenic and anticarcinogenic properties, immune system stimulation, and prevention of gastrointestinal infections are some of actions. Dairy and non-dairy products can be vehicles for probiotic strains that have been studied and developed intensively by many researchers and industries. The use of dairy matrices is limited by problems such as lactose intolerance, allergenic milk proteins and cholesterol content. In this sense, cereal, fruit juice, vegetable, and soybean beverages have demonstrated to be a suitable non-dairy alternative for delivering probiotics. Current technological innovations and some factors that influence the stability and viability of probiotic strains in these products have been discussed. In this chapter an overview about traditional and commercially available dairy and non-dairy probiotic products as well as the technological aspects for its successful production is presented.

[*] Email: soccol@ufpr.br.

1. Introduction

In 1908, Metchnikoff was the Nobel prize winner in Medicine. He asserted that if lactobacilli cultures were introduced in the large intestine in the human body they could promote health and prolong life, due to their capacity of suppressing the putrefactive-type fermentations of the intestinal flora.

Only at the end of the 20th century it became clear that intestinal microflora had several functions, including metabolic, trophic and protective ones [1] and also therapeutic effects of *L. acidophilus* milk on digestion [2]. In the 1930´s Dr. Minoru Shirota selected the strains of the intestinal tract that could survive passage through the gut and the *L. casei* Shirota was the basis for the establishment of the Yakult Honsha Company [3, 4].

The term "probiotic comes from the Greek 'pro bios`, which means 'for life' [5]. A probiotic, as originally defined by Fuller [6] is "a live microbial feed supplement, which beneficially affects the host by improving its intestinal microbial balance". Besides, Salminen et al. [7, 8], described a probiotic as, "a live microbial food ingredient that is beneficial to health". According to the properties of probiotics, Del Piano et al. [9] defined as "living microorganisms, which resist to the gastric, bile, and pancreatic secretions, attach to epithelial cells and colonize the human intestine". There are some definitions for probiotic, but the most accepted definition is that "probiotics are live microorganisms that, when administered in appropriate amounts, confers a series of health benefits" [10].

In general, probiotics are consumed in the form of dairy (yogurts, fermented milks, cheeses and others) and non-dairy foods (soy fermented milks, fermented juices) and nutraceutics (powder supplements, capsules of freeze-dried organisms), cereal bars, chocolate bars, composes infant formula for children, and others.

Interest in probiotics has been spurred by the growing abundance of modern disorders such as neoplasms, atherosclerosis, cardiac diseases, hypertension and HIV infection. Probiotic consumption is reported to exert a myriad of beneficial effects including: enhanced immune response, balancing of colonic microbiota, vaccine adjuvant effects, reduction of fecal enzymes implicated in cancer initiation, treatment of diarrhea associated with travel and antibiotic therapy, control of rotavirus and *Clostridium difficile* induced colitis and prevention of ulcers related to *Helicobacter pylori*.

The benefits of the probiotic strains can be associated to metabolic, trophic and, protective functions [5]. Metabolic functions are characterized by the fermentation of non-digestible dietary residue and endogenous mucus, production of vitamin K and absorption of ions. Trophic functions are based on the control of epithelial cell proliferation and differentiation, enhancement and homeostasis of the immune system. Finally, protective functions are related to a barrier effect and protection against pathogens [9].

Probiotics are also implicated in the reduction of serum cholesterol, the antagonism against food-borne pathogens and tooth decay organisms, amelioration of lactose absorption symptoms as well as candidiasis and urinary tract infections [11]. Probiotic foods are mostly milk-based and gained popularity worldwide since they contribute to intestinal microbial balance and play an important role in health maintenance. The regular consumption of food containing probiotic microorganisms is recommended as a preventive measure against diseases for consumers of all ages.

The incorporation of probiotic strains in milk products has led to a growth and development in the dairy chain and the diversity of products released is quite extensive and has been shown to be economically advantageous.

2. Sources of Probiotics

Sources of probiotics can be observed in Figure 1, which presents the microorganism groups such as species of *Lactobacillus*, *Bifidobacterium*, other groups of lactic acid bacteria (LAB), yeasts species.

Microorganism Group	Species
Lactobacillus	*Lactobacillus species (L. reuteri, L. acidophilus, L. casei, L. rhamnosus, L. gasseri, L. plantarum, L. paracasei, L. fermentum, L. johnsonii, L. sporogenes, L. lactis, L. GG, L. delbrueckii, L. rhamnosus, L. salivarius; L. crispatus. L. johnssonii, L. farciminis, L. curvatus*
Bifidobacterium	*Bifidobacterium adolescentis, B. bifidum, B. brevis, B. infantis, B. longum; B. lactis, B. animalis, B. thermophilus*
Other groups of lactic acid bacteria	*Enterococcus species (E. faecalis, E. faecium)* *Lactococcus lactis* *Leuconostoc mesenteroides* *Pediococcus acidilactici* *Streptococcus species (S. diacetylactis, S. intermedius, S. thermophilus)*
Yeast	*Saccharomyces boulardii* *Saccharomyces cerevisiae* *Escherichia coli strain Nissle*

Source: [12, 13, 14, 15, 16].

Figure 1. Sources of Probiotics.

3. Requirements for the Use of Probiotic Strains

The guidelines published by FAO/WHO [17] stipulate the requirements for a product to be called a probiotic. It requires that the strains must retain minimum viable count at the end of their shelf life in the product formulation, and besides must confer a proven clinical end-point. The strains must present efficacy and safety. The strain is also considered probiotic if it presents: resistance to gastric acidity; bile acid resistance; adherence to mucus and/or human epithelial cells and cell lines; antimicrobial activity against potentially pathogenic bacteria;

ability to reduce pathogen adhesion to surfaces; bile salt hydrolase activity, and resistance to spermicides (applicable to probiotics for vaginal use) [17].

Besides, some studies in literature describe complementary requirements to a microorganism to be defined effectively as a probiotic: Adherence to cells such as to host epithelial tissue to act as antagonists to potentially pathogenic microorganisms; exclude or reduce pathogenic adherence; persist, multiply, keep its viability in dairy and non-dairy products; demonstration of bile tolerance, acid tolerance; produce organic acids, hydrogen peroxide, bacteriocins, etc. [18, 19, 20].

4. Probiotic Effects on Health Benefits

Probiotics have been gaining popularity worldwide since they contribute to health. Some of the benefits are: intestinal microbial balance, important role in health maintenance. The regular consumption of food containing probiotic microorganisms is recommended as a preventive measure against diseases for consumers of all ages [4]. In general, probiotic microorganisms exert different mechanisms of action, which depends on their specie and metabolism. Some of these mechanisms are:

Competitive exclusion of enteric pathogens; Inhibit growth of potential pathogens by producing organic acids, bacteriocin, etc; triggers cytokine synthesis from enterocytes by attaching to their surface; produce toxic metabolites (e.g. hydrogen peroxide); restore the normal intestinal flora during antibiotic therapy; increase intestinal motility; production of butyric acid, which increased turnover of enterocytes; neutralization of dietary carcinogens; some fermented milk products containing probiotics had a beneficial effect on *H. pylori* infections in man; suppression of allergies; control of blood cholesterol levels; modulation of immune function; prevention of cancers of the colon [15, 21, 22, 23].

According to Kaur et al. [21], to date, interest in probiotic containing foods has centered around three health propositions: first, improving general gut health [gastrointestinal uses]; second, improving body's natural defenses; and third, lowering blood cholesterol. Others health benefits already reported are: alleviation or prevention of diverse intestinal diarrhea-inducing disorders such as lactose indigestion and cause prophylaxis of intestinal and urogenital infections [21, 24]; inhibition of the mutagenicity of intestinal contents and reduce the incidence of intestinal deseases [21, 24]; beneficial outcomes of probiotics in diverse conditions such as rotavirus infections, antiobiotic-associated diarrhea, irritable bowel syndrome, inflammatory bowel disease, atopy in at risk infants and chronic sinusitis [22, 25,26].

Lactobacillus GG (LGG), a variant of *Lactobacillus casei* ssp. *rhamnosus* reduces the severity and duration of diarrhea, and also reduces the intestinal permeability defects caused by exposure to cow´s milk or rotavirus infection [21, 27]. *Lactobacillus* LGG also facilitates antigen transport, increases the number of IgA and other Ig´s in intestine, stimulates local release of interferons, and also acts as an immunoadjuvant for oral vaccines [21].

Some probiotics strains such as species of *Lactobacillus acidophilus, L. casei, L. delbrueckii* and *Bifidobacterium* strain [infant stool] present antimicrobial activities against *Staphylococcus aureus, S. typhimurium, S. thermophilus, L. monocytogenes, S. flexneri, E.*

coli, E. clocae, Clostridium dificile, Klebsiella pneumonia, Bacillus cereus, Yersinia pseudotuberculosis [15].

5. Probiotic Dairy Products

Dairy foods are the main vehicles for probiotics delivery. The incorporation of probiotic strains in milk products has led to a growth and development in the dairy chain and the diversity of products release quite extensive and has been shown to be economically advantageous. The preference could be explained due to the fact that the manufacturing processes of these products have been continuously optimized to assure and improve the survival rate of microorganisms. In this way only minor changes have to be introduced into already consolidated manufacturing technologies and/or processes [28].

In the last ten years, when probiotic supplemented products started to assume an important rolein the scientific universe, many researches started to work with fermented milks and yogurts. That is why these the probiotic dairy products predominate in worldwide probiotic market. However, actually other milky-based probiotic products, such as cheese from different kinds, ice creams, frozen yogurt, aerated sweets as mousses and sweets like puddings have been researched and developed [29, 30, 31]. With the consumption increase of these products Japan leads the dairy products market with more than 50 different milky products containing viable probiotic microorganisms. In European countries, such as Sweden, more than 25% of the fermented market consists of products containing probiotics [31]. Actually, it is estimated that in worldwide market there are more than 80 different dairy products containing *Bifidobacteria*.

Probiotic bacteria have been used frequently as adjunct culture for fermented milks and cheeses. The usual practice is the fermentation with both starter microorganisms and one or more species of probiotic bacteria in order to obtain different products.

Microbial cultures used in dairy fermented products are usually isolated from the human intestine, since these species are recognized as appropriate due to the physiological needs of the host and their easy and high capacity to colonize the intestine. These species include the GRAS (generally recognized as safe) bacteria such as *Bifidobacterium adolescentis, B. bifidum, B. breve, B. infantis, B. longum, Lactobacillus acidophilus, L. casei* ssp. *Rhamnosus* and *Enterococcus faecium* [14, 29, 31]. However, the isolation of probiotic strains from different origins has already been related. Recently, a *Lactobacillus* species have been industrialized and launched in the Spanish market as component of the first dried milk containing probiotic bacteria isolated from breast milk [32].

In order to assure the biological and therapeutical effect of the probiotic products, several organizations established standard regulation for minimum viable counts of probiotic microorganisms. Most of the international organizations have required minimum levels of 10^7 CFU/mL for *L. acidophilus* and 10^6 CFU/mL for *Bifidobacteria* in fermented milk products at the time of sale [20]. Already in Japan, the Fermented Milk and Lactic Beverages Association has established at least 10^7 CFU/mL of viable Bifidobacteria in fermented milk drinks. In probiotic yoghurts counts of viable probiotic bacteria should be at least 10^8 CFU/mL to achieve the desired effects [10, 33]. And it is necessary the ingestion of at least 100 g of food containing 10^8 to 10^9 CFU/g of viable probiotics, in order to ensure the daily minimum dose considered therapy therapeutic [34].

It is important to emphasize that, besides promoting healthy and secure effects, the probiotic cultures must be selected for incorporation in dairy products according to some technological criteria. They should not affect the taste, the texture and remain 'stable and viable' during the whole life time of the product in the market [10, 29]. The processes of freeze and unfreeze of some products such as ice creams and frozen yogurts can endanger the probiotic bacteria promoting cell death and inhibition of the metabolism. In some cases, freezing favors a higher survival rate of the microorganism, but with a long storage time an increase in the mortality rate can be observed [35, 36].

A problem frequently observed in fermented and cold sweets such as the frozen yogurts is the low feasibility of probiotic microorganisms, which in many cases makes the final count of these microrganisms not reach the necessary values to feature them as probiotic. This fact occurs mainly due to the product acidity, in general pH 4.5, injury by the cold and oxygen toxicity, by the incorporation of the air into the product [37]. Trying to minimize detrimental effects on bacteria cells, micro-encapsulation processes where tested using particles of calcium alginate. High viability of probiotic microorganisms embedded to the fermented and cold milky sweet was observed after 12 weeks of storage [38]. However, ice creams are found as a good vehicles for probiotics. Studies have demonstrated that this kind of product has a good stability when stored for more than 6 months and with minimum microbial viability loss [39, 40].

Commercial Name	Probiotic Microorganism	Company	Type of Product
Yakult®	*Lactobacillus casei* Shirota	Yakult Honsha Co. Ltd.	Fermented milk
SVELTY® Gastro Protect	*Lactobacillus johnsonii* La1	Nestlé	Fermented milk
Actimel®	*L. casei* (Defensis)	Danone	Fermented milk
LC1 Yoghurt®	*L. johnsonii* La1 and *L. acidophilus*	Nestlé	Yogurt
Activia®	*Bifidus regularis* (*B. animalis*), *L. bulgaricus*, *S. thermophilus*	Danone	Yogurt
Heini's Yogurt Cultured Cheese	*L. acidophilus*, *L. casei*, *B. lactis*	Bunker Hill Cheese Company	Cheese
Biorich®	*L. acidophilus* LA-5 and *Bifidobacterium* BB-12	Chr. Hansen A/S	Dried bacteria for yoghurt manufacture
HOWARU® Premium Probiotics	*L. acidophilus* NCFM™	Danisco A/S	Dried bacteria for application in dairy products And frozen desserts

Source: Database from companies' websites.

Figure 2. Probiotic Dairy Products.

Other dairy products are gaining space in the probiotic market are the aerated sweets, such as mousses. Researches with probiotic and symbiotic chocolate mousses, with the presence of *L. paracasei* ssp. *paracasei* LBC82, have shown that the survival of cells during its storage was not endangered [over 10^7 CFU g^{-1}] even with a pH decrease [41, 42]. However, in another study with chocolate mousses, it was observed a reduction of 3 logarithm of the strain *L. acidophilus* population after the 20th day of storage. In the same work, the micro-encapsulation of *L. acidophilus* was tested and after the same period of storage it was found a reduction of only a logarithm of the probiotic strain population [42].

Several probiotic products have been offered in the market and are classified in three categories:

1) conventional food with addition of probiotic microorganisms;
2) food supplements or fermented milk with food formulations used for delivering probiotic bacteria and;
3) dietary supplements in the form of capsules or pills [43].

Figure 2 presents some types of probiotic dairy fermented products.

5.1. Probiotic Yogurt

Yougurt is a product elaborated with active cultures of lactic bacteria that ferment the cream and/or the milk, metabolizing part of the present lactose to lactic acid, it also occurs the pH reduction with or without coagulation (isoelectric) of the proteins. The microorganisms need to be abundant and viable until the product sell-by date [30, 44]. At the end of the process, the yogurt will present features that distinguish it from other fermented milky beverages as shown in Figure 3.

Starter cultures of *Lactobacillus delbrueckii* spp. *bulgaricus* and of *Streptococcus thermophilus* are frequently used for yogurt production. These bacteria have a symbiotic relation, in which, each one of them stimulates the multiplication of the other. In this way, *L. delbrueckii* ssp. *bulgaricus* liberates aminoacids and peptides of the milk proteins, which allow the multiplication of *S. thermophilus* in the first fermentation stages. *S. thermophilus*, on the other hand, produces formic acid, which incites the multiplication of *L. delbrueckii* sp. *bulgaricus*, decreasing the fermentation time and attributing peculiar characteristics to the product [44, 45].

There is a great discussion about the probiotic character of the yogurtmicrorganisms, once *L. delbrueckii* ssp. *bulgaricus* and *S. thermophilus* improve the lactose digestion and eliminate symptoms of intolerance to this disaccharide. Although, they do not survive to the conditions of the gastrointestinal tract in compatible concentrations to carry out probiotic effects *in vivo* [46, 47]. The use of probiotic cultures or starter cultures, their features and resistance to biological barriers are variable depending on the species and even on the strains of same species [48].

Some yogurt-based products were reformulated to include, besides yogurt conventional cultures, cultures of *L. acidophilus*, *Lactobacillus casei* ssp. and of *Bifidobacterium* ssp. [49]. These products started to be called bio-yogurts [29]. During their elaboration process, the probiotic cultures can be added before fermentation, combined with the yogurt conventional cultures, or after fermentation [33]. It is important to emphasize that each cultivation condition must be properly tested, once each probiotic strain has a peculiar behavior and that several factors can injure the multiplication of these bacteria, as well as their survival during the storage period.

Composition	Yogurt	Fermented Milk
Milk protein (% w/w)	Min. 2.7%	Min. 2.7%
Milk fat (% w/w)	Less than 15%	Less than 10%
Titrable acidity, expressed as % lactic acid (% w/w)	Min. 0.6%	Min. 0.3%
Sum of microorganisms constituting the starter culture (cfu/g, in total)	Min. 10^7	Min. 10^7
Microorganisms (cfu/g, in total)	Min. 10^6	Min. 10^6

Source: [30].

Figure 3. Composition of the Yogurt and Fermented Milks.

In the probiotic production it is necessary to take care of the fermentation media composition and the dissolved oxygen concentration [50]. With the intention of stimulate the multiplication and survival of probiotic cultures, hydrolyzed of casein or whey, yeast extract, glucose, vitamins and minerals (always in proportions compatible with the legislation) can be added and this practice helps the development of the products' texture. [18, 33, 50].

5.2. Probiotic in Beverages

The fermented milks are defined as the product obtained from milk or the reconstituted milk and/or milk derivatives, reconstituted or not, with or without addition of other ingredients, in which the lactic base represents, at least, 51% of the total product ingredients, which are fermented through the action of specific microorganisms cultivation, and/or added of fermented milk and/or other fermented dairy products [30]. At the end of the process the fermented milks should present the following features demonstrated in Figure 3.

Probiotic fermented milk brings benefits to the consumer by improving the microbial balance and by re-establishing optimum ecosystem of the intestinal microbiota. The production of probiotic fermented milk is a big challenge, once the milk is not an appropriated medium to microorganism multiplication [51, 52, 53]. Milk, despite being a rich mean from the nutritional point of view, is not propitious for probiotic bacteria cultivation mainly due to the absence of proteolytic activity [54]. The incorporation of micronutrients, as peptides and amino acids, and other growth factors can be necessary to reduce the fermentation time and to propitiate probiotic's bacteria viability [55].

Fermentable simple sugars can also be added to favor the development of strains [56]. When using *Bifidobacterium* strains, a problem arises with the adaptation of the microorganism that suffers with the oxygen presence [56]. One of the alternatives to increase the multiplication of these bacteria in milks is, besides the supplementation with growth factors, the addition of bifidogenic and probiotic factors. Another solution is the selection of strains that can survive to acidification and aerobiosis [48].

The improvement of fermentation processes during the manufacturing of fermented milks containing probiotics is linked to the addition of a starter culture containing *L. delbrueckii* ssp. *bulgaricus* and *Streptococcus thermophilus* [54].A starter culture, used in the elaboration of probiotic fermented milks, can stimulate probiotic multiplication due to oxygen consumption [50]. Aiming to overcome the problem of post-acidification, the trend is to use the mixed culture, which contains *L. acidophilus*, *Bifidobacteria* and *S. thermophilus* as starter. In this way, a fermented and probiotic dairy beverage is obtained without occurrence of great alterations of the product characteristics [57].

The regular consumption of probiotic fermented milk can help in the prophylaxis and in the treatment of several pathological conditions of the gastrointestinal tract [58], in the intestinal function regularization, colon cancer prevention and other kinds of cancer, cholesterol and plasmatic triglyceride, elimination of the symptoms of lactose intolerance among others.

Kefir is an acidic-alcoholic fermented milk product that was originated in the Balkans, Eastern Europe and the Caucasus [59]. It is a refreshing, naturally carbonated fermented dairy beverage with a slightly acidic taste, yeasty flavor and creamy consistency [60]. The starter culture used to produce this beverage is an irregularly shaped, gelatinous white/yellow grain [61]. Traditionally kefir grains have been used for centuries in many countries, for example, in Tibet, China, as the natural starter in the production of the unique self-carbonated dairy beverage [62]. *Kefir* is a consortium of microbes that is mainly used in the production of the low alcoholic beverage, where milk constitutes the initial fermenting substrate. It can be produced by fermenting milk with commercial freeze-dried kefir starter cultures, traditional kefir grains, and the product that remains after the removal of kefir grains [63].

These kefir grains have a varying and complex microbial composition that includes species of yeasts, lactic acid bacteria (LAB), acetic acid bacteria (AAB), and mycelial fungi. The grain is a symbiotic association of microorganisms (*Lactobacilli, Lactococci, Kluyveromyces, Candida* etc.) embedded in a polysaccharide matrix (kefiran), which is composed of glucose and galactose produced by *Lactobacillus kefiranofaciens* [64,65]. However cell growth and kefiran production rates have been shown to be increased when *L. kefiranofaciens* is grown in a batch mixed culture together with *Saccharomyces cerevisiae* [66] demonstrating the importance of symbiosis between bacteria and yeast in kefir.

The microbiological and chemical composition of the kefir indicates that it is a complex probiotic product [67].The beneficial action of this fermented milk can be partially attributed to the inhibition of pathogenic microorganisms by metabolic products such as organic acids produced by kefir microflora [68]. Studies about the immunomodulation capacity resulting from kefir oral ingestion were also reported [69].

6. Probiotic in Non-Dairy Products

The consumption of beverages and food that contain probiotic microorganisms is a growing worldwide trend. New products containing probiotic strains have been launched, particularly non-dairy probiotic products. They have gained importance due to the ongoing trend of vegetarianism, allergenic milk proteins and to a high prevalence of lactose intolerance in many populations around the world [31]. Even though fermented dairy products are generally good matrices for the delivery of probiotics to humans, other foods have been examined for their potential as probiotic carriers. Fruit and vegetable drinks, soymilk, meats, desserts, cereal-based products and many others may be other suitable media for delivering probiotics [70, 71].

Indeed, technological advances have made possible to alter some structural characteristics of fruit and vegetables matrices by modifying food components in a controlled way such as pH modification, fortification of culture media, among others [72]. This could make them ideal substrates for probiotic cultures, since they already contain beneficial nutrients, such as minerals, vitamins, dietary fibers, and antioxidants, while lacking the dairy allergens that might prevent consumption by certain segments of the population [73].The commercial success of dairy-free probiotic products ultimately depends on taste, appearance, price, and health claim appeal to consumers. They need to receive a comprehensible and reasonable message about the physiological effects of probiotics in humans, without appearing to be exaggerated; moreover, all these factors influence directly the consumers' attitudes toward effective purchase, which is necessary for the maintenance of industry.Therefore, the food industry takes into consideration many variables to develop or reengineer nondairy probiotic products, such as sensory acceptance, stability, price, chemical, and functional properties.

6.1. Beverages

Functional beverages make up one of the most dynamic parts of the beverage worldwide market. As a result of marketing and innovation, the level of consumer awareness of different types of probiotics has improved in the last years. They have driven research efforts into the development of alternative products that can appeal to a wider range of consumers. One emerging area of innovation has been probiotics and prebiotics in nondairy beverages. Due to this value, several food companies have begun researching possible probiotic line extensions. Nowadays, worldwide, there is a great variety of traditional and commercial nondairy beverages with presumptive probiotic features (Figure 4). Many of them are nonalcoholic beverages produced with cereals, fruit, and vegetable as main raw material. Traditional fermented beverages from cereals such as *Mahewu* and *Togwa* have an important contribution to the human diet especially in many African countries. *Mahewu* (*amahew*) is a sour beverage made from corn meal, found in Africa and some Persian Gulf countries. *Togwa* is a starch-saccharified traditional beverage consumed in Africa that has been used as a probiotic medium [74]. Besides being an inexpensive technology, the fermentation preserves the food and enhances its sensory properties [75].

Product	*Microorganisms*	*Substrate*	*References*
Boza	*Lactobacillus, Lactococcus, Pediococcus, Leuconostoc*, yeast	Cereals	[76]
Hardaliye	*Lactobacillus*, yeast, molds	Red grapes and mustard seeds	[77]
Mahewu	*Lactobacillus, Lactococcus*	Maize	[74]
Pozol	*Lactobacillus, Lactococcus, Leuconostoc, Pediococcus*, yeast, molds	Cocoa and maize	[78]
Togwa	LAB, yeast	Cereals	[74]
B.E. Wholegrain Liquid	*Lactobacillus, Saccharomyces*	Cereals	AGM Foods Pty, Grainfields Australia, Wynnum, Australia
Friscus	*L. plantarum* and *L. paracasei* 8700:2	Fruit juice	Skånemejerier, Bravo, Malmö, Sweden
Gefilus	*L. rhamnosus*	Fruit juice	Valio Ltd, ValioGefilus, Helsinki, Finland
Goodbelly	*L. plantarum* 299v	Fruit juice	NextFoods, GoodBelly, Boulder, Colorado, USA
Lemon and ginger	*Lactobacillus, Saccharomyces*	Lemon juice, honey, and cereals	AGM Foods Pty, Grainfields, Australia, Wynnum, Australia
Probiotic Soymilk	Not available	Soymilk	Pulmuone Wildwood, Wildwood Organics, Fullerton, California, USA
Silk Live	*L. bulgaricus, St. thermophilus, L. acidophilus, B. bifidum, L. casei, L. rhamnosus*	Soymilk	White Wave Foods, Silk, Broomfield, Colorado, USA
Vita Biosa	*L. acidophilus, L casei, B. bifidum*	Herbs	Biola Inc., Hamilton, Ontario, Canada

Adapted from Yamaguishi, et al. [75].

Figure 4. Traditional and Commercial Non-dairy Beverages.

6.1.1. Probiotic Fruit Juices

Fruit juices represent a big market sector as functional drinks, as they are good sources of vitamins and minerals. There is a genuine interest in the development of fruit juice based functional beverages with probiotics because they have taste profiles that are appealing to all age groups and because they are perceived as healthy and refreshing foods. However, according to Rivera-Espinoza and Gallardo-Navarro [79], unsuitable contents of aromas (perfumery, dairy) and flavors (sour, savory) have been reported when probiotics are added to juices.

LAB have been considered difficult microorganisms to cultivate due to their demand of various essential amino acids and vitamins. It has been found that some probiotic strains have the capability to grow in fruit matrices. Few studies have been conducted regarding the factors that influence the viability of probiotics during storage in fruit juices compared to dairy products, it has been shown that the storage temperature and time [80], the presence of dietary fiber in the product [81], and the type of container used [82] are likely to have an important role. However, there is very little information on the effect of the chemical composition of the fruit juices on probiotic survival. Nualkaekul et al. [83] proposed a mathematical model able to predict adequately the cell survival in orange, grapefruit, blackcurrant, and pineapple juices. The survival of *Bifidobacterium longum* was studied during refrigerated storage for 6 weeks in model solutions, based on which a mathematical model was constructed describing cell survival as a function of pH, citric acid, protein and dietary fiber. The authors observed the highest cell survival [less than 0.4 log decrease] after 6 weeks of storage in orange and pineapple, both of which had a pH of about 3.8. Although the pH of grapefruit and blackcurrant was similar [pH 3.2], the log decrease of the former was 0.5 log, whereas of the latter was 0.7 log. In another study, the survival of a potential probiotic *Lactobacillus plantarum* strain during refrigerated storage in model solutions was explained as a factor of pH, citric acid and ascorbic acid [84]. Taking into account the results from the compositional analysis of the juices and the model, the authors deduced that in certain juices, other compounds such as proteins and dietary fibers seemed to protect the cells during storage. In contrast, in certain juices, such as pomegranate, cell survival was much lower than expected; this could be due to the presence of antimicrobial compounds, such as phenolic compounds. It is important consider that cell growth is highly affected by initial pH of the culture media. Acid values difficult LAB growth.

Soccol and Prado [85] developed a coconut probiotic water fermented by *L. plantarum* AC-1, *L. plantarum* B-7, and *B. animalis* ssp. *lactis* BFL-9 in same proportion. The supplementation of the beverage with yeast extract, hydrolyzed soy protein, and sucrose produced maximum yield to mixed culture in 8 hours of fermentation. After 28 days of cold storage at 4°C, lactobacilli and *B. animalis* ssp. *lactis* presented 8 and 7 log CFU/mL, respectively. The fermented beverage added of sugar and coconut aroma presented good sensorial characteristics and acceptability from potential consumers. Furthermore, the beverage could be classified as "low calorie" delivering only 33.5 kcal/100 mL.

Lactobacillus casei NRRL B-442 was cultivated in cashew apple juice and its survivability was investigated during refrigerated storage (4°C) for 42 days [86]. Optimum conditions for probiotic cashew apple juice production were defined as initial pH 6.4, fermentation temperature of 30°C, inoculation level of 7.48 log CFU/mL (*L. casei*) and 16 h of fermentation. It was observed that the *L. casei* grew during the refrigerated storage. Viable cell counts were higher than 8.00 log CFU/mL throughout the storage period (42 days). The authors found the fermented juice with *L. casei* is a good and healthy alternative functional food containing probiotics. Cashew apple juice showed to be efficient as dairy products for *L. casei* growth.

6.2. Probiotic Soy Products

Soybean, the most important legume in the traditional Asian diet, is rich in high-quality protein [87]. The soybean products play an important role in the prevention of chronic diseases such as menopausal disorder, cancer, atherosclerosis, and osteoporosis [88]. Despite its ubiquity in the diet, soybean has some limitations such as a disagreeable bean flavor for some food processes, and its content non-digestible oligosaccharides such as raffinose and stachyose, which produces flatulence [89] causing gastrointestinal discomfort. Fermentation has been a traditional option to increase digestibility of soy products and make them more flavored [89]. Soymilk is a good substrate suitable for LAB growth. Several studies reported the production and use of fermented soymilk beverage as probiotic [90, 91, 92]. *Lactobacillus plantarum*, *Streptococcus lactis* and *Leuconostoc lactis* have been found as part of the lactic flora of soybean, but there is no information about their probiotic characteristics.

According to Granato et al. [31], the sales and marketing of probiotic soy products, such as yogurts-like, blends with fruit juices and fermented beverages, have increased during the past decade, showing a trend to the development of new products with suitable sensory and nutritional appeal, and beneficial properties, when regularly consumed. Both nondairy and soy-based probiotic products represent a huge growth potential for the food industry, and may be widely explored through the development of new ingredients, processes, and products.

The development of technologies to encapsulate probiotics contributes to improve the non-dairy probiotic products. Microencapsulation has been studied to protect probiotics from damage during food processing, storage and upper gastrointestinal tract digestion. Freeze-dried microcapsules of *L. acidophilus* were incorporated into soy protein bars. The viability of probiotic cells remained in high numbers throughout 14 weeks of storage at 4°C. Results of this work with the support of other studies on microencapsulation benefits indicate a promising use of freeze-dried α-galactosidase positive microencapsulated probiotics in a soy based food [93].

6.3. Other Non-Dairy Probiotic Products

Non-dairy probiotic beverages are the more common functional products available on the market, but there are several studies on the development of alternative probiotic foods such as sausages fermented by *Lactobacillus rhamnosus* strains, fruit pieces containing *Bifidobacterium* spp., dried fruits vacuum-impregnated with *Saccharomyces cerevisiae* and *Lactobacillus casei* ssp. *rhamnosus*, an oat-based cereal bar including *Bifidobacterium lactis*, table olives enriched with *Lactobacillus rhamnosus*, *L. paracasei*, *Bifidobacterium bifidum*, and *Bifidobacterium longum* [79].

6.3.1. Cereal Based Food

Cereal grains are considered an important source of protein, carbohydrates, vitamins, minerals and fiber for people all over the world. Furthermore, they can be used as sources of non-digestible carbohydrates that besides promoting several beneficial physiological effects can also selectively stimulate the growth of *Lactobacilli* and *Bifidobacteria* present in the colon, thereby acting as prebiotics [94].

Cereals are good substrates for probiotic strains growth and due to the presence of non-digestible components of the cereal matrix may also serve as prebiotics [95, 96]. However, due to the complexity of cereals, a systematic approach is needed in order to identify the factors, which enhance the growth of probiotic strains. Previous experiments of have established the ability of human derived strains of *Lactobacillus reuteri* to resist simulated gastric acidity and bile acid, and also to grow well in a number of cereal substrates.

6.3.2. Meat and Fish Products

Besides its nutritional content rich in protein, amino acids, vitamins and minerals, meat is now being demanded as functional meat products which have reduced contents of salt, both nitrites and nitrates, cholesterol fraction, fat and whose fatty acid profile, which has been modified. According to Khan et al. [97], some of the essential nutrients which can be considered functional in meat products are vitamins A, E, C and also minerals like magnesium, calcium and potassium. Non-essential components which can be supplemented with meat include dietary fiber, long chain omega-3 fatty acids, bioactive peptides, conjugated linoleic acid, probiotic bacteria, prebiotics and antioxidants. Meat serves as an excellent medium for probiotic growth and has also been reported to protect microbes against bile [98]. In this review the potential ingredients and strategies, utilized in crafting such functional meat products, and the developments and commercial successes in functional meat industry was presented. Dry fermented sausages meet the conditions required to carry viable probiotic microbes. Fermented sausages can be potential candidates for probiotics since they are subjected to mild heating and may enhance the survival of probiotic bacteria in the digestive system [99]. Various microorganisms are commercially used in functional food products and potential bacteria for probiotic sausage production.

The suitability of fish products as a substrate for growing probiotic bacterial strains was investigated by Speranza et al. [100]. The authors proposed the incorporation of probiotic bacteria into marinated anchovies. Two probiotic strains (*Lactobacillus plantarum* and *Bifidobacterium animalis* ssp. *Lactis*) were inoculated in the marine brine under different conditions. This new kind of functional probiotic product, besides being a novelty, adds functional features to a recognized high nutritional value product. In fact, in addition to proteins and trace elements (particularly selenium), the high nutritional value of fish is mainly due to their lipid composition: fishes contain high levels of polyunsaturated fatty acids, mainly eicosapentaenoic acid and docosahexaenoic acid, which are recognized to protect against cardiovascular diseases, to prevent or delay the clinical manifestations of certain cancers and to alleviate some affective and psychiatric disorders [101].

6.3.3. Vegetable Probiotic Products

LAB are the principal microorganisms responsible for the natural fermentation of vegetables. Considering the ability of some strains of LAB to convert starch into lactic acid, Santos [102] developed a probiotic beverage with the fermented cassava flour (liquid phase plus solid) using mixed culture of *L. plantarum* A6 and *L. casei* Shirota. The beverage maintained its microbiological (9 log CFU/mL) and physicochemical quality during 28 days of storage at 4°C. Sour cassava starch is obtained by a natural fermentation, and this product is largely appreciated, for example, in Africa and South America.

Yamaguishi et al. [75] reported other studies with vegetable-based substrates for the production of probiotic products. The suitability of red beets as substrate in the production of a probiotic juice by *L. acidophilus*, *L. casei*, *L. delbrueckii* subsp. *bulgaricus*, and *L. plantarum* was studied by Yoon et al. [103]. The study reported that all strains reduced the pH of beet juice due to their ability to produce lactic acid. Rakin et al. [104] supplemented beetroot and carrot juices with brewers' yeast autolyzed before fermentation with *L. acidophilus* NCDO1748. The addition of the autolyzed favorably affected the increase of the number of LAB during the fermentation. Also affect reduction of fermentation time and enrichment of the vegetable juices with amino acids, vitamins, minerals, and antioxidants.

Nazzarro et al. [105] developed a functional fermented carrot juice beverage with *L. rhamnosus* DSM20711 and *L. bulgaricus* ATCC 11842 supplemented with prebiotic components inulin and fructooligosaccharides. Both bacterial strains were capable of growing in carrot juice and survival at low pH during 4 weeks of storage at 4°C. *L. acidophilus* DSM 20079 was encapsulated in alginate-inulin xanthan prebiotic gum and its ability to grow in carrot juice and survive 8 weeks of storage at 4°C and subsequent exposure to artificial gastrointestinal conditions were assessed by Nazzarro et al. [106]. Encapsulation significantly enhanced cell viability after fermentation and storage and protected *L. acidophilus* from exposure to simulated gastric conditions. Naturally fermented olives [107] and the suitability of tomato juice [108] were also studied for the production of probiotic products by *L. casei*, *L. acidophilus*, *L. plantarum*, *L. casei* and *L. delbrueckii.* Yoon et al. [109] also developed a probiotic cabbage juice using LAB (*L. plantarum* C3, *L. casei* A4, *L. delbrueckii* ssp. *bulgaricus* D7). The cultures grew well on cabbage substrate and reached about 9 log CFU/mL after 48 hours of fermentation. After 4 weeks of the cold storage at 4°C, the viable cell of *L. plantarum* and *L. delbrueckii* were 7 and 5 log CFU/mL, respectively. *L. casei* did not survive at low pH and lost cell viability completely after 2 weeks of the cold storage.

7. Market of Probiotic Dairy Products

The global probiotic market is expected to reach $31.1 billion in 2015 at a compound annual growth rate of 7.6 % for the 5-year period [110,111]. Besides, probiotics represent one of the largest and fastest growing segments of the dietary supplement market. Among products commercially available, the most important genera of probiotic present in dairy products which are currently worldwide marketed are: *Lactobacillus, Lactococcus*, *Enterocococcus*, *Streptococcus*, *Pediococcus*, *Leuconostoc*, and *Bifidobacterium* [112]. Several probiotic products have been offered in the market and are classified in three categories: 1) conventional food with addition of probiotic microorganisms; 2) food supplements or fermented milk with food formulations used for delivering probiotic bacteria, and; 3) dietary supplements in the form of capsules or pills [43]. Some examples of probiotic dairy fermented products commercially available are: Yakult®, fermented milk containing *Lactobacillus casei* Shirota (Yakult Honsha Co. Ltd.); SVELTY® Gastro Protect which is a fermented milk containing *Lactobacillus johnsonii* (Nestlé); Actimel® *Lactobacillus casei (defensis)* from Danone, LC1 Yoghurt® (*L. johnsonii* La1 and *L. acidophilus)* (Nestlé), Activia® *Bifidus regularis* (B. *animalis, L. bulgaricus, S. thermophilus*) (Danone); Heini´s Yogurt (a cheese containing *L. acidophilus*, from Bunker Hill).

Consumers are ready to accept fortified products and healthy food products instead to consume those food products without differential benefits. So, the market of dairy and non-dairy probiotic products are one of the segments with great increased in the last years. Examples of these products are: probiotics chocolates, probiotic ice creams; probiotic baked products; probiotic cheese, probiotic butter, probiotic cereal bars, probiotic candies, and others.

Symbiotic combinations with associate probiotics and prebiotics are increasingly attracting consumers, and also biotechnology industries are facilitating the development in this segment. There is a great interest in using prebiotic oligosaccharides as functional food ingredients to contribute to colonic and microflora growth of probiotics. Products for population which presents intestinal disorders or even for geriatrics have a potential market, because *probiotics* help the acceleration of *digestive issues* and improve the immune system as described above. Some examples of symbiotic products and respective companies in the world market are: Nido Kinder® (Nestlé); Pro and Prebiotic (AG Adrien Gagnon); Más Vital (Leche Pascual); ActifloraBioplus® (Kendy, USA); Pro Advantage Howary™ (Whitewave Foods); Yo-plus (Yoplait); Nutraelle® (Wellness Nutrition, LLC); Culturelle®; Ultimate Flora Women´s Care® (Rene Life Formula´s, Inc.); For a Bear® for Kids (Renew Life Formula´s Inc.); Julie´s low-fat frozen yoghurt (Oregon), and other products.

8. Technological Advances of Probiotics

Good manufacturing practices (GMP) must be applied in the manufacture of probiotic foods with quality assurance, and shelf-life conditions established. Further development of methods (*in vitro* and *in vivo*) must be performing to evaluate the functionality and safety of probiotics [17]. To date, products containing probiotic cultures preferred by the consumers is still yogurts and dairy products, and besides, a part of population may not tolerate milk sources due to the lactose intolerance or another consume restriction or allergies. So, new alternatives in the market focus the development of non-dairy products sources of probiotics. Non-dairy products containing probiotics cultures may be fruit and vegetables juices, cereals and cereals bars, chocolates. It is crucial that the products present viable counts of probiotic cells after processing and during product shelf-life and storage conditions (at least 10^6 CFU/mL), a sufficient number of viable cells to exert the desired therapeutic effects [113]. As described by [34], it is necessary the ingestion of at least 100 g of food containing 10^8 to 10^9 CFU/g of viable probiotics, in order to ensure the daily minimum dose considered therapy.

Adding probiotics to the juices is more complex than formulating in the dairy products because the bacteria needs protection against the acidic conditions (and also phenolic compounds which may exert an antimicrobial effect) in the fruit juice. However, with microencapsulation technologies, the probiotics can become an important ingredient in the functional foods, expanding the probiotic application outside the pharmaceutical and food supplement industries [112,114].

Some of the important aspects for industrial production of probiotics are related to the microorganism itself, the cost of nutrient substrates, and the processes used in their production. In addition, parameters for scale optimization and scale-up are necessary. Different phases of the process must be evaluated, such as the kinetics of growth, primary or secondary metabolite production, as well as separation, recovery and formulation of the products.

Microencapsulation technologies have been developed and successfully applied using various matrices to protect the bacterial cells from the damage caused by the external environment [9]. It is the process by which small particles or droplets are surrounded by a coating to produce the capsules in the micrometer to millimeter range known as microcapsules. The microencapsulation allows the probiotic bacteria to be separated from its environment by a protective coating. Several studies have reported the technique of the microencapsulation by using gelatin, or vegetable gum to provide protection to acid-sensitive *Bifidobacterium* and *Lactobacillus* [112,114].

Trends on the development of new functional fermented products offer an alternative in medicine substitution to the treatment of several human diseases [115]. The use of natural ingredients are seen as healthy and the demand for improve health and appearance is attracting fast grow segment on the dairy and non-dairy probiotic market. Botanical sources, exotic and regional fruits and vegetables from world biodiversity will be useful sources in the near future to produce new beverages and foods [5, 112, 114].

Conclusion

Important advances were made in the sector of probiotics. Dairy products now abound their reduced-fat and nonfat versions. Probiotics and prebiotics have been developed which for geriatrics in order to promote the balance of intestinal microflora, which in most cases is poor due to age, medication use, thereby reducing the associated digestive tract disorders. These products have mainly in its composition the addition of prebiotics and can be used as dietary supplements.

Other products, the non-dairy probiotic products are a great option for those who do not tolerate lactose such as cereal bars, chocolate, fruit juices, meat products and others. These products are intended for all age groups, always with the aim to improve product characteristics but mainly for their health benefits.

Technological tools are being employed to optimize the production of a great number of probiotic, prebiotic and symbiotic products that are available in the market. In this way, new species have been studied. However, a typical poor growth of these species is observed. Therefore some efforts must be done to improve growth factors and establish the effective microorganism's survival in the different versions of products.

References

[1] Guarner F, Malagelada JR (2003) Gut flora in health and disease. *Lancet* 361:512−19.

[2] Cheplin H, Rettger L (1922) The therapeutic application of *Lactobacillus acidophilus*. *Abst Bacteriol* 6:24.

[3] Yakult Honsha Co. Ltd, *History*, Online. Avaliable HTTP: <http://www.yakultusa.com/Company/History.php> (accessed 01 November 2011).

[4] Vandenberghe LPS, Yamaguishi CT, Rodrigues C, Prado MR, Spier MR, de Carvalho JC (2013) Probiotic dairy products. In: Soccol CR, Pandey A, Larroche C. (org). Fermentation Processes Engineering in the Food Industry, 1ed, vol 1, CRC Press, New York, p 329-353.

[5] Vandenberghe LPS, Spier MR, Medeiros ABP, Yamaguishi CT, Lindner JDD, Pandey A (2010) The Potential of Probiotics: A Review. *Food Technol Biotechnol* 48: 413-434.

[6] Fuller R (1989) Probiotics in man and animals. A review. *J. Appl. Bacteriol* 66-365–378.

[7] Salminen S, Deighton MA, Benno Y, Gorbach SL (1998) Lactic acid bacteria in health and disease. In: Salminen S, von Wright A. (ed) Lactic acid Bacteria: Microbiology and Functional Aspects, second ed. Marcel Dekker Inc., New York, p. 211–253.

[8] Salminen S, Bouley C, Boutron-Ruault MC, Cummings J, Franck A, Gibson JR, Isolauri E, Moreau MC, Roberfroid M, Rowland I (1998) Functional food science in gastrointestinal physiology and function. *Br J Nutr* 80:S147- S171.

[9] Del Piano M, Morelli L, Strozzi GP, Allesina S, Barba M, Deidda F, Lorenzini P, Ballar M, Montino F, Orsello M, Sartori M, Garello E, Carmagnola S, Pagliarulo M, Capurso L (2006) Probiotics: from research to consumer. *Digest Liver Disease* 38:S248–S255.

[10] Food and Agriculture Organization / World Health Organization (FAO/WHO) (2001) Evaluation of health and nutritional properties of probiotics in food including powder milk with live lactic acid bacteria. Córdoba, 34p.

[11] Saavedra JM (2001) Clinical application of probiotic agents. *Am J Clin Nutr* 73:1147S-1151S.

[12] Fuller R (1991) Probiotics in human medicine. *Gut* 32-439–442.

[13] Gordin BR, Gorbach SL (1992) Probiotics for humans. In: Fuller R (ed) Probiotics. The Scientific Basis. Chapman and Hall, London, pp. 355–376.

[14] Cruz AG, Faria JAF, Van Dender AGF (2007) Packaging system and probiotic dairy foods. *Food Res Int* 40:951–6.

[15] Commane D, Hughes R, Shortt C, Rowland I (2005).The potential mechanisms involved in the anti-carcinogenic action of probiotics. *Mutation Res* 591:276–289.

[16] Pancheniak EFR, Maziero MT, Rodriguez-León JA, Parada JL, Spier MR, Soccol CR (2012) Molecular Characterization and Biomass and Metabolite Production of L. reuteri LPB P01-001: A Potential Probiotic. *Braz J Microbiol*: ISSN 1517-8382.

[17] Food and Agriculture Organization / World Health Organization (FAO/WHO) (2002) Guidelines for the Evaluation of Probiotics in Food..

[18] Donkor, ON, Henriksson A, Vasiljevic T Shah NP (2006) Effect of acidification on the activity of probiotics in yoghurt during cold storage. Int *Dairy J* 16:1181–9.

[19] Nighswonger, BD, Brashears MM, Gilliland SE (1996) Viability of *Lactobacillus acidophilus* and *Lactobacillus casei* in fermented milk products during refrigerated storage. *J Dairy Sci* 79:212–9.

[20] Talwalkar A, Kailasapathy K (2003) A review of oxygen toxicity in probiotic yogurts: influence on the survival of probiotic bacteria and protective techniques. *Comprehensive Rev Food Sci and Food Saf* 3:117–24.

[21] Kaur IP, Chopra K, Saini A (2002). Probiotics: potential pharmaceutical applications. *European Journal of Pharmaceutical Sciences* 15:1–9.

[22] Hamilton-Miller JMT (2003) The role of probiotics in the treatment and prevention of *Helicobacter pylori* infection. *Int J Antimicrob Agents* 22:360-366.

[23] Lorca GL, Wadstrom T, Valdez GF, Ljungh A (2001) *Lactobacillus acidophilus* autolysins inhibit *Helicobacter pylori* in vitro. *Curr Microbiol* 42:39-44.

[24] Ferencik M, Ebringer L, Mikes Z, Jahnova E, Ciznar I (1999) Successful modification of human intestinal microflora with oral administration of lactic acid bactéria. *Bratisl Lek Listy* 100-238-245.

[25] De Vrese M, Schrezenmeir J (2002) Probiotics and non-intestinal infectious conditions. *Br J Nutr* 8(Suppl. 1):ss59-66.

[26] Habermann W, Zimmermann K, Skarabis H, Kunze R, Rusch V (2002) Reduction of acute relapses in patients with chronic recurrent hypertrophic sinusitis during treatment with a bacterial immunostimulant (*Enterococusfaecalis bacteria* of human origin*/a medical probiotic). Arzneimittel Forschung, 52:622-627.

[27] De Roos NM, Katan MB (2000) Effects of probiotic bacteria on diarrhea, lipid metabolism, and carcinogenesis: a review of papers published between 1988 and 1998. *Am J Clin Nutr* 71:405-411.

[28] Heller KJ (2001) Probiotic bacteria in fermented foods: product characteristics and starter organisms. *Am J Clin* 73:S374–9.

[29] Hailu, G, Boecker A, Henson S, Cranfield J (2009) Consumer valuation of functional foods and nutraceuticals in Canada: a conjoint study using probiotics. *Appetite* 52(2):257–65.

[30] Codex Alimentarius (2011) *Milk and Milk Products*, 2nd edn, Rome: FAO/WHO. Online. Available HTTP: <ftp: //ftp.fao.org/codex/ Publications/Booklets/Milk/ Milk_2011_EN.pdf> (accessed 21 June 2012).

[31] Granato D, Branco GF, Cruz AG, Faria JAF, Shah NP (2010) Probiotic dairy products as functional foods. *Comp Rev Food Sci Food Saf* 9:455-470.

[32] Zacarias MF, Binetti A, Laco M, Reinheimer J, Vinderola G. (2011) Preliminary technological and potential probiotic characterization of bifidobacteria isolated from breast milk for use in dairy products. *Int Dairy J* 21:548-555.

[33] Lourens-Hattingh A. Viljoen BC (2001) Yoghurt as probiotic carrier food. *Int Dairy J* 11:1–17.

[34] Sanders ME (2003) Probiotics: considerations for human health. *Nutr Rev* 61:91-99.

[35] Alamprese C, Foschino R, Rossi M, Pompei C, Savani L (2002) Survival of *Lactobacillus johnsonii*La1 and influence of its addition in retail-manufactured ice cream produced with different sugar and fat concentrations. *Int Dairy J* 12:201-208.

[36] Alamprese C, Foschino R, Rossi M, Pompei C, Corti S (2005) Effects of *Lactobacillus rhamnosus* GG addition in ice cream. *Int. J. Dairy Technol.* 58:200- 206.

[37] Davidson RH, Duncan SE, Hackney CR, Eigel WN, Boling JW (2000) Probiotic culture survival and implications in fermented frozen yogurt characteristics. *J Dairy Sci* 83:666-673.

[38] Shah NP, Ravula RR (2000) Microencapsulation of probiotic bacteria and their survival in frozen fermented dairy desserts. *Aust J Dairy Technol* 55:139-144.

[39] Basyigit G, Kuleasan H, Karahan AG (2006) Viability of human-derived probiotic lactobacilli in ice-cream produced with sucrose and aspartame. *J Microbiol Biotechnol* 33:796-800.

[40] Akalin AS, Ersir D (2008) Effects of inulin and oligofructose on the rheological characteristics and probiotic culture survival in low-fat probiotic ice cream. *J Food Sci* 73(4):184-188.

[41] Aragon-Allegro LC, Konta EM, Suzuki K, Silva MG, Junior AF, Rall R, Rall VLM (2007) Occurrence of coagulase-positive *Staphylococcus* in various food products commercialized in Botucatu, SP, Brazil and detection of toxins form food and isolated strains. *Food Control* 18:630-634.

[42] Borges JQ, Ferreira SRSS, Costa GW (2004) Cinética de sobrevivência de *Lactobacillus acidophilus* microencapsulados em matriz de alginato de cálcio eveiculados em musse de chocolate. In: Congresso Brasileiro de Ciência e Tecnologia

de Alimentos. Campinas: Sociedade Brasileira de Ciência e Tecnologia de Alimentos p.66.

[43] Fasoli S, Marzotto M, Rizzotti L, Rossi F, Dellaglio F, Torriani S (2003) Bacterial composition of commercial probiotic products as evaluated by PCR-DGGE analysis. *Int J Food Microbiol* 82:59– 70.

[44] Van de water J (2003) Yogurt and immunity: the health benefits of fermented milk products that contain lactic acid bacteria. In: Farnworth ER (ed).Handbook of fermented functional foods, CRC Press, Boca Raton, p 113-144.

[45] Hols P, Hancy F, Fontaine L, Grossiord B, Prozzi D, Leblond-Bourget N, Decaris B, Bolotin A, Delorme C, Ehrlich SD, Guédon E, Monnet V, Renault P, Kleerebezem M (2005) New insights in the molecular biology and physiology of *Streptococcus thermophiles* revealed by comparative genomics. *FEMS Microbiol Rev* 29:435-463.

[46] Vinderola CG, Reinheimer JA (2003) Lactic acid starter and probiotic bacteria: a comparative in vitro study of probiotic characteristics and biological barrier Resistance. *Food Res Int* 36:895-904.

[47] Guarner F, Perdigon G, Corthier G, Salminen S, Koletzko B, Morelli L (2005) Should yoghurt cultures be considered probiotic? *Br J Nutr* 93:783-786.

[48] Sacarro DM, Tamine AY, Pilleggi A, Oliveira MN (2009) The viability of three probiotic organisms grown with yoghurt starter cultures during storage for 21 days at 4°C. *Int J Dairy Technol* 62(3):397–404.

[49] Korbekandi K, Jahdi M, Maracy M, Abedi D, Jalali M (2009) Production and evaluation of a probiotic yogurt using *Lactobacillus casei* ssp. *Int J Dairy Technol* 62(1):75–9.

[50] Svensson U (1999) Industrial perspectives. In: Tannock GW (ed) Probiotics: A critical review. Horizon Scientific Press, Wymondham, p.57-64.

[51] Vinderola CG, Gueimonde M, Delgado T, Reinheimer JA, Reye-gavilán CG (2000) Characteristics of carbonated fermented milk and survival of probiotic bacteria**.** Int Dairy J 10:213-220.

[52] Lucas A, Sodini I, Monnet C, Jolivet P, Corrieu G (2004) Probiotic cell counts and acidification in fermented milks supplemented with milk protein hydrolysates. *Int Dairy* J 14:47-53.

[53] Ostlie HM, Treimo J, Narvhus JA (2005) Effect of temperature on growth and metabolism of probiotic bacteria in milk. *Int. Dairy J* 15:989-997.

[54] Oliveira MN, Damin MR (2003) Efeito do teor de sólidos e da concentração de sacarose na acidificação, firmeza e viabilidade de bactérias do iogurte e probióticas em leite fermentado. *Ciênc Tecnol Aliment* 23:172-176.

[55] Oliveira MN, Sivieri K, Alegro JHA, Saad SMI (2002) Aspectos tecnológicos de alimentos funcionais contendo probióticos. *Rev Bras Ciênc Farm* 38:1-21.

[56] Stanton C, Ross RP, Fitzgerald GF, Vansinderen D (2005) Fermented functional foods based onprobiotics and their biogenic metabolites. *Curr Opin Biotechnol* 16:196-203.

[57] Zacarchenco PB, Massaguer-Roig S (2004) Avaliação sensorial, microbiológica e de pós-acidificação durante a vida-de-prateleira de leites fermentados contendo *Streptococcus thermophilus, Bifidobacterium longum* e *Lactobacillus acidophilus. Ciênc Tecnol Aliment* 24:674-679.

[58] Chermesh I, Eliakim R (2006) Probiotics and the gastrointestinal tract: Where are we in 2005? *World J Gastroenterol* 14:12(6):853-857.

[59] Fontán MCG, Martínez S, Franco I, Carballo J (2006) Microbiological and chemical changes during the manufacture of Kefir made from cows' milk, using a commercial starter culture. *Int Dairy J* 16:762–7.

[60] Powell JE, Witthuhn RC, Todorov SD, Dicks LMT (2007) Characterization of bacteriocin ST8KF produced by a kefir isolate *Lactobacillus plantarum* ST8KF. *Int Dairy J* 17(3):190–8.

[61] Güzel-Seydim, ZB, Seydim AC, Greene AK, Bodine AB (2000) Determination of organic acids and volatile flavor substances in kefir during fermentation. *J Food Comp Anal* 13(1):35–43.

[62] Saloff-Coste CJ (1996) Kefir.nutritional and health benefits of yoghurt and fermented milks. Dannone W Newsletter 11:1–7.

[63] Bensmira M, Nsabimana C, Jiang B (2010) Effects of fermentation conditions and homogenization pressure on the rheological properties of Kefir. LWT – *Food Sci Technol* 43:1180–4.

[64] Angulo L, Lopez, E, Lema, C (1993) Microflora present in kefir grains of the Galician region (North-West of Spain). *J Dairy Res* 60:263-7.

[65] Rea MC, Lennartsson T, Dillon P, Drinan FC, Reville WJ, Heapes M, Cogan TM (1996) Irish kefir-like grains: their structure, microbial composition and fermentation kinetics. *J Appl Bact* 81:83–94.

[66] Cheirsilp B, Shimizu H, Shioya, S. (2003) Enhanced kefiran production by mixed culture of *Lactobacillus kefiranofaciens* and *Saccharomyces cerevisiae. J Biotechnol* 100:43-53.

[67] Farnworth ER, Mainville I (2003) Kefir: a fermented milk product. In: Farnworth ER (ed) Handbook of Fermented Functional Foods, vol 1. CRC Press, Boca Raton, pp 77-112.

[68] Garrote GL, Abraham AG, De Antoni GL (2000) Inhibitory power of kefir: the role of organic acids. *J Food Prot* 63:364–9.

[69] Vinderola CG, Duarte J, Thangavel D, Perdigón, G, Farnworth E, Matar C (2005) Immunomodulating capacity of kefir. *J Dairy Res* 72:195-202.

[70] Champagne CP, Gardner NJ (2005) Challenges in the addition of probiotic cultures to foods. *Crit Rev Food Sci Nut* 45:61-84.

[71] Homayouni A, Azizi A, Ehsani, MR, Yarmand MS, Razavi SH (2008) Effect of microencapsulation and resistant starch on the probiotic survival and sensory properties of symbiotic ice cream. *Food Chem* 111:50 - 55.

[72] Betoret N, Puente L, Díaz MJ, Pagán MJ, García MJ, Gras ML (2003) Development of probiotic enriched dried fruits by vacuum impregnation. *J Food Eng* 56:273–7.

[73] Sheehan VM, Ross P, Fitzgerald GF (2007) Assessing the acid tolerance and the technological robustness of probiotic cultures for fortification in fruit juices. *Innovative Food Sci Emerg Technol* 8:279-284.

[74] Prado FC, Parada JL, Pandey A, Soccol CR (2008) Trends in non-dairy probiotic beverages. *Food Res Int* 41:111-123.

[75] Yamaguishi C, Spier MR, Lindner JDD, Soccol VT, Soccol CR (2011) Probiotics: Current Market Trends and Future Directions. In: Min-Tze Liong. (org) Probiotics. Springer-Verlag, Berlin:, v. 1, 299-320.

[76] Todorov SD, Botes M, Guigas C, Schillinger U, Wiid I, Wachsman WB, Holzapfel WH, Dicks LMT (2008) Boza, a natural source of probiotic lactic acid bacteria. *J Appl Microbiol* 104(2): 465-477.

[77] Arici M, Coskun F (2001) Hardaliye: Fermented grape juice as a traditional Turkish beverage. *Food Microbiol* 18:417–421.

[78] Wacher C, Cañas A, Bárzana E, Lappe P, Ulloa M, Owens JD (2000) Microbiology of Indian and Mestizo pozol fermentations. *Food Microbiol* 17:251-256.

[79] Rivera-Espinoza Y, Gallardo-Navarro Y (2010) Non-dairy probiotic products. *Food Microbiol* 27:1–11.

[80] Saarela M, Virkajarvi I, Alakomi HL, Mattila PS, Matto J (2006a) Stability and functionality of freeze-dried probiotic *Bifidobacterium* cells during storage in juice and milk. *Int Dairy J* 16:1477–1482.

[81] Saarela M, Virkajarvi I, Nohynek L, Vaari A, Matto J (2006b) Fibres as carriers for *Lactobacillus rhamnosus* during freeze-drying and storage in apple juice and chocolate-coated breakfast cereals. *Int J Food Microbiol* 112:171–178.

[82] Champagne CP, Raymond Y, Gagnon R (2008) Viability of *Lactobacillus rhamnosus* R0011 in an apple-based fruit juice under simulated storage conditions at the consumer level. *Food Microbiol Safe* 73:221-226.

[83] Nualkaekul S, Salmeron I, Charalampopoulos D (2011) Investigation of the factors influencing the survival of *Bifidobacterium longum* in model acidic solutions and fruit juice. *Food Chem* 129:1037–1044.

[84] Nualkaekul S, Charalampopoulos D (2011) Survival of *Lactobacillus plantarum* in model solutions and fruit juices *Int J Food Microbiol* 146:111–117.

[85] Soccol CR, Prado FC (2007) Processo tecnológico para produção de uma bebida fermentada a base de água de coco com propriedades probióticas. *Brazilian patent* PI0703244–7.

[86] Pereira ALF, Maciel TC, Rodrigues S (2011) Probiotic beverage from cashew apple juice fermented with *Lactobacillus casei*. *Food Res Int* 44:1276-1283.

[87] Wang YC, Yu RC, Yang HY, Chou CC (2006) Antioxidatives activities of soymilk fermented with lactic acid bacteria and *bifidobacteria*. *Food* Microbiol 23:128–135.

[88] Liu DM, Li L, Yang XQ, Liang SZ, Wang JS (2006) Survivability of *Lactobacillus rhamnosus* during the preparation of soy cheese. *Food Technol Biotechnol* 4:417–422.

[89] Han BZ, Rombouts FM, Nout JMJR (2001) A Chinese fermented soybean food. *Int J Food Microbiol* 65:1-10.

[90] Garro MS, De Valdez GF, De Giori GS (2004) Temperature effect on the biological activity of *Bifidobacterium longum* CRL 849 and *Lactobacillus fermentum* CRL 251 in pure and mixed cultures grown in soymilk. *Food Microbiol* 21:511-518.

[91] Tzortzis G, Goulas AK, Baillon MLA, Gibson GR, Rastall, RA (2004) *In vitro* evaluation of the fermentation properties of galactooligosaccharides synthesized by α-galactosidase from *Lactobacillus reuteri*. *Appl Microbiol Biotechnol* 64:106-111.

[92] Wang YC, Yu RC, Yang HY, Chou CC (2003) Sugar and acid content in soymilk fermented with lactic acid bacteria alone or simultaneously with *bifidobacteria*. *Food Microbiol* 20(3):333-338.

[93] Chen M, Mustapha A (2012) Survival of freeze-dried microcapsules of α-galactosidase producing probiotics in a soy bar matrix. *Food Microbiol* 30:68-73.

[94] Andersson H, Asp NG, Bruce A, Roos S, Wadström T, Wold, AE (2001) Health effects of probiotics and prebiotics. A literature review on human studies. *Scand J Nutr/Näringsforskning* 45:58–75.

[95] Charalampopoulos D, Pandiella SS, Webb C (2002) Growth studies of potentially probiotic lactic acid bacteria in cereal-based substrates. *J Appl Microbiol* 92:85-9.

[96] Charalampopoulos D, Pandiella SS, Webb C (2003) Evaluation of the effect of malt, wheat and barley extracts on the viability of potentially probiotic lactic acid bacteria under acidic conditions. *Int J Food Microbiol* 82:133 - 41.

[97] Khan MI, Arshad MS, Anjum FM, Sameen A, Rehman A, Gill WT (2011) Meat as a functional food with special reference to probiotic sausages. *Food Res Int* 4:3125-3133.

[98] Gänzle MG, Hertel C, Vossen JM, Hammes WP (1999) Effect of bacteriocin producing lactobacilli on the survival of *Escherichia coli* and *Listeria* in a dynamic model of the stomach and the small intestine. *Int J Food Microbiol* 48:21-35.

[99] De Vuyst L, Falony G, Leroy F (2008) Probiotics in fermented sausages. *Meat Sci* 80:75-78.

[100] Speranza B, Bevilacqua A, Sinigaglia M, Corbo MR (2012) Shelf life definition for Italian anchovies inoculated with *Lactobacillus plantarum* and *Bifidobacterium animalis* subsp. *lactis*, Innovative Food Sci Emerg Technol (in press) doi:10.1016/j.ifset.2012.05.009.

[101] Sidhu KS (2003) Health benefits and potential risks related to consumption of fish or fish oil. *Reg Toxicol Pharmacol* 38:336-344.

[102] Santos MCR (2001) Desenvolvimento de bebida e farinha láctea fermentada de ação probiótica a base de soro de leite e farinha de mandioca por cultura mista de *Lactobacillus plantarum* A6, *Lactobacillus casei* Shirota e *Lactobacillus acidophilus*. (MSc thesis). Curitiba, PR: Fed Univ Paraná. 106 pp.

[103] Yoon KY, Woodams EE, Hang YD (2005) Fermentation of beet juice by beneficial lactic acid bacteria. Lebensm-Wiss u -Technol 38(1):73-75.

[104] Rakin M, Vukasinovic M, Siler-Marinkovic S, Maksimovic M (2007) Contribution of lactic acid fermentation to improved nutritive quality vegetable juices enriched with brewer's yeast autolysate. *Food Chem* 100:599–602.

[105] Nazzaro F, Fratinni F, Sada A, Orlando P (2008) Synbiotic potential of carrot juice supplemented with *Lactobacillus* spp. and inulin or fructooligosaccharides. *J Sci Food Agric* 88:2271–6.

[106] Nazzaro F, Fratinni F, Coppola R, Sada A, Orlando P (2009) Fermentative ability of alginate-prebiotic encapsulated *Lactobacillus acidophilus* and survival under simulated gastrointestinal conditions. *J Funct Foods* 1:319–23.

[107] Randazzo CL, Restuccia C, Romano AD, Caggia C (2004) *Lactobacillus casei* - dominant species in naturally fermented Sicilian green olives. *Int J Food Microbiol* 90:9-14.

[108] Yoon KY, Woodams EE, Hang YD (2004) Probiotication of tomato juice by lactic acid bacteria. *J Microbiol* 42:315–318.

[109] Yoon KY, Woodams EE, Hang YD (2006) Production of probiotic cabbage juice by lactic acid bacteria. *Bioresource Technol* 97(12):1427-1430.

[110] BCC Research Market Forecasting. Probiotics market to grow to $31.1 billion globally in 2015. October 14th 2011. Available in: http://www.bccresearch.com Accessed in 06 July 2012.

[111] Nutraceutical World (2012) Probiotics Grow in Popularity, Vary in Form.Rodman Publishing. Available in: http://www.nutraceuticalsworld.com Accessed in 05 July 2012.

[112] Spier MR, Habu S, Pandey A, Soccol VT, Soccol CR (2009) Fermented Functional Foods. Chapter 29. In: Pandey A, Larroche, Soccol CR, Gnansounou E, Nigam P (eds). Comprehensive Food Fermentation and Biotechnology. Elsevier, New York, p.892-925.

[113] Shah NP (2000) Probiotic bacteria: Selective enumeration and survival in dairy foods. *J Dairy Sci* 83(4):894–907.

[114] Tiwari R, Tiwari G, Rai AK (2011) Probiotic novel beverages and their applications. *Syst Rev Pharm* 2:30-6.

[115] Lindner, JDD, Penna, ALB, Demiate, IM, Yamaguishi, CT, Prado MRM, Parada JL (2013) In: Soccol, CR, Pandey A, Larroche C. Fermentation Processes Engineering in the Food Industry, 1 ed. Taylor & Francis Inc., Bosa Roca 2013, p. 263-297.

In: Microbes in Process
Editors: Neelam Garg and Abhinav Aeron
ISBN: 978-1-63117-127-7

Chapter 11

Bacterial Survival Strategies and the Microbiological Safety of Food

Shadia M. Abdel-Aziz*
Microbial Chemistry Dept., Genetic Engineering and Biotechnology Division,
National Research Center, Dokki, Cairo, Egypt

Abstract

Microorganisms are well known to be beneficial or harmful. Knowledge of the stress factors and stress resistance by microorganisms is important for the development of biotechnology. Stress-induced proteins serve as molecular markers of the health of the starting cultures. Appearance of such markers may show that the culture has been subjected to stressors and may have adapted to these stressors. Such bacteria are called fighting hazardous bacteria which cause food deterioration. However, most of microorganisms make crucial contributions to the welfare of humanity. Such beneficial microorganisms are widely used in food industry for production of a large number of fermented food products and are also very helpful for conversion of food industrial wastes into value added products. The ability of microorganisms to convert large and complex molecules into the simplest one depends upon the type of culture and the growth requirements. The selection of a suitable culture to convert a specific type of substrate into useful products, thus, plays a vital role in fermentation technology.

1. Introduction

Knowledge of the stress factors and stress resistance by microorganisms is important for the development of biotechnology in different fields as well as in the field of industrial microbiology. The use of biological systems for the fulfillment of human needs perhaps started in 6000 BC, but the greatest revolution till now, commenced in the 1970s and 1980s, when a new term, ***Biotechnology***, arose to connect the biological science with technology.

* Email: abdelaziz.sm@gmail.com.

Biotechnology is a third wave in biological sciences and represents an interface between basic and applied sciences, where gradual transformation of science into technology can be witnessed. Biotechnology is the application of biological organisms, systems or processes to manufacturing and service industries, generating useful products and services. It enables the conversion of raw materials to final products with the involvement of biological entities at one stage or the other. Following this history, in recent years, microorganisms have found their application not only in the production of a variety of metabolites but also in the bio-transformation of several chemicals. The genetically engineered micro organisms are also being used for the commercial production of some non microbial products such as insulin, interferon, human growth hormone and viral vaccines. Microbes are also being used to meet effectively the crisis in both environment and energy sectors. They can reduce environmental pollution through a variety of processes and other means including the following:

1) recovery of metals from polluted waterways;
2) elimination of sulphur from metal ores and coal fired power; and
3) use of biofertilizers and biopesticides.

In the energy sector, they can be used for production of single cell proteins to meet food and fodder problems, and for biogas production to provide energy to electrify villages. Through the use of biofertilizers and bioinsecticide, they can also add to crop productivity. Similarly, they can provide cheap and cost effective methods of mining and metallurgy. Thus microbial biotechnology will have a great impact on industry. So, it is of importance to know more about the capability of microorganisms to adapt and survive in the extreme environments and to gain Knowledge about how to encounter any problems during processing steps. Bacteria can survive in almost all environments on Earth, including some considered extremely harsh. Under hardness abiotic conditions, microbes have been found thriving. Their tenacity to survive in such extreme and varied conditions allows them to play fundamental roles in global nutrient cycling. Microbes also cause a wide range of human diseases and can survive in hospital conditions [1, 2]. Many studies have dealt with the molecular system that bacteria and fungi use to adapt to changes in their environment. The concept of stress was advanced by Hans Selye, a Canadian physiologist, who wrote "Stress is Life and Life is Stress". Humans, animals, plants, and microorganisms frequently encounter stressful situations, this being the reason for the interest in stress and the nature of this phenomenon [3]. Stress can be defined as a change that resulted in a decrease in the growth rate or survival. This change may be in the genome, proteome, or in the environment. For microorganisms, stress response is of particular importance because their habitats are usually subjected to continual changes in different parameters such as acidic or alkaline pH, high or low temperature, osmotic pressure, desiccation, salinity, or substrate availability [4, 5]. An organism is still alive as long as: 1) its cell membrane and nucleic acids retain integrity; 2) the processes of protein folding are duly maintained; and 3) the macromolecules such as DNA and RNA are far from any damage. Microbial cells that cease to grow, grow slowly, or enter the steady state acquire stress resistance. Any decrease in growth rate leads to rapid molecular reactions that repair the stress-induced damages to protect the cells against stressors [3]. The response and survival mechanisms by which bacteria can overcome extreme conditions and challenge environmental changes are reported herein, with special reference about the safety of industrial food products against food poisoning pathogens.

2. Abiotic Stress Response

Microorganisms have an outstanding ability to adapt to most habitats.

2.1. Definition

Abiotic stress is defined as the negative impact of non-living factors on the living organisms in a specific environment. Survival of an organism from moment to moment depends on its ability to sense and respond to the changes in its environment.

2.2. Adapatation of Microrganisms

The stress response is a mechanism used by organisms to adapt to and overcome a stress stimulus in order to preserve homoeostasis. Adaptation is an evolutionary process in which traits in a population are tailored by natural selection to better face the challenges presented by the local environment. Evolution occurs as a consequence of organisms adapting to different environments over space and time. Therefore, demonstrations of such environment-fit adaptations in organisms are crucial for elucidating evolution in action. In regarding to the physiological process and its contribution to survival, stress response cannot be considered a single response to a single system. It needs to be viewed as different variables in different systems with complex interactions between the physiological and psychological [6]. Mechanisms for responding to environmental changes are universally present in living beings. For example, when mammals perceive a sudden environmental change, a rush of adrenaline precipitates the well-known "fight or flight" response. Such physiological stress responses in complex organisms require appropriately regulated interactions among numerous systems. But how do single celled organisms respond to potentially lethal threats. Such organisms have identified specific mechanisms that contribute to the microbial survival under rapidly changing conditions [6]. Short term stress responses enable organisms to adapt to and overcome short term stress stimuli. The stress response enables the organism to mobilize energy to "fight or flight" from a threat thereby enhancing its chances of survival. Depression as a stress response would appear to be a significant failure of adaptation. A prolonged stress response may result from a relatively long term stressor.

2.3. Impact on the Microbial Cell

Exposure of microbes to the stress factors resulted in:

1) protein denaturation;
2) breaks in nucleic chains; and
3) oxidation of macromolecules such as DNA, RNA, proteins, and fatty acids. However, microbes can challenge environmental changes and overcome extreme conditions using specific survival mechanisms.

Over the years, it is apparent that a combination of constitutive and inducible strategies which result in removal of protons (H^+), changes in the composition of cell envelop, production of general shock proteins and chaperones, expression of transcriptional regulators such as sigma factors, formation of biofilms and endospores, can all contribute to the survival of microbes [5].

3. Bacterial Survival Strategies

All living organisms (bacteria, fungi, plants, animals, and even humans) synthesize specific proteins in response to temperature elevation.

3.1. Heat Shock

All living organisms (bacteria, fungi, plants, animals, and even humans) synthesize specific proteins in response to temperature elevation, known as heat shock proteins (HSPs). The bacterial heat- shock response is not limited to changes in temperature and is a general stress response, as many of the HSPs are induced by other environmental changes, such as addition of ethanol, heavy metals, osmolarity, starvation, pollutants, or interactions with eukaryotic hosts [4]. Protein denaturation is the major event associated with heat shocks. Only properly folded polypeptide chains are considered functional undamaged proteins. An organism, exposed to abiotic stress, starts to induce the specific proteins (HSPs) to repair the damaged (denatured) proteins. It was found that, HSPs repair damaged proteins either by folding of newly synthesized polypeptides or by refolding polypeptides that are damaged or improperly folded [3]. Overproduction of HSPs renders the cells resistant to high temperatures and oxidative stress; this effect is observed in both mammals and bacteria [7]. The HSPs include chaperones and proteases that are presumably essential for overcoming changes that involve protein denaturation. Induction of this response improves tolerance to salts, heavy metals, as well as thermo-tolerance [8]. Moreover, in several bacterial species, HSPs have been shown to play an important role in pathogenesis [9]. Heat shock proteins are also essential for stationary phase and for bacterial differentiations in *Myxococcus xanthus* and in *Bacillus subtilis* [10].

3.1.1. Chaperones

Induction of stress proteins is of importance for adapting of network of microorganisms. A diverse range of stress factors such as heat shock, salt stress, ethanol, starvation for oxygen or nutrient etc., induce stress proteins which fall into two groups: general stress proteins and specific stress proteins. General stress proteins serve a general protective function allowing adaptation to stress and starvation, regardless of the nature of the stress factor. Therefore, these proteins are called general stress proteins [3]. In addition, each single stress factor induces its specific set of proteins which may exert a specific protective function against a unique stress factor. In order to face heat stress, all living cells increase the level of their HSPs, which include two important groups, namely chaperones and ATP-dependent proteases [3]. Chaperones are high-molecular weight proteins with a subunit structure, exhibiting

ATPase activity. While the chaperones prevent misfolding and aggregation of partially denatured proteins, proteases are involved in the degradation of denatured proteins. It is found that, damaged proteins act as a substrate for the synthesized proteins and that in case of absence of chaperones, polypeptide chains that are released from ribosomes were found to be functionally inactive. Chaperones are required at all temperatures to aid essential proteins to fold and maintain their proper conformation.

3.1.2. Toxic Effects of Malformed Proteins

Malformed proteins that are produced under stress conditions may aggregate and exert toxic effects in the cells. For this reason, organisms are equipped with a defense system, which is based on their ability to degrade malformed proteins. Malformed proteins (denatured, inactive, or unstable proteins) are recognized by proteases or by the chaperone complex [3]. Such proteins are either cleaved into small peptides by proteases (known as chaperonins) and eliminated or recovered into functional undamaged proteins by the chaperone complex (HSPs). Thus, chaperonins intervene in numerous stresses for various tasks such as protein folding, renaturation, protection of denatured proteins, and evacuation of damaged proteins. Chaperonins play an important role in the response to acid shock, may be to ensure proper functioning of DNA repair mechanisms.

3.1.3. Repair of Macromolecules

A number of proteins that play a role in the protection or repair of macromolecules, such as DNA and other proteins, are essential for optimal heat and acid resistance. In the mechanism of protein repair, ribosomes release polypeptide chains which unite with the chaperones to form a complex. This complex becomes a target to which a new chaperone will bind, and this reaction between chaperones and polypeptides continue to occur. It is found that, chaperones have a great affinity to unfolded proteins and they give the protein its folding and structure, after which dissociation of the complex and release of the folded protein occurs [3].

3.2. Cold Shock

Bacteria are exposed to all dangers of cold shock (e.g., in polar regions). Cold shock doesn't cause protein denaturing but causes:

1) inhibition of DNA replication, gene transcription and translation;
2) decrease in enzyme activity and the rate of metabolic processes;
3) decrease in membrane fluidity and hinder the trans-membrane influx; and
4) formation of ice crystals in the intracellular fluid, which damage the cell structure and causing cell death (3).

Bacteria synthesize specific proteins (cold shock proteins, CSPs) in response to the cold shock. Studies on the microbial adaptation to the low temperatures were initiated by the increasing use of food freezing for the purposes of long-term storage. In many bacteria, cold shock induces increased synthesis of small proteins (7 KDa). Psychrophilic microorganisms are characterized by increased membrane fluidity and increased functional adaptability of the enzymatic system as well as increased ability to synthesize anti-freeze proteins which prevent formation of ice crystals [3]. Bacteria which are adapted to cold shock stress have a shorter lag-phase and a higher growth rate. Cold shock in *E. coli* leads to a decrease in supercoiling of DNA. In response, *E. coli* modified the mechanism to increase supercoiling of DNA.

3.3. Low pH Tolerance

Bacteria possess an acid resistance system that can help them to overcome the challenge posed by different acidic environments. Gram-positive bacteria play vital roles in food and health, although the nature of this role can vary greatly. Gram-positive bacteria include GRAS (generally regarded as safe) for organisms used in food fermentations, ubiquitous gastric commensals, and potentially beneficial probiotic bacteria (in yoghurt and milk products). The acidic environments encountered in food and in the gastrointestinal tract provide a significant survival challenge for these diverse organisms. For example, the preferred delivery vehicles for probiotic cultures are yoghurt and fermented milk, both of which present an acidic challenge. Subsequently, the bacteria will need to survive the highly acidic gastric juice if they reach the small intestine in a viable state. Hence, acid tolerance is accepted as one of the desirable properties used to select potentially probiotic strains. In contrast, the acid tolerance of a pathogen such as *Listeria monocytogenes,* again assisting survival in the stomach, could be considered a virulence factor. Thus it is timely to consider the mechanisms and strategies used by organisms to protect themselves and challenge in the extreme conditions by which they can be aided or impeded.

Intrinsic resistance to gastric acid is a relatively rare probiotic property, with *Lactobacillus acidophilus* and bifidobacteria being among the most resistant. Gastric survival is also a highly valued property because it overcomes the limitations associated with needing to use food carriers with high buffering potential to aid gastric transit and ensure the delivery of viable cells into the small intestine [5]. While gram-positive bacteria may encounter a wide range of low-pH environments, most studies have focused on those where bacterial growth and survival have an impact on human health and economics, i.e., survival and growth in food, effect on the oral cavity, gastric transit, and intracellular survival.

Acid has been used as a food preservative for millennia, primarily in the form of food fermentations. In each case, the fermentation process involves the oxidation of carbohydrates to generate a range of products including organic acids, alcohol, and carbon dioxide. Such products have a preservative effect through limiting the growth of spoilage and pathogenic members of the flora in the food product [5]. Weak acids have potent antimicrobial activity because the undissociated form of weak acids can pass freely through the cell membrane. Addition of weak acids is unnecessary in foods fermented with, for example, the homo-fermentative *Lactococcus lactis*, which converts 90% of metabolized sugar to lactic acid [5]. An understanding of the acid resistance mechanisms used by lactic acid bacteria (LAB) to survive the by-products of their own metabolism and the responses available to spoilage and pathogenic organisms in low-pH foods is thus of great importance.

3.4. Osmotic Stress

In response to osmotic shock stress, microorganisms use two-fold mechanisms to sustain high salt concentrations: 1) discharge of excess salt and; 2) accumulation of osmolytes which are small organic molecules retaining water solubility at high concentrations. Osmolytes are not damaging to the intracellular structures and are accumulated intracellularly by microorganisms to act as osmoregular [3]. For pathogenic bacteria which pass from environmental waters to the human body for infection, osmolarity is an important criterion to distinguish between the external and host associated environments. Osmolarity of an aqueous environment is thought to be no greater than that equivalent to 0·06 M NaCl while in the intestinal lumen, the osmolarity is much higher (equivalent to 0·3 M NaCl). In the blood stream the bacteria encounter an osmolarity equivalent to about 0·15 M NaCl [11]. Thus, an increase in osmolarity is associated with expression of virulence factors in many pathogenic organisms.

3.5. Oxidative Stress

The expression of adhesion and invasion factors of several food-spoilage bacteria, for example, is regulated by oxygen concentrations. High oxygen concentration is usually repressive, whereas low oxygen induces invasiveness. During switch from aerobic to anaerobic growth conditions, a set of genes are induced while others are repressed. Oxidative stress induces the following damages: 1) breaks nucleic acid chains; 2) cleaves protein molecules; and 3) blocks the active sites of enzymes [3]. Oxidative stress is caused by Reactive Oxygen Species (ROSs) such as superoxide ($O_2^{\bullet -}$) and hydrogen peroxide (H_2O_2). These ROSs are byproducts of normal aerobic metabolism and are produced, mainly, by partial reduction of oxygen during respiration [12]. In presence of traces of metal ions, $O_2^{\bullet -}$ can react with H_2O_2 and generate the more reactive singlet oxygen ($^1O_2$). H_2O_2 can oxidize iron sulfur centers and cysteines in certain proteins or react with transition metals and produce the hydroxyl radical ($OH^{\bullet}$), which can oxidize virtually any cell molecule, causing DNA damage, protein inactivation, protein cross-linking and fragmentation, and lipid peroxidation [12]. Cells have a number of mechanisms to maintain low intracellular ROSs levels. The enzymes superoxide dismutases (dismutate $O_2^{\bullet -}$ to H_2O_2), catalases (decompose H_2O_2 to H_2O), peroxidases (decompose H_2O_2 through other substrate oxidation), glutathione peroxidases (use glutathione to decompose H_2O_2 to H_2O) and peroxiredoxins (use reducing power to decompose H_2O_2 to H_2O) are the most ubiquitous effectors of the enzyme mediated antioxidant response [13]. All these oxidants produced as an oxidative burst have a powerful microbial killing activity.

Oxidative stress response in microorganisms could be achieved by several means. On one hand, microbial cells possess enzymes that degrade ROSs by an oxidative defense system which include catalase, peroxidase, and superoxide dismulase. On other hand, microbial cells contain molecules with anti-oxidase activity such as carotenoides, polyamines, and fat-soluble vitamins. All of these factors act to protect the DNA from oxidation [3]. ROSs had been implicated in determining the life span of several organisms, cancer, chronic inflammation, cardiovascular, and neurodegenerative [14]. External environmental factors including ionizing radiation and a variety of chemicals can generate ROSs. Cells have developed antioxidant defense systems aimed at protecting against these ROSs [15]. Oxidative attack of macromolecules is a problem caused by continued metabolic activities in the growth-arrested cell.

3.6. Cell Membrane Changes

The bacterial cell envelope is the first and major line of defense against threats from the environment. It is an essential structure that gives the cell its shape and counteracts the high internal osmotic pressure. It is important for controlling transport of solutes. The cell envelope is also the target for numerous antibiotics because of its many crucial functions. Therefore, the monitoring and maintenance of cell envelope integrity is crucial for survival [16]. Gram-positive cell envelope consists only of two functional layers (compared with three in Gram-negative bacteria) that enclose the cellular contents: a cytoplasmic membrane, surrounded by a thick cell wall. Microorganisms sense and adapt to changes in their environment. When favored nutrients are exhausted, some bacteria may become motile to seek out nutrients, or they may produce enzymes to exploit alternative resources. One example of an extreme survival strategy employed by certain Gram-positive bacteria is the formation of endospores [17]. This complex developmental process is often initiated in response to nutrient deprivation. It allows the bacterium to produce a dormant and highly resistant cell to preserve the cell genetic material in times of extreme stress. Endospores can survive environmental assaults that would normally kill the bacterium. These stresses include high temperature, high UV irradiation, desiccation, chemical damage and enzymatic destruction [17]. The extraordinary resistance properties of endospores make them of particular importance because they are not readily killed by many antimicrobial treatments. A variety of different organisms form spores or cysts, but the endospores of Gram-positive bacteria are by far the most resistant to harsh conditions and can be explained by its unique cellular structure. Proteins in the outer proteinaceous coat, surrounding the spore, function to: 1) provide much of the chemical and enzymatic resistance to the cells; 2) tightly bind and condense the DNA; and 3) resist the UV light and DNA-damaging chemicals [17].

3.7. Regulators

Regulation of gene expression in response to the extracellular environment is an adaptive response that is required for bacterial replication and survival. Alternative sigma factors have been demonstrated to play an important role in the coordinate regulation of gene expression in bacteria undergoing major adaptive changes in physiology. When bacteria are subjected to DNA damage about 30 genes are coordinately induced, a reaction known as the “SOS response” [18]. In *E. coli* heat-shock induces approximately 30 genes under control of sigma factor (σ^{32}). Induction of the SOS genes occurs when their common repressor, LexA, is inactivated. LexA inactivation is a proteolytic reaction that is greatly enhanced by RecA protein bound to single-stranded DNA. Obviously, the SOS response is highly induced after cells are exposed to DNA damaging agents, such as UV-light, that produce single-stranded DNA [18].

To fully understand the biological contributions of regulatory proteins such as sigma factors, it is critically important to identify genes regulated by these proteins. Investigators have used combinations of global and more focused strategies to identify sigma factor regulons [6, 19]. Their data suggested the possibility the sigma factor may represent an important switch mechanism that facilitates bacterial transition from a free-living organism to a host-invading pathogen. Such studies provide powerful new insight into the field of microbial physiology by enabling rapid identification of genes that may appear to be unrelated in function, but that must be coordinately regulated to enable an organism to survive and respond appropriately under rapidly changing environmental conditions. These coordinately regulated genes may prove to be appropriate targets for development of novel antimicrobial strategies, thus providing tangible realization of the promise and power of the application of genomics tools for improving human health.

3.8. Sigma Factors

Bacteria are classified as Gram-negative or Gram-positive based on microscopically observed staining properties associated with different cell membrane structures. σ^S (RpoS) and σ^B (SigB) have been identified as general stress responsive alternative sigma factors in Gram-negative and in Gram-positive bacteria, respectively. σ^S was identified in both *E. coli* and in *S. typhimurium* as a Group II sigma factor that activates expression of numerous genes required to maintain cell viability as the cell leaves exponential growth conditions and moves into stationary phase [20]. In addition to helping *E. coli* and *S. typhimurium* respond to different environmental stress conditions, σ^S also contributes to expression of virulence-associated genes. Since its initial discovery, the presence of σ^S and its role in stress response has been confirmed in multiple, diverse of Gram-negative pathogenic bacteria, including *Pseudomonas aeruginosa, Legionella pneumophila, Borrelia burgdorferi, Yersinia enterocolitica, and Shigella flexneri.*

3.9. Biofilms

Understanding mechanisms of bacterial persistence will suggest novel ways to control infectious diseases. Different systems have been documented as making contribution to a pathogen survival. Biofilms consist of microorganisms typically embedded in an organic polymer matrix of microbial origin [21]. They develop on all surfaces immersed in natural aqueous environments, including both biological (aquatic plants and animals) and abiological (concrete, metal, plastics, stones). In nature, biofilms constitute a protected growth modality that allows the bacteria to survive in hostile environments. Bacterial biofilms colonize any humid surface (plaque on the teeth, slippery slime on river stones, gel-like film on the inside of a vase, and infected tissue). Formation of biofilms can be regulated by factors such as availability of surface, nutrients and environmental cues. Biofilms have shown to elicit specific mechanism for initial attachment to a surface and micro colony formation. Microorganisms within the biofilms when present in body are protected from the body's normal defense mechanisms. Moreover, biofilms are pervasive and problematic in medical, industrial and environmental settings because these communities express biofilm-specific properties such as increased resistance to antibiotics, UV light, and chemical biocides, increased rates of genetic exchange, altered biodegradability and increased secondary metabolite production.Thus the biofilm provides a source of infection by bacteria and when they infect the body, it is difficult to eradicate them with therapeutic agents [22]. Biofilms also provide an ideal niche for the exchange of extrachromosomal DNA responsible for antibiotic resistance, virulence factors and environmental survival capabilities at accelerated rates making it a milieu for emergence of drug resistance [22]. There is need to understand the role of biofilms in chronic infections and their mechanism of antimicrobial resistance so that the future drug treatments can be designed taking into consideration the antibiotic resistant microorganisms residing in biofilms [23]. Such information about the microbial biofilm can facilitate the development of potentially new antibiotics. The ability to produce a biofilm has been recognized as a survival strategy used by environmental microbes for many years but, particularly in the last decade, it has become evident that it is also important for infectious diseases.

Abiotic stress, as a natural part of every ecosystem, may be either beneficial or detrimental. While gram-positive bacteria may encounter a wide range of low-pH environments, most studies have focused on those where bacterial growth and survival have an impact on human health and economics, i.e., survival and growth in food, effect on the oral cavity, gastric transit, and intracellular survival. Acid has been used as a food preservative for millennia, primarily in the form of food fermentations. In each case, the fermentation process involves the oxidation of carbohydrates to generate a range of products including organic acids, alcohol, and carbon dioxide. Such products have a preservative effect through limiting the growth of spoilage and pathogenic members of the flora in the food product such as *Listeria monocytogenes*, *Rhodococcus equi,* and *Mycobacterium avium*. Addition of probiotic and fermentative bacteria (beneficial) that tolerate acidic environment, to milk and milk products can prevent and limit the growth of pathogenic members (detrimental) of the flora naturally occurred in human body which include the following:

3.9.1. Teeth Flora

Teeth flora such as *Streptococcus mutans* and *Streptococcus sobrinus bacteria, cause dental caries.* Following the intake of dietary carbohydrate, acidogenic (acid-producing) and aciduric (acid-resistant) bacteria constituting the oral biofilm ferment metabolizable sugars, producing acid that can lower the plaque pH from 7.0 to 4.0 within 3 min. This is significant, given that tooth demineralization can occur at pH 5.2 and below [5]. The demineralization phases are followed by periods of alkalinization that promote remineralization. Dental caries occurs when the acidification phases prevail over the alkalinization phases, allowing the establishment of a more acidogenic flora, which in turn results in lower plaque pHs and enhanced and prolonged enamel demineralization. This process ultimately results in the formation of a carious lesion. Thus, the ability of bacteria to produce acid in combination with growth at low pH is regarded as a virulence factor. So that, the capacity of oral biofilms to generate alkaline compounds, which can neutralize acids and prevent the emergence of a cariogenic microflora, is a critical factor in caries prophylaxis and in the inhibition of caries progression.

3.9.2. Gastrointestinal and Intestinal Flora

The role of gastric acid is most apparent in patients who are achlorydic, i.e., unable to produce HCl, or have undergone a partial gastrectomy. In these circumstances, the number of aerobes, anaerobes, and coliforms increases so that there is overgrowth of bacteria in the stomach. It has also been observed that neutralization of acid by food or bicarbonate decreases the infectious dose; for example, an increase in the pH of the stomach results in gastric transit of greater numbers of *Listeria [*24] while gastrectomy and chronic gastritis are risk factors for the development of intestinal tuberculosis [25].

4. Microbial Safety of Food Products

Processed food is food which has been chemically altered through additives such as flavors, flavor enhancers, binders, colors, fillers, preservatives, stabilizers, emulsifiers, etc. Food processing is the set of methods and techniques used to transform raw ingredients into food by the food processing industry. Food processing typically takes clean, harvested crops or butchered animal products and uses these to produce attractive, marketable and often long shelf-life food products. Processing even increases the bioavailability of some nutrients, such as lycopene, found in tomatoes. However, eating processed food may cause: the cancers and diseases due to the dioxins and other toxic chemicals through consumption of processed food [26]. Our bodies are designed to eat natural, raw foods. Our bodies are amazing organs; capable of extraordinary things and they are designed to filter out toxins at an astounding rate and have a highly sophisticated defense system. The key lies in providing our bodies with the necessary building blocks to accomplish what they were designed to do to keep us alive. On other hand, raw foods contain beneficial enzymes and nutrients that may destroy through food processing. Nutrients, beneficial enzymes, and other components of the foods we eat work synergistically. We really don't know how well they work when they're isolated from each other. We can opt for the apple over the apple juice. We can choose a baked potato or salad over french fries. We can choose whole-grain bread over the white fluff that is passed off as bread [26]. We can choose processed foods with a very short list of ingredients; the longer the list, the more processing involved, and the more nutrition lost.

Development of industrial procedures to ensure the microbial safety of foods is well documented [27]. The introduction of commercially processed and notably canned foods allowed foods to be stored for long time period at ambient temperatures. However, despite this development, the need for establishing means of assuring safety of food products against food-spoilage bacteria is necessary, as well as the important knowledge about their survival strategies. For instance, the USA health authorities, in 1918-1922, set up a Botulism Commission about the finding on the extent of botulism from commercially canned foods (150 cases between 1918 and 1922 alone). Then, the heat resistance of the *Clostridium botulinum* spores established the bases of microbiologically safe process and the safety of low-acid canned foods [27]. The recommendations are developed and rapidly accepted by the US canning industry and internationally to assure the safety of food industry. Since such pioneer work, industry has increasingly recognized the need to base microbiological safety on sound knowledge of the microbiological hazards associated with a food and the development of technological means for their control [27]. Development of industrial procedures against microbes, to ensure the safety of foods, is a challenge to determine whether a potential pathogen will be able to survive a particular process.

4.1. Assessment of Microbiological Safety

Microorganism kind includes both beneficial and harmful types. The microbiological risks associated with the consumption of any food depend on the types and number of microorganisms, or amounts of microbial produced toxin(s), present in a food at the time of consumption and the susceptibility of the person consuming the food to these factors. Thus any microbiological safety requirements must, on the one hand, take account of microbiological knowledge concerning the occurrence and fate of potentially hazardous microorganisms within a particular product and medical knowledge concerning the numbers of microorganisms (or amounts of toxin) able to cause illness. We have much knowledge of the growth, survival and death of food borne disease-causing microorganisms and this can be used to identify potential microbiological hazards in a particular foodstuff [27]. The development of database and associated mathematical models are able to predict microbial growth, death and survival rates.

A major problem with microbiological hazard assessments is the paucity of quantitative information concerning the likely outcome of ingestion of a food contaminated with different levels of pathogenic microorganisms: i) will infection occur, and, if so, will this lead to mild, moderate or severe illness?. We know that a large number of factors will affect the initiation of infection and disease; ii) factors associated with the organism, particularly concern viability and the developed of its virulence factors which will depend on the strain and its history prior to, and during, its association with the food, including its location in a foodstuff and effects of processing, storage and use; iii) the effects of food, including whether it is liquid or solid (including size of particles), fat and iron content, acidity, other flora, preservative factors, and storage conditions - all of which will effect microbial viability and ability to survive passage through the stomach and to establish itself in the lower intestinal tract. Finally there are the idiosyncratic factors associated with an individual consuming the food including: nutritional and physiological factors, immunological status age, status of intestinal tract at the time of consumption, and previous exposure of the individual to infection by the organism.

4.2. Industrial Food Safety

Of importance to understand more about how microbes behave in food and the extent to which they are able to respond to their environment, switch genes and particularly develop means to change their resistance to preservatives and preservation systems used by the food industry. We need to understand the disease processes and particularly the molecular basis whereby microbes initiate infection and disease in man and how man may develop resistance to disease. Traditionally the microbiological safety of foods has been established via challenge tests. These tests simulated the effects of environmental conditions on food, in terms of growth and proliferation of spoilage and pathogenic microorganisms. Challenge tests can provide data useful in determining the safety and shelf-life of food under set conditions [27].

Microbiological challenge testing has been and continues to be a useful tool for determining the ability of a food to support the growth of spoilage organisms or pathogens. Microbiological challenge tests also play an important role in the validation of processes that are intended to deliver some degree of lethality against a target organism or group of target organisms. Quite often, with this latter purpose, there is an associated performance standard that the process must deliver (for example, a 5 log reduction of *Escherichia coli* O157:H7 for fermented meats). An appropriately designed microbiological challenge test will validate that a specific process is in compliance with the pre-determined performance standard [28]. The design, implementation, and assessment of microbiological challenge studies is a complex task that depends on factors related to how the product is formulated, manufactured, packaged, distributed, prepared, and consumed. An expert microbiologist must consider the relevant factors and design a study that best assesses the food safety of the product. Failure to account for specific product and environmental factors in the design of the test could result in flawed conclusions [28]. Microbiological challenge studies are also useful in determining the potential shelf life of certain refrigerated or ambient-stored foods. The determination of whether challenge studies are appropriate or useful must be made by considering such factors as the likelihood of the product to support growth of spoilage pathogens, or knowledge of the previous history of the product. For example, it is not useful to conduct challenge studies on frozen foods that would not support growth under proper storage conditions; nor would it be especially useful to conduct challenge tests on commercially sterile retorted canned foods. However, in the canned food example, it may be appropriate to conduct inoculated pack studies as part of the protocol for process validation. Microbiological challenge testing is very useful for food products that may sustain the growth of pathogenic organisms and that are stored under refrigeration, elevated temperature, or at ambient temperature and vulnerable to the growth of microorganisms.

4.3. Evaluation and Definition of Potentially Hazardous Foods

To make decisions on whether a food requires time/temperature control for safety, the properties of the food itself must be considered. Hazardous foods are common food commodities, including added preservatives and processing steps that may affect their microbial ecology. The microbiological hazards that may occur from consuming particular food commodities or their derived products must be discussed. Consideration must also be given for processing technologies or other methods that may be useful in minimizing hazards to ensure safety of foods. Potentially hazardous food means a food that is natural or synthetic, and that requires temperature control. Food could be potentially hazardous because it is in a form capable of supporting the rapid and progressive growth of infectious or toxigenic microorganisms, growth and toxin production of *Clostridium botulinum*, as well as growth of *Salmonella enteritidis* in raw shell eggs [29]. Potentially hazardous food includes an animal food (a food of animal origin) that is raw or heat-treated, or a food of plant origin that is heat-treated or consists of raw seed sprouts. Potentially hazardous food does not include a food with a pH level of 4.6 or below, a food that is commercially processed to achieve and maintain commercial sterility under conditions of non-refrigerated storage and distribution, or a food for which laboratory evidence demonstrates that the rapid and progressive growth of infectious or toxigenic microorganisms can't occur. Australia New Zealand Food Standards Council [30] adopted three Food Safety Standards, provides further discussion of high, medium, and low risk foods.

4.3.1. High-Risk Foods

High risk foods are foods that "may contain pathogenic microorganisms and will normally support formation of toxins or growth of pathogenic microorganisms." Examples are raw meat, fish, oysters, poultry, milk, tofu, fresh filed pasta, meat pies, frankfurts, salami, cooked rice, and lasagna.

4.3.2. Medium-Risk Foods

Medium-risk foods are foods that "may contain pathogenic microorganisms but will not normally support their growth due to food characteristics; or food that is unlikely to contain pathogenic microorganisms due to food type or processing but may support formation of toxins or growth of pathogenic microorganisms." Examples are fruits and vegetables, orange juice, canned meats, pasteurised milk, dairy products, ice cream, peanut butter, and milk-based confectionery.

4.3.3. Low-Risk Foods

Low-risk foods are foods that "are unlikely to contain pathogenic microorganisms and will not normally support their growth due to food characteristics." Examples are grains and cereals, bread, carbonated beverages, sugar-based confectionery, alcohol, and fats and oils.

5. Potentially Hazardous Foods

Processing technologies or other methods for foods may be useful in minimizing hazards to ensure safety of foods.

5.1. Raw Meat and Poultry Products

These foods consist of raw products, shelf-stable, raw-salted and salted-cured products, perishable raw-salted and salted-cured products (fresh sausage). Ready-to-eat products include perishable cooked uncured products (cooked roast beef, cooked pork, cooked turkey); perishable cooked cured products (a variety of luncheon meats), canned shelf-stable cured products (Vienna sausages, corned beef, meat spreads, canned sausages with oil, dried beef), perishable canned cured products (cured meats), canned uncured products (roast beef, meat stew, chili, chicken), fermented and acidulated sausages (salamis, pepperoni, sausage), and dried meat products (beef sticks, basturma). Because of the complexity of the product/processing matrices, product parameters (moisture protein ratio, water, and pH) and processing schedules are needed to ascertain whether ready-to-eat products require time/temperature control for safety or are shelf stable.

5.2. Fish and Seafood Products

These foods include fresh and frozen fish and crustaceans, cooked crustacean products, salted and smoked seafood products, sushi and seafood products.

5.3. Fats and Oils

Fats and oils are primary components of many foods that are emulsions comprised of oil as the continuous phase and water as the discontinuous phase. Mayonnaise, salad dressings, and related products are examples where water is the continuous phase and oil (fat) is the discontinuous phase. In recent years, products such as garlic-in-oil, various herb/spices-in-oil, and flavored oils have proliferated.

5.4. Butter and Margarine

Butter must be at least 80% milk fat, can be salted or unsalted, and may contain starter cultures for additional flavor. Regular margarine includes any fat composition emulsified to at least 80% fat. A wide range of fats and oils are used to process margarine. Other ingredients in margarine include salt, emulsifiers, and preservatives, and some margarines may contain milk solids.

5.5. Fruits and Vegetables

Fruits, vegetables, and related products include foods that are sold fresh, minimally processed, canned, frozen, juiced, or dried. In addition to being sold fresh, fruits are also sold dried and packaged with preservatives. Minimally processed fruit can be sold as fruit salads or incorporated into dairy products such as yogurt, cottage cheese, or ice cream [31].

5.6. Milk and Milk Products

Milk can be obtained in a concentrated form, such as evaporated or condensed milk, or in a dry form. Bacterial cultures can be used in making other products such as cultured milk, yogurts, and cheeses. More than 500 cheeses are manufactured worldwide, with variations deriving from the procedure in the cheese making technique. Microbiological changes occurring during cheese making are complex. Cheeses types can be classified according to many different criteria, but a general classification divides cheeses into fresh or un-ripened, soft, semisoft, hard, and processed cheese. In processed cheese, heat and sanitary packaging are used to prevent microbial hazards, unless the cheese is contaminated with heat-resistant pathogenic spores. If the product is contaminated with spore-formers such as *C. botulinum*, however, germination and toxin formation can cause serious public health concerns, especially if the product is intended to be used at ambient temperature. In this case, pH, moisture content, and antimicrobials become critical parameters that may preclude pathogenic growth and toxin formation and will determine the need for time/temperature control. Pathogens of concern and control methods for the various product categories were summarized and listed in Figure 1.

Product Category	Pathogens of Concern	Types of Process Control
Meats and poultry (fermented sausage).	*Clostridium botulinum* and *Clostridium perfringens*, *Salmonella* spp., *Escherichia coli*, *Camplylobacter jejuni*, *Yersinia enterocolitica*, *Staphylococcus aureus*, *Listeria monocytogenes*.	Time/temperature, pH, preservatives, moisture protein ratio, fermentation, heat processing.
Fish and seafood (smoked fish).	*Vibrio vulnificus*, *Vibrio parahaemolyticus*, *Vibrio cholerae*, *C. botulinum*, *L. monocytogenes*, *Salmonella* spp., *Shigella* spp., *S. aureus*.	Time/temperature, harvest site control, fermentation, pH, water-phase salt, preservatives, drying, salting.
Fruits and vegetables (peeled carrots).	*Salmonella* spp., *Shigella* spp., *E. coli*, *L. monocytogenes*, *Bacillus cereus*, *C. botulinum*, *Y. enterocolitica*.	Production control (Good Agriculture Practices), time/temperature, cooking, preservation techniques.
Cereal grains and related products (fresh pasta, foccacia bread).	*Salmonella* spp., *S. aureus*, *B. cereus*, *C. botulinum*.	Cooking, pH, preservatives, time/temperature.
Fats, oils & salad dressings (garlic-in-oil).	*S. aureus*, *Salmonella* spp., *B. cereus*, *C. botulinum*.	pH, water activity, salt.
Butter and margarine (light salted butter).	*S. aureus*, *L. monocytogenes*, *Y. enterocolitica*.	Production/raw ingredient quality control, moisture droplet size in the water-in-oil emulsion, water phase salt.
Milk and milk products (yoghurt).	*Salmonella* spp., *L. monocytogenes*, *E. coli*, *S. aureus*, *B. cereus* (cells and spores), *C. botulinum* (cells and spores), *Campylobacter jejuni*	Production control, time/temperature, cooking/pasteurization, preservatives.
Cheese and cheese products (Natural Swiss cheese).	*Salmonella* spp., *L. monocytogenes*, *E. coli*, *S. aureus*, *Shigella* spp., *C. botulinum* (cells and spores).	Production control, moisture content, pasteurization, preservatives, pH.

Figure 1. **Pathogens of Concern and Control Methods for Various Product Categories.**

6. Survival Strategy of Some Food Borne Pathogens

A pathogen is a microorganism that is able to cause disease in a plant, animal, insect, or human. Pathogenicity is the ability to produce disease in a host cell. Microbes express their pathogenicity by means of their virulence, a term which refers to the degree of pathogenicity of the microbe. Hence, the determinants of virulence of a pathogen are any of its genetic or biochemical or structural features that enable it to produce disease in a host [9]. The relationship between a host and a pathogen is dynamic, since each modifies the activities and functions of the other. The outcome of such a relationship depends on the virulence of the pathogen, the degree of resistance or susceptibility of the host, presence of underlying disease, delays in diagnosis and treatment, surgical indications, and timing of surgery [9, 32].

Food is the greatest source of pathogens. Pathogenic bacteria can cause disease by two broad mechanisms: 1) Invasiveness; that is, the ability to invade tissues. It encompasses mechanisms for colonization (adherence and initial multiplication), production of extracellular substances which facilitate invasion (invasins) and ability to bypass or overcome host defense mechanisms; 2) Toxigenesis which means the ability to produce toxins. Bacteria may produce two types of toxins called exotoxins and endotoxins (9). Exotoxins are released from bacterial cells and may act at tissue sites removed from the site of bacterial growth. Endotoxins are cell-associated substance which may be release from growing bacterial cells and cells that are lysed as a result of effective host defense (e.g. lysozyme) or the activities of certain antibiotics. Hence, bacterial toxins, both soluble and cell-associated, may be transported by blood and lymph and cause cytotoxic effects at tissue sites remote from the original point of invasion or growth. The first stage of microbial infection is colonization (the establishment of the pathogen at the appropriate portal of entry). Sites of pathogens entry in human hosts include the urogenital tract, the digestive tract, the respiratory tract and the conjunctiva. Organisms that infect these regions have usually developed tissue adherence mechanisms and some ability to overcome or withstand the constant pressure of the host defenses at the surface.

The health and function of the human intestinal epitheium is dependent on the presence of a thriving and diverse intestinal microflora. Commensal microorganisms are directly involved in the immune response of the intestinal epithelium beyond their mere physical presence as competitive organisms [33]. The presence of beneficial microorganisms, such as *lactobacilli*, *bifidobacteria* and anaerobic bacteria, have been shown to occupy critical receptors on intestinal epithelial cells, which prevent colonizing pathogens from occupying a niche for persistence and proliferation. However, beyond their mere physical presence, commensal microorganisms are directly recognized by toll-like receptors on the intestinal epithelial surface, which plays a critical role in the host defense against microbial infection. Toll-like receptors are capable of recognizing both pathogenic and commensal bacteria and their toxic products such as lipopolysaccharide. Commensal microorganisms are thus an integral component of the health and normal functioning immune status of the intestinal epithelium, which preserves intestinal epithelial integrity, a critical aspect of intestinal barrier function necessary to protect the host against external toxins and injurious agents [33]. A healthy host thus remains resistant to such colonization if these systems of defense remain intact. However, microorganisms that evolved virulence systems to erode the normal intestinal defense mechanisms can overcome colonization resistance and establish infection. Examples of microorganisms that can cause infection in a presumably normal host include, enteropathogenic *E. coli, Helicobacter pylori, Salmonella spp.*

6.1. Survival Strategies of Some Food-Borne Pathogens

6.1.1. Escherichia coli

Pathogenic *E. coli*, such as Enterohemorrhagic *E. coli* (EHEC) and Enteroaggregative *E. coli* (EAEC), are globally widespread bacteria. Some may cause the hemolytic uremic syndrome. Varying strains cause epidemics all over the world. Recently, an epidemic outbreak of a multi-resistant EHEC strain in Western Europe was reported, mainly in Germany. The Robert Koch Institute reports >4300 infections and >50 deaths (in July, 2011). Farmers lost several million EUR since the origin of infection was unclear. Pauling et al. [34] have contribute to the currently ongoing research with a computer-aided study of EHEC transcriptional regulatory interactions, a network of genetic switches that control, for instance, pathogenicity, survival and reproduction of bacterial cells. This study included gene regulatory interactions, based on the identification of evolutionarily conserved regulatory sites within the DNA of a harmless *E. coli* K-12. Identifying and characterizing the genetic control mechanism network of such a pathogen will allow for a better understanding of its survival and infection strategies.

6.1.2. Salmonella T*yphimurium*

Human salmonellosis infections are usually acquired via contaminated food as a result of the ability of *Salmonella* to colonize and persist within the gastrointestinal tract of their hosts. In addition, after food ingestion and in order to cause food-borne disease in humans, *Salmonella* must be able to resist several deleterious stress conditions which are part of the host defence against infections. Following food consumption, pathogenic bacteria are exposed to the extreme stomach pH, and later to bile salts, high osmolarity and low oxygen tension in the intestine [35]. Furthermore, the competition for nutrients with gastrointestinal commensal bacteria that produce bacteriocins and metabolites with antimicrobial activity also represent a significant challenge. The acid environment of the stomach can be considered one of the host's first lines of defence against ingested bacteria. The importance of human gastric fluid as an antibacterial barrier against enteric pathogens is well-documented. Humans produce approximately 1–2 l of gastric fluid per day, with a pH as low as 1.5 and HCl levels of 150–160 mEq l^{-1} [35]. Furthermore, several organic acids can be present in the stomach, including lactic, acetic, propionic and butyric acids, which can also contribute to the antimicrobial effect. However, it has been suggested that certain solid food sources, especially those rich in fat or with a high protein content, protect *Salmonella* against stomach acidity [35], and feed composition has been shown to influence the gastric pH and the stomach organic acid content. With regard to liquid foods, such as water and juices, their short emptying time from the stomach could also facilitate *Salmonella* survival. Despite being neutrophilic bacteria, *Salmonella* may be able to survive in the hostile environment of the stomach through the induction of the so-called acid tolerance response, which can be defined as the capacity to undergo an adaptive response to moderately acidic pH that enhances the subsequent survival under conditions of lethal pH, such as those prevalent in the stomach [35]. Synthesize of acid shock proteins (ASPs), during the acid tolerance response, are also induced by *S. typhimurium* in order to protect or repair the macromolecular damage caused by acid stress. An extensive research effort has been made in recent decades to identify and characterize these stress proteins in *S. typhimurium*, and several regulatory genes controlling the expression of different subsets of ASPs have been described [35]. Most of the identified ASPs are involved in cellular regulation, molecular chaperoning, energy metabolism, transcription, translation, synthesis of fimbriae, regulation of the cellular envelopes, colonization and virulence. Alternative sigma factor σ^S (RpoS) were also identified for *Salmonella* survival in stationary phase, as well as under various stress conditions [20]. RpoS has been shown to be involved in the acid tolerance response of *S. typhimurium*, controlling the expression of at least 10 ASPs. In addition, *Salmonella* changes its membrane composition in response to the environmental signals found in their surroundings, with the purpose of maintaining a degree of membrane fluidity. The composition of fatty acids (FA) in the cell membrane is responsible for membrane fluidity, and a relationship between membrane fluidity and stress tolerance for *Salmonella* has been suggested [35]. When *S. typhimurium* cells are exposed to acid pH, a membrane adaptation, characterized by a decrease in the unsaturated FA to saturated FA ratio and in membrane fluidity, occurs.

6.1.3. Clostridium spp.

Perfringens food poisoning is the term used to describe the common food-borne illness caused by *Clostridium perfringens*. A more serious, but rare, illness termed pig-bel may be caused by ingesting food contaminated with type C strains. It is likely that the acid resistance of the organism contributes to its survival in foods and *in vivo*, since *C. perfringens* is capable of escaping from the macrophage phagosome [36]. Beginning in the year 2000 there has been a resurgence of *Clostridium difficile* infection (CDI) as a major cause of hospital morbidity and mortality in North America and Europe that was completely unanticipated [2]. The strain, *C. difficile,* has risen in frequency to become the most common healthcare-associated infection pathogen, exceeding methicillin-resistant *Staphylococcus aureus.* Protection against CDI is thought to be directed primarily at *C. difficile* toxins which are a major survival strategy. However, treatment of CDI with antimicrobials will also further disrupt the protective naturally bacterial microbiota leaving the patient susceptible to recurrent CDI. Disruption or alteration of the normal microbiota by antibiotics enables the spores of *C. difficile* to vegetate, colonize the gut, and produce the toxins that resulted in occurrence of diarrhea and colitis [2]. In addition, patients most susceptible to CDI, the advanced elderly, may already have a limited immune response and fail to increase their adaptive immune response with infection.

6.1.4. Listeria Monocytogenes

L. monocytogenes, the causative agent of listeriosis, may encounter low-pH environments at a number of stages during the course of its infectious cycle, including in low-pH foods, during gastric transit, following exposure to fatty acids in the intestine, and finally in the phagosome of macrophages. *Listeria* must therefore have a number of mechanisms for responding to, and surviving in, acidic environments [5]. When exposed to an acid-pH stress, *L. monocytogenes* can synthesize a number of proteins which reach to 47 proteins induced by pH 3.5. The ability of *L. monocytogenes* to survive acid stress was found to be growth-phase dependent, i.e., 100% of stationary phase cells can survive exposure to pH 3.5 for 60 min. Moreover, acid adaptation aids the survival of this food pathogen in acidic foods such as cottage cheese, yoghurt, whole-fat cheddar cheese, orange juice, and salad dressing and thus must be taken into consideration when using low pH as a preservative [5]. In addition to acidic foods, *Listeria* encounters low-pH stress in vivo, as the role of stomach acid is a primary barrier against listeriosis. To appreciate the consequences of the emergence of acid-resistant strains, spontaneous acid-tolerant mutants were isolated which were almost as resistant, without acid induction, as adapted wild-type cells and which could be further induced to display even greater levels of survival [37]. Compared with wild-type cells, these mutants displayed enhanced acid tolerance during all stages of growth, enjoyed improved survival in acidic foods, displayed enhanced invasiveness and growth in enterocyte-like and macrophage-like cell lines [37].

Conclusion

Bacteria can survive in almost all environments on Earth, including some considered extremely harsh. Under hardness abiotic conditions, microbes have been found thriving, and their tenacity to survive in such extreme conditions acquires them stress resistance. Any decrease in growth rate leads to rapid molecular reactions that repair the stress-induced damages to protect the microbial cells against stressors. Microbes can persist by survival mechanisms and genetic mutation to overcome extreme conditions and challenge environmental changes. Knowledge of the stress factors and stress resistance by microorganisms is important for the development of biotechnology in different fields as well as in the field of industrial microbiology. On the other hand, mankind has learned over many millennia much more about the ravages of food poisoning and also discovered the value of drying, freezing, smoking, fermentation, and acidification as a means of preserving foods.

References

[1] Alroumi F, Sarwar A, Grgurich P, Lei Y, Hudcova J, Craven D (1995) Strategies for prevention of ventilator-associated pneumonia: bundles, devices and medications for improved patient outcomes. *Hosp Pract* 40:81-92.

[2] Dale N (2012) *Clostridium difficile* infection prevention: Biotherapeutics, immunologics, and vaccines. Loyola Univ Chicago Stritch School Medicine.

[3] Vorob'eva L (2004) Stressors, stress reactions, and survival of bacteria: A Review. *Prikl. Biokhim*. 40:261-269.

[4] Ron E, Segal G, Sirkis R, Robinson M, Graur D (2000) Regulation of heat-shock response in bacteria. *Proceedings of the 8th International Symposium on microbial ecology*. Atlantic Canada Society for microbial ecology, Halifax, Canada.

[5] Paul D, Colin H (2003) Surviving the acid-test: Responses of Gram-positive bacteria to low pH. *Microbiol Mol Biology Rev* 67:429-453.

[6] Boor K (2006) Bacterial stress responses: What doesn't kill them can make them stronger. *PLoS Biol* 4: e23.

[7] Vijayabaskar P, Babinastarlin S, Shankar T, Sivakumar T, Anandapandian K (2011) Quantification and characterization of exopolysaccharides from *Bacillus subtilis* (MTCC 121). *Adv Biolog Res* 62:71-76.

[8] Kusukawa N, Yura T (1998) Heat-shock protein *GroE* of *Escherichia coli:* Key protective roles against thermal stress. *Genes Dev* 2:874-882.

[9] Todar K (2009) The mechanisms of bacterial pathogenicity. Dept. Bact. Univ. Wisconsin-Madison.

[10] Yang Z, Geng Y, Shi W (1998). A DnaK homolog in *Myxococcus Xanthus* is involved in social motility and fruiting body formation. *J Bact* 180:218-224.

[11] Bernardini M, Fontaine A, Sansonetti P (1990) The two-component regulatory system *omp*R *env*Z controls the virulence o *Shigella flexneri*. *J Bacteriol* 172:6274-6281.

[12] Jesus A, Wilhem H, Navarott R (2006) Fungal responses to reactive oxygen species. *Med Myco*l 44: 101-107.

[13] Aguirre J, Rios-Momberg M, Hewitt D, Hansberg W (2005) Reactive oxygen species and development in microbial eukaryotes. *Trends Microbiol* 13:111-118.
[14] Stadtman E (1992) Protein oxidation and aging. *Sci* 257:1220–1224.
[15] Sam D, Thomas N (1998) Bacterial senescence: stasis results in increased and differential oxidation of cytoplasmic proteins leading to developmental induction of the heat shock regulon. *Genes* 12:3431–3441.
[16] Jordan S, Matthew I, Thorsten M (2008) Cell envelope stress response in Gram-positive bacteria. *FEMS Microbiol Rev* 32:107–146.
[17] William C (2012) Bacterial endospore. Cornell Univ., College Agr. Life Sci., Dept. Microbiology.
[18] Taddei F, Matic I, Radman M (1995) cAMP-dependent SOS induction and mutagenesis in resting bacterial populations. *Proc Natl Acad Sci USA* 92:11736–11740.
[19] *Joseph C (*2012*)* Dealing with stress: New research highlights the survival skills of disease-causing *E. coli*. Arizona State University.
[20] Fang F, Libby S, Buchmeier N, Loewen P, Switala J, Harwood J, Guiney D (1992) The alternative sigma factor *katF (rpoS)* regulates *Salmonella* virulence. *Proc Natl Acad Sci USA* 89:11978–11982.
[21] Prem T, Soam P (2012) Time-dependant gene expression pattern of *Listeria monocytogenes* J 0161 in biofilms. *Adv Genomics Gen* 2:1-18.
[22] Prakash B, Veeregowda B, Krishnappa G (2003). Biofilms: a survival strategy of bacteria. *Curr Sci* 85:1299-1307.
[23] Li Y, Lau P, Tang N, Nan T, Gunnel S, Ellen R, Dennis G (2002) Novel two-component regulatory system involved in biofilm formation and acid resistance in *Streptococcus mutans*. *J Bacteriol* 184:6333-6342.
[24] Czuprynski C, Faith N (2002) Sodium bicarbonate enhances the severity of infection in neutropenic mice orally inoculated with *Listeria monocytogenes* EGD. *Clin Diagn Lab Immunol* 9:477-481.
[25] Laxman S, Rajni M (2010) Survival mechanisms of pathogenic *Mycobacterium tuberculosis* H37Rv. FEBS J 277:2416-2427.
[26] Eric T (2008) How do you define processed food? Natural Health (http://www.zijapower.com).
[27] Baird-Parker A (1995) Development of industrial procedures to ensure the microbial safety of food. *Food Cont* 6:29-36.
[28] Vestergaard E (2001) Building product confidence with challenge studies. *Dairy Food Env* 21:206-209.
[29] Food and Drug Administration [FDA] (1999) Food Code: 1999 recommendations of the United States Public Health Service, Food and Drug Administration. Available from: http://www.foodsafety. gov/~ dms/foodcode.html.
[30] ANZFA (2001) Australian New Zealand Food Authority. Food safety: the priority classification system for food business.http://www.foodstandards.gov.au/ newsroom/publications /thepriorityclassific352.cfm.
[31] ICMSF (1998) International Commission on Microbiological Specification for Foods. Microorganisms in foods. Microbial ecology of food commodities. New York: Blackie Academic & Professional.
[32] Franck T, Dominique G, Frederic C, Gilbert H, Didier R (2012) Management of infective endocarditis: challenges and perspectives. *The Lancet* 379:965-975.

[33] John A, Olga Z, Wu L (2005) The impact of stress and nutrition on bacterial–host interactions at the intestinal epithelial surface. *Curr Opin Clin Nutr Metabolic Care* 8:205–209.

[34] Pauling J, Rottger R, Neuner A, Salqado H, Collado-Vides J, Kalaqhatqi P, Azevedo V, Tauch A, Puhler A, Baumbach J (2012) On the trail of EHEC/EAEC-unraveling the gene regulatory networks of human pathogenic *Escherichia coli* bacteria. Computational Systems Biology, Max Planck Institute for Informatics, Germany.

[35] Avelino A, Máire B, Miguel P, Winy M, Lopez M, Ana B, Colin H (2011) *Salmonella* spp. survival strategies within the host gastrointestinal tract *Microbiol* 157:3268-3281.

[36] O'Brien D, Melville S (2000) The anaerobic pathogen *Clostridium perfringens* can escape the phagosome of macrophages under aerobic conditions. *Cell Microbiol* 2:505-519.

[37] Conte M, Petrone G, Biase A, Ammendolia M, Suparti F, Seqanti L (2000) Acid tolerance in *Listeria monocytogenes* influences invasiveness of enterocyte-like cells and macrophage-like cells. *Microbiol Pathog* 29:137-144.

In: Microbes in Process
Editors: Neelam Garg and Abhinav Aeron
ISBN: 978-1-63117-127-7

Chapter 12

Diversity and Biopotential of Plant-Associated Endophytic Actinomycetes

***M. Gangwar*[1,*], *G. Pandove*[1] *and Neelam Garg*[2]**
[1]Department of Microbiology, Punjab Agricultural University, Ludhiana, Punjab, India
[2]Department of Microbiology, Kurukshetra University, Kurukshetra, Haryana, India

Abstract

Endophytic actinomycetes have been found in virtually all terrestrial plants, where they colonize the internal tissues of their host plant and establish a different type of relationship including symbiotic, mutualistic and commensalistic. Most of the endophytes appear to originate from the rhizosphere or phyllosphere, however some may be transmitted through the seed. Endophytic actinomycetes can act as plant growth promoting and antimicrobial agents. Endophytic actinomycetes can also be beneficial to their host by producing a range of natural products that could be harnessed for potential use in medicine, agriculture or industry. In addition they can be used for bioremediation of contaminated soils and play a role in soil fertility through phosphate solubilization and may act as biocontrol agents against phytopathogenic fungi.There is increasing interest in developing potential bioactive compounds which might prove promising sources of novel antibiotics and plant growth regulators.

1. Introduction

Plants in all climates and in all ecosystems are associated with a diverse array of microbes. Endophytic actinomycetes have been found in large variety of plants, where they colonize the internal tissues of their host plant and establish a different type of relationship including symbiotic, mutualistic and commensalistic [1]. The internal plant tissue is a biological complex and distinct microhabitat within the terrestrial ecosystem because of its varying contents of alkaloids, terpenoids, steroids and aromatic compounds. Thus, healthy

* E-mail: madhugangwar@yahoo.com.

plant tissues represent an untapped reservoir of novel endophytic microorganisms producing bioactive metabolites [2]. Although the presence of endophytic actinomycetes has been known for long, awareness of their abundance and potential benefits has grown rapidly in recent years. Most of the endophytes appear to originate from the rhizosphere or phyllosphere, however some may be transmitted through the seed. Endophytic actinomycetes can also be beneficial to their host by producing a range of natural products In some cases, they can act as biological control or antimicrobial agents [3, 4], enhance plant growth [5, 6] and promote plant establishment under adverse conditions [7]. These actinomycetes are relatively unstudied and potential sources of novel natural products for exploitation in medicine, agriculture, and industry [8]. In addition they can be used for bioremediation of contaminated soils [9] and play a role in soil fertility through phosphate solubilization [10]. There is increasing interest in developing potential bioactive compounds which might prove promising sources of novel antibiotics and plant growth regulators.

Actinomycetes form major rhizospheric microbial community and are a diverse group of filamentous Gram-positive bacteria that are characterized by high G+C content in their genomes. The actinomycete that reside in the tissue of living plants and do not visibly harm the plants are known as endophytic actinomycetes [11].The first endophytic actinomycetes isolate belonging to genus *Frankia* is a nitrogen fixing actinomycete that forms nodules in non-leguminous woody trees [12]. Almost all vascular plants are found to harbor several endophytic actinomycetes and they are often rich in its root. Initially, they were exploited for their potential as antagonists of soil born phytopathogens, but now endophytic actinomycetes are considered as exciting novel sources for obtaining new bioactive compounds to overcome the emergence of drug-resistant microbial pathogens. Thus, there is increased demand for potent antibiotics, antioxidants that can be obtained from endophytic actinomycetes. These microbes live in different organs [roots, stems, leaves, flowers, fruits and seeds] of the host plants, mainly in inter or intracellular spaces. It is noteworthy that, of the nearly 300,000 plant species on the earth, each individual plant is considered to host one or more type of endophytes [2], creating an enormous biodiversity. However, only a few of these plants associated endophytic actinomycetes have been studied, indicating the opportunity to find interesting species and related natural products among myriads of plants in different niches and ecosystems is great. Significant variations in the populations of both indigenous and introduced endophytes have been reported. These variations are attributed to plant source, plant age, tissue type, time of sampling and environment. Generally, bacterial populations are larger in roots and decrease in the stems and leaves [13].

The most common endophytic actinomycetes belong to genera *Streptomyces* were isolated from surface sterilized plant tissues [14]. The remainder belongs to genera *Actinopolyspora, Micromonospora, Microbispora, Nocardia, Streptosporangium, Saccharopolyspora* etc.

Endophytic infected plants often grow faster than non-infected ones. This effect is due to production of phytohormones, such as Indole-3-acetic acid (IAA) [6], cytokinins and through secretion of plant growth regulators like Pteridic acid A and B which have auxin like activity [5], by phosphate solubilizing activity [15] and by production of siderophores [16]. These microbes are also important producers of secondary metabolites that have a great diversity of biological activities including antibacterial, antifungal, antihelminthic, insecticidal activities which could protect the plant against various devastating agents [17]. Endophytic actinomycetes isolated from wheat roots have shown ability to reduce the impact of take all disease on wheat by up to 70% in glass house trials using naturally infested soil. A number of endophytic actinomycetes exihibited suppression of fungal pathogens of wheat *in vitro* and *in planta* including *Rhizoctonia solani, Pythium* spp., *Fusarium oxysporum, Phytophthora spp.* and *Gaeumannomyces graminis* var. *tritici* indicating their potential use as biocontrol agents [18]. In addition, endophytic actinomycetes also synthesize an array of biodegradative enzymes including chitinases, glucanases, peroxidases and other enzymes possibly involved in mycoparasitic activity.

Medicinal plants synthesize chemical substances, providing raw material for pharmaceutical, cosmetic and fragrance industries [19]. Endophytes of medicinal plants probably participate in metabolic pathways of medicinal plants and produce analogous or novel bioactive compounds, for example taxol [20].

They have been isolated from stem and root interior of many plants, such as snakevine, *Zingiber officinale* [21], *Mentha longifolia* [22] and *Azadirachta indica* A. Juss [4]. Relative to stems and leaves, roots harboured more diverse actinomycetes populations. The endophytic *Streptomycete* NRRL 30562 obtained from the snakevine produced novel peptide antibiotics, which possessed wide-spectrum activity against many pathogenic fungi and bacteria [23]. Some of which have impressive activity against the malarial parasite [23, 24]. These studies demonstrated that endophytic actinomycetes offer some promises for the discovery of novel metabolite products with pharmacological and agricultural potentials.

2. Overview of Actinomycetes in Nature

The definition of endophytes is described as microorganisms that establish an endosymbiotic relationship with plants, whereby plants receive biological benefits, such as ability to increase tolerance to stresses or plant growth promotion [7]. On the other hand, endophytic actinomycetes are defined as bacteria that do not visibly harm the plants and can be isolated from the surface of the disinfected plant tissues or extracted from inside of the plants. However, most endophytes have been demonstrated to improve and promote the growth of the host plants as well as reduce disease symptoms caused by plant pathogens or various environmental stresses. Higher plants are now known to host an untold number of endophytes. Many endophytes produce a variety of bioactive molecules, which are currently attracting the attention of scientific institutions/ scientists towards isolation and exploration of their biotechnological promise. It is interesting that the useful natural products made by plants are also produced by certain endophytes residing within them that is why endophytes have been aptly called 'chemical factories inside plants'. Endophytes produce a plethora of substances e.g. novel antibiotics, antimycotics, immunosuppressants, and anticancer compounds of potential use to modern medicine, agriculture, and industry [2].

The association of actinomycetes with plants is found to confer many advantages to host plants. They can promote the growth of many field crops by producing plant growth-promoting substances and by fixing nitrogen from the atmosphere and potential sources of novel natural products for exploitation in medicine, agriculture and industry [25]. Endophytic infected plants often grow faster than the non-infected ones. This effect is due to the production of the phytohormone Indole-3-acetic acid (IAA) [6], Pteridic acid A and B, which has auxin like activity [5], phosphate solublizing activity [15], production of siderophores

[16], cytokinin and other plant growth promoting substances by the infecting endophytes. They also play a crucial role in the recycling of refractory biomaterials by decomposing complex mixtures of polymers in dead plant, animal and fungal materials producing several volatile sesquiterpene substances namely, geosmins, responsible for the characteristic "wet earthy odor" and in bioremediation of contaminated soils [26].

2.1. Isolation and Cultivation Methods for Endophytic Actinomycetes

The first endophytic actinomycete to be identified and studied was the *Frankia* species. *Frankia* are nitrogen fixing actinomycetes that form actinorrhizae with eight families and over 200 species of angiosperms [27]. Since the isolation of *Frankia,* a number of other biologically active endophytes and root-colonizing microorganisms belonging to actinomycete group have been isolated or detected from almost all vascular plants examined, ranging from woody species to herbaceous plants. It is evident that colonization of terrestrial plants by actinomycetes is ubiquitous and common in nature. Thus, endophytic actinomycetes are important components of microbial biodiversity because of the numerous plant species. Although, theoretically, each plant can be selected for actinomycetes isolation. Those plants with an unconventional setting and biology as well as with established ethnobotanic values would be preferred and promising sources of endophytes producing novel bioactive products. A specific protocol for isolation of actinomycetes from a given plant is important, as isolation represents the most crucial step during the process of obtaining pure cultures, and the fact that host species, sampling strategy, host-endophyte and inter-endophyte interactions, tissue type and ages, geographic and habitat distribution, culture conditions, surface sterilants and selective media all influence the detection and enumeration of endophytes [28]. Some detailed isolation methods and procedures, including plant sampling, surface sterilization and used media have been reviewed [29] and introduced by Coombs and Franco [30].

Surface sterilization is the first and obligatory step for endophyte isolation in order to kill all the surface microbes. It is usually accomplished by treatment of plant tissues with oxidant or general sterilant for a period, followed by a sterile rinse. Commonly used surface disinfectants are ethanol (70– 95%), sodium hypochlorite (3–10%) and hydrogen peroxide. Some surfactants such as Tween 20, Tween 80 and Triton X- 100 can be added to enhance surface sterilization effectiveness [29]. A common protocol involves a three-step procedure as described by Coombs and Franco [30]. After the treatment with sodium hypochlorite and sodium thiosulfate, plant tissues can be soaked in 10% NaHCO3 solution in order to inhibit the endophytic fungi, which could overgrow the samples and mask the actinomycetes [31] After each treatment, validation of surface sterilization should be checked to prove that the epiphytic microorganisms have been killed and that subsequent isolates are true endophytes. In general, the sterilization procedure should be optimized for each plant tissues, especially the sterilization time, since the sensitivity varies with plant species, age and organs. Pretreatment of plant issues is an important step for isolation of endophytic actinomycetes. Surface-sterilized plant samples often need drying at 80°C or 100°C for 15 to 30 min to kill bacteria. Commonly, materials were aseptically sectioned into small fragments about 0.2×1.0 cm [32, 33] then distributed onto isolation media. On this basis, tissues can be aseptically crumbled into smaller fragments by commercial blender [34, 35, 36], which enlarged the colonization area of samples onto the media agar, thus making it helpful for recovering endophytes. Tender and soft tissues can be crushed with pestle and homogenized in a mortar with extraction solution or buffer and then use serial dilution technique.

Growth of microbes in the laboratory is dependent on the composition of the media and the cultivation conditions that are applied. For endophytic actinomycetes, some classical media for soil actinomycetes isolation, such as humic acid vitamin (HV) [37], International *Streptomyces* Project media ISP 2 and ISP 5 [38], raffinose-histidine agar [39] and starch casein agar [40] are also available. Low nutrient medium TWYE was found effective for isolation of endophytic actinomycetes [41,42], due to the fact that high nutrient concentration allowed fast growing bacteria to overgrow slower growing actinomycetes. The addition a certain amount of plant extracts into the isolation medium is effective. This may be due to the different physiological properties of some actinomycetes in plant tissues and soils [43].

2.2. Diversity of Endophytic Actinomycetes

Endophytic actinomycetes have been isolated from a variety of healthy plant species ranging from crop plants, such as wheat, rice, potato, carrots, tomato and citrus [44-49], different woody tree species[50, 51, 52], ferns and club mosses [53]. In general, *Streptomyces* spp. were the most predominant species and *Microbispora*, *Micromonospora*, *Nocardioides*, *Nocardia* and *Streptosporangium* are the common genera. For instance, 619 actinomycetes were isolated from different cultivars of tomato, and all of them were *Streptomyces* spp. [54]. Taechowisan et al. [21] isolated 330 strains belonging to four different genera (*Streptomyces*, *Microbispora*, *Nocardia* and *Micromonospora*) from 36 medicinal plant species of Thailand. Some actinomycete genera such as *Actinoplanes, Amycolatopsis, Catenuloplanes, Dactylosporangium, Kineospora, Microbispora, Micromonospora, Nonomuraea*, which are often very difficult to isolate and cultivate due to their slow growth, are called rare actinomycetes [55]. Lee et al [56] isolated 81 endophytic actinomycetes including eight genera from Chinese cabbage roots, and *Microbispora* spp. were the most common isolates, followed by *Streptomyces* spp. and *Micromonospora* spp. Relative to stems and leaves, roots harboured more diverse actinomycete populations. An endophytic actinomycetes strain designated YIM 56134^{T} from the root of a Chinese medicinal plant, *Carex baccans* Nees, collected from Yunnan, south-west China. An analysis of 16S rRNA gene sequence similarities showed that strain YIM 56134^{T} was a member of the genus *Glycomyces,* for which the name *Glycomyces endophyticus* sp. nov. is proposed [57]. A total of 55 isolates were obtained from 20 plants of *Azadirachta indica* A. Juss. The dominant genus was *Streptomyces,* followed by *Streptosporangium*, *Microbispora*, *Streptoverticillium*, *Sacchromonospora* sp., and *Nocardia*. Overall 32 of the 55 isolates were found to have broad spectrum significant antimicrobial activity [4]. An endophytic actinomycete strain was isolated and designated YIM 60475^{T} by Chen et al [58] from the roots of *Maytenus austroyunnanenis*, a traditional medicinal plant, collected from the tropical rainforest of south-west China. Phenotypic, chemotaxonomic and phylogenetic analyzes established the affiliation of the new isolate to the genus *Streptomyces mayteni* sp. nov. is performed.

To date, more than 40 new taxa have been found by polyphasic taxonomic approaches, including four new genera, *Plantactinospora*, *Actinophytocola*, *Phytohabitans* and *Jishengella*. Most of these published species had more than 97% similarities with recognized nearest relatives based on the analysis of the 16S rRNA gene sequences with closest nonendophytic neighbours. In addition, few species possessed unique rRNA gene sequences, which probably had experienced long-time evolution. This may presume that some actinomycetes endophytes are very similar with nonendophytic isolates, especially those from soils. In fact, the endophyte population was considered to be a subset of the rhizosphere popolation and soil type, soil microbial population significantly affects the endophytic actinomycetes communities [59, 60].

3. Mechanisms of Mode of Action of Endophytic Actinomycetes

Root colonizing actinomycetes commonly referred to as actinorhiza exert beneficial impact on plant development directly or indirectly by using various mechanisms.

3.1. Plant Growth Promoting Actinomycetes [PGPA]

Nowadays, increased public concern about environmental problems caused either directly or indirectly by the use of fertilizers, pesticides, herbicides, and fungicides, has prompted researchers to consider alternatives to these established chemical strategies for facilitating plant growth in agriculture, horticulture, and silviculture [61]. Substantially more microorganisms are present near plant root surface than in bulk soil. This rhizosphere effect is caused due to release of exudates from growing root tissues and lyses of cells of older root part [62]. Endophytic actinomycetes are of special interest since they possess many properties that could benefit to plant growth. The plant growth promoting actinomycetes endophytes possess similar mechanisms for one of the following functions: production biological control agents, or production of plant growth promotion compounds, such as auxins, cytokinins and gibberellins, or producing siderophore to bind Fe^{3+} from the environment and help to improve nutrient uptake, supply of plant nutrients [nitrogen, phosphate and other mineral nutrients], or suppression of stress ethylene production by 1-aminocyclopropane-1- carboxylate (ACC) deaminase activity [63-66] Plant growth promoting actinomycetes (PGPA) improve plant growth and increased yield of crop plant and are genetically considered to be of two type i.e. symbiotic and free living. Actinomycetes strains having the most important stimulatory effect on plant growth and protective effect against phytopathogenic fungi were also those able to colonize the external surface of roots as well as to develop inside the root. Direct and indirect interactions between actinomycetes and other non-pathogenic soil microorganisms also influence plant growth.

The actinomycetes stimulated the intensity of mycorrhizal formation that resulted in improved plant growth [67]. Actinomycetes are unparalleled sources of bio-active metabolites including antibiotics, plant growth factors, and other substances [68]. Plant growth-promoting properties of endophytic actinomycetes and the recent increased understanding of some of the mechanisms suggest that this promising source merits further investigations for potential application in agriculture.

3.2. Indole Acetic Acid [IAA]

The actinomycetes can promote plant growth by producing promoters such as indole-3-acetic acid [IAA] to help growth of roots or produce siderophores to improve nutrient uptake [69]. Indole acetic acid [IAA] is a common natural auxin and is a product of L-tryptophan metabolism in microorganisms [70]. Approximately 80% of rhizosphere bacteria can secrete IAA [71]. The rich supply of substrates available in root exudates creates the potential for *Streptomycetes* to synthesize and release IAA [72]. Several *Streptomyces* species such as *S. olivaceoviridis*, *S. rimosus* and *S. rochei* were isolated from the tomato rhizosphere, have the ability to produce IAA and improve plant growth by increased seed germination, root elongation and root dry weight [73]. Pteridic acids A and B was obtained from the fermentation broth of an endophytic *Streptomyces hygroscopicus* TP-A045 (isolated from *Pteridium aquilinum*) as plant growth promoters with auxin like activity. These compounds accelerated formation of adventitious roots in hypocotyls of kidney beans at 1 nM as effectively as indole acetic acid [5]. For example, several endophytic actinomycetes isolated from winter rye produced indolyl-3-acetic acid. Treatment of winter rye seeds with auxin producing strains increased the germination capacity and enhanced an intensive seedling growth in vitro [74]. Endophytic actinomycetes from *Aquilaria crassna* Pierre ex Lec and wheat produced indole acetic acid [75].

3.3. Phosphate Solubilization

Phosphorus, one of the 17 chemical elements required for plant growth and reproduction, is often referred to as the "energizer" since it helps store and transfer energy during photosynthesis as it plays important role in capturing and converting the sun's energy into useful plant compounds. Phosphorus is one of the major plant nutrients in the soil. It is a constituent of plant cells, essential for cell division and development of the growing tip of the plant. For this reason it is vital for seedlings and young plants. Phosphorus availability to crops is subject to chemical fixation in soil with other metal cations, depending on soil pH. Biological phosphate solubilization is an alternative for improving the efficiency of natural phosphate utilization in agriculture. The main mechanism involved in the solubilization of phosphorus is related to organic acids synthesized by microorganisms, which also promote the acidification of microbial cell and its environment. Among these, gluconic acid apparently is the most frequent solubilization compound. Other acids involved in the solubilization of phosphates are 2-ketogluconic acid, lactic acid, isovaleric acid, isobutyric acid and acetic acid [76]. Besides the organic acids, the release of protons H^+ by the external cellular surfaces of ATPases, the production of chealating substances or the production of organic acids (Sulfuric, nitric and carbonic) can constitute alternative mechanisms for solubilization of organic phosphates [77]. The main phosphorus providing molecules by means of mineralization are the nucleic acids, phospholipids and phosphate sugars that are easily degradable, and phytic acid, polyphosphates and phosphates which are mineralized more slowly. This mineralization is promoted by phosphatases or phosphohydrolases and classified as acid or alkaline according to the pH at which they show optimal activity. These enzymes can be either secreted through the plasma membrane or remain in the membrane [76]. A large number of microorganisms including bacteria, fungi and actinomycetes are known to produce acidic metabolites which by change of soil pH or by direct chelation of metal cations, release fixed or insoluble phosphorus in available form [78]. Actinomycetes are of special interest in phosphate solubilization since these filamentous sporulating bacteria are able to develop in extremely different soils [79].

Hamdali et al. [10] reported high amount of phosphate solubilizing activity in *Streptomyces grieseus*, *Streptomyces cavourensis* and *Micromonospora aurantica*. Gangwar et al [80] reported that twenty isolates of endophytic actinomycetes from rice plants and seventeen from wheat plants were able to solubilize phosphate It is suggested these actinomycetes could be used as plant growth promoters. The amount of phosphate released from rock phosphate was higher in the presence of actinomycete strains than in the presence of plant alone, indicating the important contribution of rock phosphate solubilizing actinomycetes in the improvement of plant growth [81].

Actinomycetes isolated from the rhizosphere which was capable of increasing availability of phosphorus to plants either by mineralization of organic phosphate or by solubilization of rock phosphate by production of acids [82]. Among the six selected actinomycetes, *Streptomyces griseus, Streptomyces cavourensis, Micromonospora aurantiaca* strains show significantly improved wheat plant growth in test tubes as well as in rock phosphate soil.

3.4. Siderophore Production

Siderophores (from the Greek: "iron carriers") are defined as relatively low molecular weight, ferric ion specific chelating agents elaborated by bacteria and fungi growing under low iron stress. Siderophores play a vital role in the suppression of plant pathogen by chelation of iron thereby creating competition for iron. Siderophore production under iron stress conditions confers upon these antagonistic organisms as an added advantage, resulting in exclusion of pathogens due to iron starvation. Although various bacterial siderophores differ in their abilities to sequester iron, in general, they deprive pathogenic fungi of this essential element since the fungal siderophores have lower affinity [83]. In this way, the plants are benefited in two ways i.e. suppression of pathogen as well as enhanced iron nutrition that results in increasing plant growth. *Streptomyces* have been reported to produce hydroxamate type siderophore. Siderophore synthesized by microorganisms colonizing the rhizosphere could be a source of soluble iron for the host plant. Uptake of microbial siderophore by plant have been attributed to microorganism living within plant tissues or closely associated with plant surface.

Detection of siderophores is most readily achieved in iron-limited media, which generally means either a synthetic (minimal) recipe or introduction of a complexing agent that will render the iron selectively unavailable. Although most siderophores are either hydroxamates or catechols, earlier tests based on such functional groups proved unreliable since they are absent from a few siderophores. Siderophores are classified on the basis of the chemical functional groups they use to chelate iron. Catecholate-type (phenolate) siderophores bind Fe^{3+} using adjacent hydroxyl groups of catechol rings. Fe^{3+} is chelated using nitrogen atoms of thiazoline and oxazoline rings in hydroxamate-type siderophores [84]. Ferrichrome is the classic Hydroxamate-type siderophore.

Streptomyces have been reported to produce hydroxamate type siderophore which inhibit phytopathogen growth by competing for iron in rhizosphere soils [85]. Siderophore exhibited considerable structural variability and affinity for iron, which determine the growth of microbe under competitive conditions when iron availability is a limiting factor. Competition for iron is considered a possible method for phytopathogen control [3]. In addition some plants growth promoting rhizobacteria increased growth by supplying the plant with sequestered iron. Microbial siderophores are used by plants as source of iron.

Streptomyces lydicus WYEC108 originally isolated from a rhizosphere soil of linseed was found to produce hydroxamate type siderophore [86]. Iron is found in proteins such as nitrogenase, ferredoxins, cytochromes and leghemoglobin all of which are important in symbiotic relationship between root nodules and *Frankia* spp. A nodulated legume has been found to have an increased demand for iron compared to that of non-nodulated plant [87].

Endophytic actinomycetes are of special interest since they possess many properties that could benefit to plant growth, producing siderophore to bind Fe^{3+} from the environment and help to improve nutrient uptake, supply of plant nutrients nitrogen.

3.5. Nitrogen Fixation

Diazotrophs ("nitrogen eaters") are the only organisms that are able to reduce N_2 to NH_3 producing an enzyme, nitrogenase. Nitrogen fixation by symbiotic associations between soil bacteria belonging to actinomycetes and root system of a diversified group of woody dicotyledonous plants is less generally well known than that by the legume-*Rhizobium* symbiosis. The actinomycetes induced nodulation of these woody species and their role in nitrogen fixation have begun to be recognized as comprising one of the largest sources for biological fixation of atmospheric di-nitrogen.

Among endophytic actinomycetes, only *Frankia* sp. are able to fix atmospheric N_2 in non- leguminous woody plants by nodulation in roots similar to rhizobium in legume plants [88]. The non-legume angiosperms which bear nodular structures on their roots also fix nitrogen by a symbiotic endophyte, which is not fungal in nature, but is actinomycetes, which was explained in the monograph under the heading "Actinomyceten-Symbiosen"[89].

Tokala et al. [90] observed that the symbiotic relationship between *Streptomyces lydicus* WYEC108 strain and pea plants. Vegetative hyphae of *Streptomyces lydicus* WYEC108 colonizes the surface of emerging nodules and then, sporulate within root surface cell layers and it appears to develop a beneficial relationship with nodule bacteroids by promoting nodulation, enhances nodule growth, bacteroids differentiation and aids the bacteroids in assimilating iron and possibly other inorganic nutrients from the soil and the result is enhanced overall growth of the plant.

4. Secondary Metabolites Produced by Endophytic Actinomycetes

The novel actinomycetes strains are rich sources for drug discovery, as they are likely to contain new genes in theory and held promising for novel products, thus the chance of finding novel pharmaceutical bioactive compounds from endophytic actinomycetes is considerable. Endophytic actinomycetes associated with traditionally used medicinal plants especially of the tropics could be a rich source of functional metabolites [8]. Recent studies of the culturable endophytic actinomycetes have resulted in the identification of many new natural products with diverse biological activities.

4.1. Antibiotics

Many endophytic actinomycetes, especially those from medicinal plants possess the ability of inhibiting or killing a wide variety of harmful microorganisms like pathogenic bacteria, fungi and viruses. Thus, there is great application value to develop antimicrobial drugs from endophytic actinmycetes. The actinomycetes are noteworthy as antibiotic producers, making three quarters of all known products, the *Streptomyces* are especially prolific and can produce a great many antibiotics and other class of biologically active secondary metabolites. Approximately 80% of world's antibiotics are produced by actinomycetes, mostly by the genus *Sterptomyces* sp. and *Micromonospora* sp. [91]. Microbial metabolites which have application as antibiotics are synthesized from intermediates of primary metabolites such as amino acids, aliphatic acids, sugars, nucleotides and lipids. The most important groups of secondary metabolites which have found application as lead molecules are the peptides and polyketides.

A lot of new antibiotics have been isolated, such as munumbicins A-D [23], celastramycins A-B [24], kakadumycins [92] and demethylnovobiocins [5]. From the culture broth of endophytic strain *Streptomyces* spp. TP-A0595, a simple compound 6- prenylindole was isolated and exhibited significant antifungal activity against plant pathogen *Fusarium oxysporum*. 6- Prenylindole was originally isolated from the plant liverwort (Hepaticae), and this is an additional example of the isolation of the same compound from plant and endophyte [93]. Two novel compounds cedarmycins A and B were isolated from the strain *Streptomyces* spp. TP-A0456, which was isolated from a twig of cedar. Cedarmycins A showed in vitro antifungal activity against *Candida glabrata* with the MIC value of 0.4 μg/ml [93]. An endophytic *Streptomyces* spp. Tc022 isolated from roots of *Alpinia galanga* strongly inhibited *Colletotrichum musae* and *Candida albicans*. Extraction of the culture medium of *Streptomyces* spp. Tc022 afforded a major component actinomycin D, which displayed very strong antifungal activity [94]. Castillo et al [95] isolated two new chromophoric peptides antibiotics, designated munumbicins E-4 and E-5 from endophytic *Streptomyces* NRRL 30562, which also produced broad-spectrum antibiotics munumbicins A-D. Both compounds showed broad-spectrum activity against gram-positive and gram-negative bacteria. Recently, a new antimycotic compound saadamycin was isolated from endophytic *Streptomyces* spp. Hedaya48, and it exhibited significant antimycotic activity against dermatophytes and other clinical fungi [96]. These studies listed above, reinforced the assumption that endophytic actinomycetes could be a promising source of antimicrobial substances.

4.2. Enzymes

Actinomycetes can produce a variety of extracellular hydrolytic enzymes such as cellulase, chitinase, amylase which can be applied for agriculture [97]. Using a plate assay, the *S. griseoflavus*P4 was found to be capable of producing amylase enzymes. Some

Streptomyces strains are known to produce hydrolytic cell wall degrading enzymes such as hemicellulases, and glucanases [98]. Pronase is a proteolytic enzymes produced by *Streptomyces gratiae* shown to have mixture of several exopeptidases and endopeptidases valued for analytical processing [99]. *Streptosporangium* sp. isolated from leaves of maize was reported to produce glucoamylase, which is expected to improve industrial processes of starch degradation. [100]. The production of L-asparaginase has been studied in actinomycetes [101]. It was reported by El-Tarabily [102] that isolates of *Microbispora rosea, Micromonospora chalcea* and *Actinoplanes philippinensis* produced β-1,3, β -1,4 and β -1,6-glucanases *in vitro*, caused lysis of *Phytophthora aphanidermatum* hyphae and reduced damping-off disease of cucumber under controlled glasshouse conditions. These isolates were able to suppress damping-off in soil amended with or without cellulose. Endophytic actinomycetes from the genera *Streptomyces, Streptosporangium* and *Nocardiopsis* were studied for the production of enzymes related to starch hydrolysis. Endophytic actinomycete *Streptomyces* spp. shows chitinase production and the potential of this for control of phytopathogenic fungi [103].

4.3. Antitumour and Anti-Inflammatory Agents

The endophytic actinomycetes are promising pharmaceutical resources for searching the antitumour agents [104]. Endophytic actinomycetes live in close association with their host plants and the long co-evolution relationship, there is a real possibility that genes involved in natural products biosynthesis could be exchanged via horizontal gene transfer (HGT) between microbes and plants, resulting in production of plant-derived compounds by a microbe such as the paclitaxel-producing *Kitasatospora* sp. isolated from *Taxus baccata* in Italy. This is the first report of production of taxol from endophytic actinomycetes. Another example is themaytansinoids (19-membered macrocyclic lactams related to ansamycin antibiotics), extraordinarily potent antitumor agents that were originally isolated from members of the higher plant families Celastraceae, Rhamnaceae and Euphorbiaceae [105], as well as some mosses [106], and, remarkably, from plant-associated actinomycete *Actinosynnema pretiosum* [107]. One novel chlorine-containing ansamycin, namely naphthomycin K was isolated from the endophytic strain *Streptomyces* spp. CS of the maytansinoids producer medicinal plant *Maytenus hookeri*. It showed evident cytotoxic activity against P388 and A-549 cell lines at IC50 0.07 and 3.17 μM, but no inhibitory activities against *Staphylococcus aureus* and *Mycobacterium tuberculosis* [108]. Interestingly, ansacarbamitocins, a new family of maytansinoids, were reported from a soil actinomycete strain *Amycolatopsis* CP2808 which belongs to the family *Pseudonocardiaceae*, the same family as the ansamitocin-producing strain *A. pretiosum* [109]. Strain *Streptomyces* spp. CS was an outstanding bioactive compounds producer. Previously, 24-demethylbafilomycin C1, a new member of the bafilomycin subfamily and two more new bafilomycin derivatives were isolated from the this strain [110]. During the subsequent research, five new 16-membered macrolides, belonging to the bafilomycin subfamily were isolated and showed cytotoxic activity against MDA-MB-435 cell line in vitro [111]. From the strain *Streptomyces* sp. ls9131, which was also isolated from *M. hookeri*, two novel macrolides were found. Bioassay results showed that the compound dimeric had strong antineoplastic activity and antibacterial activity [112]. Two novel anthraquinones, lupinacidins A and B were isolated from the fermentation broth of a new endophytic actinomycete *Micromonospora lupine*. Lupinacidins were found to show significant inhibitory effects on the invasion of murine colon 26-L5 carcinoma cells [113]. Kim et al [114] isolated two new 6-alkylsalicylic acids, salaceyins A and B from strain *Streptomyces laceyi* MS53 Salaceyins A and B exhibited cytotoxicity against a human breast cancer cell line. A new cytotoxic compound, pterocidin, isolated from the endophytic *Streptomyces hygroscopicus* TP-A0451, showed cytotoxicity against some human cancer cell lines with IC50 values of 2.9–7.1 μM [115]. Recently, two compounds 5, 7-dimethoxy-4-phenylcoumarin and 5, 7-dimethoxy-4-p-methoxylphenylcoumarin, originally produced by numerous species of plants were isolated from endophytic *Streptomyces aureofaciens* CMUAc130 and showed to have antifungal and antitumor activity [116,117]. They were also investigated for their effects not only on the formation of nitric oxide [NO], prostaglandin E2 (PGE2) and tumour necrosis factor (TNF-α), but also on inducible nitric oxide synthase and cyclooxygenase-2 in lipopolysaccharide (LPS)-induced murine macrophage RAW 264.7 cells.

The inhibitory effects were shown in concentration dependent manners. They significantly reduced the formation of TNF-α. These findings support the application of them as anti-inflammatory agent [118]. From the endophytic strain *Streptomyces* sp. SUC1, four novel secondary metabolites, lansai A–D were isolated. Lansai B showed weakly active against the BC cell line [119], lansai C also showed significant anti-inflammatory activity in LPS-induced RAW 264.7 cells [120]. Huang et al [104] reported that 41.1% of the endophytic actinomycetes obtained from 12 tropical medicinal plants showed activity against liver cancer cell Smmc-7721.These studies indicate that endophytic actinomycetes still remain a relatively untapped source of novel natural products, presumed to push forward the frontiers of drug discovery.

4.4. Other Pharmaceutical Compounds from Endophytic Actinomycetes

Endophytic actinomycetes also produced other bioactive compounds with diverse functions different from those discussed above. An endophytic *Streptomyces* sp. (AC-2), isolated from a traditional Chinese medicinal plant *Cistanche deserticola* Y. C. Ma, produced tyrosol. This compound can promote an increase of intracellular cAMP special on GPR12 transfected cells, such as CHO and HEK293, might be a new possible ligand for GPR12 [121]. A novel peptide compound coronamycin was purified from a verticillate *Streptomyces* sp. (MSU-2110) colonizing an epiphytic vine, Monstera sp. It exhibited remarkable active against the malarial parasite, *Plasmodium falciparum*, with an IC50 of 9.0 ng/ml. In addition, showed strong activity against pythiaceous fungi and the human fungal pathogen *Cryptococcus neoformans* [122]. These natural bioactive metabolites with multifold applications expand utilization of actinomycetes. Endophytic actinomycetes in higher plants, particularly those associated with medicinal plants and being used traditional medicines, are more likely to contain bioactive compounds. This principle is supported by the examples of taxol and maytansinoids as discussed. Therefore to search for novel compounds could be directed towards plants that commonly serve indigenous populations for medicinal purposes and plants growing in unique environmental setting or interesting endemic locations as they are expected to harbour novel endophytes that may produce unique metabolites with diversified applications.

5. Biocontrol of Plant Pathogens

Endophytic actinomycetes are considered to be potential biocontrol agents as they can colonize the interior of the host plant avoiding competition by any other microbes in the soil. Endophytic actinomycetes were found to produce at least three kinds of antagonistic substances in plant tissues, including antibiotics, enzymes and siderophores. Mechanisms that have been proposed to explain how some microorganisms control plant diseases caused by fungal pathogens include mycoparasitism and antibiosis [123]. Mycoparasitism involves the production of extracellular enzymes that hydrolyse fungal cell walls, whereas antibiosis involves the production of secondary metabolites in the rhizosphere. These metabolites either inhibit the growth and/or differentiation of fungal pathogens. They have been shown to protect plants against different soil-borne plant pathogens, including *Rhizoctonia solani*, *Verticillium dahliae*, *Plectosporium tabacinum*, *Gaeumannomyces graminis* var. *tritici*, *F. oxysporum*, *Pythium aphanidermatum* and *Colletotrichum orbiculare* [124, 125, 126, 127].

Coombs [128] have tested around 60 actinomycete species isolated from the root of wheat and identified some species which can kill the fungus responsible for „take-all" disease. These species can reduce the impact of take-all on wheat by upto 70%. Taechowisan and Lumyong [97] isolated 59 endophytic actinomycets from the roots of *Zingiber officinale* and *Alpinia galanga* and tested their antifungal activity against *Candida albicans, Colletotrichum musae, Fusarium oxysporum* and other *phytopathogenic* fungi. Among them *Streptomyces aureofoliene* showed the most intense antagonism *against* these test organisms. Studies on the mechanisms of action of these endophytes have focused mainly on the production of bioactive compounds, such as antibiotics, cell wall degrading enzyme and competition for nutrients.Shekhar et al. [129] purified a bioactive compound from endophytic *Streptomyces violaceusniger* that showed a strong antagonism towards various wood-rotting fungi, and chitinase enzymes were associated with this inhibition. In general, the higher chitinase activity was correlated with higher fungal inhibition. For this reason, chitinolytic *Streptomyces* strains are a likely choice as potential biological control agents.

A commercial product containing *Streptomyces griseoviridis* mycostop was applied through the irrigation system to control important plant pathogens such as *Fusarium oxysporum*, *Botrytis cinera* and *Alternaria brassicicola* [130]. A detailed biocontrol effect of endophytic actinomycetes both in *vitro* and *in planta* was given by Coombs et al [14]. They isolated 38 strains of endophytic actinomycetes from surface sterilized wheat and barley roots and tested antagonistic activity to *Gaeumannomyces graminis, Rhizoctonia solani* and *Pythium* spp., which are all pathogenic to wheat root, and found that 17 of the isolates, displayed statically significant activity *in planta* against *Gaeumannomyces graminis* and some of them were able to control the development of disease symptoms in treated wheat plants. A total of 619 actinomycetes were isolated from the roots of tomato. All of them were *Streptomyces* spp. Many streptomycete antagonists showed the ability to degrade the *R. solanacearum* cell wall, but the proportion of strains from wilting plants was higher than that from healthy plants. Probably the density of pathogen increased when the plants became wilting, so the bacterial cell wall-degrading activity of endophytic *Streptomyces* would help them overcome the competition from pathogens. Fifty percent of all the isolated endophytic actinomycetes from rice showed antagonism against at least one rice pathogen. Fifty five percent of all the isolated *S. griseofuscus* have antagonism against rice pathogens. Sixty percent of all the isolated *S. hygroscopicus* and 50% of all the isolated *S. globisporus* showed strong antagonism. These endophytic actinomycetes may play an important role in protecting the plant host against pathogenic microorganisms. Some metabolites of endophytic actinomycetes can strongly inhibit rice pathogens and may be a resource for new fungicides [131]. In addition, endophytic actinomycetes have also been found to hold the ability of triggering plant induced systemic resistance (ISR). The 'versatile' endophytic strain *Streptomyces galbus* R-5 not only released cellulose, pectinase, produced actinomycin X2 and fungichromin to induce resistance in the rhododendron seedlings, but also triggered plant jasmonate-associated defence responses [132]. Conn et al. [133] observed that seed inoculations with endophytic *Streptomyces* sp. EN27 and *Micromonospora* sp. strain EN43 led to increased resistance in Arabidopsis thaliana leaves against pathogens, *Erwinia carotovora* and *F. oxysporum*, and triggered the expression of defence genes related to SA- or jasmonic acid/ethylene-dependent signalling pathways in the absence of a pathogen. Moreover, culture filtrates obtained from cells of *Micromonospora* sp. strain EN43 grown in either minimal or rich medium activated the system acquired resistance and jasmonates/ethylene pathways, respectively, indicating that two different sets of metabolites are synthesized by this endophytic actinomyctes for eliciting the plant defense pathway [134]. At present, whether or not these metabolites are antibiotics or other molecules is not known. The intimate association of endophytic actinomycetes with plants and born bioactive natural products offers a unique opportunity to find effective drugs or biofungicides for potential application in plant protection and biological control.

6. Future Perspectives

The microbial natural products are particularly the ones, produced by the members of the order *Actinomycetales*, will continue to represent an important route to the discovery of novel classes of bioactive compounds. As a result, the search for and discovery of lesser-known and/or novel actinomycetes is of significant interest to the industry due to a growing need for the development of new and potent therapeutic agents, mainly against drug resistant bacteria [135]. New discoveries, however, will only stem from a sound understanding and interpretation of knowledge derived from conventional studies conducted since the discovery of streptomycin, on the ecology, taxonomy, physiology and metabolism of actinomycetes, and from a combination of this knowledge with currently available and continuously advancing molecular tools. Such a powerful information platform will then inevitably reveal the whereabouts, taxonomical and chemical identities of previously undetected bioactive actinomycetes including novel species of *Streptomyces* as potential producers of novel drug candidates. Recent applied metagenomic approaches, genome scanning and whole genome sequence mining [136] provide efficient ways for screening gene cluster involved in the biosynthesis of pharmaceutical metabolites and revelation of biosynthesis mechanisms from undiscovered endophytic actinobacteria. Comprehensive cooperation among global taxonomists, ecologists, natural product chemists, agronomists and bioengineers is also suggested to better exploit their biodiversity and biotechnological potential [136]. There is increasing interest in developing the potential biotechnological applications of endophytic actinomycetes for improving phytoremediation and the sustainable production of nonfood crops for biomass and biofuel production.

References

[1] Bacon CW, White J F (2000) *Microbial endophytes*. Marcel Dekker, New York.

[2] Strobel G and Daisy B (2003) Bioprospecting for microbial endophytes and their natural products microbial. *Microbiol. Mol. Biol. Rev*. 67:491-502.

[3] Cao LX, Qiu ZQ, You JL, Tan HM, Zhou S (2005) Isolation and characterization of endophytic streptomycete antagonists of fusarium wilt pathogen from surface-sterilized banana roots. *FEMS Microbiol. Lett*. 247:147-152.

[4] Verma VC, Gond SK, Kumar A, Mishra A, Kharwar RN, Gange AC (2009) Endophytic actinomycetes from *Azadirachta indica* A. Juss: Isolation, diversity and anti-microbial activity. *Microb. Ecol*. 57:749-56.

[5] Igarashi Y, Iida T, Yoshida R and Furumai T (2002) Pteridic acids A and B, novel plant growth promoters with auxin-like activity from *Streptomyces hygroscopicus* TP-A0451. *J. Antibiot*. 55:764-67.

[6] Lee S, Flores-Encarnacion M, Contreras-Zentella M, Garcia-Flores L, Escamilla JE, Kennedy C (2004) Indole-3-acetic acid biosynthesis is deficient in *Gluconacetobacter diazotrophicus* strains with mutations in Cytocrome C biogenesis genes. *J. Bacteriol*. 186:5384-91.

[7] Hasegawa S, Meguro A, Shimizu M, Nishimura T, Kunoh H (2006) Endophytic actinomycetes and their interactions with host plants. Actinomycetologica 20:72-81

[8] Strobel G, Daisy B, Castillo U, Harper J (2004) Natural products from endophytic microorganisms. *J. Nat. Prod*. 67:257-268.

[9] Benimeli CS, Fuentes MS, Abate CM, Amarosoa M J (2008) Bioremediation of Lindanecontaminated soil by Streptomyces sp. M7 and its effects on Zea mays growth. *Int. Biodeter. Biodegr*. 61:233-239.

[10] Hamdali H, Hafidi M, Virolle M J and Ouhdouch Y (2008) Rock phosphate-solubilizing actinomycetes screening for plant growth promoting activities. *World J. Microbiol. Biotechnol*. 24:2565-2575.

[11] Stone JK, Bacon CW, White JF (2000) An overview of endophytic microbes: endophytism defined. In: Bacon CW, White JF (eds) *Microbial endophytes*. Marcel Dekker Inc., New York, pp 3-29.

[12] Benson DR, Silvester WB (1993) Biology of *Frankia* strain, actinomycetes symbionts of actinorrhizal plants. *Microbiol. Rev*. 57:293-319.

[13] Zinniel DK, Lambrecht P, Harris NB, Feng Z, Kucznarski D, Highley P, Ishimaru CA, Arunakumari A, Barletta RG, Vidaver AK (2002) Isolation and characterization of endophytic colonizing bacteria from agronomic crops and prairie plants. *Appl. Environ. Microbiol*. 68:2198-2208.

[14] Coombs J T, Michelsen P P and Franco C M M (2004) Evaluation of endophytic actinobacteria as antagonists of *Gaeumannomyces graminis* var. *tritici* in wheat. *Biol. Cont*. 29:359-366.

[15] Wakelin SA, Warren RA, Harvey PR, Ryder MH (2004) Phosphate solubilization by *Penicillium* spp. closely associated with wheat roots. *Biol. Fert. Soils* 40:36-43.

[16] Costa J M and Loper J E (1994) Characterization of siderophore production by the biological- control agent *Enterobacter cloacae*. *Mol. Plant. Microbe Interact*. **7**:440-48.

[17] Jain PK, Jain PC (2007) Isolation, characterization and antifungal activity of *Streptomyces sampsonii* GS 1322. *Ind. J. Exp. Biol*. 45:203-206.

[18] Chamberlain K, Crawford DL (1999) *In vitro* and *in vivo* antagonism of pathogenic turfgrass fungi by *Streptomyces hygroscopicus* strains YCED9 and WYE53. *J. Ind. Microbiol. Biotechnol.* 23:641-646.

[19] Karthikeyan B, Jaleel CA, Lakshmanan GMA, Deiveekasundaram M (2008) Studies on rhizosphere microbial diversity of some commercially important medicinal plants. *Colloids Surf.* B 62:143-45.

[20] Strobel GA, Ford E, Li JY, Sears J, Sidhu RS, Hess WM (1999) *Seimatoantlerium tepuiense* gen. nov., a unique epiphytic fungus producing taxol from the Venezuelan Guyana. *Syst. Appl. Microbiol*. 22:426-33.

[21] Taechowisan T, Peberdy JF, Lumyong S (2003) Isolation of endophytic actinomycetes from selected plants and their antifungal activity. *World J. Microbiol. Biotechnol.* 19:381-385.

[22] El-Shatoury S, Abdulla H, El-Karaaly O, El-Kazzaz W, Dewedar A (2006) Bioactivites of endophytic actinomycetes from selected medicinal plants in the world heritage site of Saint Katherine, Egypt. *Int. J. Bot*. 2:307-12.

[23] Castillo UF, Strobel GA, Ford EJ, Hess WM, Porter H, Jensen JB, Albert H, Robinson R, Condron MA M, Teplow DB, Stevens D, Yaver D (2002) Munumbicins, wide spectrum antibiotics produced by *Streptomyces* NRRL 30562, endophytic on *Kennedia nigriscans. Microbiol.* 148:2675-2685.

[24] Pullen C, Schmitz P, Meurer K, Bamberg DD, Lohmann S, Franc SDC, Groth I, Schlegel B, Möllmann U, Gollmick F, Gräfe U, Leistner E (2002) New and bioactive compounds from *Streptomyces* strains residing in the wood of *Celastraceae*. *Planta.* 216:162-67.

[25] Kumar U, Singh A, SivaKumar T (2011) Isolation and screening of endophytic actinomycetes from different parts of *Emblica officinalis*. *Ann. Biol. Res.* 2:423-434.

[26] Fuentes MS, Benimeli CS, Cuozzo SA, Amoroso MJ (2010a) Isolation of pesticide-degrading actinomycetes from a contaminated site: Bacterial growth, removal and dechlorination of organochlorine pesticides. *Int. Biodeter. Biodegr*. 64:434-441.

[27] Provorov NA, Borisov AY, Tikhonovich IA (2002) Developmental genetics and evolution of symbiotic structures in nitrogen-fixing nodules and arbuscular mycorrhiza. *J. Theor. Biol.* 214:215-232.

[28] Zhang HW, Song YC, Tan RX (2006) Biology and chemistry of endophytes. *Nat. Prod. Rep*. 23:753-771.

[29] Hallmann J, Berg G, Schulz B (2006) Isolation procedures for endophytic microorganisms. In: Schulz BJE, Boyle CJC, Sieber TN (eds) *Microbial. Root endophytes*. Springer, New York, pp 299-314.

[30] Coombs JT, Franco CM (2003) Isolation and identification of actinobacteria from surface sterilized wheat roots. *Appl. Environ. Microbiol*. 69:5603-5608.

[31] Qin S, Xing K, Jiang JH, Xu LH, Li WJ (2011) Biodiversity, bioactive natural products and biotechnological potential of plant- associated endophytic actinobacteria. *Appl. Microbiol. Biotechnol.* 89:457-473.

[32] Cao L, Qiu Z, You J, Tan H, Zhou S (2004) Isolation and characterization of endophytic *Streptomyces* strain from surface sterilized tomato (*Lycopersicon esculentum*) roots. *Lett. Appl. Microbiol.* 39:425-30.

[33] Fialho de Oliveira M, Germano da Silva M, Van Der Sand ST (2010) Anti-phytopathogen potential of endophytic actinobacteria isolated from tomato plants (*Lycopersicon esculentum*) in southern Brazil, and characterization of *Streptomyces* sp. R18(6), a potential biocontrol agent. *Res. Microbiol.* 161:565-572.

[34] Qin S, Li J, Zhao GZ, Chen HH, Xu LH, Li WJ (2008a) *Scharopolyspora endophytica* sp. nov., an endophytic actinomycete isolated from the root of *Maytenus austroyunnanensis*. *Syst. Appl. Microbiol.* 31:352-357.

[35] Qin S, Wang HB, Chen HH, Zhang YQ, Jiang CL, Xu LH, Li WJ (2008b) *Glycomyces endophyticus* sp. nov., an endophytic actinomycete isolated from the root of Carex baccans Nees. *Int. J. Syst. Evol. Microbiol.* 58:2525-2528.

[36] Li J, Zhao GZ, Qin S, Zhu WY, Xu LH, Li WJ (2009d) *Streptomyces sedi* sp. nov., isolated from surface-sterilized roots of Sedum sp. *Int. J. Syst. Evol. Microbiol.* 59: 1492-1496.

[37] Hayakawa M (1990) Selective isolation methods and distribution of soil actinomycetes. *Actinomycetologica* 4:103-112.

[38] Shirling EB, Gottlieb D (1966) Methods for characterization of *Streptomyces* species. *Int. J. Syst. Bacteriol.* 16:313-340.

[39] Vickers JC, Williams ST, Ross GW (1984) A taxonomic approach to selective isolation of streptomycetes from soil. In: Ortiz-Ortiz L, Bojalil LF, Yakoleff V (eds) *Biological, biochemical and biomedical aspects of actinomycetes*. Academic, Orlando, pp 553-561.

[40] Küster E, Williams ST (1964) Media for the isolation of *Streptomycetes*: starch casein medium. *Nature* 202:928-929.

[41] Qin S, Li J, Chen HH, Zhao GZ, Zhu WY, Jiang CL, Xu LH, Li WJ (2009b) Isolation, diversity, and antimicrobial activity of rare actinobacteria from medicinal plants of tropical rain forests in Xishuangbanna, China. *Appl. Environ. Microbiol.* 75:6176-6186.

[42] Li J, Zhao GZ, Huang HY, Qin S, Zhu WY, Li WJ (2009a) *Kineosporia mesophila* sp. nov., isolated from the surfacesterilized stem of *Tripterygium wilfordii*. *Int. J. Syst. Evol. Microbiol.* 59:3150-3154.

[43] Okazaki T (2003) Studies on actinomycetes isolated from plant leaves. In: Kurtböke DI (ed) Selective isolation of rare actinomycetes. *Queensland Complete Printing Service,* Australia, pp 102-121.

[44] Nejad P, Johnson PA (2000) Endophytic bacteria induce growth promotion and wilt disease suppression in oilseed rape and tomato. *Biol. Cont.* 18:208-215.

[45] Araujo WL, Marcon J, WJr M, Van Elsas JD, JWVan V, Azevedo JL (2002) Diversity of endophytic bacterial populations and their interaction with *Xylella fastidiosa* in citrus plants. *Appl. Environ. Microbiol.* 68:4906-4914.

[46] Surette MA, Sturz AV, Lada RR, Nowak J (2003) Bacterial endophytes in processing carrots (Daucus carota L. var. sativus): their localization, population density, biodiversity and their effects on plant growth. *Plant Soil* 253:381-390.

[47] Sessitsch A, Reiter B, Berg G (2004) Endophytic bacterial communities of field-grown potato plants and their plant-growth promoting and antagonistic abilities. *Can. J. Microbiol.* 50:239-249.

[48] Tian XL, Cao LX, Tan HM, Han WQ, Chen M, Liu YH, Zhou SN (2007) Diversity of cultivated and uncultivated actinobacterial endophytes in the stems and roots of rice. *Microb. Ecol.* 53:700-707.

[49] Velazquez E, Rojas M, Lorite MJ, Rivas R, Zurdo-Pineiro JL, Heydrich M, Bedmar EJ (2008) Genetic diversity of endophytic bacteria which could be find in the apoplastic sap of medullary parenchym of the stem of healthy sugarcane plants. *J. Basic Microbiol.* 48:118-124.

[50] Zin NM, Sarmin NIM, Ghadin N, Basri DF, Sidik NM, Hess WM, Strobel GA (2007) Bioactive endophytic streptomycetes from the Malay Peninsula. *FEMS Microbiol. Lett.* 274:83-88.

[51] Yuan HM, Zhang XP, Zhao K, Zhong K, Gu YF, Lindstrom K (2008) Genetic characterisation of endophytic actinobacteria isolated from the medicinal plants in Sichuan. *Ann. Microbiol.* 58(4):597- 604.

[52] Zhao GZ, Li J, Huang HY, Zhu WY, Zhao LX, Tang SK, Xu LH, Li WJ (2010a) Pseudonocardia artemisiae sp. nov., a novel actinobacterium isolated from surface-sterilized Artemisia annua L. *Int. J. Syst. Evol. Microbiol.* doi:10.1099/ijs.0.021931-0.

[53] Janso JE, Carter GT (2010) Biosynthetic potential of phylogenetically unique endophytic actinomycetes from tropical plants. *Appl. Environ. Microbiol.* 76:4377-86.

[54] Tan HM, Cao LX, He ZF, Su GJ, Lin B, Zhou SN (2006) Isolation of endophytic actinomycetes from different cultivars of tomato and their activities against Ralstonia solanacearum in vitro. *World J. Microbiol. Biotechnol.* 22:1275-80.

[55] Hayakawa M (2008) Studies on the isolation and distribution of rare actinomycetes in the soil. *Actinomycetol.* 22:12-19.

[56] Lee SO, Choi GJ, Choi YH, Jang KS, Park DJ, Kim CJ, Kim JC (2008) Isolation and characterization of endophytic actinomycetes from Chinese cabbage roots as antagonists to Plasmodiophora brassicae. *J. Microbiol. Biotechnol.* 18:1741-1746.

[57] Qin S, Wang HB, Chen HH, Zhang YQ, Jiang CL, Xu LH, Li W J (2008) Glycomyces endophyticus sp. nov., an endophytic actinomycete isolated from the root of Carex baccans Nees. *Int. J. Syst. Evol. Microbiol.* 58:2525-28.

[58] Chen HH, Qin S, Lee JC, Kim CJ, Xu LH, Jiang CL, Li WJ (2009a) Streptomyces mayteni sp. nov., a novel endophytic actinomycete isolated from Chinese medicinal plant. *Ant. Leeuw.* 95:47-53.

[59] Conn VM, Franco CMM (2004) Analysis of the endophytec actinobacterial population in the roots of wheat (Triticum aestivum L.) by terminal restriction fragment length polymorphism and sequencing of 16S rRNA clones. *Appl. Environ. Microbiol.* 70: 1787-1794.

[60] Compant S, Clément C, Sessitsch A (2010) Plant growth-promoting bacteria in the rhizo- and endosphere of plants: their role, colonization, mechanisms involved and prospects for utilization. *Soil Biol. Biochem.* 42:669-678.

[61] Glick BR, Todorovic B, Czarny J, Cheng Z, Duan J, McConkey B (2007) Promotion of plant growth by bacterial ACC deaminase. *Crit. Rev. Plant Sci.* 26:227-242.

[62] Lynch J M, Wilson K L, Ousley M A and Whipps J M (1991) Response of lettuce to Trichoderma treatment. *Lett. Appl. Microbiol.* 12:59-61.

[63] Compant S, Duffy B, Nowak J, Clément C, Barka EA (2005) Use of plant growth-promoting bacteria for biocontrol of plant diseases: principles, mechanisms of action, and future prospects. *Appl. Environ. Microbiol.* 71:4951-4959.

[64] Kannan V, Sureendar R (2008) Synergistic effect of beneficial rhizosphere microflora in biocontrol and plant growth promotion. *J. Basic Microbiol.* 49:158-164.

[65] Sun Y, Cheng Z, Glick BR (2009) The presence of a 1- aminocyclopropane-1-carboxylate (ACC) deaminase deletion mutation alters the physiology of the endophytic plant growth promoting bacterium Burkholderia phytofirmans PsJN. *FEMS Microbiol. Lett.* 296:131-136.

[66] Nimnoi P, Pongsilp N, Lumyong S (2010) Endophytic actinomycetes isolated from Aquilaria crassna Pierre ex Lec and screening of plant growth promoters production. *World J. Microbiol. Biotechnol.* 26:193-203.

[67] Abdel-Fattah GM, Mohamedin AH (2000) Interaction between a vesicular-arbuscular mycorrhizal fungus (Glomus intraradices) and Streptomyces coelicolor and their effects on sorghum plants grown in soil amended with chitin of brawn scales. *Biol. Fertil. Soils* 32:401-409.

[68] Shahidi BGH, Fooladi MH, Mahdavi MJ, Shahghasi A (2004) Broadspectrim, a Novel Antibacterial from Streptomyces sp. *Biotechnol.* 3:126-130.

[69] Merckx R, Dijkra A, Hartog AD, Veen JAV (1987) Production of root-derived material and associated microbial growth in soil at different nutrient levels. *Biol. Fert. Soils* 5:126-132.

[70] Bhavdish N, Johri A, Sharma J, Virdi S (2003) Rhizobacterial diversity in India and its influence on soil and plant health. *Adv. Biochem. Eng. Biotechnol.* 84:49-89.

[71] Leinhos V, Vacek O (1994) Biosynthesis of auxins by phosphate solubilizing rhizobacteria from wheat and rye. *Microbiol. Res.* 149:31-35.

[72] Martens DA, Frankenberger WT Jr (1994) Assimilation of exogenous 2‟-14C-indole acetic acid and 3”-14C-tryptophan exposed to the roots of three wheat varieties. *Plant Soil* 166:281-90.

[73] El-Tarabily KA (2008) Promotion of tomato (Lycopersicon esculentum Mill.) plant growth by rhizosphere competant 1-amino-cyclopropane-1-carboxylic acid deaminase producing Streptomycete actinomycetes. *Plant Soil* 308:161-74.

[74] Merzaeva OV, Shirokikh IG (2010) The production of auxins by the endophytic bacteria of winter rye. *Appl. Biochem. Microbiol.* 46:44-50.

[75] Gangwar M, Rani S, Sharma N (2012a) Diversity of endophytic actinomyetes from wheat and its potential as plant growth promoting and biocontrol agents. *J. Adv. Lab. Res. Bio.* 3:15-23.

[76] Rodriguez H, Fraga R (1999) Phosphate solubilizing bacteria and their role in plant growth promotion. *Biotechnol. Adv.* 17:319-339.

[77] Oliveira MF (2003) *Identificação e caracterização de actinomicetos isolados de processo de compostagem.* Diss. Universid Ade Federal Do Rio Grande DoSul.

[78] Reyes I, Bernier L, Antoun H (2002) Rock phosphate solubilization and colonization of maize rhizosphere by wild and genetically modified strains of Penicillium rugulosum. *Microb. Ecol.* 44:39-48.

[79] Pathom-Aree W, Stach JE, Ward AC, Horikoshi K, Bull AT, Goodfellow M (2006) Diversity of Actinomycetes isolated from Challenger Deep sediment (10,898 m) from the Mariana Trench. *Extremo* 10:181-89.

[80] Gangwar M, Rani S, Sharma N (2012b) Investigating endophytic actinomycetes diversity from Rice for plant growth promoting and antifungal activity. *Int. J. Adv. Life Sci.* 1:10-21.

[81] Welch SA, Taunton AE, Banfield JF (2002) Effect of microorganism and microbial metabolites on apatite dissolution. *Geomicrobiol. J.* 19:343-67.

[82] Hinsinger P, Plassard C, Tang C, Jaillard B (2003) Origin of root mediated pH changes in the rhizosphere and their response to environmemtal constraints: a review. *Plant Soil* 248:43-59.

[83] Loper JE, Henkels MD (1997) Availability of iron to *Pseudomonas fluorescens* in rhizosphere and bulk soil evaluated with an ice nucleation reporter gene. *Appl. Environ. Microbiol.* 63:99-105.

[84] Crosa JH, Walsh CT (2002) Genetic assembly line enzymology of siderophore biosynthesis in bacteria. *Microbiol. Mol. Biol. Rev.* 66:223-49.

[85] Khamna S, Yokota A, Lumyong S (2009) Actinomycetes isolated from medicinal plant rhizosphere soil: diversity and screening of antifungal compound, indole-3-acetic acid and siderophore production. *World J. Microbiol. Biotechnol.* 25:649-655.

[86] Hamby MK (2001) *M. Sc thesis*. University of Idano. Moscow.

[87] Deryto M, Skorupska A (1992) Rhizobial siderophore as an iron source for clover. *Physiol. Plant.* 85:549-53.

[88] Knowlton S, Berry A, Torrey JG (1980) Evidence that associated soil bacteria may influence root hair infection of actinorhizal plants by Frankia. *Can. J. Microbiol.* 26:971-77.

[89] Bond G (1967) Fixation of nitrogen by higher plants other than legumes. *Ann. Rev. Plant Physiol.* 18:107-26.

[90] Tokala RK, Strap JL, Jung CM, Crawford DL, Salove DLA, Bailey JF, Morra MJ (2002) Novel plant microbe rhizosphere interaction involving Streptomyces lydicus WYEC108 and the pea plant (Pisum salivum). *Appl. Environ. Microbiol.* 68:2161-2171.

[91] Pandey B, Agrawal VP (2004) Studies on the antibacterial activity of actinomycetes isolated from the Khumbu region of Mt. Everest. *A paper presented in the International Conference on the Great Himalayas: Climate, Health, Ecology, Management and Conservation, Kathmandu.* January 12–15.Organized by Kathmandu University and the Aquatic Ecosystem Health and Management Society, Canada.

[92] Castillo U, Harper JK, Strobel GA, Sears J, Alesi K, Ford E, Lin J, Hunter M, Maranta M, Ge H, Yaver D, Jensen JB, Porter H, Robison R, Miller D, Hess WM, Condron M, Teplow D (2003) Kakadumycins, novel antibiotics from Streptomyces sp. NRRL 30566, an endophyte of Grevillea pteridifolia. *FEMS Microbiol. Lett.* 234:183–190.

[93] Igarashi Y (2004) Screening of novel bioactive compounds from plant-associated Actinomycetes. *Actinomycetologica* 18:63-66.

[94] Taechowisan T, Wanbanjob A, Tuntiwachwuttikul P, Taylor WC (2006) Identification of Streptomyces sp. Tc022, an endophyte in Alpinia galanga and the isolation of actinomycin D. *Ann. Microbiol.* 56:113-117.

[95] Castillo UF, Strobel GA, Mullenberg K, Condron MM, Teplow DB, Folgiano V, Gallo M, Ferracane R, Mannina L, Viel S, Codde M, Robison R, Porter H, Jensen J (2006) Munumbicins E-4 and E-5: novel broad-spectrum antibiotics from Streptomyces NRRL 3052. *FEMS Microbiol. Lett.* 255:296-300.

[96] El-Gendy MMA, EL-Bondkly AMA (2010) Production and genetic improvement of a novel antimycotic agent, saadamycin, against dermatophytes and other clinical fungi

from endophytic Streptomyces sp. Hedaya48. *J. Ind. Microbiol. Biotechnol.* 37: 831–841.

[97] Taechowisan T, Lumyong S (2003) Activity of endophytic actinomycetes from roots of Zingiber officinale and Alpinia galanga against phytopathogenic fungi. *Ann. Microbiol.* 53:291-298.

[98] Trejo-Estrada SR, Paszczynski A, Crawford DL (1998) Antibiotics and enzymes produced by the biological control agent Streptomyces violaceusniger YCED-9. *J. Ind. Microbiol. Biotechnol.* 21:81-90.

[99] Neihans GA, Kratzke RA, Froberg MK, Aeppli DM, Nguyen PL, Geradts J (1999) G1 checkpoint protein and p53 abnormalities occur in most invasive transitional cell carcinomas of the urinary bladder. *Br. J. Cancer* 80:1175-1184.

[100] Stamford TLM, Stamford NP, Coelho LC, Araunjo JM (2002) Production and characterization of a thermostable glucomylase from Streptosporangium sp. endophyte of maize leaves. *Bioresour. Technol.* 83:105-109.

[101] Narayana KJP,Prabhakar P,Vijayalakshmi M,Vekateswarlu Y, Krishna PSJ (2007) L-Asparaginase production by Streptomyces Albidoflavus. *Indian J. Microbiol.* 48:331-336.

[102] El-Tarabily KA, Sivasithamparam K (2006) Non-Streptomyces actinomycetes as biocontrol agents of soil-borne fungal plant pathogens and as plant growth promoters. *Soil. Biol. Biochem.* 38:1505-1520.

[103] Quecine MC, Araujo WL, Marcon J, Gai CS, Azevedo JL, Pizzirani-Kleiner AA (2008) Chitinolytic activity of endophytic Streptomyces and potential for biocontrol. *Lett. Appl. Microbiol.* 47:486-4891.

[104] Huang XL, Zhuang L, Lin HP, Li J, Goodfellow M, Hong K (2012) Isolation and bioactivity of endophytic filamentous Actinobacteria from tropical medicinal plants. *Afr. J. Biotechnol.* 11:9855-9864.

[105] Powel RG, Weisleder D, Smith CR, Kozlowski J, Rohwedder WK (1982) Treflorine, trenudine, and Nmethyltrenudone: nove maytansinoids tumour inhibitors containing two fused macrocyclic rings. *J. Am. Chem. Soc.* 104:4929-4934.

[106] Suwanborirux K, Chang CJ, Spjut RW, Cassady JM (1990) Ansamitocin P-3, a maytansinoid from Claopodium crispifolium and Anomodon attenuatus or associated actinomycetes. *Experientia* 46:117-120.

[107] Higashide E, Asai M, OotsuK TS, Kozay Y, Hasegawa T, Kishi T, Sugino Y, Yoneda M (1977) Ansamitocins, a group of novel maytansinoid antibiotics with antitumour properties from Nocardia. *Nature* 270:721-722.

[108] Lu CH, Shen YM (2007) A novel ansamycin, naphthomycin K from Streptomyces sp. *J. Antibiot.* 60:649-653.

[109] Snipes CE, Duebelbeis DO, Olson M, Hahn DR, Dent WH, Gilbert JR, Werk TL, Davis GE, Lee-Lu R, Graupner PR (2007) The ansacarbamitocins: polar ansamitocin derivatives. *J. Nat. Prod.* 70:1578-1581.

[110] Lu CH, Shen YM (2004) Two new macrolides produced by Streptomyces sp. CS. *J. Antibiot.* 57:597-600.

[111] Li J, Lu CH, Shen YM (2010) Macrolides of the bafilomycin family produced by Streptomyces sp. CS. *J. Antibiot.* doi:10.1038/ ja.2010.95.

[112] Zhao PJ, Fan LM, Li GH, Zhu N, Shen YM (2005) Antibacterial and antitumor macrolides from Streptomyces sp. ls9131. *Arch. Pharm. Res.* 28:1228-1232.

[113] Igarashi Y, Trujillo M E, Martínez-Molina E, Miyanaga S, Obata T, Sakurai H, Saiki I, Fujita T, Furumai T (2007) Antitumor anthraquinones from an endophytic actinomycete Micromonospora lupini sp. nov. *Bioorg. Med. Chem. Lett.* 17:3702-3705.

[114] Kim N, Shin JC, Kim W, Hwang BY, Kim BS, Hong YS, Lee D (2006) Cytotoxic 6-alkylsalicylic acids from the endophytic Streptomyces laceyi. *J. Antibiot.* 59:797-800.

[115] Igarashi Y, Miura S, Fujita T, Furumai T (2006) Pterocidin, a cytotoxic compound from the endophytic Streptomyces hygroscopicus. *J. Antibiot.* 59:193-95.

[116] Taechowisan T, Lu C, Shen Y, Lumyong S (2005) Secondary metabolites from endophytic Streptomyces aureofaciens CMUAc130 and their antifungal activity. *Microbiol.* 151:1691-1695.

[117] Taechowisan T, Lu CH, Shen YM, Lumyong S (2007a) 4- arylcoumarin inhibits immediate-type allergy. *Food Agric. Immunol.* 18:203-211.

[118] Taechowisan T, Lu CH, Shen YM, Lumyong S (2007b) Antitumor activity of 4-arylcoumarins from endophytic Streptomyces aureofaciensCMUAc130. *J. Cancer Res. Trer.* 3:86-91.

[119] Tuntiwachwuttikul P, Taechowisan T, Wanbanjob A, Thadaniti S, Taylor WC (2008) Lansai A–D, secondary metabolites from Streptomyces sp. SUC1. *Tetrahedron.* 64: 7583-7586.

[120] Taechowisana T, Wanbanjob A, Tuntiwachwuttikul P, Liu JK (2009) Anti-inflammatory activity of lansais from endophytic Streptomyces sp. SUC1 in LPS-induced RAW 264.7 cells. *Food Agric. Immunol.* 20:67-77.

[121] Lin ZJ, Lu XM, Zhu TJ, Fang YC, Gu QQ, Zhu WM (2008) GPR12 Selections of the metabolites from an endophytic Streptomyces sp. asociated with Cistanches deserticola. *Arch. Pharm. Res.* 31:1108–1114.

[122] Ezra D, Castillo UF, Strobel GA, Hess WM, Porter H, Jensen J B, Condron M A M, Teplow D B, Sears J, Maranta M, Hunter M, Weber B, Yaver D (2004) Coronamycins, peptide antibiotics produced by a verticillate Streptomyces sp. (MSU-2110) endophytic on Monstera sp. *Microbiol.* 150:785-793.

[123] Van Driesche RG, Bellows TS (1996) Biological Control. *Chapman and Hall U S A.* ISBN 0-414-02861-1.

[124] Krechel A, Faupel A, Hallmann J, Ulrich A, Berg G (2002) Potato associated bacteria and their antagonistic potential towards plantpathogenic fungi and the plant-parasitic nematode Meloidogyne incognita (Kofoid & White) Chitwood. *Can. J. Microbiol.* 48: 772– 786.

[125] El-Tarabily KA (2003) An endophytic chitinase-producing isolate of Actinoplanes missouriensis, with potential for biological control of root rot of lupine caused by Plectosporium tabacinum. *Aust. J. Bot.* 51:257–266.

[126] El-Tarabily KA, Nassar AH, GEStJ H, Sivasithamparam K (2009) Plant growth promotion and biological control of Pythium aphanidermatum, a pathogen of cucumber, by endophytic actinomycetes. *J. Appl. Microbiol.* 106:13–26.

[127] Shimizu M, Yazawa S, Ushijima Y (2009) A promising strain of endophytic Streptomyces sp. for biological control of cucumber anthracnose. *J. Gen. Plant. Pathol.* 75:27–36.

[128] Coombs JT (2002) Antibiotics for wheat. Science Now! *The national science forum,* 20–22 August, 2001. http://www.abc.net.au/science/news/stories/s351442.htm (viewed 24–3-06).

[129] Shekhar N, Bhattacharya D, Kumar D, Gupta RK (2006) Biocontrol of wood-rotting fungi with Streptomyces violaceusniger XL-2. *Can. J. Microbiol.* 52:805-808.

[130] Tuomi T, Heino M, Nordstrom K, Laakso S (2001) Fiber fractions from processing of barley in production and conservation of a biologic control agent. *Appl. Biochem. Biotechnol.* 94:135-145.

[131] Tian XL, Cao LX, Tan HM, Zeng QG, Jia YY, Han WQ, Zhou SN (2004) Study on the communities of endophytic fungi and endophytic actinomycetes from rice and their antipathogenic activities in vitro. *World J. Microbiol. Biotechnol.* 20:303-309.

[132] Shimizu M, Suzuki T, Mogami O, Kunoh H (2005) Disease resistance of plants induced by endophytic Actinomycetes. In: Tsuyumu S, Leach JE, Shiraishi T, Wolpert T (eds) *Genomic and genetic analysis of plant parasitism and defense*. APS, St. Paul, pp 292–293.

[133] Conn VM, Walker AR, Franco CMM (2008) Endophytic Actinobacteria induce defense pathways in Arabiodopsis thaliana. *Mol. Plant Microbe Interact.* 21:208-218

[134] Hirsch AM, Valdes M (2010) Micromonospora: an important microbe for biomedicine and potentially for biocontrol and biofuels. *Soil Biol. Biochem.* 42:536-542.

[135] Kurtböke Dİ (2012) Biodiscovery from rare Actinomycetes: an eco-taxonomical perspective. *Appl. Microbiol. Biotechnol.* 93:1843-52.

[136] Hornung A, Bertazzo M, Dziarnowski A, Schneider K, Welzel K, Wohlert SE, Holzenkämpfer M, Nicholson GJ, Bechththold A, Sussmuth RD, Vente A, Pelzer S (2007) A genomic screening approach to the structure-guided identification of drug candidates from natural sources. *Chem. biochem.* 8:757-766.

In: Microbes in Process
Editors: Neelam Garg and Abhinav Aeron
ISBN: 978-1-63117-127-7

Chapter 13

Biosurfactant Production from Agro-Industrial By-Products and Wastes

Atipan Saimmai[1,*] and Suppasil Maneerat[2]
[1]Faculty of Agricultural Technology, Phuket Rajabhat University, Phuket, Thailand
[2]Department of Industrial Biotechnology, Faculty of Agro-Industry, Prince of Songkla University, Hat Yai, Songkhla, Thailand

Abstract

Biosurfactants are a structurally diverse group of surface-active molecules synthesized by microorganisms. These amphiphilic compounds contain a hydrophobic and a hydrophilic moiety, and have the ability to reduce interfacial tension between different fluid phases. The interest in biosurfactants has increased considerably in recent years, as they are potential candidates for many commercial applications in the biomedical, pharmaceuticals, petroleum and food processing industries. Biosurfactants have several advantages over their synthetic counterparts including high ionic strength tolerance, high temperature tolerance, higher biodegradability and less toxicity and higher surface activity. In spite of the advantages, fermentation must be cost competitive with chemical synthesis surfactants, and many of the potential applications that have been considered for biosurfactants depend on whether they can be produced economically. In this review, we describe the current knowledge and latest advances in the search for cost effective renewable agro-industrial alternative substrates for their production.

1. Introduction

It has been generally known that most of the surfactant used in the commercial or industrial sector nowadays is produced by chemical substances especially petrochemical substances. Although surfactants give great benefits to many organizations in several industries such as food, cosmetics, oil recovery and medical treatment, a severely dangerous

* Tel.: +66 76215133, Fax: +66 76215133. E-mail: s4680108@hotmail.com.

consequence is the overall environmental damage they cause. The greater the surfactant produced by inorganic substances is continuously utilized; the more the environment is gradually damaged. Accordingly, this is enough reason for us to try to find new ways of making surfactants nature friendly. It is time to change the old materials for green-friendly ones by relying upon natural materials such as substrates. Serious study is needed to find the kinds of natural materials to be used as suitable sources of microorganisms. These mediums must be improved to give good growth and to effectively produce biosurfactants on a large scale. Not only do the biosurfactants possess the advantages of lower toxicity and biodegradability, they also have the effectiveness of a wider range of pH and temperature values than their counterparts [1, 2]. These are the reasons why this kind of attractive microbial compound is much better than the chemically synthesized compounds.

There are many kinds of problems in biosurfactant production, such as the small yield and high production cost of materials. The use of many kinds of waste generated from the processing of many agricultural products in processing plants could ease the high cost of biosurfactant production significantly. They could also help it to be produced on a large scale. Those wastes which include some parts of agricultural product such as peels, seeds and husks left during the process in the processing plants, are good substrates for the growth of microorganisms and biosurfactant production. Furthermore, there are also many kinds of by-products produced during the processing of agricultural products and some vegetable oils which are not suitable for human consumption. These can also be used to help reduce the cost of microbial surfactant production. All the useful materials remarked above can be used as cheap substrates for biosurfactant production; this proposal is reviewed in this paper.

2. Alternative Substrates for Biosurfactant Production to Overcome the High Cost of Production

The high cost of production is considered to be the main obstacle to biosurfactant production and the material used as substrate accounts for the high cost. However, currently new kinds of substrate which can resolve this problem are being continuously reported. These include some parts of agricultural production which leave unused material and by-products. Produced during agricultural processing, these can be used as alternative substrates for biosurfactant production to overcome the high costs of production. These are identified below.

2.1. Cheap Substrates from Unwanted Parts of Agricultural Products

Each year there are a million tons of unused remnants of agriculture products left in every country; these unwanted parts become waste which needs more management. It would be much better to make them available as material as low cost substrates for biosurfactant production. These parts of agricultural products are mostly remnants of tropical agronomic crops including: the peels of cassava; the hulls of soybean [3]; the peels of sugar beet [4]; the peels and the stalks of sweet potato and potato; the peels of sweet sorghum [5]; the stem husks of rice and wheat [6]; the bran and the straw of rice and wheat [6]; the bagasse of sugarcane and cassava; the pulp of coffee; and the husks of coffee. The use of these wastes has been reported as substrates for biosurfactant production [7] and they are listed with their advantages in Figure 1.

Source	Microorganism Strains	Products	Reference
Animal fat waste	*Candida bombicola*	Sophhorolipids	[35]
Babassu oil	*Candida lipolytica* IA 1055	Sophorolipids	[87]
Cashew apple juice	*Bacillus subtilis* LAM1005	Surfactin	[17]
Cashew apple juice / Pomace	*Pseudomonas aeruginosa*	Rhamnolipids	[13]
Cassava / Flour	*Bacillus* sp.	Biosurfactant	[33]
Cassava flour wastewater	*Bacillus subtilis* LB5a	Lipopeptide	[26], [73]
Cheese whey	*Streptococcus thermophilus* A or *Lactococcus lactis* 53	Glicolipid or glicoprotein	[79]
Citrus fruits / Peel	*Pseudomonas aeruginosa* MTCC 2297	Rhamnolipids	[9]
Curd whey and distillery wastes	*Pseudomonas aeruginosa* BS2	Rhamnolipids	[59]
Dairy wastewater	*Cryptococcus curvatus* ATCC20509	Sophorolipids	[61]
Dairy whey / Whey	*Pseudomonas aeruginosa* BS2	Rhamnolipids	[59]
Fish oil	*Pseudomonas aeruginosa* BYK-2 KCTC 18012P	Rhamnolipids	[86]
Ligoncellulosic waste	*Lactococcus lactis* 53	Biosurfactant	[79]
Mixed vegetable oils (Sunflower seed oil and oleic acid)	*Thermophhilic eubacterium* HB8	Rhamnolipids	[51]
Molasses waste	*Pseudomonas aeruginosa* GS3	Rhamnolipid	[88]
Oil refinery wastes	*Candida antarctica* and/or *Candida apicola*	Glycolipids	[32]
Okara waste	*Bacillus subtilis* NB22	Lipopeptide and iturin	[22-23]
Palm oil decanter cake	*Inquilinus limosus* KB3	Lipopeptide	[50]
Palm oil mill effluent	*Nevskia ramosa* NA3	Biosurfactant	[44]
Potato process effluents	*Bacillus subtilis*	Lipopeptide	[55] [56]
Rapeseed oil	*Pseudomonas* sp. DSM 2874	Rhamnolipids	[52]
Rice and wheat bran / Stem Husk	*Pseudomonas aeruginosa*	Biosurfacctant	[89]
Simple alcohol glycerol	*Pseudomonas aeruginosa*	Rhamnolipids	[28]
Soap stock waste	*Pseudomonas aeruginnosa* LB1	Rhamnolipids	[31]
Soy molasses waste	*Candida bombicola* ATCC 22214 and *Cryptococcus curvatus* ATCC 20509	Sophorolipids	[61]
Soybean oil	*Candida* sp. SY16	Mannosylerythritol lipid	[90]
Soybean oil refinery wastes	*Pseudomonas aeruginosa* AT10	Rhamnolipids	[91]
Soybean oil / Seeds	*Pseudomonas aeruginosa* PACL	Rhamnolipid	[3]
Soybean soapstock waste	*Pseudomonas aeruginosa* LBI	Rhamnolipid	[92]
Starchy waste	*Bacillus subtilis*	Surfactin	[56]
Sugar beet / Peel	*Pseudomonas* spp.	Rhamnolipid	[4]
Sugarcane bagasse /Stem Husk	*Pseudomonas aeruginosa*	Biosurfactant	[6]
Sunflower and soybean oil	*Pseudomonas aeruginosa* DS 10-129	Rhamnolipid	[27]
Sunflower oil	*Serratia marcescens*	Lipopeptide	[93]
Sunflower oil soapstock waste	*Pseudomonas aeruginosa* LBI	Rhamnolipid	[31]
Sweet potato / Peel	*Bacillus subtilis* B6-1	Lipopeptides	[94]
Sweet sorghum / Peel	*Bacillus subtilis* NB22	Lipopeptides	[23]
Turkish corn oil	*Candida bombicola* ATCC 22214	Sophorolipids	[95]
Vegetable fat waste	*Candida glabrata* UCP1002	Biosurfactant	[38]
Waste frying oils	*Pseudomonas aeruginosa* 47T2 NCBI 40044	Rhamnolipid	[36]

Figure 1. Wastes of Agricultural Products as Substrate for Biosurfactant Production.

Citrus fruit is one of the most valuable agricultural products which makes high profits in the international market and is mostly used to produce orange juice. Consequently, much of its unwanted peel and waste are left after the processing [8]. There was a report about using orange fruit peel as a sole carbon source for rhamnolipid production by *Pseudomonas aeruginosa* MTCC 2297 [9].

2.2. Cheap Substrates from Agricultural Product Processing Plants

There has been much research on using agro-industrial wastes from agricultural processing plants such as oil refinery wastes, molasses, starch-rich wastes, cassava waste water and distilled grape marc [10, 11, 12]. These unwanted wastes can be renewable low cost starting substrates in the production process of biosurfactant and can be used to produce it at an industrial level and on a large scale. Rice water and water from the processing of cereals, pulses and molasses are very good for use as useful substrates for the production of microbial surfactant. It was also reported that cashew apple juice can be utilized as substrates where *Pseudomonas aeruginosa* ATCC 10145 grew. It gave superb surfactants after peptone of 5.0 g/L and nutritive broth had been added as supplements. In addition, compared to other amino acid sources, surface tension during the fermentation was significantly reduced by 41% when the microorganism culture was performed in peptone-supplemented cashew apple juice [13]. In India, there have been several efforts in the production biosurfactant by using some valuable agro-industrial wastes as substrates.

2.2.1. Substrate of Cashew Apple Juice

Cashew apples, the by-product of the cashew industry, are mostly grown in tropical countries, especially Brazil. They can provide another practical substrate for the production of microbial surfactant because of their richness in reducing sugar, vitamins and minerals salts; they are also cheap (US $ 0.50/kg). These advantages make them attractive and easy for use as a culture medium [14, 15, 16]. Cashew apple juice and mineral mediums, with added peptone and nutritive broth, were utilized as substrates to produce surface active agents by *Pseudomonas aeruginosa* ATCC 10145. The reduction of surface tension of the medium confirmed that this kind of low cost substrate with its supplements could be utilized as a medium for microbial growth and biosurfactant production [13]. A successful feasibility study for producing microbial surfactant from clarified cashew apple juice by *Bacillus subtilis* LAM1008 was reported in 2009. A group of researchers from Brazil used a mineral medium containing clarified cashew apple juice (MM-CAJC) for the medium of *B. subtilis* LAM1008. The results indicated that the highest reduction of surface tension from the various combinations of supplements was obtained with the culture in MM-CCAJC, with added yeast extract.

Furthermore, the microbial surfactant (surfactin) produced showed great surface and emulsifying performances and a yield of 3.5 mg/L was gained when MM-CCAJCC, supplemented with yeast extract, was utilized [17]. In the same year there was another study on using cashew apple juice supplemented with a mineral medium (MM-CCAJ) to get biosurfactant by *B. subtilis* LAMI005 and a surfactin product of 123 mg/l was reported. These kinds of microbial surfactants gave a significant surface performance and emulsifying activity confirming the possibility of this substrate to produce biosurfactant.

2.2.2. Substrate of Okara Waste

In Japan, there have been reports of the use of okara to synthesize microbial surfactant. This kind of waste is the residue from ground soy beans when the extraction of the water from the extractable fraction used to produce soy milk and tofu is performed. Okara is an industrial waste from the processing plant and is mostly disposed by incineration [18]. It is reported that each year approximately 700,000 tons of okara are generated from the production of tofu in Japan alone [19]. About 1.1 kg of fresh okara could be obtained after each kilogram of soybeans is processed into soy milk. It consists of valued components, namely water (81.1%), protein (4.8%), fat (3.6%), starch and sugar (6.4%), fiber (3.3%), and ash (0.8%). It can thus be used as substrate for the simultaneous production of lipopeptide, surfactin and iturin in solid state fermentation by *Bacillus subtilis* NB22 [20-23].

2.2.3. Substrate of Simple Alcohol Glycerol

Simple alcohol glycerol has been claimed as a possibility for use as sole carbon and energy source for three producers of surface active agents. This kind of waste is the main byproduct obtained from trans-esterification of vegetable oils and animal fats [24]. Furthermore, glycerol is a component of lipids that are abundant in nature. It is known that many microorganisms could grow naturally using a sole carbon and energy source like glycerol. This substrate is usually used as a substitute for common carbohydrates, such as sucrose, glucose and starch [25]. As biodiesel production has increased continuously, it makes for more accumulation of glycerol much more as a byproduct of this industry. This is a good opportunity to use this cheap and valuable substrate as a practical medium for biosurfactant production. Low cost glycerol could be specifically used as a water soluble substrate in the production process. For example, it was reported that this kind of substrate was obtained as a sole carbon source in the production process of rhamnolipid by *Pseudomonas aeruginosa* [26]. Though the results showed that its yield was less than that of the traditional hydrophobic substrates, this low cost substrate can overcome this disadvantage in yield. There are many other reports about utilizing glycerol as a substrate for the production of attractive microbial surfactants as can be seen below.

A group of researchers claimed that they could gain 1.77 g/l rhamnolipid biosurfactant from the production by *P. aeruginosa* DS10-129 [27]. In 2005, 15.4 g/l of rhamnolipids was obtained by using *P. aeruginosa* cultivated in a basal mineral medium containing glycerol as the sole carbon source [28]. Basidiomycete yeast and *Pseudozyma antarctica* JCM 10317 were used to show that they could efficiently synthesize mannosylerythritol lipids (MELs), a glycolipid biosurfactant, by using glycerol as a substrate. Then 16.3 g/l MELs was obtained by the intermittent adding of glycerol [29].

A mineral medium formulated with glycerol (93%) was used to synthesize biosurfactant by *Pseudomonas aeruginosa* UCP0992. A product of 8.0 g/l with a surface tension reduction of 27.4 mN/m was achieved after 96 hours. In addition, the researchers obtained a high emulsification index value of 80% and a critical micelle concentration of 700 mg/l after 72 hours of microbial growth. This research indicates that the production of biosurfactant from glycerol is possible and there is potential for application in the environment [24]. Biodiesel-derived crude glycerol was used to produce glycolipids by the fungi, *Ustilago maydis*. The researchers learned that *U. maydis* is a superb host for the bioconversion of crude glycerol to add value to the products [30]. The research results above are enough to confirm that glycerol has suitable properties to be a good substrate or carbon source for the growth microorganism and biosurfactant production.

2.2.4. Substrate of Oil Waste

Oil wastes from the food industry and vegetable oil refineries can also be used as a good substrates for biosurfactant production. In addition, there are many other kinds of industrial oil wastes such as soap-stock, tallow lard and free fatty acids which are good for helping microorganisms to grow and generate metabolitic production. It has been found that the waste oils generated from vegetable oil refineries and soap industries can be used as substrate for the production of active microbial agents [31, 32, 33]. Apart from the oil wastes above, vegetable fat wastes and animal fat wastes can be suitable as cheap substrates [34, 35] and are also readily available in large quantities. Researchers found that *Pseudomonas aeruginosa* 47 T2 could give two important rhamnolipid homologs, Rha-C10-C10 and Rha-Rha- C10-C10 during cultivating in olive oil waste water or in waste frying oils obtained from olive/sunflower (50:50; v/v). The rhamnolipids produced were as much as 8.1 g/l [28, 36, 37]. These wastes can be considered useful alternative substrates and can help ease the high cost of biosurfactant production.

2.2.4.1. Substrate of Animal Fat Waste

Each year animal fat and tallow are produced in large amounts in meat processing plants throughout the world. It was found that they could be used as a valuable and cheap substrate for the biosurfactant production of sophorolipids by the yeast, *Candida bombicola* [35]. The researchers used fat as a sole carbon source and the results showed that the yeast developed quite poorly. However, after 10% of the glucose and 10% of the fat were mixed together, the highest level of growth was gained. This indicated that additional carbon sources in the medium are necessary. It was also reported that using the available abundant and inexpensive animal fat for the surface active agent production of sophorolipids by *C. bombicola* could obtain 120 g/l of the product [35].

2.2.4.2. Substrate of Vegetable Fat Waste

The first use of a vegetable fat waste as the substrate for the production of microbial surfactant was reported in 2010. This was after a group of researchers had investigated the use of vegetable fat waste to synthesize biosurfactant by *Candida glabrata.* A factorial design was applied to investigate the effects and the interactions of waste, yeast extract and glucose on the surface tension after 144 hours of cultivation. From the experiment, they obtained the maximum surface activity by using vegetable fat waste at 5% and yeast extract at 0.2%. Furthermore, the biosurfactant containing cell-free broth retained its surface-active performance after incubation at high temperatures (120°C) at a wide range of pH values [2-12] and salt concentrations (0-10%, w/v of NaCl) [38]. From a structural examination, it could be a mixture of carbohydrates, proteins and lipids and may be suitable for use in bioremediation and oil recovery.

2.2.4.3. Substrate of Olive Oil Mill Waste Effluent

Olive oil mill waste effluent, which can be obtained from the process of olive oil extraction, results in a large amount of black liquid waste that can be effectively used as substrates. It has a high content of organic matter (20-60 kg COD/m^3), depending upon the olive oil extraction procedure, which is superb for bacteria to grow in [39]. In addition, this kind of waste can be obtained easily in large amounts and especially in Mediterranean countries where more than 98% of the world's olive oil is produced. It is estimated that each year over 2.5 million metric tons of olive oil is produced in these countries and up to 75% is produced in the European Union alone. That means much byproduct is produced as well. This kind of byproduct is toxic [40], so human beings cannot consume it in a raw or processed form. However, it has useful organic substances such as residual oils, organic acids, nitrogen compounds and sugars which can support the growth of microorganisms very well.

The production of rhamnolipids biosurfactant by *Pseudomonas aeruginosa* 47T2 growing in olive oil mill effluent (OOME) as the sole carbon source was reported for the first time in 1993 [41]. From their work, the researchers obtained a conversion yield of 0.058 g of rhamnolipid per g of substrate (OOME).

It was also found that COD of OOME was reduced by about 50% in 72 h. A group of researchers from Turkey also used OOME as the sole carbon source for rhamnolipids production from *Pseudomonas* sp. Several strains of *Pseudomonas* sp. were able to grow on OOME as the substrate to produce rhamnolipids with a highest yield of 0.875 g of rhamnolipid per g of OOME. The COD of OOME was also reduced about 70% from the initial concentration after 72 h of cultivation [42].

2.2.5. Cheap Substrate from the Palm Oil Industry

The palm oil industry is another attractive source of material for good substrate for microbial surfactant production. Each year much unwanted waste is generated and disposed of from the palm oil processing plants in many countries, especially Malaysia, where many palm oil mills are located. Not only does the palm oil production process give palm oil products, it also generate three main kinds of waste namely palm oil mill effluent, palm oil sludge and palm oil decanter cake.

These organic wastes from the agricultural product processing plants cause severe environment problems. It would be much better if they were used them as an effective substrate for biosurfactant production.

2.2.5.1. Palm Oil Mill Effluent

Palm oil mill effluent (POME) is a colloidal suspension composted of 95-96% water, 0.6-0.7% oil and 4-5% total solids. These include 2-4% suspended solids originating in the mixing of sterilizer condensate separator sludge and hydrocyclone wastewater [43]. There has been a study on using POME as a new substrate for biosurfactant production by the newly isolated biosurfactant-producing strain *Nevskia ramose* NA3. Commercial sugar (CS), $FeCl_2$, glucose, K_2HPO_4, $MgCl_2$, and $NaNO_3$ were identified by the Plackett-Burman design as important parameters for improving biosurfactant production by this strain. $FeCl_2$ and $NaNO_3$ showed good effects, while CS, glucose, K_2HPO_4 and $MgCl_2$ had mainly an adverse effect. The results suggested that a concentration of $FeCl_2$ and $NaNO_3$ higher than 1.0 and 5.0 g/l, a CS and glucose concentration lower than 1 g/l, and a K_2HPO_4 and $MgCl_2$ concentration lower than 1.0 g/l, should be used for further improvement of biosurfactant production by *N. ramose* NA3. By using POME at 20% (v/v), *N. ramose* NA3 could reduce the surface tension of culture supernatant from 67.2 to 29.5 mN/m with 60% of emulsification activity after 30 h of cultivation [44].

2.2.5.2. Sludge Palm Oil

Sludge palm oil (SPO) is the floating residual oil separated during the initial stage of POME discharge. Its quality is low grade, so it is not suitable for wide application. However, for biosurfactant production, it is considered a novel substrate. It is composed of 83.5% triglycerides, 8% diglycerides, 0.5% monoglycerides and 8% free fatty acids. SPO can perform as a carbon source for microbial growth and biosurfactant production [45].

A study was made of using SPO as a substrate for biosurfactant production. The production medium was composed of (g/l): glucose; 5, meat extract; 0.5, K_2HPO_4; 2.2, KH_2PO_4; 0.14, $NaNO_3$; 0.5, NH_4NO_3; 3.3, $CaCl_2$; 0.04, NaCl; 0.04, $FeSO_4$; 0.2, $MgSO_4$; 0.6 and 2% (v/v) of SPO. Only 22 bacterial strains were used and during the centrifugation of the fermentation broth, and only a few strains produced microbial biomass in the form of a pellet. That means they could grow in the production medium supplied. The researchers used the growth of bacterial strains in the medium as a variable during the initial screening. This was because most of the biosurfactants are secondary metabolites and this increasing coincides with the growth of microorganisms. For example, the microbial growth on water-immiscible aliphatic hydrocarbon has been associated with the production of surfactant. It was also reported that several environmental factors influenced biosurfactant production by growing cells. In most studies the lowest value of surface tension was gained in the stationary phase [46, 47, 48]. The strains that grew very well in the production medium which had SPO alone could reduce the surface tension of fermentation broth to 36.2 mN/m.

2.2.5.3. Palm Oil Decanter Cake

Palm oil decanter cake is another kind of waste produced during the palm oil milling decantation process. It is considered as a low cost substrate from agricultural products and it can be collected in large amounts from several countries in Southeast Asia especially Malaysia and southern Thailand. Palm oil decanter cake is rich in oil residues, various vitamins and mineral elements which are very suitable for biosurfactant production [49].

Biosurfactant production by bacterium *Inquilinus limosus* KB3 growing in a mineral salts medium with palm oil decanter cake was reported on by a group of researchers in Thailand [50]. They found that *I. limosus* KB3 could grow and produce biosurfactant, lowering the surface tension in the medium broth to 28.2 mN/m. It gave a crude surfactant concentration of 5.13 g/l after 54 h of cultivation. It was also found that the biosurfactant could reduce the surface tension of pure water to 25.5 mN/m with the critical micelle concentration of 9 mg/l. It also retained its properties during exposure to elevated temperatures (121°C). This microbial compound was also able to form stable emulsions with various hydrocarbons and had the ability to enhance oil recovery, polyaromatic hydrocarbons solubility and antimicrobial activity.

2.2.6. Vegetable Oil

Several kinds of vegetable oils such as sunflower and soybean oil [27] could be used to process microorganisms to obtain biosurfactant. For example, it was reported that sunflower seed oil and oleic acid were used for producing rhamnolipids by *Thermus thermophilus* HB8.

In addition, it was found that the potential production of rhamnolipids had been induced by utilizing *Thermophhilic eubacterium* [51]. Palm oil, another substrate, which is found easily and abundantly, can be a very good raw material for producing simultaneous products of polyhydroxyalkanoates and rhamnolipids by *Pseudomonas aeruginosa.*

Rapeseed oil can be used to produce rhamnolipids and L - (+) - rhamnose by *Pseudomonas* sp. DSM 2874 [52]. Jatropha oil, mesua oil, castor oil, ramtil oil and jojoba oil are all plant-extracted oils that are not suitable for human consumption because of their unfavorable odor, color and toxic composition. However, they can be effectively used for biosurfactant production. Sunflower seed oil which is directly hydrolyzed by secretion of lipase from the microorganism can be utilized as a good carbon source for rhamnolipid production. Another study reported that *Tsukamurella* sp. DSM 44370 could give approximately 30 g/l of glycolipid when it grew in 110 g/l sunflower oil. It showed high surface and interfacial tension reduction and some antimicrobial activities against some bacteria and a fungal strain [53].

2.2.7. Cassava Wastewater

Cassava wastewater produced during the preparation of cassava flour at processing plants is another carbohydrate-rich residue generated in large amount. It is a good alternative substrate in fermentation process for biosurfactant production by *Bacillus subtilis* [10]. Cassava wastewater substrate can also be used for the simultaneous production of surfactants, polyhydroxyalkanoates and rhamnolipids by *Pseudomonas aeruginosa* [54].

There was a test of the use of five cassava flour wastewater preparations as culture mediums for biosurfactant production by a wild-type *Bacillus* sp. isolate. The results showed that the microbe could grow and produce biosurfactant in all preparations. Later, the same research group applied two *Bacillus subtilis* strains for biosurfactant production in cassava effluent. The results indicated that both *B. subtilis* ATCC 21332 and *B. subtilis* LB5a exhibited a good surface activity and produced similar yields of surfactin [26, 33]. In addition, this group investigated the production and properties of biosurfactant obtained from the *B. subtilis* LB5a strain with a cassava wastewater substrate. They found that the crude surfactant of lipopeptide with a concentration of 3.0 g/l could withstand exposure to elevated temperatures (100°C), high salinity (20% NaCI) and a wide range of pH values and form stable emulsions with various hydrocarbons. From the repots, we can conclude that this kind of waste could be effective as a suitable substrate for practical surface-active agent production [10].

2.2.8. Starchy Waste

A main source of starchy-waste material is potato processing plants. It was found that starchy waste can be another practical substrate in the fermentation process because this kind of byproduct is inexpensive and can be obtained easily. Potatoes are generally available with 80% water, 17% carbohydrates, 2% protein, 0.1% fat and 0.9 vitamins. They are a big source of carbon in the form of starch and sugars, nitrogen and sulfur from protein, inorganic minerals, trace elements and vitamins. There was a report on using high solids and low solids of potato effluents as substrates for surface active agent production by three formulas. These were: effluents diluted 1:10; not diluted; and diluted with trace minerals or corn steep liquor by using *Bacillus subtilis* 21332. The results showed that this strain could grow in all three kinds of formula which was not affected by any addition of exogenous nutrients. There was also a comparison between the growth rate of the microorganisms using all high solid and low solid mediums to that with the B-PS (biotic purified starch) control. The results showed that the growth rate of the two former ones is significantly higher than that of the latter [55]. In addition, the starchy waste could also be used to produce surfactant by *B. subtilis* ATCC 21332 [56].

A study was made on reducing the high cost in the biosurfactant production process by Das and Mukherjee [57]. The potato associated waste was used to produce biosurfactants by *B. subtilis* ATCC 21332. They compared growth, surface activity and carbohydrate utilization of the microorganism in established potato, simulated liquid and solid potato waste mediums and a commercially prepared potato starch in a mineral salts medium. The utilization of potato substrate and the production of surfactant was indicated by high surface tension reduction [58]. There was also the investigation into the efficiency of two *Bacilllus subtilis* strains for the production of biosurfactants in two fermentation systems using powdered potato peels as a substrate [57]. The researchers immersed potato peels in very hot water before drying them in the oven and they ground the dried peels into powder and stored them at 4°C before further use. The results showed that both of the fermentation processes gave lipopeptide biosurfactant with great surface activity and yield.

2.2.9. Dairy Wastewater

Dairy wastewater or dairy effluent is the whey from processing plants and can be a low cost substrate for biosurfactant fermentation as it is cheap. It has been found that much of whey from the organized dairy sector is not utilized and is being disposed through the effluent treatment systems. It is preferable if a large portion of that effluent which supports microbial growth could be used to produce biosurfactant. Disposed effluents have many nutrients such as proteins, peptides, amino acids, lipids, minerals and vitamins which are all favorite mediums for the microorganism [55, 59]. There was also a report on using dairy wastes as substrates by Daniel et al. [60, 61]. They gained a high concentrations of about 422 g/l sophorolipids by a two-stage cultivation process with the yeast *Cryptoccoccus curvatus* ATCC 20509. In the first state, deproteinized whey concentrate, which is rich in lactose, was used for the culture of *C. curvatus* ATCC 20509. In the second stage, biomass gained from the first stage was homogenized under high pressure and autoclaved to generate a crude cell extract composted of single cell oil. This was used as substrate for the growth of *Candida bombicola* ATCC 22214 for sophorolipid production.

Another study was reported on the use of whey mixed with glucose for sophorolipids production with the yeast *Candida bombicola* in the medium. The researchers obtained a product up to 34 g/l under experimental conditions in an optimized medium formulation [62].

2.2.10. Molasses

Molasses is a co-product which can be obtained from sugar processing plants. It is produced during the sugar manufacturing process from sugarcane or sugar beet. This kind of material is useful because much carbon is available. The molasses from sugarcane is very suitable for culturing bacteria because cane molasses (75% dry matter) consist of 48-56% total sugar, 9-12% organic matter excluding sugar, 2.5% protein, 1.5-5.0% potassium, 0.4-0.8% calcium, 0.06% magnesium, 0.06-2.0% phosphorus, 1.0-3.0 mg/kg biotin, 15-55 mg/kg pantothenic acid, 2,500-6,000 mg/kg inositol and 1.8 mg/kg thiamine [63]. These nutrients are essential for microbial growth. *Bacillus licheniformis* TR7 and *Bacillus subtilis* SA9 was cultivated in molasses as a whole medium to produce lipopeptides biosurfactant. The results showed that the microbe gave the maximum production of biosurfactant–after 4% (w/v) molasses was used. The yields of TR7 and SA9 biosurfactant were found to be 3.30 and 3.78 g/l, respectively. It could reduce the surface tension of pure water to 28.5 and 29.5 mN/m, with critical micelle concentrations of about 10 and 30 mg/l, respectively [64].

2.2.11. Soy Molasses

Soy molasses is rich in potentially fermentable sugar content and many other valuable growth factors that help microbial growth. It can also be a cheap substrates because it is a low-value co-product from soybean processing plants. Nowadays there are many more health-conscious consumers and they demand soy protein-based foods and drinks. These demands could help the soy protein industry which has reach a sustained 10% annual growth for the past several years with a market value of nearly $4 billion [65]. Accordingly, it offers a good opportunity to make use this increasing agricultural waste from soy processing plants. It is a valuable raw and cheap material for the production of surface active agents. Soy molasses is mainly composed of soluble carbohydrates, namely sucrose, raffinose and stachyose. These are effective in microbial fermentation and biosurfactant production. For example, a soy molasses-based medium was used to reduce the cost of production of sophorolipids biosurfactant by *Candida bombicola* [66]. The yield of the biosurfactant obtained was 21 g/l. Solaiman et al. [67] have also shown that soy molasses have a potential for use as carbon and nitrogen sources for sophorolipids production by *C. bombicola* with yields of 55 g/l

2.2.12. Soap Stock

Soap stock is generated when the oilseeds are extracted and refined to make edible oil in the processing plant by using hexane and other chemicals. This gummy, amber colored by-product has been utilized to produce emulsan and biodispersan. The former can form and stabilize oil-in-water emulsion while the latter can disperse large solid limestone granules, making micrometer-size water suspension [68]. The use was made of a mineral salt medium with soap stock as the sole carbon source to produce surface-active rhamnolipids biosurfactant (RLLBI). This was done by the method of batch fermentation with *Pseudomonas aeruginnosa* strain LBI, isolated from petroleum-contaminated soil [31]. It was found that biosurfactant production of the microbes significantly increased after nitrogen depletion and gave the best rhamnolipids concentration of 15.9 g/l. RLLBI produced stable emulsions with hydrocarbons (crude oil, kerosene, toluene, *n*-alkanes (C12-C14) and mineral oil) and vegetable oils (linseed oil, almond oil) [30].

2.2.13. Mixed Substrate from Vegetable Industry

Mixed substrate from the vegetable industry is another interesting choice that could be utilized as an economical substrate for surface active agent production. It was claimed that *Candida bombicola* could produce sophorolipid biosurfactant when cultivated in a medium which consisted of two different carbon sources and a nitrogen source. The carbon sources were sunflower oil and the readily available sugar to maximize biomass. The results showed that 120 g/l sophorolipid was obtained in 8 days under the best optimal operational conditions [69]. Haba et al. [36] found that nine *Pseudomonas* strains and two *Bacillus* strains could reduce the surface tension (32-36 mN/m) and give stable emulsions with kerosene oil. In addition, *Pseudomonas aeruginosa* 47T2 could grow in frying cooking oil waste, the mixed substrate of sunflower and olive oil, and produce 2.7 g/l of rhamnolipid with a production yield of 0.34 g/g.

A report was made of using oleic acid and rapeseed oil respectively as additional carbon sources with glucose in both optimized feed-batch and continuous culture processes. This could ease the high cost of production. *Candida bombicola* ATCC 22214 could produce a high yield of 300 g/l sophorolipid. It also increased the productivity of of the feed-batch to 57 g/l/d culture and to 76 g/l/d for the continuous culture under the optimized culture conditions [70]. Using low-cost raw materials of soybean oil, safflower oil and glycerol respectively as substrate for the culture of oil-degrading strains, *Pseudomonas aeruginosa* DS10-129, achieved yields of 4.31, 2.98 and 1.77 g/l rhamnolipid biosurfactant, respectively [71].

Researchers used rape seed oil as a substrate for culturing *Pseudomonas* sp. DSM 2874 with a developed biotechnological process for the production of rhamnolipids. They obtained a 45 g/l yield of mixtures of mono and di-rhamnolipids. Furthermore, after enzymatic modification of the substrate by direct addition of naringinase to the resting cells, they obtained the production of rhamnolipid [1-4], L-(+) rhamnose and (R,R)-3- (3-hydroxydecanoyloxy) decanoic acid [52]. This could be considered one of the interesting integrated microbial/enzymatic processes for the production of pure rhamnolipid. Bednarski et al. [32] also reported that *Candida antarctica* and *Candida apicola* could produce glycolipids biosurfactant when using oil refinery waste (either with soap stock or post-refinery fatty acids) as a substrate. The formula of the medium enriched by the oil refinery waste gave a 7.5-8.5 fold higher concentration of glycolipid in comparison to the medium without the enrichment of oil refinery waste [32].

Many types of commercial sugar such as sugarcane juice and cane molasses, sugarcane juice alcohol stillage, glycerol, mannitol and soybean oil were tested as a cheap substrates with *Bacillus subtilis* ATCC 6633. The biosurfactant produced from the formulas can reduce the surface tension and increase emulsification. This indicates that renewable and low-cost carbon sources can be effectively used as substrates to produce biosurfactants [72].

In research in Brazil, researchers tested biosurfactant production by culturing *Pseudomonas aeruginosa* LBI in oil from buriti (*Mauritia flexuosa*), cupuacu (*Theobroma grandiflora*), passion fruit (*Passiflora alata*), andiroba (*Carapa guianensis*), Brazilian nut (*Bertholletia excelsa*) and babassu (*Orbignya* spp.) to obtain rhamnolipid. The results showed that the highest rhamnolipid concentrations were obtained from the oils of the Brazilian nut at 9.9 g/l and of passion fruit at 9.2 g/l [73]. Inanother Brazilian study, the researchers used different cheap carbon sources such as soybean oil, soybean oil soapstock, fish oil and glycerol. These were used to grow *P. aeruginosa* LBM10 isolated from a southern coastal area in Brazil to produce a rhamnolipid-type surface active agent. There was a notably high maximum yield which was obtained by using soybean oil as substrate, and the surface active agent was stable at a wide range of pH. They also found that the rhamnolipid obtained was suitable for the bioremediation of spills in marine and estuarine environments because of its salt tolerance [74].

A combination of molasses with three different oils (soybean, sunflower or olive oil) as a substrate for sophorolipids (SLs) production by the yeast *Candida bombicola* was evaluated. *C. bombicola* could produce biosurfactant of approximately 24 g/l in comparison to a single medium. In addition, this yield was also comparable to the costly conventional method using a synthetic medium containing yeast extract, urea, soybean oil and glucose. Furthermore, further study was made of the various parameters' effect on sophorolipid production by the same microbe. A yield of 60 g/l was obtained in the fermentor under the optimal conditions using sugarcane molasses 50 g/l, soybean oil 50 g/l, inoculum size 5% (v/v), temperature 30°C, inoculum aged 2 days and agitation of 200 rpm [62, 75, 76].

3. Substrates from Other Sources

Various substrates can be obtained from other sources

3.1. Lignocellulosic Waste

Lignocellulosic materials are the main components of different waste materials from various industries, municipalities, agriculture and forestry. These waste materials are mostly burned and CO_2 is released to contribute to the greenhouse effect [77]. Lignocellulose is composed of three main types of polymer namely cellulose, hemicellulose and lignin which are strongly intermeshed and chemically bonded by both non-covalent forces and covalent cross-linkages. There have been many studies of the microbial degradation of these macromolecules by fungi and bacteria [78].

Lignocellulosic, a rich agriculture waste, can be a particularly economical substrate for the production of surface active compounds. The residues could be used to produce biosurfactants by *Lactobacilli* spp. [79, 80]. The hemicellulosic hydrolyzates from various agricultural wastes with *Lactobacillus* sp. were used for the first time to synthesize the simultaneous products of biosurfactants and lactic acid. The dual production strategy of the researchers made biosurfactant more economically viable on the market. It also reduced the overall environmental damage from the effects of the burning of waste. From their research, they obtained a biosurfactant product of 0.71 g/g of biomass after hemicellulosic hydrolyzates from trimmed vine shoots was added [81]. It can be concluded that the hemicellulosic sugars from the agricultural wastes are practical substrates for the low cost production of biosurfactants. It also made clear that the large amount of pruning wastes of vine-stocks produced worldwide and the resulting monomeric sugar solutions are the potential renewable sources. They can be used for the production of other biomolecules like lactic acid as well.

Recently a study was made into another abundant byproduct of distilled grape marc from wine industry in Spain which is composed of complex lignocellulosic material [82]. This abundant grape marc is generated when the crushed grapes are pressed during the process of wine production. Some of this grape marc is usually distilled in winery's to get ethanol to produce spirituous liquors. However, a large amount of the unused distilled grape marc has much hemicelluloses. An organic acid and could be used for the production of valuable biomolecules like lactic acid and biosurfactants. This research group also evaluated the liquors containing sugars from hydrolyzates of distilled grape marc as mediums for the lactic acid and biosurfactants production. This could make biosurfactant production more economical [83]. They considered the effects of variables temperatures, reaction times and H_2SO_4 concentration on hydrolysis by using an incomplete factorial design. The experiment yielded 4.8 mg/l intracellular biosurfactant which was equivalent to 0.60 mg of intracellular biosurfactant per gram of sugar that used *Lactobacillus pentosus*. The studies on stability and emulsifying quality on obtained biosurfactant showed much better volume and more stable emulsion than with commercial surfactin and other surfactants [84]. This is another indication that using low cost agriculture wastes as substrates for microbial surfactant production with its economical implications is possible.

3.2. Municipal Solid Waste

Currently in urban societies people's activities produce a large amount of municipal solid waste. This kind of waste is another rich source of energy as it has high content of organic matter. It usually ends up in landfills, so it may be preferable better to use it for biosurfactants production instead. In addition, it is increasingly difficult to find out more space for new landfills and urban populations are continuously increasing. Accordingly, it is time to convert a huge amount of this waste to useful substrate. It has been reported that one such process for converting this waste to useful molecules like biosurfactants or biophotosensitizer for a range of chemical applications has been developed. The results of the process showed that this biomass waste can be used as a valuable source for the chemical industry [85]. The biosurfactants obtained from using this biomass conversion process can give a real economic return in terms of energy saving and thus enhance the net profits compared with synthetic surfactant production.

In the study, the researchers made a compost isolated humic acid-like (cHAL) material obtained from a 1:1 (v/v) mixture of municipal solid and lignocellulosic wastes composted for 15 days. This was applied to textile dyeing (TD) and soil remediation by washing (SW) and compared the performance to that of sodium dodecylsulfate (SDS), the commercial synthetic surfactant.

The performances were rated in the TD experiments based on the fabric color intensity and uniformity. In the SW experiments, the performances were based on the total hydrocarbon concentration (CWPAH) and on the residual surfactant (Cre) concentrations in the washing solution equilibrated with the two contaminated soils. The results showed that in the TD case the cHAL biosurfactant at 0.4g/l yields good color intensity and equal color uniformity as SDS at 5g/l. In the SW case, the humic acid-like was found to enhance CWPAH by a factor of 2-4 relative to SDS with one soil. However, it was found that the two surfactants performed similarly in the other soil. The Cre data showed that both soils absorbed far more SDS (68-95%) than cHAL (12-54%). These results indicate intriguing technological and environmental perspectives deriving from the use of compost isolated biosurfactants in place of synthetic surfactants [85].

3.3. Fish Oil Substrate

Fish oil is not usually known or reported on for use as substrate for the production of surface active agents. However, it has been claimed to be able to serve this objective as well. It could be a promising carbon source. Furthermore, a study showed a yield of 17 g/l rhamnolipid. The culture medium for the *Pseudomonas aeruginosa* BYK-2 KCTC 18012P was optimized for enhancing production by using 25 g/l fish oil as a carbon source. This study is considered as a unique report on fish oil as a practical substrate for microbial surfactant production [86].

Conclusion

Based on the outcomes of the many studies of the researchers discussed above there is hope of reducing the high cost of materials. The production of greater amounts of biosurfactant production with good properties is possible and feasible. There are effective and cheap materials waiting to be converted to useful and practical substrates. This can be accompanied by the production of suitable microorganisms through optimal processes for producing valuable microbial surfactants. Many industries depend on these at present and will do so much more in the future. In addition, not only can invaluable wastes be converted into valuable wastes, reducing the cost of biosurfactant production, these processes will also help improve the overall environment.

Acknowledgment

We are grateful to Phuket Rajabhat University for providing a scholarship to Saimmai A. This work was supported by the Office of the Higher Education Commission and International Foundation for Science No. F/5204-1.

Conflict of Interest

The authors declare that they have no conflict of interest.

References

[1] Banat I, Franzetti A, Gandolfi I, Bestetti G, Martinotti M, Fracchia L, Smyth T, Marchant R (2010) Microbial biosurfactants production, applications and future potential. *Appl. Microbiol. Biotechnol*. 87:427-444.

[2] Cameotra SS, Makkar RS, Kaur J, Mehta SK (2010) Synthesis of biosurfactants and their advantages to microorganisms and mankind. *Adv. Exp. Med. Biol*. 672:261-280.

[3] Lima de CJB, Ribeiro EJ, Sérvulo EFC, Resende MM, Cardoso VL (2009) Biosurfactant production by *Pseudomonas aeruginosa* grown in residual soybean oil. *Appl. Biochemi. Biotechnol*. 152:156-168.

[4] Onbasli AB (2009) Biosurfactant production in sugar beet molasses by some *Pseudomonas* spp. *J. Envir. Biol.* 30:161-163.

[5] Makkar RS, Cameotra SS (2002) An update on use of unconventional substrates for biosurfactants production and their new applications. *Appl. Microbiol. Biotechnol.* 58: 428-434.

[6] Krieger N, Doumit C, David AM (2010) Production of microbial biosurfactants by solid-state cultivation. *Adv. Exp. Med. Biol.* 672:203-210.

[7] Pandey A, Soccol CR, Mitchell D (2000) New developments in solid state fermentation: I Bioprocesses and products. *Process Biochem.* 35:1153-1169.

[8] Adalgisa B, Marzia GDL, Mario M, Rosalena T (2005) Effects of industrial orange wastes on soil characteristics and on growth and production of durum wheat. *Agron. Sustain. Dev.* 25:129-135.

[9] George S, Jayachandran K (2008) Analysis of rhamnolipid biosurfactants produced through submerged fermentation using orange fruit peelings as sole carbon source. *Appl. Biochem. Biotechnol.* 158:694-705.

[10] Nitschke M, Pastore G (2006) Production and properties of a surfactant obtained from *Bacillus subtilis* grown on cassava wastewater. *Bioresour. Technol.* 97:336-341.

[11] Rivera OMP, Moldes AB, Torrado AM, Dominguez JM (2007) Lactic acid and biosurfactants production from hydrolyzed distilled grape marc. *Process. Biochem.* 42:1010-1020.

[12] Sobrinho HBS, Rufino RD, Luna JM, Salgueiro AA, Campos-Takaki GM, Leite LFC, Sarubbo LA (2008) Utilization of two agro industrial by-products for the production of a surfactant by *Candida sphaerica* UCP0995. *Process. Biochem.* 43:912-917.

[13] Rocha M, Souza M, Benedicto S, Bezerra M, Macedo G, Pinto G, Gonçalves L (2007) Production of biosurfactant by *Pseudomonas aeruginosa* grown on cashew apple juice. *Appl. Biochem. Biotechnol.* 137-140:185-194.

[14] Campos DCP, Santos AS, Wolkoff DB, Matta VM, Cabral LMC, Couri S (2002) Cashew apple juice stabilization by microfiltration. *Desalination* 148:61-65.

[15] Honorato T, Rabelo M, Gonçalves L, Pinto G, Rodrigues S (2007) Fermentation of cashew apple juice to produce high added value products. *World J. Microbiol. Biotechnol.* 23:1409-1415.

[16] Rabelo MC, Fontes CPML, Rodrigues S (2009) Enzyme synthesis of oligosaccharides using cashew apple juice as substrate. *Bioresour. Technol.* 100:5574-5580.

[17] Rocha P, Barreto G, Melo V, Gonçalves B (2009) Evaluation of cashew apple juice for surfactin production by *Bacillus subtilis* LAMI008. *Appl. Biochem. Biotechnol.* 155: 63-75.

[18] O'Toole DK (1999) Characteristics and use of okara, the soybean residue from soy milk production: a review. *J. Agric. Food Chem.* 47:363-371.

[19] Kuan YH, Liong MT (2008) Chemical and physicochemical characterization of agrowaste fibrous materials and residues. *J. Agric. Food Chem.* 56:9252-9257.

[20] Ohno A, Ano T, Shoda M (1993) Effect of temperature change and aeration on the production of the antifungal peptide antibiotic iturin by *Bacillus subtilis* NB22 in liquid cultivation. *J. Ferment. Bioeng.* 75:463-465.

[21] Ohno A, Ano T, Shoda M (1993) Production of the antifungal peptide antibiotic, iturin by *Bacillus subtilis* NB22 in solid state fermentation. *J. Ferment. Bioeng.* 75:23-27.

[22] Ohno A, Ano T, Shoda M (1995) Production of a lipopeptide antibiotic, surfactin, by recombinant *Bacillus subtilis* in solid state fermentation. *Biotechnol. Bioeng.* 47:209-214.

[23] Ohno A, Ano T, Shoda M (1996) Use of soybean curd residue, okara, for the solid state substrate in the production of a lipopeptide antibiotic, iturin A, by *Bacillus subtilis* NB22. *Process. Biochem.* 31:801-806.

[24] da Silva GP, Mack M, Contiero J (2009) Glycerol: A promising and abundant carbon source for industrial microbiology. *Biotechnol. Adv.* 27:30-39.

[25] Bognolo F (1999) Biosurfactants as emulsifying agents for hydrocarbons. *Colloid. Surface A* 152:41-52.

[26] Nitschke M, Ferraz C, Pastore GM (2004) Selection of microorganisms for biosurfactant production using agroindustrial wastes. *Braz. J. Microbiol.* 35:81-85.

[27] Rahman KS, Rahman TJ, McClean S, Marchant R, Banat IM (2002) Rhamnolipid biosurfactant production by strains of *Pseudomonas aeruginosa* using low cost raw materials. *Biotechnol. Prog.* 18:1277-1281.

[28] Zhang GL, Wu YT, Qian XP, Meng Q (2005) Biodegradation of crude oil by *Pseudomonas aeruginosa* in the presence of rhamnolipids. *J. Zhejiang. Univ. Sci.* 6: 725-730.

[29] Morita T, Habe H, Fukuoka T, Imura T, Kitamoto D (2007) Convenient transformation of anamorphic basidiomycetous yeasts belonging to genus *Pseudozyma* induced by electroporation. *J. Biosci. Bioeng.* 104:517-520.

[30] Liu Y, Koh CMJ, Ji L (2011) Bioconversion of crude glycerol to glycolipids in *Ustilago maydis*. *Bioresour. Technol.* 102:3927-3933.

[31] Benincasa M, Abalos A, Moreira I, Manresa A (2002) Rhamnolipid production by *Pseudomonas aeruginosa* LB1 growing on soapstock as the sole carbon source. *J. Food Eng.* 54:283-288.

[32] Bednarski W, Adamczak M, Tomasik J (2004) Application of oil refinery waste in the biosynthesis of glycolipids by yeast. *Bioresour. Technol.* 95:15-18.

[33] Nitschke M, Pastore GM (2003) Cassava flour wastewater as a substrate for biosurfactant production. *Appl. Biochem. Biotechnol.* 106:295-302.

[34] Carolina Gusma De AB, Rufino RD, Sarubbo LA (2010) Laboratory production and characterization of a new biosurfactant from *Candida glabrata* UCP1002 cultivated in vegetable fat waste applied to the removal of hydrophobic contaminant. *World J. Microbiol. Biotechnol.* 26:1683-1692.

[35] Deshpande M, Daniels L (1995) Evaluation of sophorolipid biosurfactant production by *Candida bombicola* using animal fat. *Bioresour. Technol.* 54:143-150.

[36] Haba E, Espuny M, Busquets M, Manresa A (2000) Screening and production of rhamnolipids by *Pseudomonas aeruginosa* 47T2 NCIB 40044 from waste frying oils. *J. Appl. Microbiol.* 88:379-387.

[37] Zhang H, Xiang H, Zhang G, Cao X, Meng Q (2009) Enhanced treatment of waste frying oil in an activated sludge system by addition of crude rhamnolipid solution. *J. Hazar. Mater.* 167:217-223.

[38] Gusmao C, Rufino R, Sarubbo L (2010) Laboratory production and characterization of a new biosurfactant from *Candida glabrata* UCP1002 cultivated in vegetable fat waste applied to the removal of hydrophobic contaminant. *World J. Microbiol. Biotechnol.* 26:1683-1692.

[39] Marques IP (2001) Anaerobic digestion treatment of olive mill wastewater for effluent re-use in irrigation. *Desalination* 137:233-239.

[40] Hamman OB, de la Rubia T, Martinez J (1999) Decolorization of olive oil mill wastewater by *Phanerochaete flavido-alba*. *Envir. Toxicol. Chem.* 18:2410-2415.

[41] Mercade ME, Manresa MA, Robert M, Espuny MJ, de Andres C, Guinea J (1993) Olive oil mill effluent (OOME) New substrate for biosurfactant production. *Bioresour. Technol.* 43:1-6.

[42] Sidal U, Kolankaya N, Kurtonur C (2000) Obtaining biosurfactant from olive oil mill effluent (OOME) using *Pseudomonas* sp. *Turk. J. Biol.* 24:611-626.

[43] Ma AN (2000) Environmental management for the palm oil industry. *Palm Oil Dev.* 30:1-9.

[44] Chooklin CC, Phertmean S, Cheirsilp B, Maneerat S, Saimmai A (2013) Utilization of palm oil mill effluent as a novel substrate for biosurfactant production by *Nevskia ramosa* NA3. *Songklanakarin. J. Sci. Technol.* 35:167-176.

[45] Chow MC, Ho CC (2002) Chemical composition of oil droplets from palm oil mill sludge. *J. Oil Palm. Res.* 14:25-34.

[46] Desai JD, Banat IM (1997) Microbial production of surfactants and their commercial potential. *Microbiol. Mol. Biol. Rev.* 61:47-64.

[47] Tuleva BK, Ivanov GR, Christova NE (2001) Biosurfactant production by a new *Pseudomonas putida* strain. *Z. Naturforsch.* 57:356-360.

[48] Priya T, Usharani G (2009) Comparative study forbiosurfactant production by using *Bacillus subtilitis* and *Pseudomonas aeruginosa*. *Botany Res. Intl.* 2:284-287.

[49] Yahya A, Sye CP, Ishola TA, Suryanto H (2010) Effect of adding palm oil mill decanter cake slurry with regular turning operation on the composting process and quality of compost from oil palm empty fruit bunches. *Bioresour. Technol.* 101:8736-8741.

[50] Saimmai A, Rukadee O, Onlamool T, Sobhon V, Maneerat S (2013) Production and characterization of biosurfactant from marine bacterium of *Inquilinus limosus* KB3 grown on low-cost raw materials. *Ann. Microbiol.* doi 10.1007s13213-012-0592-7.

[51] Anastasia A, Pantazaki MI, Dimopoulou OM, Simou OM, Pritsa AA (2010) Sunflower seed oil and oleic acid utilization for the production of rhamnolipids by *Thermus thermophilus* HB8. *Appl. Microbiol. Biotechnol.* 88:939-951.

[52] Trummler K, Effenberger F, Syldatk C (2003) An integrated microbial/enzymatic process for production of rhamnolipids and L-(+)-rhamnose from rapeseed oil with *Pseudomonas* sp. DSM 2874. *Eur. J. Lipid. Sci. Technol.* 105:563-571.

[53] Vollbrecht E, Rau U, Lang S (1999) Microbial conversion of vegetable oils into surface-active di-, tri-, and tetrasaccharide lipids (biosurfactants) by the bacterial strain *Tsukamurella* spec. *Eur. J. Lip. Sci. Tech.* 101:389-394.

[54] Siddhartha GV, Costa AO, Lépine F, Milot S, Deziel E, Nitschke M, Contiero J (2009) Cassava wastewater as a substrate for the simultaneous production of rhamnolipids and polyhydroxyalkanoates by *Pseudomonas aeruginosa*. *J. Ind. Microbiol. Biotechnol.* 36: 1063-1072.

[55] Thompson DN, Fox SL, Bala GA (2000) Biosurfactants from potato process effluents. *Appl. Biochem. Biotechnol.* 84:917-930.

[56] Noah KS (2005) Surfactin production from potato process effluent by *Bacillus subtilis* in a chemostat. *Appl. Biochem. Biotechnol.* 122:465-474.

[57] Das K, Mukherjee AK (2007) Comparison of lipopeptide biosurfactants production by *Bacillus subtilis* strains in submerged and solid state fermentation systems using a cheap carbon source: some industrial applications of biosurfactants. *Process. Biochem.* 42:1191-1199.

[58] Fox SL, Bala GA (2000) Production of surfactant from *Bacillus subtilis* ATCC 21332 using potato substrates. *Bioresour. Technol.* 75:235-240.

[59] Dubey K, Juwarkar A (2004) Determination of genetic basis for biosurfactant production in distillery and curd whey wastes utilizing *Pseudomonas aeruginosa* strain BS2. *Indian J. Biotechnol.* 3:74-81.

[60] Daniel HJ, Otto RT, Reuss M, Syldatk C (1998) Sophorolipid production with high yields on whey concentrate and rapeseed oil without consumption of lactose. *Biotechnol. Lett.* 20:805-807.

[61] Daniel HJ, Reuss M, Syldatk C (1998) Production of sophorolipids in high concentration from deproteinized whey and rapeseed oil in a two stage fed batch process using *Candida bombicola* ATCC 22214 and *Cryptococcus curvatus* ATCC 20509. *Biotechnol. Lett*. 20:1153-56.

[62] Daverey A, Pakshirajan K (2010) Sophorolipids from *Candida bombicola* using mixed hydrophilic substrates: production, purification and characterization. *Colloids Surf. B* 79:246-253.

[63] Saharan BS, Sahu RK, Sharma D (2011) A Review on biosurfactants: fermentation, current developments and perspectives. *Genet. Eng. Biotechnol. J. GEBJ*-29.

[64] Saimmai A, Sobhon V, Maneerat S (2011) Molasses a whole medium for bosurfactants production by *Bacillus* strains and their application. *Appl. Biochem. Biotechnol.* 165: 315-335.

[65] Deak NA, Johnson LA (2006) Functional properties of protein ingredients prepared from high-sucrose/low-stachyose soybeans. *J. Am. Oil Chem. Soc.* 83:811-818.

[66] Solaiman DK, Ashby RD, Nuñez A, Foglia TA (2004) Production of sophorolipids by *Candida bombicola* grown on soy molasses as substrate. *Biotechnol. Lett.* 26:1241-1245.

[67] Solaiman DK, Ashby RD, Zerkowski JA, Foglia TA (2007) Simplified soy molasses-based medium for reduced-cost production of sophorolipids by *Candida bombicola. Biotechnol. Lett.* 29:1341-1347.

[68] Rosenberg E, Rubinovitz C, Gottlieb A, Rosenhak S, Ron EZ (1988) Production of biodispersan by *Acinetobacter calcoaceticus* A2. *Appl. Environ. Microbiol*. 54:317-322.

[69] Casas J, Garcia-Ochoa F (1999) Sophorolipid production by *Candida bombicola*: medium composition and culture methods. *J. Biosci. Bioeng.* 88:488-494.

[70] Rau U, Hammen S, Heckmann R, Wray V, Lang S (2001) Sophorolipids: a source for novel compounds. *Ind. Crops Pro.* 13:85-92.

[71] Rahman KS, Banat IM, Thahira J, Thayumanavan T, Lakshmanaperumalsamy P (2002) Bioremediation of gasoline contaminated soil by a bacterial consortium amended with poultry litter, coir pith and rhamnolipid biosurfactant. *Bioresour. Technol.* 81:25-32.

[72] Reis FA, Servulo EF, De Franca FP (2004) Lipopeptide surfactant production by *Bacillus subtilis* grown on low-cost raw materials. *Appl. Biochem. Biotechnol.* 115: 899-912.

[73] Costa SG, Nitschke M, Haddad R, Eberlin MN, Contiero J (2006) Production of *Pseudomonas aeruginosa* LBI rhamnolipids following growth on Brazilian native oils. *Process. Biochem*. 41:483-488.

[74] Prieto LM, Michelon M, Burkert JFM, Kalil SJ, Burkert CAV (2008) The production of rhamnolipid by a *Pseudomonas aeruginosa* strain isolated from a southern coastal zone in Brazil. *Chemosphere* 71:1781-1785.

[75] Daverey A, Pakshirajan K (2009) Production, characterization, and properties of sophorolipids from the yeast *Candida bombicola* using a low-cost fermentative medium. *Appl. Biochem. Biotechnol*. 158:663-674.

[76] Daverey A, Pakshirajan K (2010) Kinetics of growth and enhanced sophorolipids production by *Candida bombicola* using a low-cost fermentative medium. *Appl. Biochem. Biotechnol.* 160:2090-2101.

[77] Kukhar V (2009) Biomass - feedstock for organic chemicals. *Kem. Ind.* 58:57-71.

[78] Perez J, Munoz-Dorado J, la Rubia T, Martinez J (2002) Biodegradation and biological treatments of cellulose, hemicellulose and lignin: an overview. *Int. Microbiol*. 5:53-63.

[79] Rodrigues LR, Teixeira JA, van der Mei HC, Oliveira R (2006) Physicochemical and functional characterization of a biosurfactant produced by *Lactococcus lactis* 53. *Colloids Surf.* 49:79-86.

[80] Velraeds M, van der Mei H, Reid G, HJ B (1996) Inhibition of initial adhesion of uropathogenic *Enterococcus faecalis* by biosurfactants from *Lactobacillus* isolates. *Appl. Environ. Microbiol*. 62:1958-1963.

[81] Portilla-Rivera OM, Moldes Menduiña AB, Torrado Agrasar AM, Domínguez González JM (2007) Biosurfactants from grape marc: stability study. *J. Biotechnol.* 131:S136-S136.

[82] Portilla-Rivera OM, Moldes AB, Torrado AM, Domínguez JM (2007) Lactic acid and biosurfactants production from hydrolyzed distilled grape marc. *Process. Biochem.* 42: 1010-1020.

[83] Portilla-Rivera O, Torrado A, DomiÌ·nguez JM, Moldes AB (2008) Stability and emulsifying capacity of biosurfactants obtained from lignocellulosic sources using *Lactobacillus pentosus*. *J. Agric. Food Chem*. 56:8074-8080.

[84] Portilla-Rivera OM, Rivas B, Torrado A, Moldes AB, Domínguez JM (2008) Revalorisation of vine trimming wastes using *Lactobacillus acidophilus* and *Debaryomyces hansenii*. *J. Sci. Food Agri.* 88:2298-2308.

[85] Montoneri E, Boffa V, Savarino P, Perrone DG, Musso G, Mendichi R, Chierotti MR, Gobetto R (2009) Biosurfactants from urban green waste. *Chem. Sus. Chem*. 2:239-247.

[86] Lee K, Hwang SH, Ha S, Jang JH, Lim DJ, Kong JY (2004) Rhamnolipid production in batch and fed-batch fermentation using *Pseudomonas aeruginosa* BYK-2KCTC 18012P. *Biotechnol. Bioproc. Eng.* 9:267-273.

In: Microbes in Process
Editors: Neelam Garg and Abhinav Aeron
ISBN: 978-1-63117-127-7

Chapter 14

Control of Corrosion Caused by Sulfate-Reducing Bacteria

Ashraf M. El-Shamy*
Physical Chemistry Department, National Research Centre,
Cairo, Egypt

Abstract

This chapter presents an overview of the microbiology influenced corrosion caused by sulfate-reducing bacteria (SRB) and how to determine its effects on metal structures in oil and gas industries. Sulfate reducing bacteria are a group of specialized microorganisms that occur in aqueous environments in the absence of oxygen. The nutrition for SRB depends on the simple organic acids as carbon source in presence of molecular hydrogen (H_2) from decomposing natural organic matter. The hydrogen sulfide, H_2S is produced as metabolites from the process of metabolism as a result of reduction of sulfate to sulfide.

The formed H_2S is the main factor in the dangerous effects caused by SRB as well as it contaminates gas and stored oil, precipitates ferrous sulfide that plugs injection wells and promotes corrosion. The corrosion of iron and steel will occur exactly in the absence of oxygen (anaerobic corrosion). The mechanism by which SRB are involved in corrosion is concerned with their ability to depolarize iron surfaces by consumption of cathodically formed hydrogen and it is briefly explained. The measure of growth of the SRB is carried out by serial dilution method and the most probable number was used to detect the bacterial counts presented.

On the basis of measured growth characteristics of the SRB, possibilities for controlling their activity are discussed. Possibilities for controlling their activity are discussed. The control of SRB by using biocides is well discussed and briefly done by using alternatively injection of oxidized and none oxidized biocides.

* elshamy10@yahoo.com.

1. Introduction

The biofouling and microbially influenced corrosion (MIC) are mediated by micro-organisms attached to the metal surface and/or embedded in a gelatinous organic matrix (the biofilm), so it is very important to study and discuss the nature of microorganisms and how to identify its mode of action. The processes of microbial adhesion lead to an important modification of the metal/solution interface, inducing changes in the type and concentrations of ions, pH, oxygen levels, flow velocity and buffering capacity of the liquid microenvironment or the interface. The electrochemical interface used in corrosion studies should also be mentioned in this chapter.

1.1. Nature of Sulfate Reducing Bacteria

In 1997 Madigan, Martinenko and Parkerix was defined the nature of Sulfate reducing bacteria (SRB). Sulfate reducing bacteria have long been known to be involved in bio-corrosive processes due, in part, to the generation of hydrogen sulfide that then initiates the electrolytic corrosion of metals. The SRB form an integral part of the group referred to as "Sulfur Bacteria" is considered to be nuisance bacteria in a number of ways [1] included in the sulfur bacteria group are:

- Sulfate Reducing Bacteria
- Sulfur Oxidizing Bacteria
- Sulfur Reducing Bacteria

Generally sulfur occurs in living matter chiefly as a component of certain amino acids. In the soil sulfur is present primarily as a part of proteins and is taken up by plants in its oxidized form, sulfate (SO_4^{2-}). The sulfur containing proteins are first degraded into their constituent amino acids by enzymes and then excreted by numerous soil organisms as hydrogen sulfide (H_2S). Hydrogen sulfide (H_2S), immediately identified by its pungent odor, is highly toxic to most biological systems, if it was ever to accumulate. Under aerobic conditions, H_2S oxidizes spontaneously to sulfate, which its most readily utilized form by sulfate bacteria. However, under anaerobic conditions, sulfate reducing bacteria can reduce sulfate to H_2S. The effect of SRB is well noticed in cooling systems, corrosion and fouling are frequently attributed to SRB. Figure 1 shows the biofilm formed from SRB on the metal surface. The organism is often found in sludge in the basins of cooling towers. In these systems the bacterium, using sulfate as a hydrogen acceptor, produces hydrogen sulfide and ferrous sulfide, which in combination form a malodorous black slime. In severe cases of biological fouling, bubbles of hydrogen sulfide rise through the water in the basin causing objectionable odor in the area.

Generally microbially Induced Corrosion (MIC) proved to be caused by different types of microorganisms [2]. In water systems microorganisms are found to be suspended and floating (planketoinc) and settling and adhering to immersed surfaces (sessile) the latter group forms what is called (Biofilms). Biofilm forming microorganisms have capability of forming extracellular biopolymers (e.g slime) helping them, to adhere metallic and non metallic surface. The biofilm forming microorganisms include aerobic, facultative anaerobic and aerobic bacteria [3-5].

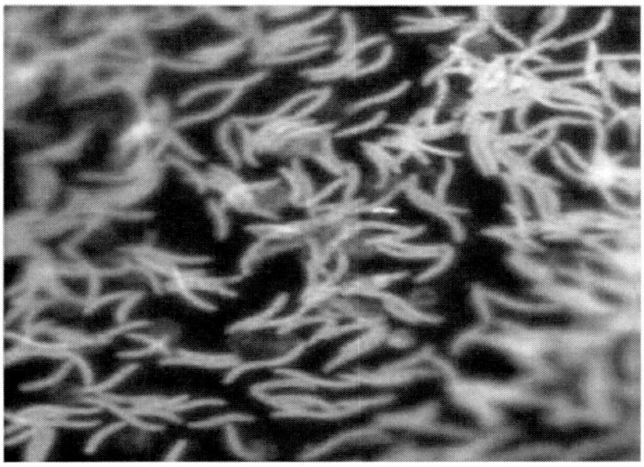

Figure 1. Biofilm of Sulfate-Reducing Bacteria.

It is well known that SRB is more destructive than other kinds of microorganisms. SRB corrosion occurs under adherent biofilms [5]. Biocides are not as effective against sessile organisms within biofilms as against a planktonic population [8]. SRB cause corrosion of cast iron, carbon and low alloy steels, stainless steels, high nickel alloys and copper alloys [6]. The biofilm of SRB causes corrosion problems for oil industry especially the underground pipelines, and it is also noticed in stagnant conditions or low flow rate of river or sea water [7-9].

The formed biofilm can also cause damage to cooling systems [10]. Under anaerobic conditions the SRB that can reduce the sulphate anion into sulfide in contact with ferrous metal surface that can use the cathodic hydrogen and cause cathodic depolarization i.e. promote corrosion process [11, 12]. Naturally derived mixed culture is used in studying corrosion behaviour in off sure structural steels [13]. Naturally SRB mixed culture was developed by the Lactate containing media. The black precipitate of ferrous sulfide is an indication of SRB attack, so it is used as visual inspection of presence of SRB in the samples. Sulphate reducing bacteria (SRB) are a group of genetically similar anaerobic organisms that were first discovered by Beijerinck in 1895. At present over 33 species of SRB are known most having been discovered in the last 20 years. They posses very similar characteristics and are found in many diverse habitats across the world from the anaerobic mud found at the bottom of the ocean to the intestines of humans. There are many species of organisms included in environment such as: bacteria, algae, and fungi. The microorganisms present in the water have the ability to attach and grow on the immersed surfaces of structural materials, resulting in the formation of a biological film, or biofilm. Larger, macroscopic organisms, such as the well known barnacles and mussels, are also present in many environments and also can concerned in the formation of biofilm. There is a voluminous amount of literature on the formation of biofilms and their adverse effects [14-17].

The emphasis of this article on microbiologically influenced corrosion (MIC) is on the effects of the microscopic organisms and the byproducts they produce on the electrochemical corrosion of steel. The biofilms that form on the surface of metals and its alloys immersed in aqueous environments have the capability to influence the corrosion of those metals and alloys. There are many environmental factors which affect on the activity and ability of the microorganisms such as: pH, oxidizing power, velocity of flow, concentration of chemical species at the metal surface. Thus, the value of a given parameter at the metal-water interface within the biofilm may be quite different from that in the bulk electrolyte away from the interface. The result can be the initiation of corrosion under conditions in which there would be none in the absence of the film, a change in the mode of corrosion (that is, from uniform to localized), or an increase or decrease in the corrosion rate, as shown in Figure 2. It is important to note, however, that the presence of a biofilm does not necessarily mean that there will always be a significant effect on corrosion.

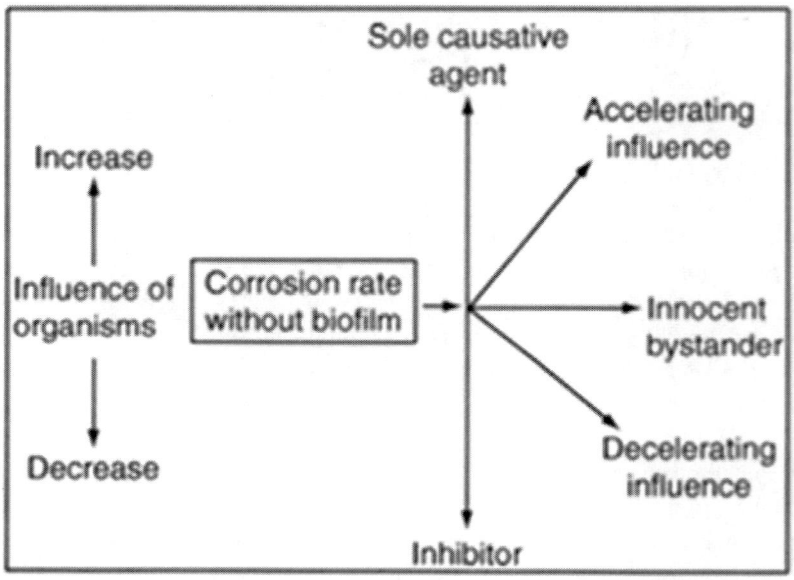

Figure 2. Schematic Showing that the Influence Microorganisms can have on Corrosion Ranges from Sole Causative Agent to Inhibitor. The Most Usual Influence, However, is to Change the Mode of Corrosion from General to Localized and Increase the Rate of Penetration [18].

1.2. Identification of SRB

The presence of sulfate reducing bacteria is not easily identified through microbiological tests. Even the determination of total anaerobic bacteria suddenly provides precise early warning, before the population will rise to unacceptable levels and remarkable irreversible damage already had taken place to the system. Being SRB an anaerobic and sensitive organism, it is very common to be eliminated through the sampling and cultivation procedure, and being present in the original sample, cannot finally be determined. The recognition of the biological nature of sulphate reduction in natural environments, and identification of the microbial species involved back to the latter part of the nineteenth century, and the seminal work of such giants of the early days of microbiology as Beijerinck and Winogradsky. The central role of environmental studies in highlighting the issues to be addressed and the problems to be solved; has remained to this day a constant theme in microbiological analyses of the sulphate reducers. The modern era of such analyses, however, can be said to date from the period around 1960 when the demonstrations by Postgate and Peck, respectively, of the presence of cytochromes and of phosphorylation linked to anaerobic respiration in sulphate reducing bacteria, fundamentally altered our view of the biochemical nature of these organisms and, in particular, of their mechanisms of energy conservation.

There then followed a period of intense activity centered on: elucidation of the metabolic pathways of substrate utilization and the mechanisms of energy generation; cultural techniques and the identification of an ever increasing number of new species; and the appreciation of their significant role in maintaining, or disrupting, the biological balance of many natural and man-made ecosystems. These themes of biochemistry and cell physiology, phylogeny, and ecology remain central to the understanding of SRB themselves, and of their interactions with other components of the biosphere. During the nineteenth century, there was a growing acceptance of the ubiquity of microbial activity and the realization of the impacts that these microorganisms did have directly and indirectly on human society. In the latter part of this century there was initially an equal interest in environmental and medical aspects but the major discoveries by Pasteur, Koch and many others shifted the focus to pathogenic microorganisms.

1.3. Biofilm Formation

The bacteria implicated in corrosion may begin their lives on a metal surface as a scatter of individual cells, as shown in Figure 3.

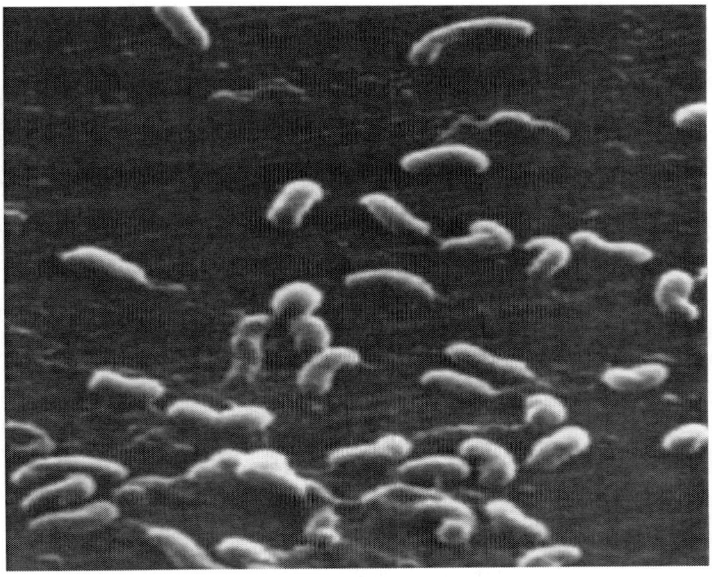

Figure 3. Various Forms of Bacterial Film that can Influence Corrosion Started as Scatter of Individual Cells on the Metal Surface (6050X).

As the biofilm matures, however, the organisms are usually found as individuals or in colonies embedded in the matrix of a semi-continuous and highly heterogeneous biofilm Figures 4a, b.

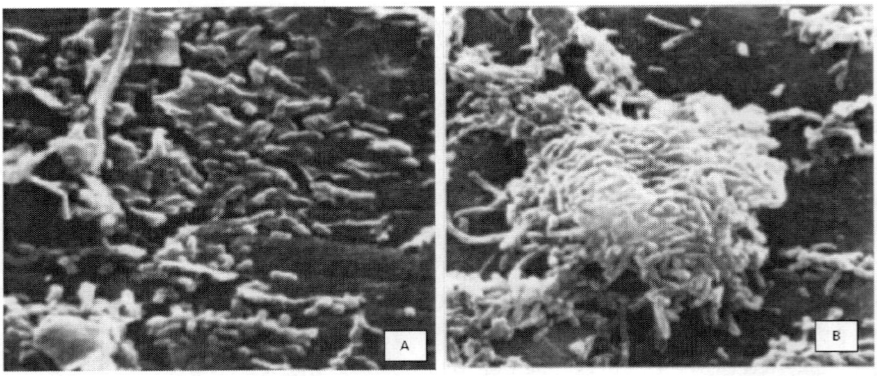

Figure 4. Various Forms of Bacterial Film that can Influence Corrosion. (a) Semi-continuous Film of Bacteria in Slime (3150X) (b) Bacterial Cells in a Colony (2700X).

The process of colonization begins immediately on immersion with the adsorption of a nonliving organic conditioning film. This conditioning film is nearly complete within the first 2 h of immersion, at which time the initially colonizing bacteria begin to attach in substantial numbers. The bacterial, or primary, slime film develops over a period of 24 to 72 h in most natural waters, depending on the temperature, although further changes in the film can often be observed over more than a 2 month period [19,20] as shown in figure 5.

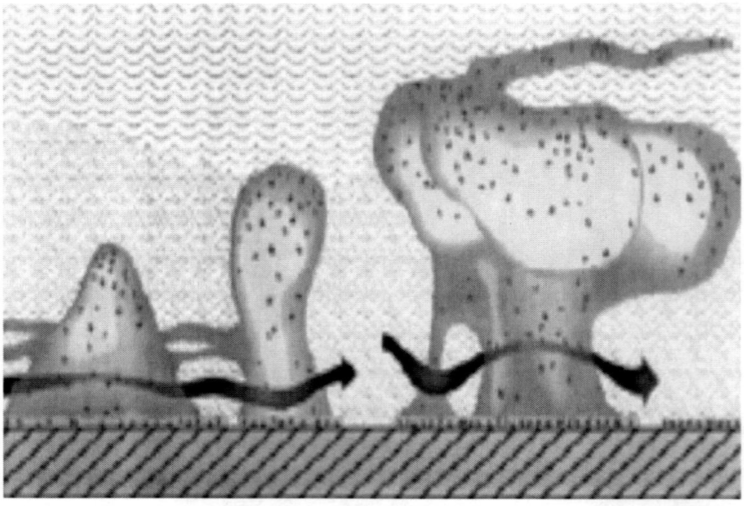

Figure 5. Conceptual Model of Biofilm Structure as Determined by Confocal Scanning Laser Microscopy. The Metal is Shown Cross-Hatched. Note the Nearly Continuous Base Film Upon Which Sits a Series of Cauliflower like Cell Clusters Separated by Voids Through Which Bulk Water can Flow (arrows) [21].

The sulfate reducing bacteria (SRB) are implicated in the corrosion of iron-base alloys in a variety of environments [22-24]. The traditional view is that most SRB are obligate anaerobes, yet they are known to accelerate corrosion in aerated environments. This becomes possible when the aerobic organisms in the outer portions of the columnar structure create an anaerobic microenvironment with the organic acids and nutrients necessary for growth of the SRB close to the metal surface (Figure 6).

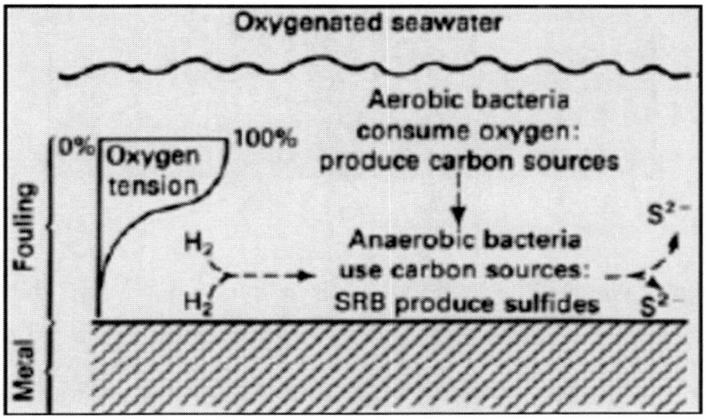

Figure 6. Variations Through the Thickness of a Bacterial Film. Aerobic Organisms near the Outer Surface of the Film Consume Oxygen and Create a Suitable Habitat for the Sulfate-Reducing Bacteria at the Metal Surface [25].

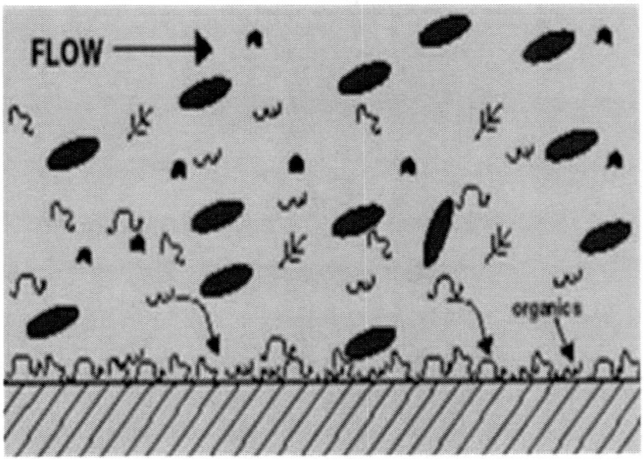

Figure 7. Adsorption of Organic Molecules on a Clean Surface forms a Conditioning Film [6].

The adsorption of organic molecules on the clean surface of pipelines form a conditioning film and this step illustrates the initiation of biofilm formation is clearly shown in figure 7 while figure 8 shows the adhesion of pioneer bacteria in the pipe of flowing water, some of the planktonic (free-floating) bacteria will approach the pipe wall and become entrained within the boundary layer, where flow velocity falls to zero. Cells begin to form structures which may permanently adhere to the cell surface. Transport of bacteria cells to the conditioned surface.

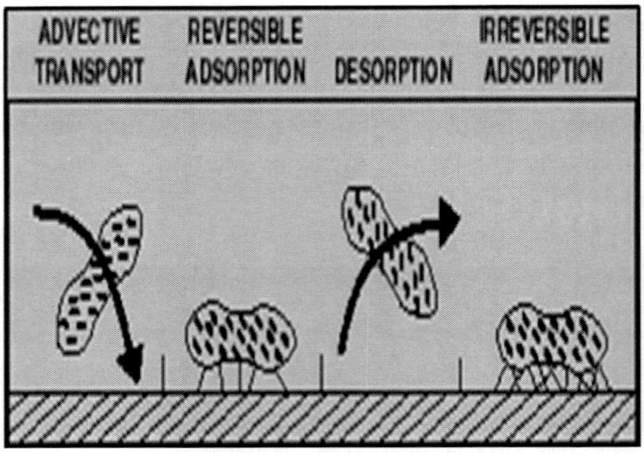

Figure 8. Adhesion of Pioneer Bacteria.

The formation of extracellular polysaccharides (slime formation) is illustrated in Figure 9, since the biofilm bacteria excrete extracellular polysaccharides, or sticky polymers, which hold the biofilm together and cement it to the pipe wall. These polymers strands, much like the structure of a spider's web, trap scarce nutrients and protect bacteria from biocides. This extracellular slime acts as an ion-exchange system for trapping and concentrating trace nutrients from the overlying water. The slime acts as a protective coating for the attached cells which mitigates the effects of biocides and other toxic substances.

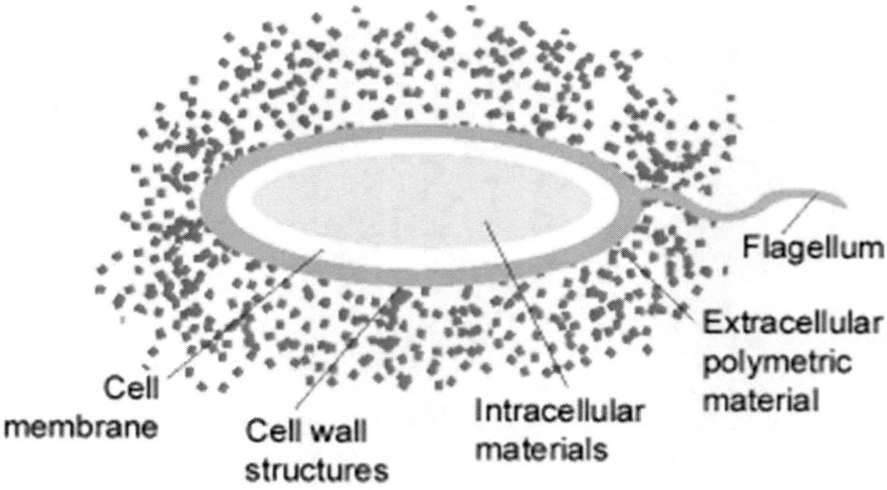

Figure 9. Extracellular Polysaccharides (Slime) Formation.

The bacteria and other microorganisms develop cooperative colonies or (consortia) within the biofilm as illustrated in Figure 10. An anaerobic biofilm may develop underneath the aerobic layer. It is to be noted that the coliforms are facultative i.e., they can live in both anaerobic or aerobic conditions.

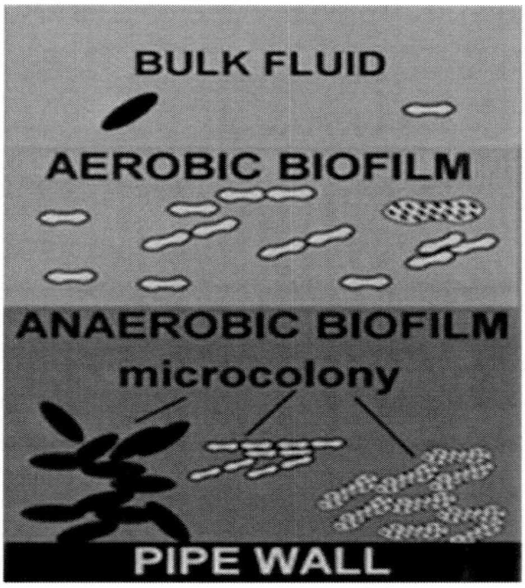

Figure 10. Biofilm Cross-Section.

Thus, the organisms influencing corrosion can often flourish at the corrosion site by associating with other organisms in a microbial colony or consortium, even when the bulk environment is not conducive to their growth. This traditional view is currently being challenged by recent research showing many SRB are micro-aerophilic and able to function under low oxygen conditions at any depth of the biofilm. The anaerobic corrosion is resulting in the presence of sulfate reducing bacteria and the stability of biofilm is depending on the flow rate of electrolyte and the concentration of produced microbial cells. Even though the sulfate-reducing bacteria are obligatory anaerobes, they may survive but not actively grow if exposed to oxygen. The bacterium is present in most natural waters including fresh, brackish and seawater, and most soils and sediments contain sulfate reducers. Sulfate or sulfite must be present for active growth. The bacteria may tolerate temperatures as high as 176°F and a pH from about 5 to 9.

2. Metabolic By-Products and Corrosion Phenomenon

Many of the microorganisms implicated in corrosion are able to have an influence on the electrochemical reactions involved by virtue of chemical compounds consumed and the products produced during the course of their metabolism. A large percentage of them can form extracellular polymeric materials termed simply *polymer,* or *slime.* The slime is involved in attaching the organisms to the surface, trapping and concentrating nutrients for the microbes to use as food, and often, in shielding the organisms from the toxic effect of biocides. Corrosion occurs where biological matter settles or attaches to surfaces. Biologically produced chemicals such as ammonia, hydrogen sulfide and acids can increase general corrosion rates in a system, near and away from the generating organisms. The corrosion of iron by SRB is rapid, and unlike ordinary rust, is not self-limiting. Tubercles formed by sulfate-reducing bacteria consist of an outer shell of red ferric oxide mixed with black magnetite iron oxide. Beneath the shell is a soft black center of ferrous sulfide, under which there is usually a deep pit. When cleaned, the interior surface of the pit usually has a characteristic bright silvery appearance. The slime film can influence corrosion by trapping or complexing heavy-metal ions near the surface. Some species of microbes can produce organic acids, such as formic and succinic, or mineral acids, such as H_2SO_4. These chemicals are corrosive to many metals. One series of bacteria is involved in metabolizing nitrogen compounds. As a group, they can reduce nitrates (NO^{-3}) (often used as corrosion inhibitors) to nitrogen (N_2) gas. Others can convert nitrates to nitrite, or vice versa, or they can break it down to form ammonia (NH_3). Still other series of bacteria are involved in the transformation of sulfur compounds. They can oxidize sulfur or sulfides to sulfates (SO_4^{2-}) (or H_2SO_4), or they can reduce sulfates to sulfides, often producing corrosive H_2S as an end product. The presence of a biological film on a corroding metal surface does not introduce some new type of corrosion, but it may influence the occurrence and/or the rate of all currently known types of corrosion. The net effect of both micro and macro-biofouling film formation on corrosion is variable. The most usual effect of both types of fouling is to make corrosion of the metal substratum more localized than it would be in water with the same bulk chemistry but without microorganisms. This effect is a consequence of the heterogeneity of the biofilm, which in turn leads to an increase in the variability of the chemical environment from place to place along the metal surface. Microbial biofilms change the chemistry at the metal-liquid interface in a number of ways that have an important bearing on corrosion. As the biofilm grows, the bacteria in the film produce a number of by-products. Among these are organic acids, hydrogen sulfide, and protein-rich polymeric materials commonly known as slime. The first effect of the composite film of bacteria and associated polymer is to create a diffusion barrier between the metal-liquid interface and the bulk seawater. The barrier itself is over 90% water, so it does not truly isolate the interface. Instead, it supports strong concentration gradients, both parallel and perpendicular to the metal surface [26]. Thus, the water chemistry at the interface may be different from that in the bulk water, although the two are closely coupled through diffusive processes. These biological influences can be divided into four general categories :Production hydrogen sulfide, Production of sulfides under oxygen-free (anaerobic) conditions, Production of organic and inorganic acids as metabolic by-products, Introduction of new redox reactions, Production of oxygen or chemical concentration cells.

2.1. Production of Hydrogen Sulfide

The SRB is capable of generating hydrogen sulfide from the reduction of sulfate. A number of problems are related to the generation of H_2S that then initiates the electrolytic corrosion of metals [26, 27]. There are three major events that can result from the presence of this gas: Odour generation resembling rotten eggs, electrolytic corrosive processes and generation of black iron sulfides [28]

The activities of SRB in natural and man-made systems are of great concern to many different industrial operations. This primarily concerns the petroleum industries, which use immense amounts of seawater in their technologies while recovering oil from under the sea bed caused severe economic losses due to corrosion of equipment [29, 30]. Oil, gas and shipping industries are seriously affected by the sulfides generated by SRB [31]. A large amount of SRB may cause the oil and gas to acidify, the piping to corrode and technical installations to become clogged [32].

Owing to their quite considerable ecological and economic importance, SRB have become recently a popular subject of scientific investigation. In recent years, and by using experimental techniques and analytical approaches of molecular biology, sulphate reducing prokaryotes are now the chosen organisms of study in many of the major microbiological laboratories worldwide. Additionally, there is an extensive literature covering their several unique characteristics which, in many cases, may help to shed light on certain issues of fundamental importance to our understanding of the evolution and development of life processes. Many authors accepted this challenge, and given so willingly of their time, effort and insight. It has been a conscious decision to include in the same volume studies of: genomic and proteomic analyses; phylogenetic diversity; molecular characterization of enzymes and respiratory systems; thermodynamic analyses of metabolic processes, including anaerobic oxidation of hydrocarbons; response to stress, most particularly with regard to oxygen and other alternative electron acceptors; extreme and specialized micro-environments, including biofilms; environmental impact in, for example, bioremediation and corrosion. These apparently disparate subject areas nevertheless form an intellectual continuum, within which it is possible to see the interdependence of the techniques and thought processes of one study area impacting directly on another. MIC occurs in many environments and on most alloy types, including some non-metallic materials. When a piece of metal is exposed in water, whether potable or not, it almost immediately starts to build up an inorganic layer, which consists of charged inorganic molecules. After a further period of time, organic matter starts to attach onto this surface and the thickness of the layer increases.

Aerobic species such as slime forming bacteria then start to colonize the layer, which is referred to as a biofilm. Thereafter, once the biofilm has reached a certain thickness, SRB colonize the biofilm and locate themselves at the interface between the biofilm and the metal substrate. Pipelines are used extensively in industry to convey various products. The internal corrosion of pipelines can lead to perforation and leakage of the product. A growing problem, not often accurately recognized, is internal corrosion of pipelines caused by bacteria. This form of corrosion is not widely recognized in many industries, although it is widespread and causes many corrosion problems in pipelines, both internal and external. Sulphate reducing bacteria (SRB) are responsible for the bulk of the corrosion problems caused by MIC. Corrosion caused by SRB is found extensively in water pipelines conveying raw, potable and waste water, as well as in pipelines conveying slurries. This problem also occurs in pipelines conveying crude oil. This paper describes internal corrosion problems of slurry pipelines caused by bacteria and outlines mitigation methods to improve slurry pipeline integrity, thereby extending the life of such pipelines.

2.2. Production of Sulfides under Oxygen-Free (Anaerobic) Conditions

The most thoroughly documented case in which microbes are known to cause corrosion is that of iron and steel under anaerobic conditions in the presence of SRB. Based on electrochemistry, de-aerated soils of near-neutral pH are not expected to be corrosive to iron and steel. However, if the soil contains SRB and a source of sulfates, rapid corrosion has been found to occur. The classical mechanism originally proposed for this corrosion involved the removal of atomic hydrogen from the metal surface by the bacteria using the enzyme hydrogenase [33]. The removed hydrogen was then supposedly used by the bacteria in the reduction of sulfates to sulfides. The following set of equations was proposed to explain this mechanism:

$4Fe \rightarrow 4Fe^{2+} + 8e^-$ Anodic reaction (1)

$8H_2O \rightarrow 8H^+ + 8OH^-$ Dissociation of water (2)

$8H^+ + 8e^- \rightarrow 8H$ Cathodic reaction (presence of SRB) (3)

$SO_4^{2-} + 8H \rightarrow S^{2-} + 4\ H_2O$ Cathodic depolarization (4)

$Fe^{2+} + S^{2-} \rightarrow FeS$ Corrosion product (5)

$3Fe^{2+} + 6\ (OH)^- \rightarrow 3Fe(OH)_2$ (corrosion product) (6)

Without SRB, the mechanism would stop after Eq 3, when the surface became covered by a monolayer of hydrogen. According to the theory, this hydrogen is stripped off by the bacteria, a process known as cathodic depolarization; this process allows corrosion to continue. It is now recognized that this original mechanism, although it undoubtedly plays an important role, does not represent the entire process (Figure 11).

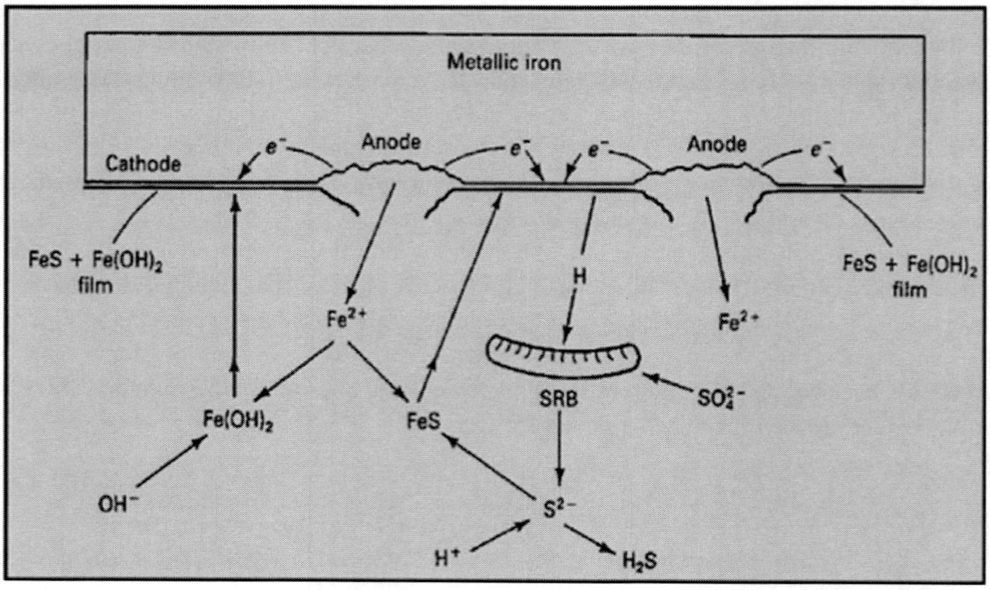

Figure 11. Schematic of the Anaerobic Corrosion of Iron and Steel Showing the Action of Sulfate-Reducing Bacteria (SRB) in Removing Hydrogen from the Surface to form FeS and H2S.

It has been shown that the iron sulfide (FeS) film produced is protective if continuous, but that it causes galvanic corrosion of the bare iron underneath if defective. The bacteria thus transform a benign environment into an aggressive one in which steel corrodes quite rapidly. Other corrosive substances, such as H_2S, can also be produced. The SRB also have been identified as contributors to the corrosion of stainless, copper, and aluminum alloys [34].

2.3. Production of Organic and Inorganic Acids

The acids produced by these organisms can also contribute to corrosion by aiding the breakdown of coatings systems. Alternatively, other organisms that have no direct influence on corrosion may be involved in the breakdown of coatings. The breakdown products are then sometimes usable as food by the acid producing bacteria, ultimately leading to accelerated corrosion of the underlying metal. Organic acids produced by fungi were identified as the cause of pitting failures in painted carbon steel holds on a bulk carrier and aluminum fuel storage tanks [35]. Bacterial action on decaying organic matter in the slime film can also result in the production of ammonia and sulfides. Ammonia causes stress-corrosion cracking of copper alloys [36], and sulfides have been implicated in accelerated localized and/or uniform corrosion of both copper alloys and steels [37].

2.4. Redox Reactions

Biofilms are able to accumulate ions, such as the heavy metals manganese and iron, in concentrations far above those in the surrounding bulk water. This behaviour makes new redox reactions available to the corrosion process, particularly to the cathodic reaction. These reactions have been proposed for the acceleration of corrosion reactions by manganese redox cycling within biofilms as follows:

$$MnOOH + 3H^+ + e^- \rightarrow Mn^{2+} + 2H_2O \quad (7)$$

Or by disproportionation:

$$MnOOH + 2H^+ \rightarrow MnO_2 + Mn^{2+} + 2H_2O \quad (8)$$

The presence of manganese oxidizing bacteria keeps the redox cycle going to allow reoxidation. The manganese redox cycle can include both inorganic and biological processes and the reaction path way follow these equations.

$$4H^+ + MnO_2 + 1e^- \rightarrow Mn^{3+} + 2H_2O \quad (9)$$

$$2H_2O + 2Mn^{3+} \rightarrow MnO_2 + Mn^{2+} + 4H^+ \quad (10)$$

$$2H_2O + Mn^{2+} + O_2 \rightarrow 4OH^- + 1e^- \quad (11)$$

Equation 9 describes the one-electron reduction of MnO_2 to a soluble Mn^{3+} species, which is meta stable and disproportionate quickly to Mn^{4+} and Mn^{2+} in seawater (Eq. 10) unless complexed by strong organic ligands [38-40].

2.5. Oxygen and Chemical Concentration in Cells

The areas with relatively high oxygen concentrations within the biofilm are cathodic relative to areas with less oxygen. Beneath a microbial colony, oxygen is depleted as it is used by the organisms in their metabolism. Oxygen from the bulk electrolyte is unable to replenish those areas because of a combination of effects. First, oxygen migration through the film is slowed by the diffusion barrier effect, and second, oxygen that does penetrate the film is immediately used by the microbial metabolism. Formation of such a corrosion cell, as shown in Figure 12, causes a pit to form at the anodic area under the bacterial colony. As the pit grows, iron dissolves according to the anodic reaction:

$$Fe \rightarrow Fe^{2+} + 2e^- \qquad (12)$$

The cathodic reaction is reduction of dissolved oxygen outside the pit to form OH^- according to:

$$O_2 + 2H_2O + 4e- \rightarrow 4OH^- \qquad (13)$$

The insoluble ferrous hydroxide corrosion product forms by the reaction:

$$3Fe^{2+} + 6OH^- \rightarrow 3Fe(OH)_2 \qquad (14)$$

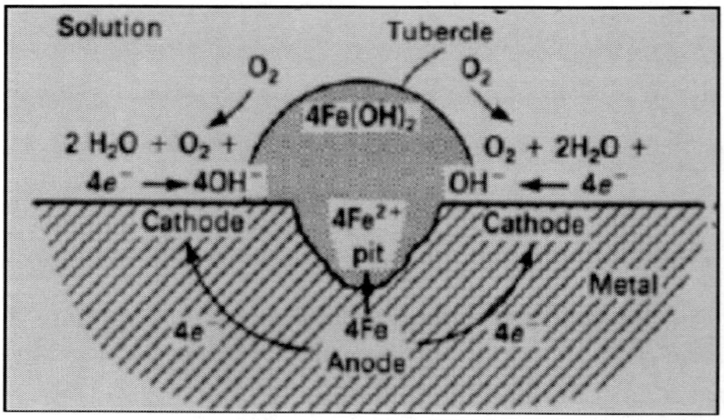

Figure 12. Schematic of Pit Initiation and Tubercle Formation due to an Oxygen Concentration Cell under a Biological Deposit [41].

Conversely, the anode should form where SRB reside at the metal surface, leading to a neutral pH reaction such as:

$$HS^- + Fe \rightarrow FeS + H^+ + 2e^- \qquad (15)$$

3. Testing for SRB

The presence of microorganisms on a metal surface, even in the early stages of biofilm formation, can change the chemistry at the interface where corrosion reactions take place. Testing for microbiologically influenced corrosion (MIC) is used to determine if the microbes are the cause of specific types of corrosion failures or if the microbes in a given environment result in an unacceptably high corrosion rate for the metal structures and how can I determine whether the corrosion in my system is MIC or not?. In order to prove that the organisms present were the sole cause of corrosion, one would have to show first that the organisms can reproduce the attack when introduced into a sterile system and second, that the effect does not happen when the organisms are absent from an otherwise identical environment. There are four types of evidence metallurgical, microbiological, chemical, and electrochemical are generally used to determine if the influence of the organisms is important enough that it must be addressed for a corrosion control program to be successful.

Figure 13. Monitoring of Bacterial Counts by Taking Samples from the System Before and After Injection of Biocide.

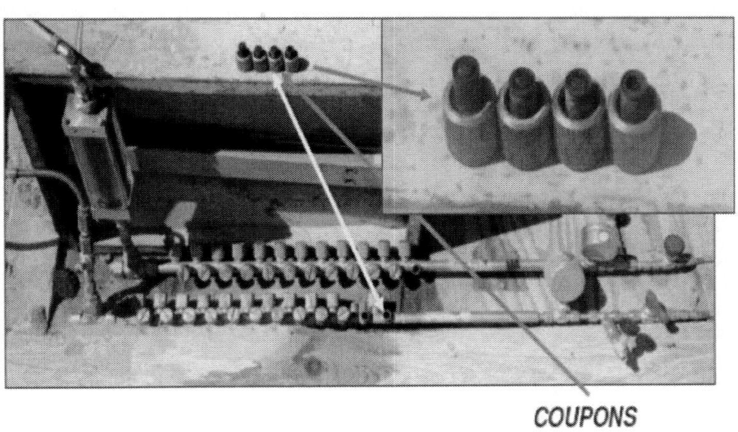

Figure 14. Monitoring of Corrosion Caused by SRB, by Using Coupons Inside the System and Calculating the Weight Loss and Determining the Corrosion Rate.

3.1. Serial Dilution Method

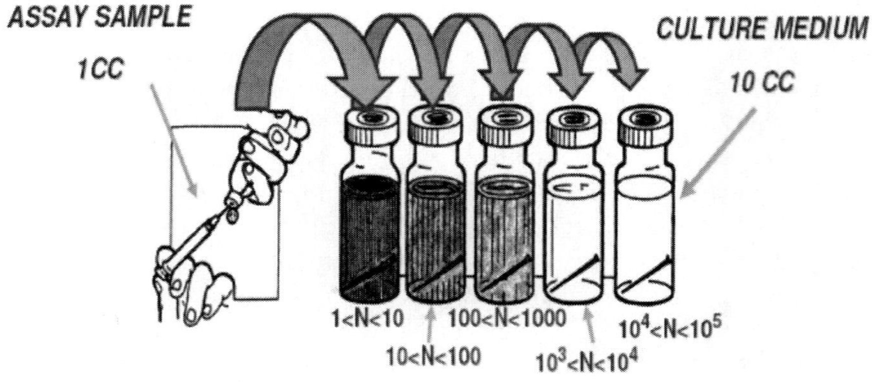

Figure 15. Method of Detecting Bacterial Counts in the Field.

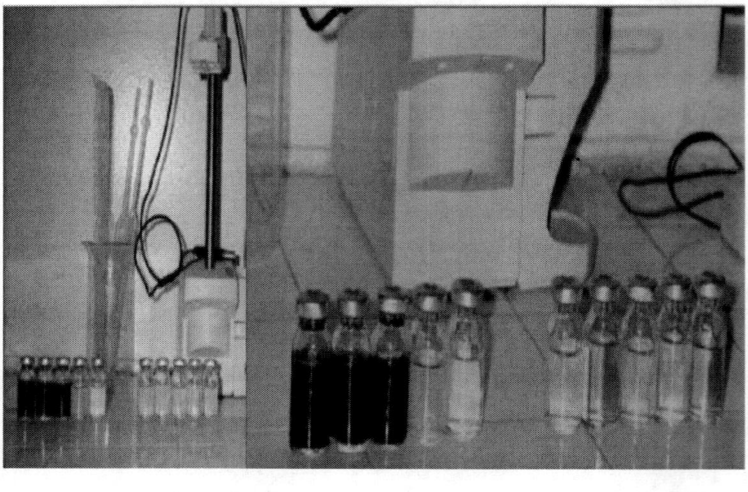

Figure 16. Method of Detecting SRB Counts in the Laboratory.

3.2. Evidence of SRB

Sulphate reducing bacteria are considered as prominent aggressive micro-organisms causing unexpected failure of materials. These micro-organisms are mainly concerned with the effect of sulfide ion produced, by dissimilatory reduction of sulfate, by bacterial metabolic activity and by-products from its oxidation and reduction. The iron sulfides (FeS_x), formed by the precipitation of biogenic sulfide with ferrous ion, can act as a cathode in the galvanic couple and enhance corrosion. The enzyme, hydrogenase, present in bacterial cell membrane, can consume protons to generate molecular hydrogen and accelerate the direct electron transfer between the hydrogenase and zero valent metal. The enhanced cathodic activity of the FeS corrosion product can significantly enhance corrosion in anaerobic condition even in the absence of an efficient oxygen cathode. All of previous statements are considered as evidences of SRB which could be summarized in particularly as follows:

3.2.1. Metallurgical Evidence

The metallurgical history of the alloy in use is important, as is the morphology of the corrosive attack. Some types of MIC are recognizable, in part, by the pattern of corrosion on the surface, as detailed elsewhere [42, 43]. The distribution and morphology of the corrosive attack should be recorded photographically.

3.2.2. Microbiological Evidence

Sampling of biological materials and corrosion products must be done using proper procedures, moreover; the data must be gathered while the corrosion site is still wet and the corroded component is removed from the system, discarded, and allowed to dry before it is examined. It is important to photographically record the initial appearance of the corrosion site in color as soon as possible, so that the organisms are still alive [44, 45].

3.2.3. Chemical Evidence

Any localized inputs of chemicals or pollutants not normally present in the environment should be noted [45]. Detailed chemical information should also be gathered for the corrosion products and any biological mounds that may be present at or near the corrosion site. The color, texture, odor, and distribution of the materials as well as their organic and inorganic chemistries should be included in the gathered information.

3.2.4. Electrochemical Evidence

The electrochemical tests have proven useful within certain limitations for MIC testing [46]. There are many conventional techniques such as variation of corrosion potential and corrosion rates at different periods. Noise electrochemical techniques should be very fruitful for monitoring corrosion evolution [47]. Open circuit corrosion potential measurements have been used in MIC studies for many years. It has been often reported that the microbial films changed the corrosion potential or the open circuit corrosion potential of passive metals that immersed in natural seawater [48-53]. Measurement of OCP changing with time is important for estimating the effect of depolarizers on corrosion reactions. A plot of potential as a function of time could be useful to detect the initiation or an accelerated attack of SRB [54]. An increase of OCP means depolarization of cathode and increase corrosion, a drop in potential is evidence for decreased corrosion. Also rapid changes in the corrosion potential can be used to indicate a depolarization or enhancement of the anodic reaction, or to the formation of a semi-protective film [55]. Suggested mechanisms of corrosion in anaerobic environment by SRB are mainly concerned with the effect of sulfide ion produced, by dissimilatory reduction of sulfate, bacterial metabolic activity and products from its oxidation and reduction [56]. Microbiologically induced corrosion (MIC) is referred to the accelerated deterioration of the materials, owing to the metabolism or the presence of bacteria. MIC has been inviting quite a lot of attention of many research groups in the recent ten years [57, 58].

4. Control of SRB Corrosion

The general approaches to maintaining a system free of biocorrosion problems vary with the materials of construction, environment, economics, and duty cycle of the equipment. The most common approaches involve the use of sterilization, coatings, cathodic protection and appropriate selection of materials. The most important step in prevention is to start with a clean system and to keep it clean.

4.1. Physical Methods

The physical method by which, we can clean the infected systems by irradiation using gamma or UV for disinfection of materials and environments.

4.2. Chemical Methods

The chemical methods are concerned with using of chemical compounds have the ability to kill or suppress the growth of microorganisms called biocides. The biocidal action has been widely used for many years to control biofilm formation in closed systems, such as heat exchangers, cooling towers, and storage tanks.

4.3. Alkali Wash

We have found that an alkali wash after pigging is very effective in killing SRB. The high-pH kills the bacteria and it is also usually easier to dispose of alkali solutions than to dispose of acid wash effluent. We would therefore not usually recommend an acid clean of slurry pipelines, particularly when the slurries that are conveyed by the pipeline are basic in nature such as lime.

4.4. Biocide Injection

Biocide treatments are widely used to kill or inhibit the growth of microorganisms and decreasing the biofouling and MIC in steel pipes and in closed systems. The cathodic protection is also used to prevent MIC when used with coatings. On the other hand, the use of biocides and cathodic protection techniques are very expensive for industrial applications. Some of the used biocides cause environmental pollution due to its higher toxicity, and its harmful effects on metal structures if they are used at concentrations in excess of a few ppm. The traditional method for biocides evaluation is the minimum inhibitory concentration (mic) but it does not tell much about the particular biocide actually protect against (SRB) [59, 60].

Therefore, in the present work the electrochemical methods has been used to commercial preparations against a stabilized mixed culture containing SRB (SMC–SRB) and biofilm bacteria isolated from mild steel coupon immersed for one month in the sea water. Regular treatment with biocides and corrosion inhibitors, in addition to mechanical cleaning should be used to control bacterial growth and thereby biocorrosion. Control of sessile bacteria, mostly SRB is important for controlling SRB. Carefully planned and well executed biocide programs can be highly cost effective if they can prevent the problems created by bacteria [60,61]. Although it is possible to use biocides to kill and/or control the numbers of bacteria and inhibitors to reduce the rate of corrosion, their use in once-through systems is limited for the following reasons:

- The volumes of chemicals required in once-through systems can make treatment prohibitively expensive
- Treatment and disposal of the chemicals at the end of the pipeline may be problematic and expensive
- SRB can adapt to biocides. This necessitates that biocide treatment regimes are rotated between oxidizing and non-oxidizing biocides. It is therefore more practical to batch biocides, if they are going to be used, between cleaning pigs, following hard cleaning of the pipeline first to break down the tubercles.

4.5. Mechanical Methods

Microbial colonization of metals and alloys caused by the formation of biofilms made of bacteria, extracellular polymeric substances (EPS) and mainly water. These biological deposits can drastically modify the corrosion behavior of metals and alloys and specifically enhancing localized corrosion depending upon alterations in the type and concentrations of ions, pH, and oxygen levels. The biofilm formed have to subject for treatment to minimize the problems due to biocorrosion and biofouling of industrial systems range from heavy microbiological contamination with consequent energy and efficiency losses to structural failures owing to corrosion. Mechanical method for removing biofilm is considered as one of the most important method for cleaning systems. Pigging is the most famous mechanical method used for get rid the fouling and biofilm from the piplines.

4.5.1. Pigging

It is difficult to remove SRBs once they have become established. Simply injecting biocides into the slurry will not work because they bacteria are located beneath the tubercles. In order to get to the bacteria, the scale and tubercles need to be removed. This can be achieved using hard pigs and the degree of abrasion can be designed to suit the application. Once the scale and tubercles have been removed, it is possible to pump appropriate chemicals through the pipeline to kill the bacteria.

5. Industries Affected by Microbial Corrosion

Not only the petroleum companies which affected by SRB corrosion but also various chemical and petrochemical industries have been affected by SRB corrosion problems. Figure 1 illustrate the problems in the chemical processing, nuclear power, oil field (both onshore and offshore), and underground pipeline industries are most common in the published literature. This does not necessarily mean that these industries are more prone to problems of this type than other industries. It does mean that these industries have been quick to recognize the biological causes of some of their corrosion problems, to take steps to solve these problems, and to make the results public. More recently, biological corrosion represents a growing challenge to the fire protection industry.

Industry or structure	Problem areas
Chemical processing	Stainless steel tanks, pipelines, and flanged joints, particularly in welded areas after hydro-testing with natural river or well waters
Power generation: nuclear, hydro, fossil fuel, and geothermal	Carbon and stainless steel piping and tanks; copper-nickel, stainless, brass, and aluminum-bronze cooling water pipes and tubes, especially during construction, hydro-test, and outage periods
Onshore and offshore oil and gas	Mothballed and water flood systems; oil- and gas-handling systems, particularly in those environments soured by sulfate-reducing-bacteria (SRB)-produced sulfides
Underground pipeline	Water-saturated clay-type soils of near-neutral pH with decaying organic matter and a source of SRB
Water treatment	Heat exchangers and piping
Sewage handling and treatment	Concrete and reinforced-concrete structures
Highway maintenance	Culvert piping
Buildings and stone works	Ancient and modern lime stones, marbles, sandstones, granites, and mortars
Pulp and paper	Raw, process, and cooling water systems; paper machine head boxes and suction rolls
Hydraulic systems	Increased corrosivity of hydraulic oils with entrained water; fouling and corrosion of filters
Fire protection systems	Pinhole leaks in piping systems; plugging of piping and sprinkler heads
Metalworking	Increased wear from breakdown of machining oils and emulsions

Figure 17. Industries Affected by SRB Corrosion.

6. Cost of SRB Corrosion

The petroleum companies suffer from SRB corrosion and it is estimated that in one of the Egyptian Companies (Gulf of Suez Petroleum Co., GUPCO) is more than $ 1 million per year [62] but it is estimated about $ 4–6 billion/year in the in United States.. The reason for the occurrence of these problems is that microorganisms are found in the treated seawater used for injection into the producing wells to enhance oil recovery [63]. The anaerobic sequencing batch reactor (ASBR) has received much attention in recent years, mainly in industrial wastewater treatment. Compared to similar continuous reactors a larger useful volume for the treatment is generally required; however it does away with the use of a separated sedimentation, [64,65] The action of the SRB is not limited to mild steel and has been reported for copper and its alloys and stainless steel [66,67]. The literature survey of MIC problems proved that, not all produced corrosion caused by bacteria but in many cases bacteria existing in natural and man-made environments do not cause or accelerate corrosion.

Bacteria may exist at the corrosion site only because the electrochemistry of corrosion creates a favorable environment for their growth.

7. Recent Work on Sulfate-Reducing Bacteria

Sulfate reducing bacteria play significant role in the corrosion of stainless steels exposed to marine environment in several industrial plants especially in the extraction and processing oil and gas industry [68]. The electrochemical characteristics of 1Cr18Ni9Ti in sulphate-reducing bacteria (SRB) solutions and the biofilm of SRB on the surface of the 1Cr18Ni9Ti electrode were studied by electrochemical, microbiological, and surface analysis methods [69]. The sulphate reducing bacteria was isolated from coastal samples, by using Starkey's medium and detection of main groups of SRB were carried out using commercially kits [70]. The development of biofilm contained a bacterial population with black and rusty appearance. The biofilm bacteria were collected from the mild steel surfaces and a stabilized mixed culture containing sulphate reducing bacteria (SMC-SRB) was obtained through enrichment on Postage medium B [71]. A stabilized mixed culture containing sulphate reducing bacteria (SMC-SRB) was isolated and used to study the effect of two alternated biocides and two commercial corrosion inhibitors on the production of H_2S [72]. The corrosion potential of AISI 304 stainless steel coupons influenced by (SRB) has been studied [73]. The MIC is used to address the corrosion enhanced by micro-organisms [74]. The growth and Mineralization of sedimentary biofilms (microbial mats) through precipitation of calcium carbonate may have been the first step in the preservation of stromatolites, fossil forms of the oldest life on Earth [75-77].

Conclusion

Conventional techniques typically used by slurry pipeline operators have severe limitations in assessing the actual integrity of slurry pipelines and all integrity plans based solely on such techniques result in a false sense of security. Sulphate-reducing bacteria are responsible for the bulk of the microbial corrosion problems experienced in slurry pipelines. Microbially influenced corrosion (MIC) leads to perforation and leakage of slurry pipelines and therefore, is the biggest contributor to losses arising out of biofilms made on metal surface by such bacteria. The products of electrochemical reactions provides a conducive environment for microbial growth because of liberating minerals utilized as nutritional sources by the growing bacteria on metal surface. There are many techniques which are used for monitoring and controlling MIC, but foremost, it is very important to make mechanical removal of biofilm and then washing the surface by alkali before injection of biocide. More research and focus is needed to avoid damage to man-made structures from such biofilm forming metal eating bacteria.

Acknowledgment

I would like to express my very great appreciation to Dr Smar Mouneir for his valuable and constructive suggestions during the planning and development of this chapter. His willingness to give his time so generously has been very much appreciated. I would also like to thank Dr Shadia for her advice and assistance in keeping my progress on schedule of this work. I also wish to thank all members in physical chemistry department, National Research Centre for their assistance and guidance. My grateful thanks are also extended to my family my wife Samer, my daughter Sara and my son Omar for their help in saving time for the completion of this work.

References

[1] American Water Works Association (1995) *Sulfur Bacteria.* Chapter (3) in AWWA Manual M7, AWWA Denver, Co. pp. 671-676

[2] Shams El-Din AM, Saber TMH, Hammoud AA (1996) Biofilm formation on stainless steels in Arabian Gulf Water. *Desalinat* 107:251-264

[3] Costerton JW, Geesey GG, Ladd TJ, Nickel JC, Dasgupta M, Marrie TJ (1987) Bacterial biofilms in nature and disease. *Ann Rev Microbiol* 41:435-464

[4] Charaklis WG, Marshal KC (1990) Biofilms: a basis for an interdisciplinary approach. *John wiley & Sons,* Inc. New York

[5] Tantal RE (1988) Fundamentals of Bacteria Induced Corrosion. *Materials Performance* 20:32-38

[6] Hamilton WA (1985) Sulphate-reducing bacteria and anaerobic corrosion. *Annu Rev Microbiol* 39:195-217

[7] Licina GJ (1988) Sourcebook for microbiologically influenced corrosion in nuclear power plants RP2812-2. Electric Power Research Institute, Palo Alto, Calif

[8] Beloglazov SM, Dzhafarov ZI, Polyakov VN, Demushin NN (1991) Quaternary ammonium salts as corrosion inhibitors of steel in the presence of sulfate reducing bacteria. *Prot Met* (USSR) 27:810-813

[9] Miller JDA (1981) Metals. In: Rose AH Ed. Microbial biodeterioration. *Academic Press,* New York, 149-202

[10] Borenstein SW (1994) Microbiologically Influenced Corrosion Handbook. *Woodhead,* Cambridge, England

[11] Postgate J (1984) 2nd Edn. The Sulfate Reduced Bacteria. *Cambridge Univ. Press*

[12] Iverson WP (1972) Biological Corrosion. *Adv Corr Sci Technol* 2:1-41

[13] Edyvean RGJ, Thomas CJ, Brook R (1986) The Use of Biologically Active Environments for Testing Corrosion Fatigue properties of Offshore Steels. In *Biologically Induced Corrosion.* 245-267. Ed; S. C. Dexter, NACE-8

[14] Tiller AK (1990) Biocorrosion in civil engineering. In *Microbiology in Civil Engineering (Ed.).* London: E. & F.N. Spon

[15] Smart JS, Smith GL (1998) Pigging and Chemical Treatment of Pipelines. Paper Presented At the Pipeline Pigging Conference, Clarion Technical Conferences, Houston, TX, February

[16] Costlow JD, Tipper RC (1984) Ed., Marine Bio-deterioration: An Interdisciplinary Study, Proceedings of the Symposium, *Naval Institute Press*

[17] Marshall KC (1976) Interfaces in Microbial Ecology, *Harvard University Press*

[18] Savage DC, Fletcher M (1985) Ed., Bacterial Adhesion, *Plenum Press*

[19] Evans LV (2000) Ed., Biofilms: Recent Advances in Their Study and Control, *Harwood Academic Publishers*

[20] Dexter SC (2003) Microbiological Effects, Corrosion Tests and Standards, 2nd ed., R. Baboian, Ed., *ASTM Manual* 20, ASTM

[21] De Beer D, Stoodley P (1995) Relation between the Structure of an Aerobic Biofilm and Transport Phenomena. *Water Sci Technol* 32(8):11–18

[22] De Beer D, Stoodley P, Lewandowski Z (1996) Liquid Flow and Mass Transport in Heterogeneous Biofilms. *Water Res* 30(11):2761–2765

[23] Miller JDA (1970) Ed., *Microbial Aspects of Metallurgy,* Elsevier

[24] Microbial Corrosion, Proceedings of the Conference, (1983). *National Physical Laboratory,* The Metals Society

[25] Hamilton WA, Maxwell S (1986) Biological and Corrosion Activity of SRB in Natural Biofilms, *Proceedings of Biologically Induced Corrosion,* NACE International, p 131–136

[26] Dexter SC (1986) Ed., Biologically Induced Corrosion, *Proceedings of the Conference,* National Association of Corrosion Engineers

[27] Maxwell S, Hamilton WA (1986) The Activity of Sulphate Reducing Bacteria on Metal Surfaces in an Oilfield Situation. pp: 284-290, in Biologically Induced Corrosion, NACE 8, Ed., S. C. Dexter

[28] Chen G, Clayton CR (1997) Influence Of Sulphate Reducing Bacteria on the Passivity of Type 304 Austenitic Stainless. *J Electrochem Soc* 144:3140-3146

[29] Cullimore R (2000) Practical Atlas for Bacterial Identification. *Boca Raton:* CRC Press.

[30] Odom JM (1990) Industrial and Environmental Concerns with Sulphate Reducing Bacteria. *ASM News* 56:473-476

[31] Odom JM, Singleton R (1992) The Sulfate Reducing Bacteria: Contemporary Perspectives. New York: *Springer Verlag*

[32] Hamilton WA (1994a) Biocorrosion: The Action of Sulphate Reducing Bacteria. In Biochemistry of Microbial Degradation. C. Ratledge. Dordrecht: *Kluwer Academic*

[33] Battersby PA (1988) Sulphate reducing bacteria. In methods on aquatic bacteriology, B. Austin (Ed.). New York: *John Wiley & Sons*. pp. 269-299

[34] Peng CG, Park JK, Patenaude RW (1994) Statistics Based Classification of Microbially Influence Corrosion In Freshwater Systems. *Wat Res* 28:951-959

[35] Gibson GR (1990) Physiology and Ecology of the Sulfate Reducing Bacteria. *J Appl Bacteriol* 69:769-797

[36] Little B, Wagner P, Mansfeld F (1991) Microbiologically Influenced Corrosion of Metals and Alloys, *Int Mater Rev* 36(6):253-272

[37] Linhardt P (1996) Failure of Chromium-Nickel Steel in a Hydroelectric Power Plant by Manganese-Oxidizing Bacteria, Microbially Influenced Corrosion of Materials, E. Heitz et al., Ed., *Springer-Verlag*

[38] Olesen BH, Avci R, Lewandowski Z (2000) Manganese Dioxide as a Potential Cathodic Reactant in Corrosion of Stainless Steels. *Corros Sci* 42:211–227

[39] Xu K (2000) Effect of Biofilm Heterogeneity on Corrosion Behavior of Passive Alloys in Seawater. Ph.D. dissertation, University of Delaware

[40] Borenstein SW, Lindsay PB (1994) MIC Failure Analyses, A Practical Manual on Microbiologically Influenced Corrosion, G. Kobrin, Ed., *NACE International,* p 189

[41] Little BJ, Wagner PA, Mansfeld F (1997) Microbiologically Influenced Corrosion, Corrosion Testing Made Easy, Vol. 5, B. C. Syrett, Series Ed., *NACE International*, p 23

[42] Little B, Wagner P (1994) Indicators for Sulfate-Reducing Bacteria in Microbiologically Influenced Corrosion, Biofouling and Biocorrosion in Industrial Water Systems, G.G. Geesey, Z. Lewandowski, and H.-C. Flemming, Ed., Lewis Publishers, CRC Press Inc., *Boca Raton,* FL, p 213–230

[43] Videla HA (1996) Manual of Biocorrosion, *CRC Press, Inc.,* p 13–45, 121–133, 160–173, 179–185, 228-237, 232

[44] Videla H A, Bianchi F, Freitas MMS, Canales CG, Wilkes JF (1994) Monitoring Biocorrosion and Biofilms in Industrial Waters: A Practical Approach, Microbiologically Influenced Corrosion Testing, Special Technical Publication 1232, J.R. Kearns and B.J. Little, Ed., *ASTM,* p 128

[45] Ruseski I, Robbins J, Costerton JW (1982) Biocide Testing Against Corrosion-Causing Oil-Field Bacteria Helps Control Plugging. *Oil Gas J* 80:253

[46] Sanders PF, Hamilton WA (1986) Biological and Corrosion Activities of Sulphate-Reducing Bacteria in Industrial Process Plant, Biologically Induced Corrosion, S. C. Dexter, Ed., *NACE International*, p 47

[47] Videla HA, deMelle MFL, Silva RA, Bianchi F, Gonzales Canales C (1990) A Practical Approach to the Study of the Interaction Between Biofouling and Passive Layers on Mild Steel and Stainless Steel in Cooling Water. Paper 124, *Corrosion/* 90, *NACE International*

[48] Franklin MJ, Nivens DE, Vass AA, Mittelman MW, Jack RF, Dowling NJE, White DC (1991) Effect of chlorine and chlorine/bromine biocide treatments on the number and activity of biofilm bacteria and on carbon steel corrosion. *Corr* 47:128-134

[49] Fontana MG (1986) Corrosion engineering, 3rd edn. *McGraw-Hill,* New York

[50] Pankhurst ES (1968) Significance of sulphate-reducing bacteria to the gas industry: a review. *J Appl Bacteriol* 31:179-193

[51] Cord-Ruwisch R, Kleinitz W, Widdel F (1987) Sulfate-reducing bacteria and their activities in oil production. *J Petrol Technol* 39:97-105

[52] Boivin J (1995) Oil Industry Biocides. *Mat Perf* Feb:65-68

[53] Flemming HC, Wingender J (2001) Relevance of Microbial Extracellular Polymeric Substance (EPS). Part II Technical Aspects. *Wat Sci Technol* 43(6):9-16

[54] El-Raghy SM, Ghazal HH, Abo El- Liel HM () Material Induced Corrosion of Subsea Pipeline in the Gulf of Suez. *SEP* 37791

[55] Ateya BG, El-Raghy SM, Abdelsamie ME, Mohamoud MN, Bayoumie RS (2008) Microbial Induced Corrosion (MIC) II-Effect of Some Biocides and Corrosion Inhibitors on a Stabilized Mixed Culture Containing Sulphate Reducing Bacteria (SMC-SRB) *J App Sci Res* 4(12): 1805-1810

[56] Dague RR, Habben CE, Pidaparti SR (1992) Initial Studies on The Anaerobic Sequencing Batch Reactor. *Wat Sci Technol* 26:2429-2432

[57] Sung S, Dague RR (1995) Laboratory Studies On The Anaerobic Sequencing Batch Reactor, *Wat Environ Res* 67:294-301

[58] Angell P, Luo JS, White DC (1995) Microbially Sustained Pitting Corrosion of 304 Stainless Steel in Anaerobic Seawater. *Corr Sci* 37:1085-1096

[59] Sarioglu F, Javaherdashti R, Aksoz N (1997) Corrosion of Drilling Pipe Steel in an Environment of Containing Sulphate Reducing Bacteria. *Int J Pres Ves Piping* 73:127-131

[60] Johnsen R, Bardal E (1985) Cathodic Properties of Different Stainless Steels in Natural Seawater. *Corr* 41:296-304

[61] Scotto V, Di Cintio R, Marcenaro G (1985) The Influence of Marine Aerobic Microbial Film on Stainless Steel Corrosion Behavior. *Corr Sci* 25(3):185-194

[62] Dexter SC, Gao GY (1988) Effect of Seawater Biofilms on Corrosion Potential and Oxygen Reduction of Stainless Steel. *Corr NACE* 44(10):717-723

[63] Scotto V (1989) Electrochemical Studies of Biocorrosion of Stainless Steel in Seawater. *Proc. EPRI Workshop, Microbial Corrosion: 1988*. Electric Power Research Institute, Palo Alto, CA, pp. 1-36

[64] Angell P, Urbanic K (2000) Sulphate Reducing Bacterial Activity as a Parameter to Predict Localized Corrosion of Stainless Alloys. *Corr Sci* 42:897-912

[65] Werner SE, Johnson CA, Laycock NJ, Wilson PT, Webster BJ (1998) Pitting of type 304 Stainless Steel In The Presence of Biofilm Containing Sulphate Reducing Bacteria. *Corr Sci* 40:465-480

[66] Tuovinen OH, Cragnolino G (1986) *Proceedings of the Conference on Corrosion Monitoring in Industrial Plants Using Nondestructive Testing and Electrochemical Methods, Montreal, Canada,* May 1986 pp.413-432

[67] Little BJ, Wagner P, Hart K, Ray R, Lavoie D, Nealson K, Aguilar C (1997) The Role of Metal Reducing Bacteria In Microbiologically Influenced Corrosion. Paper No. 215, *Proc. NACE Corrosion'97*. Houston, TX: National Association of Corrosion Engineers International.

[68] Videla HA, Characklis WG (1992) Biofouling and Microbially Influenced Corrosion. *Int J Biodeg Biodet* 29:195-207

[69] Nagiub A, Mansfeld F (2002) Evaluation of Microbiologically Influenced Corrosion Inhibition (MICI) with EIS and ENA. *Electrochem Acta* 47(13-14):2319

[70] Crolet JL (1992) From Biology and Corrosion to Biocorrosion, in *Proceedings of the 2nd European Federation of Corrosion Workshop*, Portugal, p. 50

[71] Lewandowski Z, Dickinson W, Whonchee L (1997) Electrochemical Interactions of Biofilms with Metal Surfaces. *Wat Sci Technol* 36(1):295

[72] Xu LC, Wong K, Chan Y (2002) Application of Atomic Force Microscopy in the Work Of Microbiologically Influenced Corrosion. *Mat Charact* 48(2-3):195

[73] Iwona BB, Sunner J (2004) Biocorrosion: Towards Understanding Interactions between Biofilms and Metals. *Curr Opin Biotechnol* 15(3):181

[74] Sekaran G, Chongdar S, Gaonkar SN (2004) Influence of Bacteria on Film Formation Inhibiting Corrosion. *Corr Sci* 46(8):1953

[75] Popova A, Sokolova E, Raicheva S (2003) AC and DC Work of the Temperature Effect on Mild Steel Corrosion in Acid Media in The Presence of Benzimidazole Derivatives. *Corr Sci* 45(1):33

[76] Franklin M J (1988) Effect of Electrochemical Impedance Spectroscopy on the Microbial Biofilm Cell Numbers. *Viab Act Corr* 47(7):519

[77] Roberge PR, Sastri VS (1991) Online Corrosion Monitoring with Electrochemical Impendence Spectroscop. *Corr* 50(10):744

In: Microbes in Process
Editors: Neelam Garg and Abhinav Aeron
ISBN: 978-1-63117-127-7

Chapter 15

Production and Application of Siderophore

***Kyoung-Ja Kim*[1,*] *and Young-Joon Yang*[2]**
[1]Dept. of Life Science and Biotechnology, University of Soonchunhyang, Korea
[2]Department of Plant Science and Technology, Sangmyung University, Cheonan, Korea

Abstract

Iron is an essential micronutrient for all microorganisms. Although abundant, iron is mostly in the oxidized insoluble form under physiological conditions and the reduced form is highly toxic for most macromolecules and in biological systems. During evolution, micro-organisms have developed strategies to acquire iron from both the environment and superior organisms, including direct uptake of iron ions from exogenous iron/heme sources and the synthesis of siderophores. Siderophores are defined as low molecular weight, ferric ion specific chelating agent produced by bacteria and fungi growing under iron stressed conditions. Siderophores have been related to virulence mechanisms in bacteria and fungi pathogenic to both animals and plants. The present review paper aims at presenting and discussing the latest achievements in siderophore production, isolation and structure, as well as applications of these molecules in therapies against iron-related diseases, and their applications as antimicrobial-, anti-parasite- and anti-tumor agents and those in the field of agriculture.

1. Introduction

Iron is one of the most abundant elements in the crust of the earth. It can exist in two oxidation states but in the presence of high concentrations of oxygen it is exclusively present in the ferric (Fe^{3+}) form, which is not highly soluble. The ferrous (Fe^{2+}) ion has a much better solubility, but it only exists under anoxic conditions or at very low pH [1]. Iron is an essential

* kyoungjakim@hotmail.com.

element involved in many important cellular processes such as DNA synthesis, for formation of heme, reduction of oxygen for synthesis of ATP, respiration and detoxification of free radicals [2]. In the environment iron exists as complex hydrated oxides of Fe (III), many of which are nonstoichiometric and quantitatively insoluble. The low availability of iron in the presence of oxygen and neutral or basic pH limits growth of microorganisms. Therefore, efficient systems for iron transport and acquisition are important for the survival of nearly all microorganisms [3]. In order to survive, most of the microorganisms have developed numerous mechanisms to scavenge iron [4]. To obtain and transport the necessary iron, the majority of aerobic and facultative anaerobic bacteria secrete siderophores, which are highly efficient and specific iron chelators [5]. These small organic molecules have a very high affinity for iron and can create very stable complexes with ferric iron. These complexes are more soluble in water and therefore make iron available for fungi or bacteria. In pathogenic microorganisms, siderophores have been suggested as virulence factors since they can interact with some host iron-containing proteins like transferrin. Siderophores are able to remove iron from these proteins directly [6] or in combination with a proteolytic process [7]. Furthermore, it has been demonstrated that siderophore biosynthesis is crucial for virulence in several human, animal or plant pathogenic microorganisms [8, 9], including the opportunistic fungus *Aspergillus fumigatus* [10, 11] and plant pathogenic bacteria *Erwinia carotovora* and *Erwinia chrysanthemi.* Iron is particularly important for photosynthetic organisms, such as cyanobacteria, as it plays a key role in photosynthetic electron transfer systems [12]. In the present review paper, the recent achievements in siderophore production will be presented and discussed, as well as clinical applications including the use of siderophores as antimicrobial-, antiparasitic-, and anti-tumor agents and their application in the field of agriculture.

2. Production of Siderophores

Microbial siderophores obtained from fungi and several bacterial genera are iron-chelating secondary metabolites which show great variability in their chemical structures, and sequester iron

2.1. Siderophore and Iron

Siderophores are low molecular weight iron binding compounds (200–2000 Da) produced by bacteria, fungi and plants in response to the low levels of this essential element available in the environment [13, 14]. These low levels of available iron stem from its profound insolubility at neutral pH in an aerobic atmosphere. The siderophores serve not only to solubilize Fe^{3+} but also to transport, and in some cases store, it in microbial cells. Most of the known siderophores employ either the hydroxamate or catecholate groups in forming very specific and extremely stable octahedral complexes with the ferric ion. Iron is essential for the growth and development of almost all living organisms, owing to the efficiency with which it converts between its Fe (II) and Fe (III) redox states [15]. Thus, through appropriate ligation, a range of redox potentials are utilized for cellular metabolism. In this way, Fe participates in oxygen transport, DNA synthesis, oxygen sensing, electron transport, and many other biological processes [16, 17]. Some notable exceptions are *Lactobacilli*, *Legionella*, *Neisseria*, and the fungus *Sacchoromyces cervisiae* [18]. For example, *Lactobacilli* have no heme enzymes and use the cobalt form of ribonucleotide reductase, and so do not require iron [19]. In most aerobic microbial habitats, iron is found in insoluble oxide hydrate complexes. During infections, iron is similarly not freely accessible for the pathogen, as it is tightly sequestered by proteins or haems in the host. Therefore, the free iron concentrations are always extremely low in the environment of microorganisms (10^{-7} to 10^{-24} M) [20]. The range of siderophore concentrations found in soil is wide, from tens of micromoles to a few millimoles per litre [21]. The siderophores play an important role in extracellular solubilization of iron from minerals so as to make it available to microorganism(s). In pathogenic microorganisms, siderophores have been suggested as virulence factors since they can interact with some host iron-containing proteins like transferrin. Siderophores are able to remove iron from these proteins directly [6] or in combination with a proteolytic process [7]. Furthermore, it has been demonstrated that siderophore biosynthesis is crucial for virulence in several human, animal or plant pathogenic microorganisms [8, 9], including the opportunistic fungus *Aspergillus fumigatus* [10, 11].

2.2. Detection and Isolation of Siderophore

Detection of siderophores is most readily achieved in iron limited media, which generally means either a synthetic (minimal) recipe or introduction of a complexing agent that will render the iron selectively unavailable. Although most siderophores are either hydroxamates or catechols, earlier tests based on such functional groups proved unreliable since they are absent from a few siderophores. The chrome azurol sulfonate (CAS) assay [22] has become widely used since it is comprehensive, exceptionally responsive, and more convenient than microbiological assays which, although sensitive, may be rigidly specific. CAS assay may be applied on agar surfaces or in solution. It is based on the color change that accompanies transfer of the ferric ion from its intense (extinction coefficient of at least 100,000) blue complex to the siderophore. A positive reaction consisted in a color change of CAS reagent from blue to orange, leading to a clear visible red-orange halo around the microbial colony. A detergent must be present in order to achieve the intense color; otherwise, only a charge-transfer extinction of a few thousand is realized. Gram-negative bacteria are impervious to detergents and hence CAS reagent can be incorporated in the agar media where it has the potential for isolation of biosynthetic, regulatory, and transport mutants. For Gram-positive bacteria and fungi, the toxicity of the reagent must be mitigated in some way [23]. These qualifications, however, do not compromise use of the dye in solution or as a spray on chromatograms [22]. Overlaid CAS (O-CAS) method [24] was reported to solve the problem of CAS assay method, which presents a problem of growth inhibition for Gram positive bacteria and fungi, due to the presence of hexadecyltrimetyl ammonium bromide (HDTMA) in the cultivation medium, which, at high concentrations, is toxic to these microorganisms. Since siderophores differ substantially in structure, no uniform procedure is available for their isolation. A preliminary examination by paper electrophoresis should reveal the charge profile as a function of pH, following which appropriate exchange resins can be applied for retention and elution of the compound(s). Most are water-soluble, and it is thus usually expedient to drive the siderophore into an organic solvent, such as benzyl alcohol or phenol-chloroform, in order to eliminate salt. The siderophore may be isolated *per se* or as its iron chelate. The latter has the advantage of visual color, but the iron must be removed before any natural product can be characterized. Vigorous hydrolysis in the presence of iron will destroy oxidizable moieties, and direct NMR analysis is ruled out by the paramagnetism of the ferric ion. In general, after the fermentation step, cells are usually removed by filtration, or centrifugation [25-29] allowing the processing of the siderophore-rich liquid phase. Prior to further processing, the filtrate can be concentrated, viz. by evaporation under vacuum [27-30] occasionally after extraction with organic solvent which already involves a purification step [29, 31] and by lyophilization [25] for more efficient manipulation. In any case, the supernatants are eluted through chromatographic columns, with adequate resins and mobile phase composition. Affinity column chromatography for catechol type- or hydroxamate type siderophores were also available [32, 33]. Some examples of siderophore purification steps are given in Figure 01.

Process (Reference)	Comments
Production of corynebactin (34)	ion exchange resin AG1X2 washing with 2.5% ammonium acetate, pH6.5, followed by of 6% ammonium acetate, pH 6.5 lyophilized, and dissolved in deionized water Source15Q ion exchange column chromatography Elution with ammonium acetate, pH 6.5 (from 2.5% to 7.5%, final with 15%)
Production of chrysobactin (32)	XAD-4 column chromatography Elution with 50% methanol Evaporation and lyophilization Boronate affinity column chromatography Elution with water (pH 5.0)
Production of exochelin (28)	Cation-exchange column chromatography Elution with 1 M NH_40H/NH_4C1, pH 9.5, Ion exchange column chromatography (high-resolving sulphonic acid ion-exchange resin) Elution with a linear gradient of 0.1 M NH_40H /acetic acid, pH 9.0,
Production of legiobactin (35)	Anion-exchange resin DEAE Sephadex A-50 Elution with MOPS buffer (NaCl gradient) Evaporation and desalting HPLC (anion exchange column) Elution with 100% A (double-distilled water, pH 5.8), decreased to 40% A while B (400 mM NaCl in ddH_2O, pH 5.8) increased to 60%
Production of staphyloferrin A (36)	Adsorption on anion exchanger IRA 401 S elution with 3M formic acid Neutralization and precipitation with methanol Gel chromatography on Biogel P2 Gel chromatography on Fractogel TSK HW 40 Acidification with TFA (pH 2) and immediate lyophilization
Production of vibrioferrin (37)	Dowex 1x8 ($HCOO^-$) resin Wash with water and elution with 3 N formic acid Evaporation Bio-gel P2 column chromatography Elution with water and evaporation TSK-gel Toyopearl HW-40 F column chromatography Elution with water-methanol (9:1) Amberite CG-120(H^+ form) Column Prep TLC (n-butanol-formic acid-water = 20:20:1)

Figure 1.

2.3. Structure of Siderophore

Siderophores have a variety of chemical structures and form a family of at least 500 different compounds [13, 14]. They possess up to six oxygen or nitrogen electron donor atoms, which can bind metal cations. Microbial siderophores are classified (Figure 2.) as hydroxamates, catecholates, carboxylates, as a mixed type, and as the recently discovered amphiphilic siderophores [38] based on the chemical nature of their coordination sites. Most siderophores are biosynthetically produced by large multienzyme synthases that resemble the eukaryotic fatty acid synthases [1, 39, 40]. They often have a peptidic backbone, with modified amino acid side chains creating the iron-coordinating ligands, such as the catecholate, hydroxamate and hydroxycarboxylate groups. Each group provides two oxygen ligands in a fixed orientation, and a total of six ligands normally surround the ferric iron in an octahedral fashion. Chemical assays were performed on concentrated culture filtrates. Production of phenolate/catecholate type siderophores was investigated by the Arnow test [41] and absorption at 318 nm [42]. Hydroxamate-type siderophores were quantified by the periodate oxidation assay [43] and the ferric perchlorate test [44]. Hydroxamates of bacteria are made up of hydroxylated and acylated alkylamines. Desferrichrome is produced by *Aspergillus* is a hydroxamate siderophore like ferrioxamine produced by *Ustilago sphaerogena* [45, 46]. Fungal hydroxamates are hydroxylated and acylated ornithine based. Based on the number of hydroxamate groups, siderophores can be categorized as mono-, di-, and tri-hydroxamates. Shenker assay was used to detect carboxylates [47]. The binding properties of siderophores are used to identify their ligand denticity as bi-, tetra-, and hexa-dentates, based on the number of bonds that the ligand forms with the metal ion [48]. Dimerumic acid and N^{α}-methyl coprogen B produced from *Scedosporium apiospermum* were hydroxamate type siderophore. It is well known that siderophores secreted by fungi principally belong to the hydroxamate class [49]. Polycarboxylates are exclusively found in Zygomycetes [50] and strikingly, phenolates/catecholates have never been reported in fungi. Catecholates, produced only by bacteria, have catecholate and hydroxy groups. Enterobactin [51, 52, 53] is one of the representatives of the catecholate siderophores [20]. It is produced by enterobacteria like *Escherichia coli* and *Salmonella*. It is a triscatechol derivative of a cyclic triserine lactone and consists of three 2, 3-dihydroxy benzoic acid metal binding units attached via amide linkage to a trilactone backbone. Breakdown product is 2,-dihydroxy benzoyl serine (2, 3-DHBS) as linear monomer, dimer, and trimer [20]. Mycobactins from *Mycobacterium avium*, *M. intracellulare*, *M. scrofulaceum*, *M. paratuberculosis,* Anguibactin from *Vibrio anguillarum*, 2, 3-dihydroxybenzoyl glycine from *Bacillus* spp. ST13 and 2, 3-dihydroxybenzoylserine from *Escherichia coli* ST2 are also catecholate type siderophore. Carboxylates are produced by fungi belonging to the Phylum *Zygomycota* (*Mucorales*) and by a few bacteria (*Rhizobium meliloti* and *Staphylococcus hyicus*). Carboxylates coordinate iron with the carboxy and hydroxyl groups [54]. Mixed siderophores, such as azotobactin and yersiniabactin, are produced by *Azotobacter vinelandii* and *Yersinia pesti,* respectively [31, 55, 56], and pseudobactin and pyoverdine, which carry both hydroxamate and catecholate groups, are widespread among the pseudomonades [57, 58]. Amphiphilic siderophores, such as marinobactins, are a recently discovered suite of marine siderophores of the marine bacteria *Marinobacter* [38]. They are amphiphilic lipopeptide siderophores having a polar head group and a long fatty-acid chain. The rest of the siderophores are linear or cyclic hexapeptides. Amphiphilic siderophores such as aquachelins, marinobactins and amphibactins are distinct families of amphiphilic peptidic siderophores produced by different genera of marine bacteria [59]. These siderophores contain a peptidic head group that coordinates iron(III) as well as one of a series of different fatty acids that is appended at the amine terminus [38, 59]. The amphiphilicity of each of these siderophores is defined by both the peptide size (e.g., six, seven or four amino acids, respectively) and the fatty acid chain length (e.g., C12–C18). The amphibactins with a short peptide head group of only four amino acids and relatively long C18 fatty acids remain cell associated and are extracted from the bacterial pellet during isolation in contrast to the marinobactins and aquachelins which are isolated from the supernatant following centrifugation of the bacterial culture to pellet the cells. Another smaller suite of amphiphilic peptide siderophores are the ornibactins, isolated from *Burkholderia cepacia*, a terrestrial bacterium [60, 61]. The ornibactins, like the marinobactins and aquachelins contain β-hydroxy aspartic acid, but are distinguished by their quite short fatty acid appendages: C4, C6, and C8. Relatively few terrestrial bacteria have been found to produce suites of amphiphilic siderophores, although in addition to the ornibactins produced by *B. cepacia* [60, 61], mycobacteria produce both suites of lipophilic siderophores and amphiphilic siderophores [62, 63]. In addition to the mycobactins, other known cell-associated siderophores include the structurally related formobactins [64], nocobactins [65], and amamistatins [66]. Acinetoferrin and rhizobactin 1021 are two amphiphilic siderophores that also contain a citrate backbone. Rhizobactin 1021 is produced by the nitrogen fixing alfalfa symbiont *Rhizobium meliloti* 1021; it is reported also as a single siderophore with one C10 fatty acid appendage [67]. Acinetoferrin and rhizobactin 1021 are each composed of citrate, 1, 3-diaminopropane, and monounsaturated acyl appendages. Acinetoferrin is produced by the opportunistic pathogen *Acinetobacter haemolyticus* [68], however it is reported as a single amphiphilic siderophore with two short C8 fatty acid groups. Corrugatin produced by *Pseudomonas corrugata* [69] is one other amphiphilic peptide siderophore that is reported as a single siderophore, as opposed to a suite like the marinobactins, aquachelins, amphibactins and ornibactins.

Catecholate Type

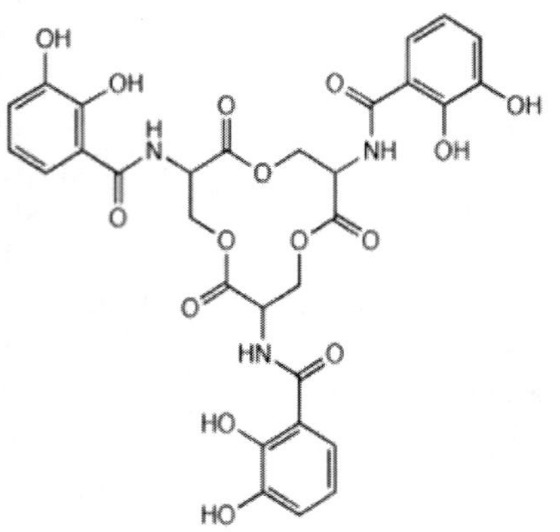

Enterobactin

(*E. coli, Salmonella typhimurium*)

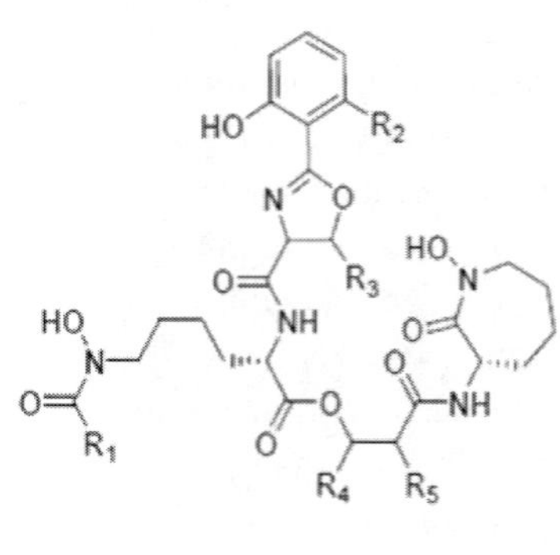

Mycobactin

(*Mycobacterium avium, M. paratuberculosis*)

Figure 2.1.

Hydroxamate Type

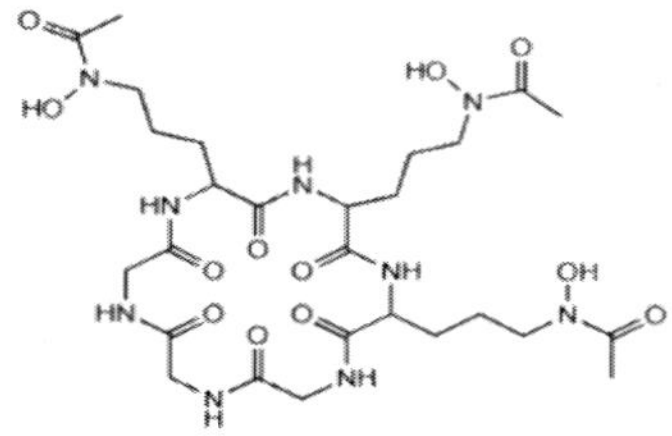

Desferrichrome
(*Aspergillus* sp.)

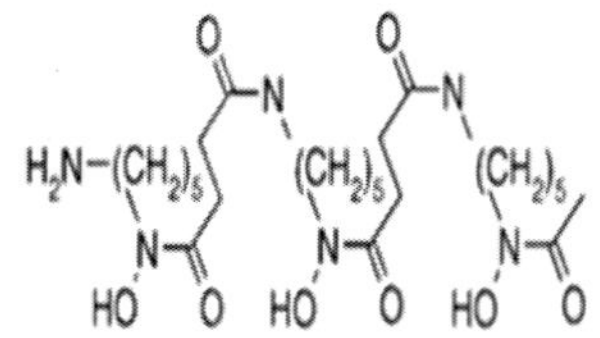

Desferrioxamine
(*Streptomyces pilosus*)

Figure 2.2.

Carboxylate Type

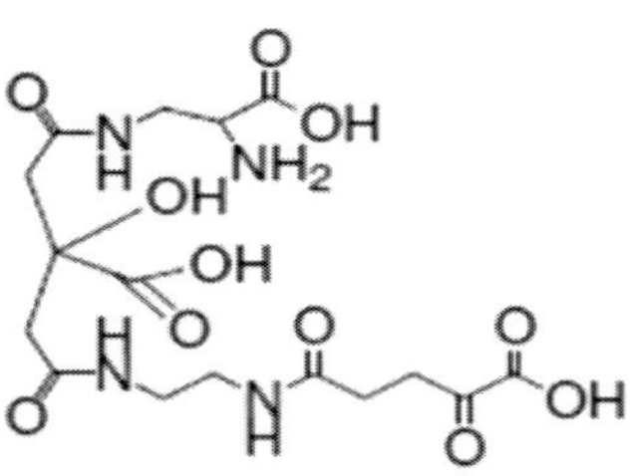

Staphloferrin B
(*Staphylococcus aureus*)

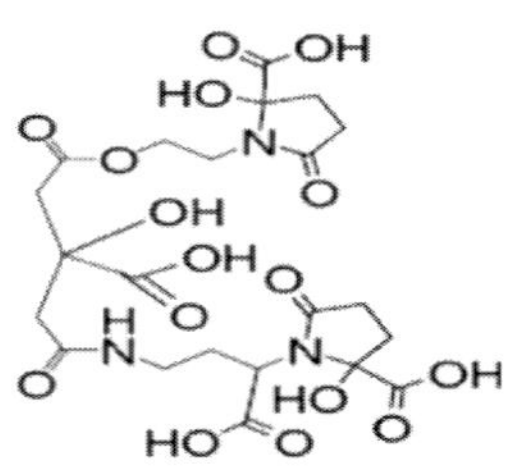

Achromobactin
(*Erwinia chrysanthemi*)

Figure 2.3.

Mixed Type

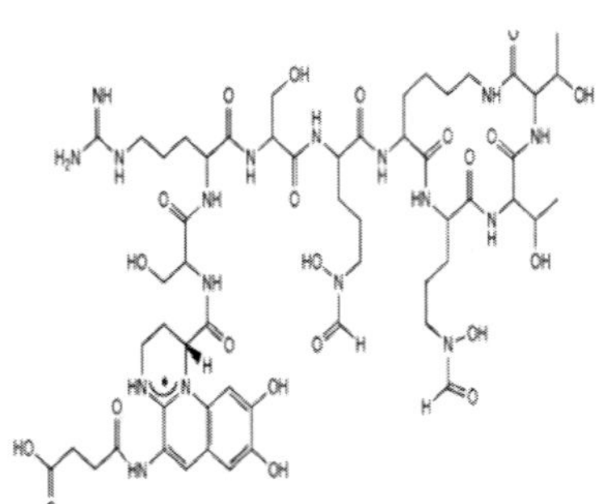

Pyoverdine
(*Pseudomonas fluorescens*)

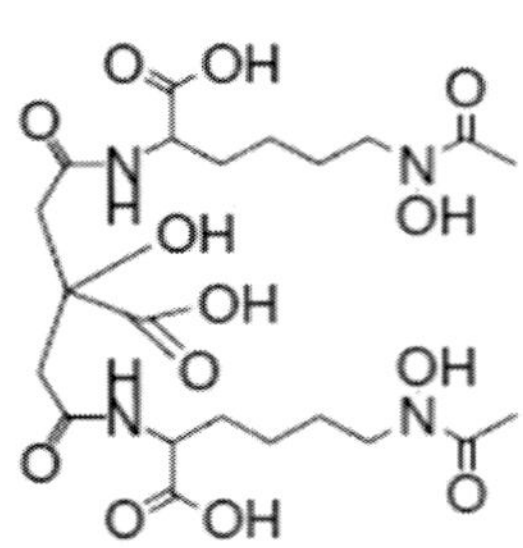

Aerobactin
(*Enterobacter spp., E.coli*)

Figure 2.4.

Amphiphilic type

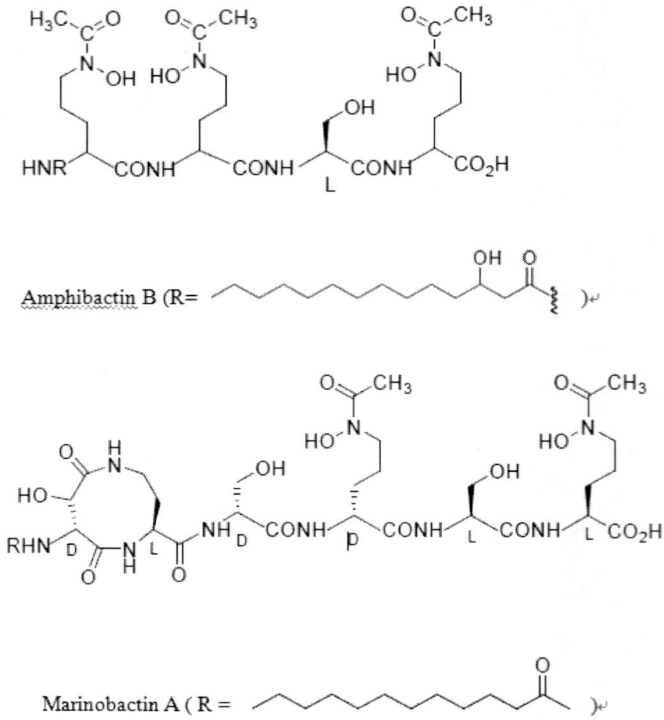

(a)

Amphiphilic type

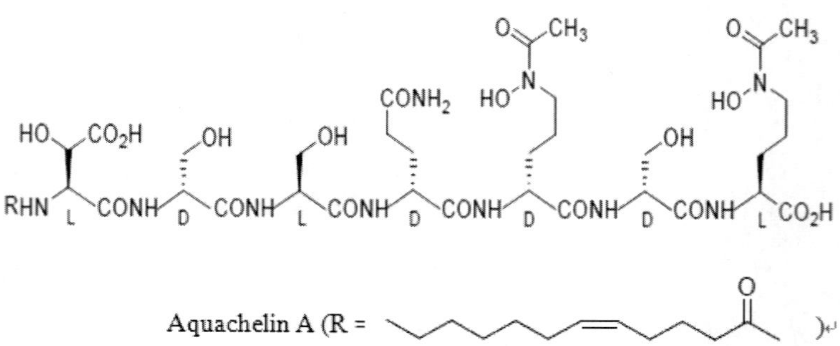

(b)

Figure 2.5.

2.4. Effect of Environmental Factors on the Production of Siderophore

Since siderophore biosynthesis occurs under iron-limiting conditions [17] above limited concentrations of iron, biosynthesis of siderophores is repressed, the threshold for repression differing according to the strain. It was observed, while addressing the production of desferrioxamine E and analoges, by *Streptomyces olivaceus* cells, that the requirements for low iron concentration were not constant throughout a batch run, being more critical in the early stages, where the addition of 10 μM iron (III) decreased production, whereas similar addition at later stages was hardly inhibitory of siderophore production.

Production of siderophore in *Frankia* strain was enhanced in the EDTA replenish condition in contrast to the iron supplementation medium suggesting that EDTA reduces the availability of other free metals and hence creates the stress condition for which the production of siderophore is enhanced [70]. Other relevant parameters that influence production are the nature of the carbon and nitrogen sources, pH, Temperature and dissolved oxygen tension, these being typically found in microbial based processes [71]. Thus, the nature of the carbon source used (glycerol or glucose) proved influential in siderophore production by *Escherichia coli* strain Nissle 1917 cells. Since most aerobactin was produced with glycerol as the carbon source at 37°C, glucose might have a catabolite repressing effect. A producer of four siderophores, enterobactin and salmochelin production was favored when glycerol was the carbon source. The use of well defined media for coprogen production by *Penicillium citrinum* was suggested by Gaeumann and Vischer, where glucose and L-asparagine were the carbon and nitrogen sources, respectively [72]. Wang and co-workers claimed that sucrose and ammonium nitrate were the most adequate carbon and nitrogen sources for siderophore production by *Aureobasidium pullulans*. Production could be increased by the addition of L-ornithine, possibly acting as a precursor [73]. This stimulation in siderophore production due to exogenous amino-acids was also reported by Calvente and co-workers when addressing the production of rhodotorulic acid by *Rhodotorula* strains. Supplementing the media with ornithine or with glutamic acid could increase production in up to 30%, although arginine had no similar effect [74]. These authors also established an optimal C: N ratio for the siderophore production system evaluated. The presence of L-ornithine and L-serine was also shown to enhance the fermentative production yield of albomycin, a siderophore antibiotic, structurally related to ferrichrome, obtained from *Streptomyces* sp. WS 116. Differentiation in siderophore production using suitable amino-acids and other precursors was also reported by Meiwes and coworkers, who ascribed such effect to the interaction with the metabolic pathways involved [75]. These authors furthermore established that desferrioxamine E production was dependent on the dissolved oxygen Tension, a level of 80% was suggested, since higher values required increased aeration with undesirable foaming and no productivity increase. Under basic conditions, ferric iron is poorly soluble and forms ferric hydroxide precipitates [76, 77], thus increasing iron stress and stimulating siderophore secretion. Conversely, acid conditions favor the solubility of ferric iron, facilitating iron acquisition by direct reduction mechanisms.

At acid pH (like pH 4), iron stress is therefore lower, thus resulting in a reduced siderophore production. Salmochelin and yersiniabactin were preferably produced under neutral to slightly alkaline conditions, whereas aerobactin production was maximal in a slightly acidic environment [78]. Cyclic catecholate siderophores have very high iron binding constants at high pH, and lower binding constants at more acidic pH values. Hydroxamate iron complexes generally have a slightly higher stability in the acidic pH range than catecholates. Yersiniabactin has a high complex formation constant at pH 8.35, but the pH dependence is not known [79]. Since yersiniabactin is unstable under acidic conditions [31, 54], one may assume that this siderophore is adapted to neutral and alkaline conditions. Yersiniabactin production was of greater advantage at neutral and slightly alkaline conditions than at pH 5.6. Agitation of the culture medium also largely influenced siderophore production since cultivation of the fungus without agitation resulted in higher production compared to that observed with continuous stirring [80].

Temperature was influential in siderophore production by *Azospirillum brasilense*, since spirilobactin production was clearly inhibited at temperatures where microbial growth was not affected [81]. Hydroxamate siderophore of *Vibrio salmonicida* was produced only at 10°C or below [82]. Metals other than iron are also able to stimulate or inhibit siderophore production in a number of bacteria, even in the presence of high iron concentrations. For instance, the presence of molybdenum in the environment of *Azotobacter vinelandii* regulates the production of azotochelin, a catecholate siderophore able to chelate this metal. At concentrations up to 100 μM of molybdenum, azotochelin production is activated, whereas at higher metal concentrations the synthesis of the siderophore is completely repressed [83]. High concentrations of aluminum increase the production of schizokinen and *N*-deoxyschizokinen (two hydroxamate siderophores) in iron-limited cultures but not iron-replete cultures of *Bacillus megaterium* [84]. Under iron-limited conditions, exposure to 10 mM Cu^{2+} upregulates genes involved in the synthesis of pyoverdine and down regulates those involved in the synthesis of pyochelin [85]. Interestingly, even in the presence of iron (100 μM), pyoverdine production is increased by 10 and 100 μM Cu^{2+} and Ni^{2+} and to a lesser extent by 100 μM Cr^{2+} (86). No activation of pyochelin production was observed, but a repression for 10 μM Co^{2+} and 100 μM Cu^{2+}, Mo^{+6} and Ni^{2+} in iron-limited medium [87]. When culture media were supplemented with 100 μM Fe^{3+}, Cd^{2+}, Cu^{2+}, Eu^{3+}, Ga^{3+}, Mn^{2+}, Ni^{2+} and Tb^{3+} decreased pyochelin production [86].

Probably, as for iron, siderophore production can be both activated and inhibited by many of these metals, depending on their concentrations. Therefore, it would be useful to test large ranges of concentrations of each metal to establish their exact roles in regulation of siderophore production in more detail. It is still unclear how metals other than iron stimulate siderophore production. One possible explanation is that the free siderophore concentration in the medium is reduced in the presence of other metals as a result of complex formation. Such a decrease in the siderophore concentration may be sufficient to activate secretion of additional siderophore into the medium. Stimulation of siderophore production by metals other than iron was reported for pyoverdine [88] in *P. aeruginosa*, azotochelin [83] in *A. vinelandii* and schizokinen [84] in *B. megaterium.* A bacterial cell must adapt to its environment in order to grow, and the production and utilization of siderophores is an important factor for its success [89]. Some examples of effect of environmental factors on the production of siderophore are given in Figure 3.

Environmental factors	Comments (References)
Iron	Iron starvation slowed *Paenibacillus polymyxa* bacterial growth and at all iron levels (0, 2, 20 μM), pH of liquid culture was decreased, but maximum decrease was observed at highest iron level (90).
EDTA	As the concentration of EDTA was gradually increased up to 40 μM, the production of hydroxamate type siderophore of *Frankia* strain showed a proportionate increase (70).
Carbon /Nitrogen source	Catecholate siderophore production of *Escherichia coli* strain Nissle 1917 was higher with glycerol than with glucose as the carbon source (78) Production of rhodotorulic acid by *Rhodotorula* strains was increased with supplementing the media with ornithine or with glutamic acid (74).
pH	Siderophore production and growth of *Scedosporium apiospermum* was fivefold higher at pH 8 compared to that observed at neutral or acid pH (Bertrand et al., 2009). Aerobactin of *Escherichia coli* strain Nissle 1917 was maximally produced at a more acidic pH (pH 5.6) (78).
Free metals	At concentrations up to 100 μM of molybdenum, azotochelin of *Azotobacter vinelandii* production is activated (Duhme *et al.*, 1998). Pyochelin production of *Pseudomonas aeruginosa* was repressed for 10 μM Co^{2+} and 100 μM Cu^{2+}, Mo^{+6} and Ni^{2+} in iron-limited medium (Visca *et al.*, 1992).
Agitation	Without agitation resulted in a 20-fold higher siderophore production compared to that observed with continuous stirring (Bertrand et al., 2009).

Figure 3.

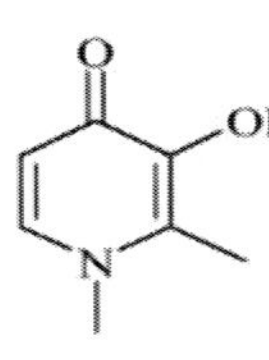

Deferiprone

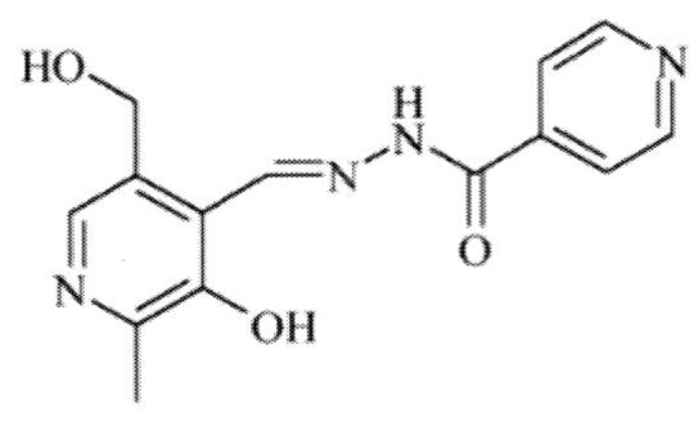

Pyridoxal isonicotinoyl hydrazone (PIH)

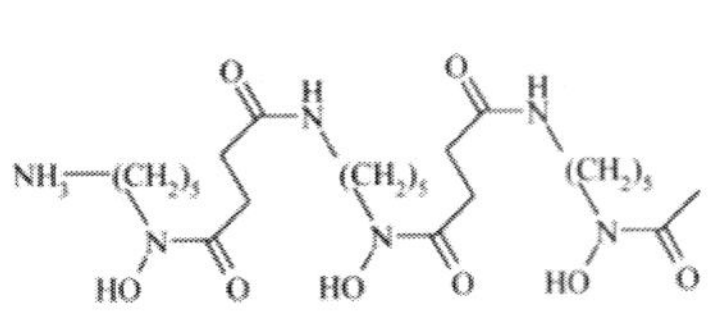

Desferrioxamine (DFO)

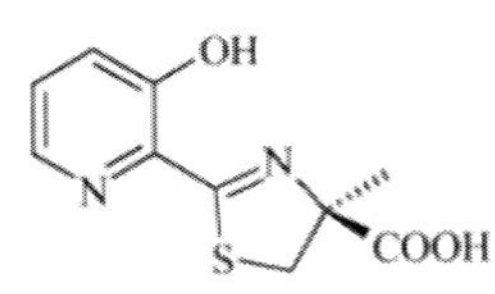

Desferrithicin (DFT)

Figure 4.

3. Application of Siderophore

In addition to their natural role in iron acquisition, siderophores have a number of applied uses. They have a role as bioremediation agents and also have novel clinical applications [91]. Clinical applications of iron chelation can be divided into three broad categories. First, removal of excess iron in order to prevent the formation of ROS (reactive oxygen species) and circumvent tissue damage (thalassemia major, neurodegenerative diseases, prevention of photo damage). Second, removal of iron form non-heme iron-containing enzymes that play a crucial role in a disease state (ribonucleotide reductase inhibitors in the treatment of cancer or malaria, protein hydroxylase inhibitors in the treatment of fibrosis). Third, attachment of a chelator to drugs [92] for their transport into the cells (sideromycins).

3.1. Clinical Application

Siderophores have various applications in clinical medicine and agricultural field

3.1.1. Treatment for the Iron Overload Disease

Patients that suffer from the disease β-thalassemia have a genetic defect in the biosynthesis of hemoglobin, which leads to the production of non-functional erythrocytes. The lack of effective hemoglobin synthesis leads to anaemia that is corrected by blood transfusions, and this along with increased gastrointestinal Fe uptake leads to Fe overload. As our bodies have a poor capacity for secreting excess iron, these patients quickly start to suffer from a significant iron overload, which, if left untreated, can cause serious damage to various organs. Consequently they have to be routinely treated with iron-chelation therapy to remove the excess iron. The siderophore Desferrioxamine (DFO) is an effective treatment for β-thalassemia because it can induce a negative Fe balance [93]. This compound binds Fe^{3+} tightly in serum and promotes its secretion into the urine. DFO is trihydroxamate siderophore, produced by *Streptomyces pilosus* and like most siderophores, this molecule provides three bidentate ligands to preferentially bind ferric iron. While Fe^{3+} has a strong preference for a six-coordinate 'octahedral' surrounding, especially if it is made up mostly of oxygen atoms as the ligands, other metal ions such as Zn^{2+} and Fe^{2+} actually prefer a four-coordinate tetrahedral surrounding with several nitrogen and sulfur ligands. Thus siderophores such as desferrioxamine B are designed to bind Fe^{3+} extremely tightly and they seem ideally suited to remove the ferric ion from the body. Unfortunately they are relatively large molecules and they are poorly absorbed if taken orally. Currently two agents are available for clinical treatment that have met with considerable success: deferiprone (tradename Ferriprox, produced by Apotex) and deferasirox (marketed as Exjade by Novartis). However, these compounds have a lower affinity for Fe^{3+} than desferrioxamine. However, in part because of their smaller size they appear to be well-tolerated and easily absorbed in the intestine and they are effective as oral agents. In addition to deferiprone, pyridoxal isonicotinoyl hydrazone (PIH, [94]), bishydroxyphenyl thiazole (ICL670, [95]), and desferrithiocin (DFT, [96]) are screened as interesting compounds for the clinical use in the iron overload disease. *Friedreich's Ataxia* is an inherited condition characterized by neurodegeneration and cardiomyopathy [97, 98]. In this condition, iron overload affects the central nervous system, which results in a gradual loss of motor skills leading to paralysis and death [99, 100]. Deferiprone is currently involved in clinical use for the treatment of the *Friedreich's Ataxia* and Parkinson's disease. Iron-mediated oxidation leads to the loss of mitochondrial DNA, greatly impairing the function of this organelle [101, 102]. This free radical damage is believed to be a secondary event after the disruption of iron homeostasis within the mitochondria. Therefore, iron chelators that are able to penetrate the mitochondria and facilitate iron removal provide a potential route of treatment for this iron overload condition [103, 104]. PIH is a tridentate iron chelator that has been shown to protect DNA against oxidative stress [105] and its analog 2-pyridylcarboxaldehyde isonicotinyl hydrozone has shown high activity in mobilizing mitochondrial iron [106]. Some examples of siderophores for the treatment of iron overload diseases are given in Figure 4.

3.1.2. Antimicrobial and Antiparasite Agents

Humans and animals acquire iron from their diet in the ferrous form by the divalent ion transporter proteins in the digestive tube, but once in the blood stream, iron is rapidly oxidized into the ferric form. The largest amount of iron (about two-thirds of iron in the heme-bound state) is found in hemoglobin in circulating red blood cells, whilst in the plasma the ferric iron is usually bound to the circulating transferrin protein [107]. Since bacteria require iron for growth, pathogenic bacteria have found ways of accessing eukaryotic iron-containing proteins from the host, including hemoglobin, transferrin and lactoferrin, besides the production of siderophores [108, 109]. One of the disadvantages of direct uptake is the necessity of a specific receptor for each of these iron sources [107]. In gram-negative bacteria, highly specific transport proteins present in the outer membranes can extract the iron sources from the environment, concentrate them on the bacterial surface, and after extraction of Fe (III), transport heme, iron and Fe (III)-siderophore complexes across the outer membrane [108]. Gram-positive bacteria are able to use the same range of substrates, to use ABC transporters to conduct iron into the cytoplasm and even present functional redundancy in iron-transporter mechanisms [110]. Micro-organisms that are able to acquire iron through different systems presented an evolution advantage over others. However, host organisms have also an array of solutions to battle microbial invasions, including constitutive components (e.g. siderophilins; transferrin in plama, lymph and cerebrospinal fluid; and lactoferrins in lachrymal secretions and mucosas) and infection induced responses (e.g. increased hepatic synthesis of hepcidin and macrophage ferritin; synthesis of lipocalins to bind siderophores) [109]. In 1944, Schade and Caroline found that ovotransferrin, which is the anti-infective agent in egg white, suppresses a wide number of micro-organisms by chelating essential iron for microbial growth [111]. Strategies have been developed to take advantage of bacterial necessities of iron: The use of iron antagonists and iron-chelators to decrease the available iron and the use of chemical compounds to inhibit the enzymatic steps of microbial iron metabolism [92, 110]. Conjugates of siderophores and antibiotics or other bactericidal compounds are often referred to as 'Trojan-horse' antibiotics, because the siderophore uptake systems are used to actively recognize and transport these compounds into the bacteria. For example, albomycin is an adduct of the hydroxamate ferrichrome siderophore that is chemically linked to a thioribosyl pyrimidine antibiotic that can inhibit the tRNA synthetase.

Another examples of siderophore-antibiotic conjugates are danomycins and salmycins produced by actinomycetes contain the trishydroxamate danoxamine as a siderophore unit that is connected with an aminoglycoside moiety [112, 113]. The compounds are highly active protein synthesis inhibitors in gram-positive bacteria such as *staphylococci* and *streptococci*. Ferrimycins are hybrids of ferrioxamine B and an antibiotically active group. Their transport is antagonized by ferrioxamines [114], and their activity is directed against gram-positive bacteria such as *S. aureus* and *Bacillus* spp. by altering their protein biosynthesis [115]. The main advantages of targeting microbial siderophore biosynthesis or iron-acquisition systems are that it could be a way to circumvent bacterial adaptation to drugs and it does not affect the cells of the infected human or animal cells. The efficacy of traditional antibacterials, in particular the reduced permeation of aminopenicillins through the outer membrane of Gram-negative pathogens, can be increased by conjugation of the antibiotic to a siderophore moiety allowing facilitated transport via essential, species specific Fe^{3+} siderophore transport pathways. In addition to the active uptake by the bacteria itself, the siderophore aminopenicillin conjugates are not substrates for efflux pumps in *P.aeruginosa*. It was recently reported that microcin E422m, an antimicrobial peptide produced by *Klebsiella pneumonia*, contained several catecholate groups that allowed binding of Fe^{+3}. This siderophore–peptide complex was shown to be actively taken up by Gram-negative bacteria through their siderophore outer membrane transporters [116]. A novel method to treat infections caused by multi-drug resistant *S. aureus* may involve the inhibition of siderophore mediated iron uptake, e.g. staphyloferrin-mediated iron uptake, which could be achieved by inhibition of the expression or function of at least one of the Sbt polypeptides required for the production of staphyloferrin [117]. Miller and Xu proposed pharmaceutical formulations to treat tuberculosis containing siderophoric mycobactins which are of paramount importance for the transport of iron into *M. tuberculosis* cells [118].

A new class of antimicrobial peptides carrying a siderophore like post-translational modification, called siderophore-peptides, showed high activity against enterobacteria including *E. coli*, *S. enteritidis* and *Enterobacter cloacae*. The antagonistic activity of a siderophore produced by a certain bacterium may be even used against a different bacterial species. Siderophore-containing extracts of nonpathogenic *Staphylococcus* strains inhibited considerably the vegetative growth of *Bacillus subtilis* strains [119]. Siderophore synthesis is also essential for fungal virulence, in particular for *Aspergillus* species. Moore and coworkers proposed the use of an agent to inhibit siderophore biosynthesis, secretion or binding capacity to prevent and treat fungal infections caused by *Aspergillus fumigatus* e.g. in immunocompromised patients [120]. Siderophore produced from yeast *Aureobasidium pullulans* is able to inhibit the growth. While the development and therapeutical application of sideromycins are one of the basic strategies for siderophore-dependent bacterial and fungal infection treatment, some iron chelators have also been found to be of value as antimalarial agents. Gysin and co-workers proposed a siderophore compound consisting of at least one hydroxamate group and one peptide bond (such as an aerobactin from pathogenic enterobacteria) as antiparasitic agent in diseases such as malaria [121]. Desferrioxamine B (deferoxamine [DFO]) is a linear trishydroxamate siderophore that is conventionally produced by large-scale fermentation of *Streptomyces pilosus* A21784 as methane sulfonate salt (Desferal; Norvatis Inc.). A fluorescent derivative of desferrioxamine B developed by Lytton et al. [122] displayed high inhibition of intra erythrocytic *Plasmodium falciparum* growth, with demonstrable Fe (III) scavenging within parasitized red cells. Salicylaldehyde isonicotinoyl hydrazone together with DFO significantly protected mice from infection with *Plasmodium vinckei*; however, they also inhibited the in vitro proliferation of B and T cells and allogeneic mixed-lymphocyte reactions to a certain extent [123]. Mode-of-action studies suggested that DFO enters the parasite for intracellular iron depletion, displaying an in vitro activity [124, 125, 126]. DFO conjugated with methylanthranilic acid showed 10-fold-higher in vitro activity against *P. falciparum* [127, 124], which could even be improved by using nalidixic acid as a conjugate against multidrug-resistant *P. falciparum* [128]. The mode of action in the latter case was suggested to involve metal-catalyzed oxidative DNA damage upon intercalation of the drug. Salicylaldehyde isonicotinoyl hydrazone and benzoyl hydrazone derivatives were found to be active at similar low micromolar concentrations in vitro [129]. Figure 5 shows some examples of iron-chelating compounds involved in natural microbial defense and therapeutics for treatment of infection.

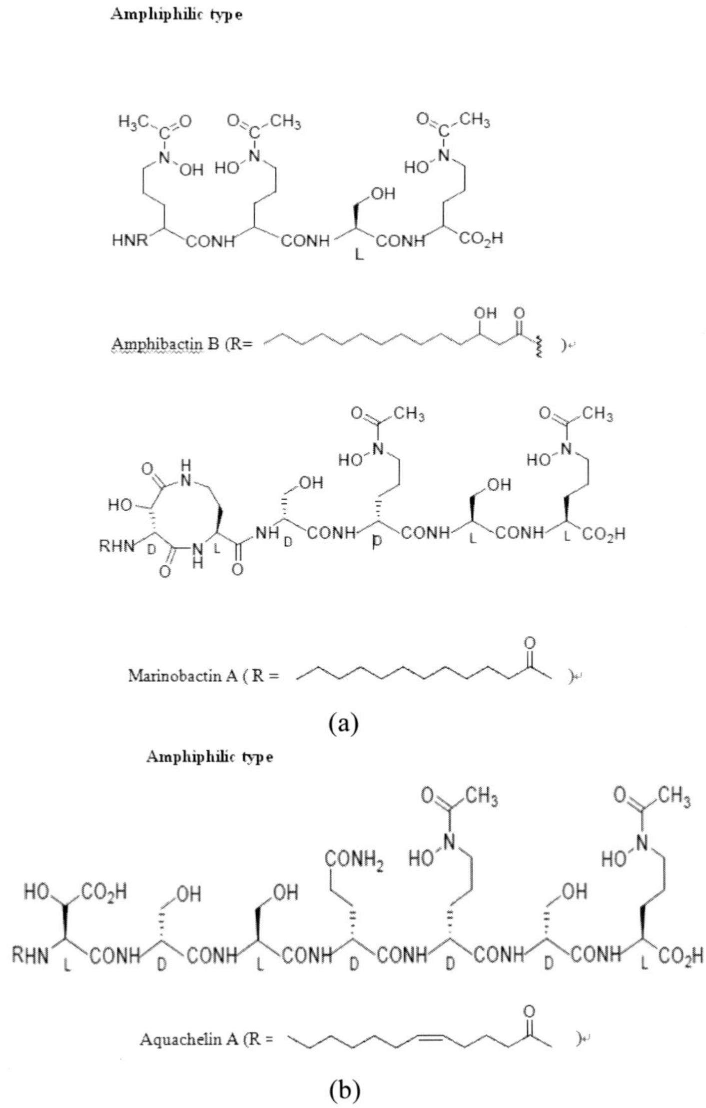

Figure 5.

3.1.3. Anti- tumor Agents

Many *in vitro* and *in vivo* studies have shown that, compared with normal cells, cancer cells are more sensitive to Fe deprivation because of their marked Fe requirements [17, 130, 131]. To facilitate rapid replication, neoplastic cells have significantly higher levels of ribonucleotide reductase (RR) and the transferrin receptor 1 (TfR1) [130, 131]. The higher Fe utilization by cancer cells than their normal counterparts provides a rationale for the selective antitumor activity of chelators. Iron chelators may be of value as therapeutic agents in the treatment of cancer. Mycobacterial desferriexochelins as well as synthetic iron chelators such as dexrazoxane and isonicotinoyl hydrazone derivatives are also used in cancer therapy [130, 132]. Depriving cancer cells of the essential nutrient iron (Fe) is a novel approach for cancer treatment [17, 130, 131]. Alternatively or additionally, iron (Fe) may form redox-active metal complexes that cause oxidative stress via production of reactive oxygen species, damaging critical intracellular targets and thereby eliciting a cytotoxic response. Fe-containing proteins perform key reactions involved in energy metabolism and DNA synthesis. Indeed, the rate-limiting step of DNA synthesis is catalyzed by the Fe-containing enzyme RR [133]. RR is up-regulated in cancerous cells, facilitating the production of deoxyribonucleotides for DNA replication and subsequent cell division [134]. Production of the iron storage protein, ferritin, is also increased in tumor tissue. Chelators with high lipophilicity theoretically possess advantages over less lipid-soluble ligands as these can easily enter cells and subsequently deplete iron from intracellular pools necessary for iron incorporation into the R2 subunit. Alternatively, these ligands can interact directly with the hydrophobic iron center of RR [135]. In addition, chelators with the highest antiproliferative activity not only chelate iron, decreasing the available iron in the LIP (labile iron pool), but may also enhance its redox activity. Examples of these ligand types include thiosemicarbazones such as Triapine [136, 137] and some aroylhydrazones such as those of the 2-pyridylketone isonicotinoyl hydrazone (PKIH) class [136]. Recently, the Fe chelator Triapine (3-aminopyridine-2-carboxaldehyde thiosemicarbazone), which inhibits RR activity and tumor growth *in vitro* and *in vivo* [138], has entered phase I and II clinical trials [130, 131]. Chelators such as desferrioxamine (DFO) which is a linear trishydroxamate siderophore that is conventionally produced by large-scale fermentation of *Streptomyces pilosus* A21784 as methane sulfonate salt (Desferal; Norvatis Inc.) arrest cells at the G_1/S interface, inhibiting cell-cycle. Chelators of the thiosemicarbazone class have been investigated for over 30 years as RR inhibitors [139-144]. Additionally, some chelators of the pyridoxal isonicotinoyl hydrazone class [145] possess potent antitumor activity, e.g., 2-hydroxy-1-naphthylaldehyde isonicotinoyl hydrazone [94]. The cytotoxic mechanism of action of this chelator involved not only Fe chelation but also redox cycling of its Fe complex to generate reactive oxygen species (ROS) [137]. Dexrazoxane and its racemate, Razoxane are the prodrugs of iron chelators designated as antitumor agents [146]. Dexrazoxane is used clinically in combination with antitumor agent doxorubicin for the treatment of advanced breast cancer. Tachpyridine is a hexadentate ligand based on a cis, cis-1, 3, 5-triaminocyclohexane scaffold, where chelation occurs via three pyridyl and three secondary amine nitrogens. As a consequence of its ability to bind cellular Fe, tachpyridine was demonstrated to inhibit ferritin synthesis [147] and also induce apoptosis by a p53-independent pathway [148]. As the majority of human cancers have a defective p53 gene, anti-cancer agents with the ability to activate p53-independent apoptosis are of great clinical interest [148]. Some examples of iron-chelators for the treatment of cancer are given in Figure 6.

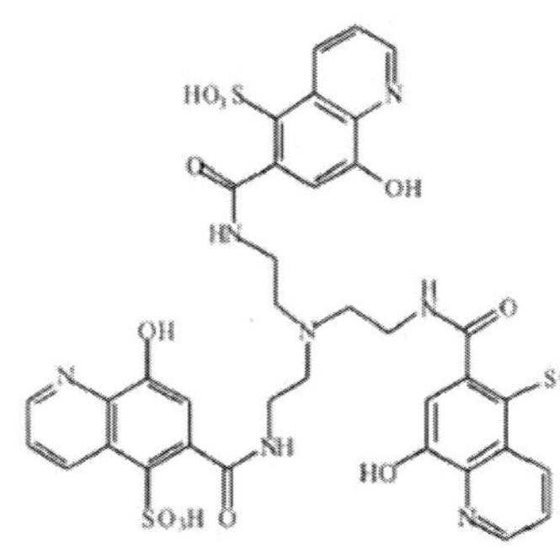

O-Trensox

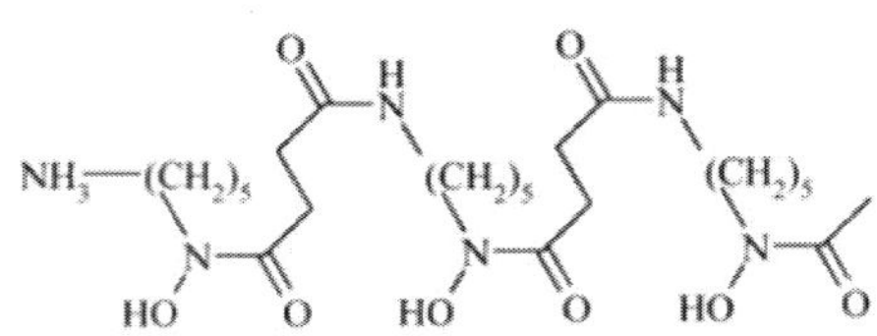

Desferrioxamine (DFO)

Figure 6.1.

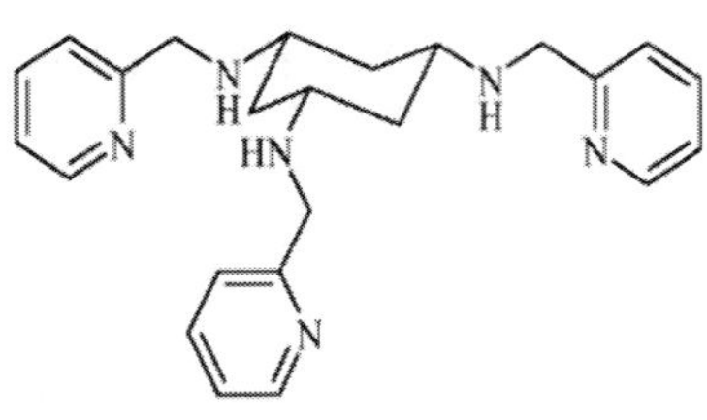

Tachpyridine

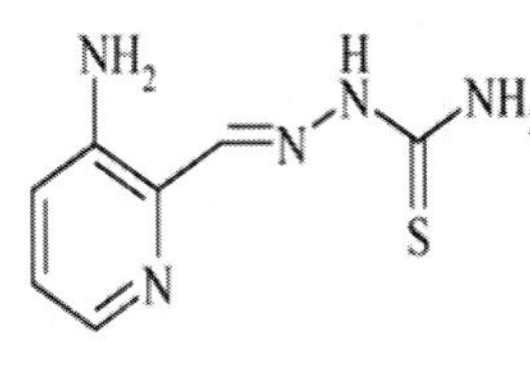

Triapine

Figure 6.2.

3.2. Agricultural Application

Fluorescent pseudomonads form a line of siderophores comprised of a quinoline moiety, responsible for the fluorescence, and a peptide chain of variable length bearing hydroxamic acid and a-hydroxy acid functions. Capacity to form these pseudobactin or pyoverdine type siderophores has been associated with improved plant growth either through a direct effect on the plant, through control of noxious organisms in the soil, or via some other route. Nitrogenase can be said to be an iron-intensive enzyme complex and the symbiotic variety, as found in *Rhizobium* spp., may require an intact siderophore system for expression of this exclusively prokaryotic catalyst upon which all life depends [149]. Microbial metabolite supply in soil was shown to increase the metal mobility [150]. In particular, bacterial siderophores can be helpful for soil remediation [151, 152] and can be taken up by plant [153, 154]. Enhanced phytoextraction of an agricultural Cr- and Pb-contaminated soil by bioaugmentation with siderophore-producing bacteria, *Pseudomonas aeruginosa, Pseudomonas fluorescens* or *Ralstonia metalliduran* was reported [88]. It was reported that siderophores from endophytic *Streptomyces* strain AzR-051 play vital role in promotion of plant growth [155].

Conclusion

Siderophores are common products of bacteria and of fungi. Since the first siderophores were discovered, several hundred distinct biologically active iron-chelator compounds such as hydroxamate-, catechol-, carboxylate- and mixed-type have been purified and characterized. The efficacy of antibiotics can be increased by conjugation of the antibiotic to a siderophore moiety (sideromycines) allowing facilitated transport via siderophore transport pathway. Essential requirement of iron for propagation of microorganisms has afforded exciting opportunities for the inhibition of antibiotic-resistant pathogens with iron chelators or sideromycines. Considering the essential role of iron in cellular proliferation and its potential to mediate oxidative damage when in excess, iron-chelating agents provide a promising form of treatment for iron overload disease. Applications of iron chelators to iron overload conditions as well as other therapeutic areas, including Parkinson's and other neurodegenerative disease, cancer and infection have the largest market potential. It must be pointed out that Fe chelators, like all anti-cancer agents, will affect both normal and neoplastic cells. Further studies are essential in order to examine the selectivity of new chelators against tumor cells compared to their normal counterparts. It is clear that iron chelator will play an appreciable role in the further treatment of various forms of neurodegeneration. The abnormal distribution of iron is probably not the primary casuse of the various forms of neurodegeneration, but is associated with disease progression. Due to the limitations in the activity of the clinically used siderophore, DFO, a vast array of iron chelators have been developed from both natural and synthetic sources in an attempt to replace it. Due to the tight relationship between siderophore utilization and virulence, the strategy of iron-dependent pathogen control is a promising field for future investigations and offers a broad array of possible therapeutic applications. While the siderophore-antibiotic strategy uses Fe-siderophore uptake systems as gateways for established antibiotics, the siderophore pathway inhibition strategy tries to abolish siderophore utilization in order to starve the pathogens out for iron. Formation of reactive oxygen species (ROS) is catalyzed by iron. ROS damage can lead to a large number of unrelated disease states. Iron chelators can reduce oxidative stress by the removal of iron from target tissues. Equally as important, removal of iron from the active site of enzymes that play key roles in various diseases can be of considerable benefit to the patients. Because of the importance of iron chelators, the application of clinically approved chelators to thalassemia major, bacterial and fungal infection, as well as the potential application of new chelators designed for specific disease states, a number of review articles on the subject of therapeutic applications of these compounds have appeared in recent literature. As demonstrated in this review, Fe is a worthy target to investigate in terms of developing novel anti-tumor agents. There are already a number of candidate chelators which show high anti-proliferative activity being far more active than DFO. Considering the many suboptimal properties of Fe chelation therapy with DFO, but its pronounced anti-tumor activity, the development of more effective chelators against cancer is worthwhile. Bioaugmentation-assisted phytoextraction is a relevant mean to reduce soil treatment duration, one of the major limitations of phytoremediation technologies.

The development of disease-specific iron chelators is an urgent and a challenging area of research.

Acknowledgments

This article was supported by Soonchunhyang University.

Conflict of Interest

There's no conflict of interest.

References

[1] Barry SM, Challis GL (2009) Recent advances in siderophore biosynthesis. *Curr Opin Chem Biol* 13:205–215.

[2] Crichton R (2009) *Iron metabolism, from molecular mechanisms to clinical consequences*, 3rd edn. Wiley Interscience, Chichester, West Sussex.

[3] Andrews SC, Robinson AK, Rodriguez-Quinones F (2003) Bacterial iron homeostasis. *FEMS Microbiol Rev 27*:215-237.

[4] Howard DH (1999) Acquisition, transport, and storage of iron by pathogenic fungi. *Clin Microbiol Rev* 12:394–404.

[5] Neilands JB (1981) Microbial iron compounds. *Annu Rev Biochem 50*:715–731.

[6] Hissen AHT, Chow JMT, Pinto LJ et al (2004) Survival of *Aspergillus fumigatus* in serum involves removal of iron from transferrin: the role of siderophores. Infect *Immun* 72:1402–1408.

[7] Wolz C, Hohloch K, Ocaktan A, Poole K, Evans RW, Rochel N, Albrecht-Gary MA, Abdallah MA, Doring G (1994) Iron release from transferrin by pyoverdin and elastase from *Pseudomonas aeruginosa. Infect Immun* 62:4021–4027.

[8] Haas H, Eisendle M, Turgeon BG (2008) Siderophores in fungal physiology and virulence. Annu Rev Phytopathol 46:149–187.

[9] Johnson L (2008) Iron and siderophores in fungal-host interactions *Mycol Res* 112:170-83.

[10] Schrettl M, Bignell E, Kragl C (2004) Siderophore biosynthesis but not reductive iron assimilation is essential for *Aspergillus fumigatus* virulence. *J Exp Med* 200:1213–1219.

[11] Hissen AHT, Wan ANC, Warwas ML, Pinto LJ, Moore M (2005) The *Aspergillus fumigatus* siderophore biosynthetic gene sidA, encoding L-ornithine N5-oxygenase, is required for virulence. *Infect Immun* 73:5493–5503.

[12] Zouni A, Witt HT, Kern J, Fromme P, Krauss N, Saenger W, Orth P (2001) Crystal structure of photosystem II from *Synechococcus* elongatus at 3.8 A resolution. *Nature* 409:739-43.

[13] Chu BC, Garcia-Herrero A, Johanson TH, Krewulak KD, Lau CK, Peacock RS, Slavinskaya Z, Vogel HJ (2010) Siderophore uptake in bacteria and the battle for iron with the host; a bird's eye view. *Biometals* 23:601-11.

[14] Hider RC, Kong X (2010) Chemistry and biology of siderophores. *Nat Prod Rep* 27:637–657.

[15] Aisen P, Enns C, Wessling-Resnick M (2001) Chemistry and biology of eukaryotic iron metabolism. *Int J Biochem Cell Biol* 33:940–959.

[16] Andrews NC (1999) Disorders of iron metabolism. *N Engl J Med* 341:1986–1995.

[17] Le NT, Richardson DR (2002) The role of iron in cell cycle progression and the proliferation of neoplastic cells.*Biochim Biophys Acta* 1603:31-46.

[18] Neilands, J. B., Konopka, K., Schwyn, B., Coy, M., Francis, R. T., Paw, B. H., and Bagg, A. (1987) in *Iron Transport in Microbes, Plants and Animals* (Winkelmann, G., van der Helm, D., and Neilands, J. B., eds) pp. 3–33, VCH Press, Weinheim.

[19] Wayne R and Neilands, J. B. (1975) Evidence for common binding sites for ferrichrome compounds and bacteriophage phi 80 in the cell envelope of *Escherichia coli*. *J Bacteriol* 121:497–501.

[20] Raymond KN, Dertz EA, Kim SS (2003) Enterobactin: an archetype for microbial iron transport. *Proc Natl Acad Sci USA* 100:3584–3588.

[21] Hersman L, Lloyd T, Sposito G (1995) Siderophore promoted dissolution of hematite. *Geochimica Cosmochimica Acta* 59:3327–3330.

[22] Schwyn B, Neilands JB (1987) Univesal chemical assay for the detection and determination of siderophore. *Anal Biochem* 160:47–56.

[23] Fekete FA, Chandkoke V, Jellison J (1989) Iron binding Compounds produced by Wood-Decaying Basidiomycetes. *App Environ Microbiol* 55:2720-2722.

[24] Pérez-Miranda S, Cabirol N, George-Téllez R, Zamudio-Rivera LS, Fernández FJ (2007) O-CAS, a fast and universal method for siderophore detection *J Microbiol Method* 70:127-131.

[25] Chan K-G (2009) Exochelin production in *Mycobacterium neoaurum*. *Int J Mol Sci* 10:345-53.

[26] Benz G, Metzger KG, Pfitzner J, Schmidt D, Zeiler H-J (1984) *Antibiotic compounds, their production and their medicinal use.* US4474762.

[27] Dhungana S, Michalczyk R, Boukhalfa H, Lack J, Koppisch A, Fairlee J, Johnson M, Ruggiero C, John S, Cox M, Browder C, Forsythe J, Vanderberg L, Neu M, Hersman L (2007) Purification and characterization of rhodobactin: a mixed ligand siderophore from *Rhodococcus rhodochrous* strain OFS. *Biometals* 20:853-867.

[28] Sharman GJ, Williams DH, Ewing DF, Ratledge C (1995) Isolation, purification and structure of exochelin MS, the extracellular siderophore from *Mycobacterium smegmatis*. *Biochem J* 305:187-96.

[29] Patel AK, Deshattiwar MK, Chaudhari BL, Chincholkar SB (2009) Production, purification and chemical characterization of the catecholate siderophore from potent probiotic strains of *Bacillus* spp. *Bioresour Technol* 100:368-73.

[30] Sayyed RZ, Chincholkar SB (2006) Purification of siderophores of *Alcaligenes faecalis* on Amberlite XAD. *Bioresour Technol* 97:1026-1029.

[31] Haag H, Hantke K, Drechsel H, Stojiljkovic I, Jung G, Zahner H (1993) Purification of yersiniabactin: a siderophore and possible virulence factor of *Yersinia enterocolitica. J Gen Microbiol* 139:2159-2165.

[32] Barnes HH, Ishimaru CA (1999) Purification of catechol siderophores by boronate affinity chromatography: identification of chrysobactin from *Erwinia carotovora* subsp. *Carotovora Biometals* 12(1):83-7.

[33] Braich N, Codd R (2008) Immobilised metal affinity chromatography for the capture of hydroxamate-containing siderophores and other Fe(III)-binding metabolites directly from bacterial culture supernatants. *Analyst* 133:877–880.

[34] Zajdowicz S, Haller JC, Krafft AE, Hunsucker SW, Mant CT, Duncan MW, Hodges RS, Jones DN, Holmes RK (2012) Purification and structural characterization of siderophore (corynebactin) from *Corynebacterium diphtheriae. PLoS One* 7(4):e34591 doi: 10.1371.

[35] Allard KA, Dao J, Sanjeevaiah P, McCoy-Simandle K, Chatfield CH, Crumrine DS, Castignetti D, Cianciotto NP (2009) Purification of Legiobactin and Importance of This Siderophore in Lung Infection by *Legionella pneumophila. Infection and Immunity* 77(7):2887–2895.

[36] Konetschny-Rapp S, Jung G, Meiwes J, Zahner H. (1990) Staphyloferrin A: a structurally new siderophore from *staphylococci. Eur J Biochem* 191:65-74.

[37] 37. Yamamoto S, Okujo N, Yoshida T, Matsuura S, Shinoda S (1994) Structure and Iron Transport Activity of Vibrioferrin, a New Siderophore of *Vibrio parahaemolyticus. J Biochem* 115(5):868-874.

[38] Martinez JS, Carter-Franklin JN, Mann EL, Martin JD, Haygood MG, Butler A (2003) Structure and membrane affinity of a suite of amphiphilic siderophores produced by a marine bacterium. *Proc Natl Acad Sci* 100:3754-3759.

[39] Crosa JH, Walsh CT (2002) Genetics and assembly line enzymology of siderophore biosynthesis in bacteria. *Microbiol Mol Biol Rev* 66:223–249.

[40] Chan DI, Vogel HJ (2010) Current understanding of fatty acid biosynthesis and the acyl carrier protein. *Biochem J* 430:1-19.

[41] Arnow LE (1937) Colorimetric determination of the components of 3, 4-dihydroxyphenylalanine–tyrosine mixtures. *J Biol Chem* 118:531–537.

[42] Young IG, Cox GB, Gibson F (1967) 2,3-Dihydroxybenzoate as a bacterial growth factor and its route of biosynthesis. *Biochim Biophys Acta, Gen Subj* 141:319–331.

[43] Emery TF, Neilands JB (1962) Further observations concerning the periodic acid oxidation of hydroxylamine derivatives. *J Org Chem* 27:1075–1077.

[44] Atkin CL, Neilands JB, Phaff HJ (1970) Rhodotorulic acid from species of *Leucosporidium*, *Rhodosporidium*, *Rhodotorula*, *Sporidiobolus*, and *Sporobolomyces*, and a new alanine-containing ferrichrome from *Cryptococcus melibiosum. J Bacteriol* 103:722–733.

[45] Leong SA, Winkelmann G (1998) Molecular biology of fron transport in fungi. *Met Ions Biol Syst* 35:147–186.

[46] Braun V (2003) Iron uptake by *Escherichia coli. Front Biosci* 8:1409–1421.
[47] Shenker M, Oliver I, Helmann M, Hadar Y, Chen Y (1992) Utilization by tomatoes of iron mediated by a siderophore produced by *Rhizopus arrhizus*. *J Plant Nutr* 15:2173–2182.
[48] Matzanke BF (1990) Structure, coordination and function of microbial iron. In: Winklemann G (ed) *Handbook of microbial iron chelates.* CRC, Boca Raton, p 15.
[49] Renshaw JC, Robson GD, Trinci APJ, Wiebe MG, Livens FR, Colllison D, Taylor RJ (2002) Fungal siderophores: structures, functions and applications. *Mycol Res* 106:1123–1142.
[50] Helm D, Winkelmann G (1994) Hydroxamates and polycarboxylates as iron transport agents (siderophores) in fungi. In: Winkelmann G, Winge D, editors. *Metal Ions in Fungi.* Marcel Dekker; New York. 39–98.
[51] O'Brien IG, Cox GB, Gibson F (1970) Biologically active compounds containing 2,3-dihydroxybenzoic acid and serine formed by *Escherichia coli. Biochim Biophys Acta* 201:453–460.
[52] O'Brien IG, Cox GB, Gibson F (1971) Enterochelin hydrolysis and iron metabolism in *Escherichia coli. Biochim Biophys Acta* 237:537–549.
[53] Pollack JR, Neilands JB (1970) Enterobactin, an iron transport compound from *Salmonella typhimurium. Biochem Biophys Res Commun* 38:989–992.
[54] Drechsel H, Tschierske M, Thieken A, Jung G, Zahner H, Winkelmann G (1995a) Siderophore rhizoferrin and its analogs produced by direct fermentations. *J Ind Microbiol* 14:105–112.
[55] Drechsel H, Stephan H, Lotz H, Haag H, Zähner H, Hantke K, Jung G (1995b) Structure elucidation of yersiniabactin, a siderophore from highly virulent Yersinia strains. *Liebigs Ann* 1995: 1727–1733.
[56] Knosp O, von Tigerstrom M, Page WJ (1984) Siderophore-mediated uptake of iron in *Azotobacter vinelandii. J Bacteriol* 159:341–347.
[57] Abdallah MA, Pattus F (2000) Siderophores and iron transport in microorganisms. *J Chin Chem Soc* 47:1–20.
[58] Boukhalfa H, Crumbliss AL (2002) Chemical aspects of siderophore mediated iron transport. *Biometals* 15(4):325-39.
[59] Martinez JS, Zhang GP, Holt PD, Jung HT, Carrano CJ, Haygood MG, Butler A (2000) Self-assembling amphiphilic siderophores from marine bacteria. *Sci* 287:1245-1247.
[60] Stephan H, Freund S, BeckW, Jung G, Meyer JM, Winkelmann G (1993) Ornibactins – a new family of siderophores from *Pseudomonas*. *Biometals* 6:93–100.
[61] Meyer JM, Van VT, Stintzi A, Berge O, Winkelmann G (1995) Ornibactin production and transport properties in strains of Burkholderia vietnamiensis and Burkholderia cepacia (formerly *Pseudomonas cepacia*). *Biometals* 8:309–317.
[62] Gobin J, Horwitz MA (1996) Exochelins of *Mycobacterium tuberculosis* remove iron from human iron-binding proteins and donate iron to mycobactins in the *M. tuberculosis* cell wall. *J Exp Med* 183:1527-32.
[63] Ratledge C, Dale J (1999) *Mycobacteria: molecular biology and virulence.* Blackwell Science Ltd. Oxford, UK.
[64] Murakami Y, Kato S, Nakajima M, Matsuoka M, Kawai H, Shin-Ya K, Seto H (1996) Formobactin, a novel free radical scavenging and neuronal cell protecting substance from *Nocardia* sp. *J Antibiot* (Tokyo) 49:83-89.

[65] Ratledge C, Patel PV (1976) Isolation, properties and taxonomic relevance of lipid-soluble, iron-binding compounds (nocobactins) from Nocardia. *J Gen Microbiol* 93:141–152.

[66] Suenaga K, Kokubo S, Shinohara C, Tsuji T, Uemura D (1999) Structures of amamistatins A and B, novel growth inhibitors of human tumor cell lines from an actinomycete. *Tetrahedron Lett* 40: 1945–1948.

[67] Persmark M, Pittman P, Buyer JS, Schwyn B, Gill PR, Neilands JB (1993) Isolation and structure of rhizobactin 1021, a siderophore from the alfalfa symbiont *Rhizobium meliloti* 1021. *J Am Chem Soc* 115:3950–3956.

[68] Okujo N, Sakakibara Y, Yoshida T, Yamamoto S (1994) Structure of acinetoferrin, a new citrate-based dihydroxamate siderophore from Acinetobacter haemolyticus. *Biometals* 7:170–176.

[69] Risse D, Beiderbeck H, Taraz K, Budzikiewicz H, Gustine D (1998) Corrugatin, a lipopeptide siderophore from *Pseudomonas corrugata*. Z Naturforsch, C: *Bioscience* 53:295–304.

[70] Singh A, Mishra AK, Singh SS, Sarma HK, Shukla E (2008) Influence of iron and chelator on siderophore production in *Frankia* strains nodulating *Hippophae salicifolia* D. *Don J Bas Microbiol* 48:104–111.

[71] Villegas MD (2007). In: Eds. *Microbial siderophores.* Berlin: Springer-Verlag 219-231.

[72] Gaeumann E, Vischer E (1967) *Process for producing coprogen and desferricoprogen* US 3297526.

[73] Wang WL, Chi ZM, Chi Z, Li J, Wang XH (2009) Siderophore production by the marine-derived *Aureobasidium pullulans* and its antimicrobial activity. *Bioresour Technol* 100:2639-2641.

[74] Calvente V, de Orellano ME, Sansone G, Benuzzi D, Sanz de Tosetti MI (2001) Effect of nitrogen source and pH on siderophore production by Rhodotorula strains and their application to biocontrol of phytopathogenic moulds.*J Ind Microbiol Biotechnol* 26(4):226-9.

[75] Meiwes J, Fiedler H-P, Zähner H, Konetschny-Rapp S, Jung G (1990) Production of desferrioxamine E and new analogues by directed fermentation and feeding fermentation. *Appl Microbiol Biotechnol* 32:505-510.

[76] Raymond KN, Mu¨ller G, Matzanke BF (1984) Complexation of iron by siderophores—a review of their solution and structural chemistry and biological function. *Top Curr Chem* 123:49–102.

[77] Kraemer SM (2004) Iron oxide dissolution and solubility in the presence of siderophores. *Aquat Sci* 66:3–18.

[78] Valdebenito M, Crumbliss AL, Winkelmann G, Hantke K (2006) Environmental factors influence the production of enterobactin, salmochelin, aerobactin, and yersiniabactin in *Escherichia coli* strain Nissle 1917. *Int J Med Microbiol* 296:513-520.

[79] Perry RD, Balbo PB, Jones HA, Fetherston JD, DeMoll E (1999) Yersiniabactin from *Yersinia pestis*: biochemical characterization of the siderophore and its role in iron transport and regulation. *Microbiology* 145:1181-1190.

[80] Bertrand S, Larcher G, Landreau A, Richomme P, Duval O, Bouchara JP (2009) *Hydroxamate siderophores of Scedosporium apiospermum Biometals* 22:1019-1029.

[81] Bachhawat AK, Ghosh S (1989) Temperature Inhibition of Siderophore Production in *Azospirillum brasilense. J Bacteriol* 171:4092-4094.

[82] Colquhoun DJ, Sørum H (2001) Temperature dependent siderophore production in *Vibrio salmonicida. Microbial Pathogenesis* 31:213–219.

[83] Duhme RC, Hider MJ, Naldrett MJ, Pau RN (1998) The stability of the molybdenum-azotochelin complex and its effect on siderophore production in *Azotobacter vinelandii. J Biol Inorg Chem* 3: 520–526.

[84] Hu X, Boyer GL (1996) Siderophore-mediated aluminum uptake by *Bacillus megaterium* ATCC 19213. *Appl Environ Microbiol* 62:4044–4048.

[85] Teitzel GM, Geddie A, De Long SK, Kirisits MJ, Whiteley M, Parsek MR (2006) Survival and growth in the presence of elevated copper: transcriptional profiling of copper-stressed *Pseudomonas aeruginosa. J Bacteriol* 188(20):7242-7256.

[86] Braud A, Geoffroy V, Hoegy F, Mislin GLA, Schalk IJ (2010) The siderophores pyoverdine and pyochelin are involved in *Pseudomonas aeruginosa* resistence against metals: Another biological function of these two siderophores. *Environ Microbiol Report* 2:419–425.

[87] 87.Visca P, Colotti G, Serino L, Verzili D, Orsi N, Chiancone E (1992) Metal regulation of siderophore synthesis in *Pseudomonas aeruginosa* and functional effects of siderophore-metal complexes. *Appl Environ Microbiol* 58:2886–2893.

[88] Braud A, Hoegy F, Jezequel K, Lebeau T, Schalk IJ (2009) New insights into the metal specificity of the *Pseudomonas aeruginosa* pyoverdine-iron uptake pathway. *Environ Microbiol* 11:1079–1091.

[89] Winkelmann G (2004) *Iron Transport in Bacteria,* ed. J. H. Crosa, A. R.Mey and S. M. Payne, ASM Press, Washington,, ch. 28, pp.437–450.

[90] Raza W, Shen Q (2010) Growth, Fe^{3+} reductase activity, and siderophore production by *Paenibacillus polymyxa* SQR-21 under differential iron conditions. *Curr Microbiol* 61(5):390-5.

[91] Bergeron RJ, Brittenham GM (eds) (1994) *The Development of Iron Chelators for Clinical Use,* CRC Press, Boca Raton, FL.

[92] Wencewicz T, Möllmann U, Long T, Miller M (2009) Is drug release necessary for antimicrobial activity of siderophore-drug conjugates? Syntheses and biological studies of the naturally occurring salmycin “Trojan Horse” antibiotics and synthetic desferridanoxamine-antibiotic conjugates. *Biometals* 22:633-648.

[93] Giardina PJ, Grady RW (1995) Chelation therapy in β-thalassemia: the benefits and limitations of desferrioxamine. *Semin Hematol* 32:304–312.

[94] Richardson DR, Tran EH, Ponka P (1995) The potential of iron chelators of the pyridoxal isonicotinoyl hydrazone class as effective antiproliferative agents. *Blood* 86:4295–4306.

[95] Galanello R, Piga A, Alberti D, Rouan MC, Bigler H, Séchaud R (2003) Safety, tolerability, and pharmacokinetics of ICL670, a new orally active iron-chelating agent in patients with transfusion-dependent iron overload due to beta-thalassemia. *J Clin Pharmacol* 43(6):565-572.

[96] Bergeron RJ, Huang G, Weimar WR, Smith RE, Wiegand J, McManis JS (2003) Desferrithiocin analogue based hexacoordinate iron(III) chelators. *J Med Chem* 46(1):16-24.

[97] Pandolfo M (2002) Iron metabolism and mitochondrial abnormalities in Friedreich ataxia. *Blood Cells Mol Dis* 29:536–547.

[98] Napier I, Ponka P, Richardson DR (2005) Iron trafficking in the mitochondrion: novel pathways revealed by disease. *Blood* 105:1867–1874.

[99] Puccio H, Koenig M (2000) Recent advances in the molecular pathogenesis of Friedreich ataxia. *Hum Mol Genet* 9:887–892.

[100] Lieu PT, Heiskala M, Peterson PA, Yang Y (2001) The roles of iron in health and disease. *Mol Aspects Med* 22:1–87.

[101] Rotig A, de Lonlay P, Chreiten D, Foury F, Koenig M, Sidi D, Munnich A, Rustin P (1997) Aconitase and mitochondrial iron-sulphur protein deficiency in *Friedreich ataxia. Nat Genet* 17:215–217.

[102] Lodi R, Rajagopalan B, Bradley JL, Taylor DJ, Crilley JG, Hart PE, Blamire AM, Manners D, Styles P, Schapira AH, Cooper JM (2002) Mitochondrial dysfunction in Friedreich's ataxia: from pathogenesis to treatment perspectives. *Free Radic Res* 36(4):461-466.

[103] Richardson DR, Mouralian C, Ponka P (2001) Development of potential iron chelators for the treatment of Friedreich's ataxia: ligands that mobilize mtochondrial iron. *Biochem Biophys Acta* 526:133-40.

[104] Richardson DR (2003) Friedreich's ataxia: iron chelators that target the mitochondrion as a therapeutic strategy? *Expert Opin Investig Drugs* 12:235–245.

[105] Hermes-Lima M, Nagy E, Ponka P, Schulman HM (1998) The iron chelator, pyridoxal isonicotinoyl hydrazone (PIH) protects plasmid pUC-18 DNA against *OH-mediated strand breaks. *Free Radic Biol Med* 25:875–880.

[106] Darnell G, Richardson DR (1999) The potential of iron chelators of the pyridoxal isonicotinoyl hydrazone class as effective antiproliferative agents III: the effect of the ligands on molecular targets involved in proliferation. *Blood* 94:781-92.

[107] Miethke M, Marahiel MA (2007) Siderophore-based iron acquisition and pathogen control. *Microbiol Mol Biol Rev* 71:413-451.

[108] Braun V, Braun M (2002) Active transport of iron and siderophore antibiotics. *Curr Opin Microbiol* 5:194-201.

[109] Weinberg ED (2009) Iron availability and infection. Biochimica et Biophysica Acta (BBA) - *General Subject* 1790:600-605.

[110] Ballouche M, Cornelis P, Baysse C (2009) Iron metabolism: a promising target for antibacterial strategies. *Recent Pat Anti-Infect Drug Discovery* 4:190-205.

[111] Schade AL, Caroline I (1944) Raw hen egg white and the role of iron in growth inhibition of *Shigella dysenteriae*, *Staphylococcus aureus*, *Escherichia coli* and *Saccharomyces cerevisiae*. *Sci* 100:14-15.

[112] Huber P, Leuenberger H, Keller-Schierlein W (1986) Danoxamin, der eisenbindende Teil des Sideromycin-Antibioticums Danomycin. Helv. *Chim. Acta* 69:236–245.

[113] Vertesy L, Aretz W, Fehlhaber HW, Kogler H (1995) Salmycin A-D, Antibiotika aus *Streptomyces violaceus*, DSM 8286, mit Siderophor-Aminoglycosid-Struktur. *Helv. Chim. Acta* 78:46–60.

[114] Zaehner H, Diddens H, Keller-Schierlein W, Nageli HU (1977) Some experiments with semisynthetic sideromycins. *Jpn J Antibiot* 30:201–206.

[115] Braun PC (1999) Nutrient uptake by Candida albicans: the influence of cell surface annoproteins. *Can J Microbiol* 45(5):353-359.

[116] Thomas X, Destoumieux-Garzon D, Peduzzi J (2004) Siderophore peptide, a new type of post-translationally modified antibacterial peptide with potent activity. *J Biol Chem* 279:28233-242.

[117] Heinrichs DE, Beasley F, Vines E, Cheung J (2010) *Siderophore mediated iron uptake in bacterial infection.* WO2010020036 A1.

[118] Miller MJ, Xu Y (2001) *Antimycobacterial agents.* US 6,310,058 B1.

[119] Raaska L, Alakomi HL, Salkinoja-Salonen M, Mattila-Sandholm T (1999) Antagonistic activity of *Staphylococcus* siderophores and chemical biocides against *Bacillus subtilis* in a paper-machine environment. *J Ind Microbiol Biot* 22:27-32.

[120] Moore MM, Hissen AHT, Wan NCA (2010) *Production of mutant strain of Aspergillus fumigatus, method of assay for inhibiting siderophore biosynthesis and diagnostic method for detecting likely Aspergillus fumigatus infection.* US2006079580-A1; US20107794712-B2.

[121] Gysin J, Crenn Y, Silva LPD, Breton C (1993) *Siderophores as antiparasitic agents.* US20035192807.

[122] Lytton SD, Cabantchik ZI, Libman J, Shanzer A (1991) Reversed Siderophores as Antimalarial Agents .2. Selective Scavenging of Fe (III) from Parasitized Erythrocytes by a Fluorescent Derivative of Desferal. *Mol Pharmacol* 40:584-90.

[123] Golenser J, Domb A, Mordechai-Daniel T, Leshem B, Luty A, Kremsner P (2006) Iron chelators: correlation between effects on *Plasmodium* spp. and immune functions. *J Parasitol* 92:170–177.

[124] Loyevsky M, Lytton SD, Mester B, Libman J, Shanzer A, Cabantchik ZI (1993) The antimalarial action of desferal involves a direct access route to erythrocytic (*Plasmodium falciparum*) parasites. *J Clin Investig* 91:218–224.

[125] Loyevsky M, John C, Dickens B, Hu V, Miller JH, Gordeuk VR (1999) Chelation of iron within the erythrocytic *Plasmodium falciparum* parasite by iron chelators. *Mol Biochem Parasitol* 101:43–59.

[126] Scott MD, Ranz A, Kuypers FA, Lubin BH, Meshnick SR (1990) Parasite uptake of desferroxamine: a prerequisite for antimalarial activity. *Br J Haematol* 75:598–602.

[127] Glickstein H, Breuer W, Loyevsky M, Konijn AM, Shanzer A, Cabantchik ZI (1996) Differential cytotoxicity of iron chelators on malariainfected cells versus mammalian cells. *Blood* 87:4871–4878.

[128] Ghosh M, Lambert LJ, Huber PW, Miller MJ (1995) Synthesis, bioactivity, and DNA-cleaving ability of of desferrioxamine B-nalidixic acid and anthraquinone carboxylic acid conjugates. *Bioorg Med Chem Lett* 5:2337–2340.

[129] Tsafack A, Loyevsky M, Ponka P, Cabantchik ZI (1996) Mode of action of iron (III) chelators as antimalarials. IV. Potentiation of desferal action by benzoyl and isonicotinoyl hydrazone derivatives. *J Lab Clin Med* 127:574–582.

[130] Buss JL, Greene BT, Turner BT, Torti FM, Torti SV (2004) Iron chelators in cancer chemotherapy. *Curr Top Med Chem* 4:1623–1635.

[131] Kalinowski DS, Richardson DR (2005) The Evolution of Iron Chelators for the Treatment of Iron Overload Disease and Cancer. *Pharmacol Rev* 57:547–583.

[132] Chong TW, Horwitz LD, Moore JW, Sowter HM, Harris AL (2002) A mycobacterial iron chelator, desferri-exochelin, induces hypoxiainducible factors 1 and 2, NIP3, and vascular endothelial growth factor in cancer cell lines. *Cancer Res* 62:6924–6927.

[133] Kicic A, Chua AC, Baker E (2001) Effect of iron chelators on proliferation and iron uptake in hepatoma cells. *Cancer* 92:3093–3110.

[134] Elford HL, Freese M, Passamani E, Morris HP (1970) Ribonucleotide reductase and cell proliferation. *J Biol Chem* 245:5228–5233.

[135] Hodges YK, Antholine WE, Horwitz LD (2004) Effect on ribonucleotide reductase of novel lipophilic iron chelators: the desferri-exochelins. *Biochem Biophys Res Commun* 315:595–598.

[136] Chaston TB, Lovejoy DB, Watts RN, Richardson DR (2003) Examination of the antiproliferative activity of iron chelators: multiple cellular targets and the different mechanism of action of Triapine compared with desferrioxamine and the potent pyridoxal isonicotinoyl hydrazone analogue 311. *Clin Cancer Res* 9:402–414.

[137] Yuan J, Lovejoy DB, Richardson DR (2004) Novel di-2-pyridyl-derived iron chelators with marked and selective antitumor activity: in vitro and in vivo assessment. *Blood* 104:1450–1458.

[138] Finch RA, Liu M, Grill SP, Rose WC, Loomis R, Vasquez KM, Cheng Y, Sartorelli AC (2000) Triapine (3-Aminopyridine-2-carboxaldehydethiosemicarbazone): A Potent Inhibitor of Ribonucleotide Reductase Activity with Broad Spectrum Antitumor Activity. *Biochem Pharmacol* 59:983–991.

[139] Finch RA, Liu MC, Cory AH, Cory JG, Sartorelli AC (1999) Triapine (3-aminopyridine-2-carboxaldehyde thiosemicarbazone; 3-AP): an inhibitor of ribonucleotide reductase with anti-neoplastic activity. *Adv Enzyme Regul* 39:3–12.

[140] Agrawal KC, Sartorelli AC (1978) The chemistry and biological activity of β-(N)-heterocyclic carboxaldehyde thiosemicarbazones. *Prog Med Chem* 15:321–356.

[141] Booth BA, Moore EC, Sartorelli AC (1971) Metabolic effects of some tumor-inhibitory pyridine carboxaldehyde thiosemicarbazones. *Cancer Res* 31:228–234.

[142] Cory JG, Cory AH, Rappa G (1994). Inhibitors of ribonucleotide reductase. Comparative effects of amino- and hydroxy-substituted pyridine-2-carboxaldehyde thiosemicarbazones. *Biochem Pharmacol* 48:335–344.

[143] Garcia-Tojal J, Garcia-Orad A, Diaz AA (2001) Biological activity of complexes derived from pyridine-2-carbaldehyde thiosemicarbazone. *J Inorg Biochem* 84:271–278.

[144] Sartorelli AC, Agrawal KC, Moore EC (1971) Mechanism of inhibition of ribonucleoside diphosphate reductase by β-(N)-heterocyclic aldehyde thiosemicarbazones. *Biochem Pharmacol* 20:3119–3123.

[145] Ponka P, Borova J, Neuwirt J, Fuchs O (1979) Mobilization of iron from reticulocytes: Identification of pyridoxal isonicotinoyl hydrazone as a new iron chelating agent. *FEBS Lett* 97:317–321.

[146] Creighton AM (1971) *Piperazine derivatives* (1,234,935) U.K..

[147] Torti SV, Torti FM, Whitman SP (1998) Tumor cell cytotoxicity of a novel metal chelator. *Blood* 92:1348-8.

[148] Abeysinghe RD, Greene BT, Haynes R (2001) p53-independent apoptosis mediated by tachpyridine, an anti-cancer iron chelator. *Carcinogenesis* 22:1607–1614.

[149] Barton LL, Hemming BC (eds) (1993) *Iron Chelation in Plants and Soil Microorganisms*, Academic Press, New York.

[150] Gadd GM (2004) Microbial influence on metal mobility and application for bioremediation. Geoderma 122:109–119.

[151] Diels L, van der Lelie N, Bastiaens L (2002) New developments in treatment of heavy metal contaminated soils. *Rev Environ Sci Biotechnol* 1:75–82.

[152] Belimov AA, Hontzeas N, Safronova VI, Demchinskaya SV, Piluzza G, Bullitta S, Glick BR (2005) Cadmium-tolerant plant growth-promoting bacteria associated with the roots of Indian mustard (*Brassica juncea* L. Czern.). *Soil Biol Biochem* 37:241–250.

[153] Wang Y, Brown HW, Crowley DE, Szaniszlo PJ (1993) Evidence for direct utilization of a siderophore ferroxamine B, in axenially grown cucumber. *Plant Cell Environ* 16:579–585.

[154] Jurkevitch E, Hadar Y, Chen Y (1988) Involvement of bacterial siderophores in the remedy of lime-induced chlorosis in peanut. *Soil Sci Soc Am J* 52:1032–1037.

[155] Verma VC, Singh SK, Prakash S (2011) Bio-control and plant growth promotion potential of siderophore producing endophytic *Streptomyces* from *Azadirachta indica* A. Juss. *J Basic Microbiol* 51(5):550-556.

In: Microbes in Process
Editors: Neelam Garg and Abhinav Aeron
ISBN: 978-1-63117-127-7

Chapter 16

Versatile Biological Macromolecules: Chitin and Chitosan

Neelam Garg [*] ***and Sonu Rani Kashyap***
Department of Microbiology, Faculty of Life Sciences,
Kurukshetra University, Haryana, India

Abstract

Chitin, the second most abundant natural polysaccharide after cellulose on earth, is widely distributed in nature, mainly as the structural component of the exoskeletons of crustaceans, insect cuticles, in marine diatoms, algae and fungal cell walls. Chitin is a polymer of N-acetyl-D-glucosamine monomers. Shellfish waste is the major source of chitin worldwide. Biosynthesis of chitin occurs in a series of reactions and is studied in different organisms. Chitin is insoluble in water and various other solvents but it can be converted by deacetylation in to its soluble form called chitosan. Chitosan is mainly found in the cell wall of some fungi. Commercial chitosan is derived from the shells of shrimp and other sea crustaceans. These are nontoxic, biocompatible, biologically safe, biodegradable and renewable biopolymers. These remarkable properties of chitin and chitosan make them versatile biomolecules with number of applications in various fields like biotechnology, chemistry, food processing, textile, agriculture, veterinary, environmental protection, and medicine production. The aim of this review is to have a look at the source, structure, preparation and properties of the chitin and chitosan. In view of this, the present review is to attract the attention of various biologists on these underutilized biological macromolecules.

1. Introduction

Chitin and chitosan are among the novel families of renewable, biological macromolecules, importance of which is increasing rapidly. Chitin is known from ancient

[*] nlmgarg@yahoo.com.

times and has been found in the shells of 100 megayear old Pterygotus (ancient scorpion) [1]. It was first discovered by prof. Henri Braccanot of France in 1811 in the fungal species of *Agaricus, Hydnum* and *Boletus* and he called it "Fungine" [2]. In 1823, Odier found this polymer in insect cuticle [3]. Chitin, which is converted into chitosan by different methods, is a poly-β-(1, 4)-N-acetyl-D-glucosamine. It is the most ubiquitous naturally occurring polysaccharide after cellulose on earth [4, 5, 6,7]. Chitin is estimated to be produced annually almost as much as cellulose[1]. The annual production of cellulose (by plants, algae and phytoplankton) and chitin (by marine zooplankton and invertebrates as well as by terrestrial arthropods) was estimated in the range of 10^{11} and 10^{10} tons, respectively [8, 9]. Chitin is often considered as cellulose derivative with an acetamido group at C-2, even though it does not occur in organisms producing cellulose [4,6]. Both of these polymers mainly serve as structural components supporting cell and body surfaces, as cellulose strengthens the cell wall of plant cells whereas chitin strengthens the cell walls of fungi and exoskeletons of arthropods [1].

Chitin is produced in enormous quantities in the biosphere. The total global estimate of accessible chitin amounts to 10 gigatons per year[1].The interest in chitin is escalating not only as an underutilized source but also as a new functional biomaterial of high potential. Commercially, chitin and chitosan are of great importance owing to their relatively high percentage of nitrogen (6.89%) compared to synthetically substituted cellulose [5]. But utilization of chitin is limited due to its insolubility in H_2O, dilute acids and most organic solvents. Its solubility may be improved by various chemical modifications, the simplest being deacetylation. The end product of chitin deacetylation is chitosan. Chitosan is a linear copolymer of β-(1, 4)-linked-N-acetyl-D-glucosamine (GlcNAc) and glucosamine (GlcN) units. Chitosan has unique functional and biological properties favorable for a variety of industrial and biological applications [6, 10]. Chitin and chitosan are the naturally abundant and renewable polymers and have excellent properties such as biodegradability, biocompatibility, non-toxicity and adsorption. Due to these properties, chitin and chitosan have a number of applications in various fields like biotechnology, agriculture, medicine, environment protection, fiber and paper Industry, cosmetic and foods etc.

2. Chitin

Chitin - a Greek term for tunic /envelope was first used by Odier in 1823 [3]. According to Lower (1984) it was derived from Greek word 'chiton' meaning a coat of mail. Chitin is colorless to off-white in colour. Chitin is hard, inelastic, nitrogenous polysaccharide found in the exoskeleton as well as in the internal parts of invertebrates[5]. Its degrees of deacetylation usually range from 5-15% and the average molecular weight ranges from 1.03×10^6 to 2.5×10^6 Dalton[11]. Chitin is highly hydrophobic and is insoluble in water and most various organic solvents. It is soluble in hexafluoro-isopropanol, hexafluoroacetone, chloroalcohols in conjugation with aqueous solutions of mineral acids and dimethylacetamide containing 5% lithium chloride [12]. Chitin is crystalline in nature and its tensile strength is much stronger than most artificially produced materials. The strength of this polymer is due to extensive hydrogen bonding along the chains. Chitin can be processed into several casting products including flakes, fine powders, beads, membranes, sponges, cottons, fibers and gels etc [13].

2.1. Sources of Chitin

Fungi, marine invertebrates and insects are the chief sources of chitin. Approximately 1.2×10^5 tons of chitin are annually accessible from shellfish waste on a worldwide basis [14] whereas the average Indian landing of shellfish has chitinous solid waste fraction ranging from 60,000 to 80,000 tonn [15]. Approximately 75% of the total weight of shellfish, such as shrimp, crab, and krill, is considered waste, and chitin comprises 20 to 58% of the dry weight of the solid waste [16]. India, Poland, Japan, US, Norway and Australia are now commercially producing chitin and chitosan. The US department of commerce reported in 1973 that they have produced over 1, 50,000 mt of chitin as processing waste from shellfish, krill, calms, oyster, squid and fungi [5].

In nature, it exists in the exoskeleton of animals such as insects [17] and crustaceans. It is also found in microorganisms e.g. in the cell walls and structural membranes of fungal mycelia (mushrooms), yeast and green algae [18, 19, 20]. Chitin is the structural component of the shells of crustaceans i.e. shrimp, crab, lobster and prawn [21]. It has also been reported in Protozoans, Arachnids, Nematodes, and Tunicates [22, 23]. The exoskeleton of crustaceans consists of 25-30% chitin by dry weight [21]. Some important sources of chitin and its location in the various organisms are given in Figure 1.

Organism	Structural component
Crayfish	Shell
Shrimp	Shell
Fungi	Cell wall
Squid	Beak, pen & stomach lining
Crab	Shell
Krill	Exoskeleton
Prawn	Shell
Lobster	Shell
Oyster	Shell
Insects	Cuticle, trachea & Midgut peritrophic memberane.
Protozoans	Cyst wall
Arachnids	Exoskeleton
Nematodes	Egg shells
Bacteria	Prostheca & Streptomycete spores.
Algae	Lorica wall
Birds	Beak & Pen

Figure 1. Sources of Chitin.

In Fungi, chitin appears to be ubiquitous [24] with exception of *Schizosaccharomyces* which has small amounts of chitin. Several zygomycete species of fungi chitin exits in its deacetylated form referred to as chitosan. But latter is found in the cell walls of fungi, to a lesser extent than Chitin [14, 25, 26]. Chitin is present in most of the fungi. Amount of chitin present in some significant fungi are given in Figure 2.

Fungus	**Division**	**Chitin content (%)**
Aspergillus flavus	Ascomycota	13.2-25.6
Aspergillus niger	Ascomycota	42.0
Aspergillus oryzae	Ascomycota	10.4
Aspergillus parasiticus	Ascomycota	17.1-24.9
Aspergillus phoenicis	Ascomycota	23.7
Blastomyces dermatitidis	Ascomycota	13.0
Glomerella cingulata	Ascomycota	6.3
Histoplasma capsulatum	Ascomycota	25.8-26.4
Histoplasma farciminosum	Ascomycota	40.0
Lentinula edodes	Basidiomycota	25.08–36.72
Mucor rouxii	Zygomycota	9.4
Neurospora crassa	Ascomycota	8.0-11.9
Neurospora tetrasperma	Ascomycota	14.7
Paracoccidioides brasiliensis	Ascomycota	11.0
Penicillium chrysogenum	Ascomycota	19.5- 42.0
Penicillium notatum	Ascomycota	24.9
Saccharomysis gutulata	Ascomycota	2.3
Rhizopus nigricans	Zygomycota	12.2
Tremella mesenterica	Basidiomycota	3.7
Trichoderma viridis	Ascomycota	12.0-22.0

Figure 2. Relative Amount of Chitin in Various Fungi (Knorr [14];Ruiz[25]; Yen and Mau[26]; Blumenthal and Roseman [27].

Chitin is present as a structural component in most existing species today. Its phylogenetic distribution varies in different types of organisms. In Prokaryotes, chitin has only been reported as a possible component of streptomycete spores and the stalks of some prosthecate bacteria. In eukaryotic microorganisms, it is present in tough structural material e.g. in cyst walls of some ciliates and amoebae; in the lorica walls of some ciliates and chrysophyte algae; in the flotation spines of centric diatoms and in the walls of some chlorophyte algae and oomycete fungi [9]. *Pneumocystis carinii*, of uncertain affinity, has chitin in the walls of its cysts and trophozoites [28]. In Animals, chitin is present in most invertebrates [29, 30] but absent from vertebrates. It is probably absent from plants, but polymers rich in (1, 4)-β-linked N-acetylglucosamine have been reported in plants. Apart from these sources, byproducts from citric acid and antibiotic manufacturing plants may also become an important source for chitin.

2.2. Structure of Chitin

Both chemical structure and biological function of chitin has similarity with cellulose [25].The crystalline structure of chitin has been shown to be similar to cellulose in the arrangements of inter and intrachain Hydrogen bonding [31]. Chitin is a linear polysaccharide consisting of β-(1-4)-linked-2-amino-2-deoxy-D-glucopyranose units, its basic repeating unit in the chitin chain is chitobiose, whereas cellulose is a linear homopolymer of anhydroglucose units linked by β-(1-4) glycosidic bonds, its basic repeating unit is disaccharide cellobiose [32]. Chitin may be regarded as a cellulose derivative (figure 3) with an acetamido group at C_2 [33]. Cellulose contains hydroxyl groups at the C_2 position of the monomers. Both cellulose and chitin strengthen the cell walls of plants and supporting structures of invertebrate animals respectively.

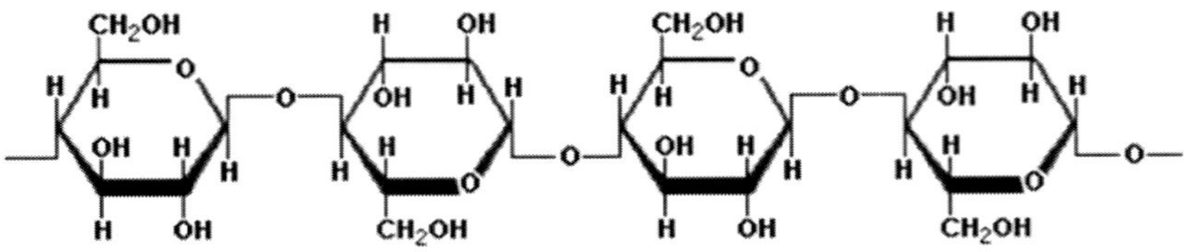

Cellulose

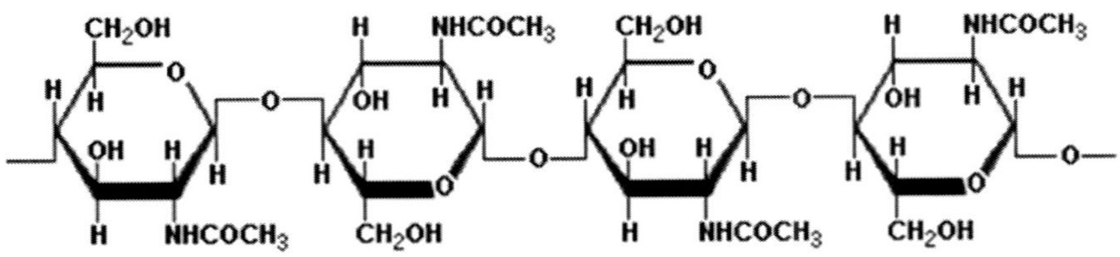

Chitin

Figure 3. Structures of Cellulose and Chitin.

The molecular configuration of a single polymer chain is helicoidal (6 aminosugar residues per turn) which is stabilized by intramolecular hydrogen bonds. The crystalline state of chitin, in which a number of molecules (~20) coalesce to form microfibrils (~3 nm in diameter), is generated by intermolecular hydrogen bonds formed between the amine and carbonyl groups. Three highly ordered crystalline structures of chitin polymorphs were described largely by X-ray diffraction studies [17, 34, 35].

In the chitin crystal structure, the chains form hydrogen-bonded sheets linked by C=O and -NH groups. In addition, intramolecular hydrogen bonds are present between the neighbouring sugar rings of each chain: the carbonyl group bonds to the hydroxyl group on C_6. There is also a second hydrogen bond between the OH-group on C_3 and the ring oxygen, similar to that in cellulose [35]. This extensive hydrogen bonding enhances the stiffness of the chitin chain. (Figure 4)

Chitin has a highly ordered, crystalline structure. There is a wide variety of types of chitin but only three hydrogen-bonded crystalline forms have been characterized: α-chitin, β-chitin, γ-chitin (Figure 5). In α-chitin, the chains are arranged in antiparallel fashion. In β-chitin, the chains are arranged in parallel fashion and γ-chitin has a three chain repeating unit, two 'up' & one 'down'.

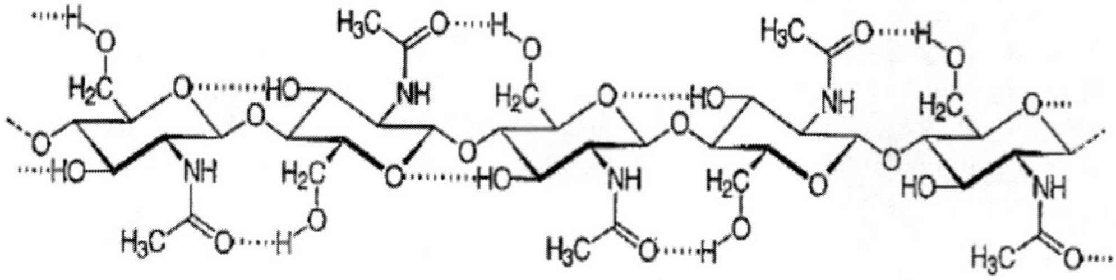

Figure 4. Chemical Structure of Chitin with its Intramolecular Hydrogen Bonds (dotted lines) between the Neigbouring Sugar Rings in Chitin.

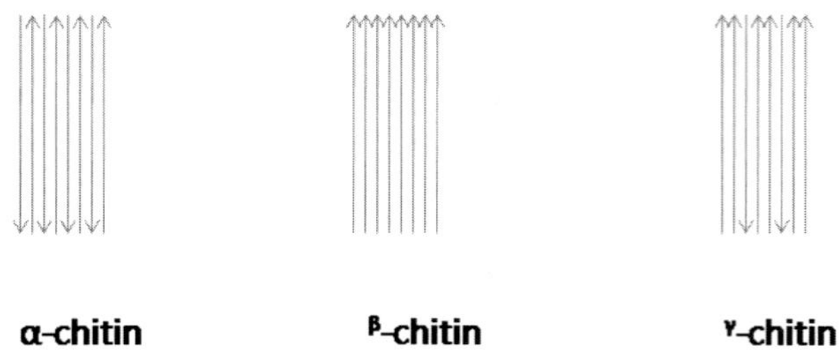

Figure 5. Arrangement of the Polymer Chains in the Three Forms of Chitin.

The three chitin variants differ in their degree of hydration, in the size of their unit cell and in the number of chitin chains per unit cell. α- chitin is the most common type. It is found in most eukaryotic microorganisms and exoskeletons of invertebrates. Most thermodynamically stable crystalline polymorph is the α-chitin in which one polymer chain is anitiparallel to the other. It has been speculated that such arrangement is due to either folding back of nascent chitin polymers or by virtue of in situ antiparallel biosynthesis at the catalytic sites [22].

The distribution of the polymorphic forms is not related to a particular category of organism, as different forms may occur in one organism, providing different functional properties. Different physical structures are responsible for performing different functions e.g. squid, *Loligo*, have α-chitin in its tough beak and radula, β-chitin in its rigid pen (endoskeleton), the polymer chains are packed in a parallel fashion [36, 37] with reducing ends pointing to the same direction and γ-chitin in its flexible stomach lining. α-chitin is by far the most abundant form, and is usually found where extreme hardness is required [17]. β and γ-chitin seem to provide toughness, flexibility and mobility and may have physiological functions other than support [31]. β-chitin is detected in *Eufolliculina uhligi*, a Protozoan and spines of diatoms, *Thalassiosira fluviatilis* whereas γ-chitin is found in coccon of beetle [17, 38]. The distribution of chitin and its different forms is shown in figure 6.

The inability of α -chitin to swell upon soaking in water is explained by the extensive intermolecular hydrogen bonding [35]. Due to lack of these interchain hydrogen bonds in β-chitin, it swells readily in water [39]. Due to the close packing of chains and its strong inter- and intramolecular bonds among the hydroxyl and acetamide groups (Figure 4), chitin shows poor solubility [40]. However, it can be converted into chitosan which is soluble in mild acidic solutions. Because chitosan is soluble in dilute acids e.g. formic acid and acetic acid and its salts are soluble in water therefore it finds wider application in industry than Chitin [41].

Phylum	**Occurrence**	**Chitin fraction [%]**	**Polymorphic form**
Arthopoda	Calcified cuticles, intersegments membranes, wings, legs and tentorium	20.0 - 85.0	α
Pogonophora	Tubes	33.0	β
Mollusca	Shell plates, mantle bristles redula, shell, redula, jaws, stomacal plates, calcified shell, pen, jaws, redula, shells.	0.1 - 36.8	α, β, γ
Annelida	Chaetae, jaws.	0.28 - 38.0	β
Fungi	Cell walls and structural membranes of mycelia stalk and spores.	traces - 45.0	α
Algae	Cell wall components.	Traces	α
Protozoa	Cyst wall, shell, perisarc, coenosteum, skeleton, podocyst.	3.2 - 30.3	α
Brachipoda	Stalk cuticle, stalk cuticle, shell.	3.8 - 29.0	β, γ

Figure 6. Occurrence of Chitin in Nature Muzzarelli [31], Roberts [42] and Shahidi [43].

Chitin chains folded in one of the forms, aggregate and produce microfibrils in living systems. Except the chitin of diatoms [44], chitin is found in nature cross-linked to other structural components. The chitin microfibrils in combination with other sugars, proteins, glycoproteins and proteoglycans form fungal septa and cell walls as well as arthropod cuticles and peritrophic matrices especially in crustaceans and insects [45]. In animals, chitin is associated with proteins, while in fungal cell wall it is associated with glucans, mannans or other polysaccharides. In fungal walls, it is found covalently bound to glucans, either directly or via peptide bridges [52]. In insects and other invertebrates, the chitin is always associated with specific proteins, with both covalent and non-covalent bonding. Chitin has also been found in varying degrees of mineralization, particularly calcification and sclerotisation involving phenolic and lipid molecules. In both fungi and invertebrates, varying degrees of deacetylation have been determined, giving an array of structure between chitin (fully acetylated) and chitosan (fully deacetylated) [46].

2.3. Extraction of Chitin

Chitin was first isolated and extracted as the alkali-resistant fraction from fungi by French chemist and pharmacist Henri Braccanot [2]. Procedures for isolation of high-molecular weight pure chitin are very important for its use as a raw material to make high quality chitosan. Since proteins, minerals, lipids, and pigments are closely associated with chitin (Figure 7), so all these components have to be removed to achieve the high purity necessary for biological applications.

Organism	**Chitin**	**Protein**	**Ash**	**Lipids**
Crab	25-30	15	40-60	2-5
Crayfish	13.2	29.8	46.6	5.6
Fungi	15-40	5-10	-	5-10
Honey bee	10-12	50-80	2-3	-
Krill	24.0	41.0	23.0	11.6
Prawn	33.0	61.6	29.4	1.4
Shrimp	30-40	35	30	5-10

Figure 7. Chitin Content in Various Organisms (%) (Synowiecki and Al-Khateeb [10], Nemtsev [47]).

Chitin from the diatom spines such as *Cyclotella cryptica* and *Thalassiosira fluviatilis* are the only form reported to be 100% poly N-acetylglucosamine that is not associated with proteins and is termed Chitan. The common procedure for extracting chitin from crustacean shells involves demineralisation, deproteinisation and decolouration. Demineralization is frequently carried out by treatment with dilute acid or chelating agent; while deproteinization involves treatment with dilute alkali or Proteolytic enzymes [31, 48]. The residues are decolourised using solvents and/or oxidants [49].

In most of the procedures, chitin is further degraded during extraction. Therefore, a great deal of interest still prevails in exploration of an efficient extraction procedure so that during this process, chitin degradation can be avoided and impurities be minimized to make the process commercially viable. In most instances, deproteinization precedes the demineralization [50] without any adverse effects on product [51, 52]. Deproteinization of chitin is usually performed by alkaline treatments, although other effective reagents have also been reported. Typically, raw chitin is treated with approximately 1.0 M aqueous solution of NaOH for 1h to 72 h at temperatures ranging from 65 to 100 °C [49]· An alternative method involves the enzymatic degradation of proteins, where the residual protein in the produced chitin often remains relatively high and the reaction time is longer compared to chemical deproteination. However, Manni [53] from Tunisia demonstrated that the crude protease from *B. cereus SV1* could be used effectively in the deproteinization (88.8%) of shrimp wastes to produce chitin.

Demineralisation is generally performed by acids including HCl, HNO_3, H_2SO_4, CH_3COOH, and HCOOH; however HCl seems to be the preferred acid and is applied at a concentration between 0.2M and 2.0M for 1h to 48 h at temperatures varying from 0°C to 100°C. Subsequently/Afterwords, the product may be decolorized with acetone and/or H_2O_2 or NaOCl [51], or Peroxide bleaches [54]. However, preparation of chitin/chitosan involves demineralization and deproteinization of shellfish waste by the use of strong acids or bases [55, 56].

Chitin has been extracted from six different local sources in Egypt. The obtained chitin was converted into the more useful soluble chitosan by steeping into solutions of NaOH of various concentrations and for extended periods of time, then the alkali chitin was heated in an autoclave which dramatically reduced the time of deacetylation. Chitin from squid pens did not require steeping in sodium hydroxide solution and showed much higher reactivity towards deacetylation in the autoclave that even after 15 min of heating a degree of deacetylation of 90% was achieved [57].

2.4. Biosynthesis of Chitin

Chitin is naturally occurring in most of the fungi and insects. Glaser and Brown [58] were the first to demonstrate in vitro chitin polymerization in a cell free system, using the filamentous fungus *Neurospora crassa.* Later, formation of chitin from UDP-GlcNAc was demonstrated in a cell-free system extracted from the southern armyworm *Prodenia eridania* also[59]. An orderly sequence of many complex reactions is involved in the process of chitin formation from glucose or with insect typical disaccharide trehalose.

The sequence of reactions (Figure 8) leading to the formation of various metabolites includes:

1. Hydrolysis of trehalose to glucose
2. Phosphorylation of glucose
3. Isomerisation of phosphorylated glucose to phosphorylated fructose
4. Transamination to form glucosamine-6-phosphate

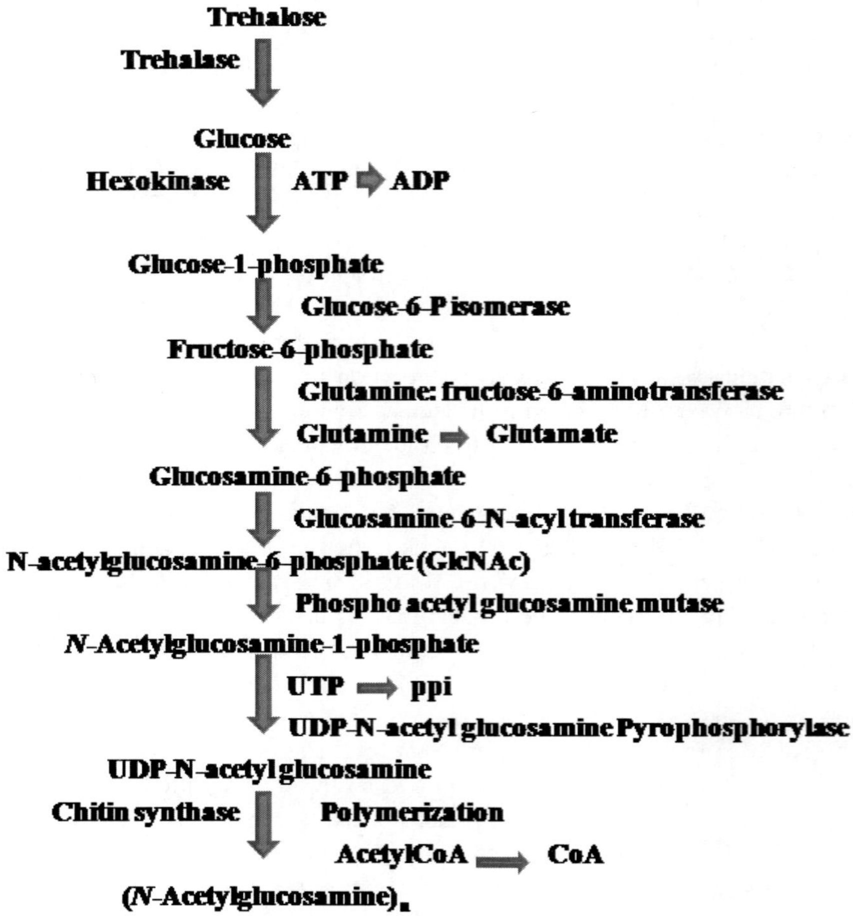

Figure 8. Flow Chart of Chitin Biosynthesis (Cohen personal communication) [64].

5. Acetylation forming N-acetylglucosamine 6-phosphate
6. Conversion of N-acetylglucosamine 6-phosphate to N-acetylglucosamine- 1-phosphate
7. Formation of the nucleotide amino-sugar conjugate UDP-GlcNAc.

The above catalytic steps and the various subsequent metabolites related to the generation of chitin were identified in yeast and filamentous fungi [60,61] as well as in insects and in crayfish [62].

The elongation of chitin chain takes place by the repeated addition of Diphosphate (UDP) – GlcNAc. The chitin biosynthesis requires UDP-*N*-acetylglucosamine as an obligate substrate and divalent cations i.e Mg^{2+} or Mn^{2+} as co-factors. Clusters of chitin synthase (CS) molecules integrated into cell membranes polymerize N-acetylglucosamine (GlcNAc) units into chitin chains.

Chitin chain elongation occurs in three distinct steps [63]:

1. The catalytic domain of chitin synthase facing the cytoplasmic site forms the polymer.
2. Translocation of the nascent polymer across the membrane and its release into the extracellular space.
3. Spontaneous assemblage of single polymers to form crystalline microfibrils of varying diameters and lengths.

2.5. Properties of Chitin

Chitin is not soluble in common solvents because of the strong intermolecular hydrogen bonding. It is soluble only in a few solvents such as hexafluoroacetone and N, N-dimethylacetamide (DMAc) containing 5–8% LiCl and methanol saturated with calcium chloride dihydrate. Chitin is natural, non-toxic, non-allergenic, anti-microbial, and biodegradable. Most of the naturally occuring polysaccharides like pectin, alginic acid, cellulose, agar, agarose and carragenas are acidic in nature whereas chitin and chitosan are very basic in nature[5]. It has a strong positive charge, which allows it to bind with negatively charged surfaces or materials, including metals, skin, and macromolecules such as proteins.

3. Chitosan

Removal of most of the acetyl groups of chitin by treatment with strong alkali yields chitosan [65] which is 2-amino-2-deoxy-β-D-glucose. A discrete nomenclature with respect to the degree of *N*-deacetylation has not been defined between chitin and chitosan [31]. In general, chitin with a degree of deacetylation of above 70% is considered as chitosan [66]. Since the production of chitosan, a deacetylated form of chitin, discovered by Rouget in 1859 it has become the subject of research for its distinct properties and its scope in commercial use. The history of chitosan begins with its first description by Braccanot in 1811. In 1894, Hope Seyle named it chitosan [10]. The term chitosan refers to a family of copolymers with various fractions of acetylated units. Many fungi including *Aspergillus niger*, *Mucor rouxii*, *Rhizopus oryzae* and *Absidia coerulea* have chitosan in their mycelial wall [67,68,69,70,71]. Chitosan provides the fungus an additional resistance to lysis by chitinolytic organisms [72].

3.1. Sources of Chitosan

Chitosan has been found in a wide range of natural sources such as crustaceans, fungi, insects and some algae. Chitosan is produced from chitin by thermochemical deacetylation. This involves boiling of chitin in concentrated alkali for several hours. However enzymatic procedures for chitin deacetylation by chitin deacetylase are also in use. In nature, Chitosan is also found in cell walls of fungi of the class *Zygomycetes,* in the green algae *Chlorella* sp., yeast and protozoa as well as in insect cuticles [73]. The cultivation of fungi (*Aspergillus niger*) can provide an alternative source of chitosan. However, the acetyl groups in chitosan produced from crustacean chitin are uniformly distributed along the polymer chain, but chitosan of similar degree of deacetylation isolated from fungal cell walls possess acetyl residues that are grouped into clusters [74]. Although, a wide range of sources of chitosan is available, but it is commercially prepared mainly from crustaceans waste such as crab, shrimp, krill, and crayfish. It is because a large amount of crustacean exoskeleton is available as a byproduct of sea food processing. Annually India produces more than 100,000 tones of shrimp processing waste [75]. Out of which only an insignificant amount is utilized for the extraction of chitin and the rest is discarded. Disposal of crustacean shell waste has been a challenge for sea food processors. The unique, abundantly renowned properties of chitin and chitosan make it a formidable biopolymer, much too precious to be dumped as waste into the sea or in landfills. Thus effective utilization of such a waste had become an important issue both from the point of clean environment and the economic value of this unique bio-waste.

3.2. Structure of Chitosan

The chemical name of chitosan is Poly-(1-4)-2-amino-2-deoxy-ß-D-glucopyranan and its molecular formula is $(C_6H_{11}O_4N)_n$ [76]. Chitosan forms extensive intra and intermolecular H-bonding due to the presence of one primary amino and two free –OH groups (one primary hydroxyl group at C-6and one secondary hydroxyl group at C-3) for each repeating unit (Figure 9). Its nitrogen content varies from 5 to 8% depending on the extent of deacetylation; it is mostly in the form of primary aliphatic amino groups.

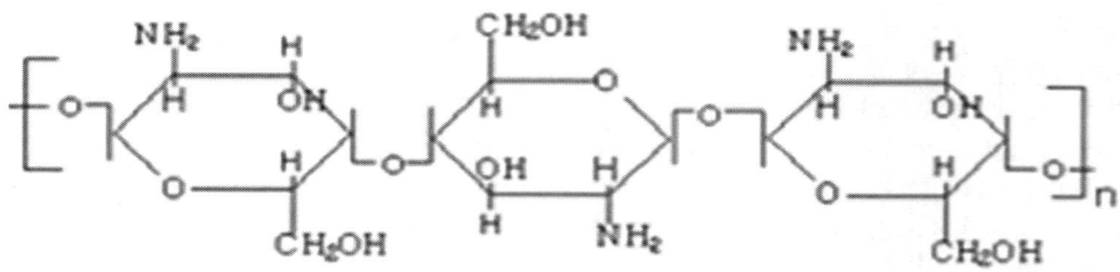

Figure 9. Structure of Chitosan.

3.3. Chitosan Preparation

Commercial chitosan is derived from the shells of shrimp and other sea crustaceans, including *Pandalus borealis* which have chitin, as the structural element in their exoskeleton. Chitosan can be prepared from fungi or it can be prepared from chitin by two methods viz. thermochemical deacetylation and enzymatic deacetylation. In thermochemical process (Figure 10), chitin is treated with highly concentrated (40-50%) solutions of NaOH or KOH at high temperatures (100-150°C) and exclusive of air for about an hour [77,78].

The thermochemical deacetylation of chitin from crustacean shells is further of two types, heterogeneous deacetylation and homogeneous deacetylation [78,79]. The extent of deacetylation of chitin depends on concentration of alkali used, particle size, density of chitin and previous pretreatment. There are two methods of thermochemical deacetylation:

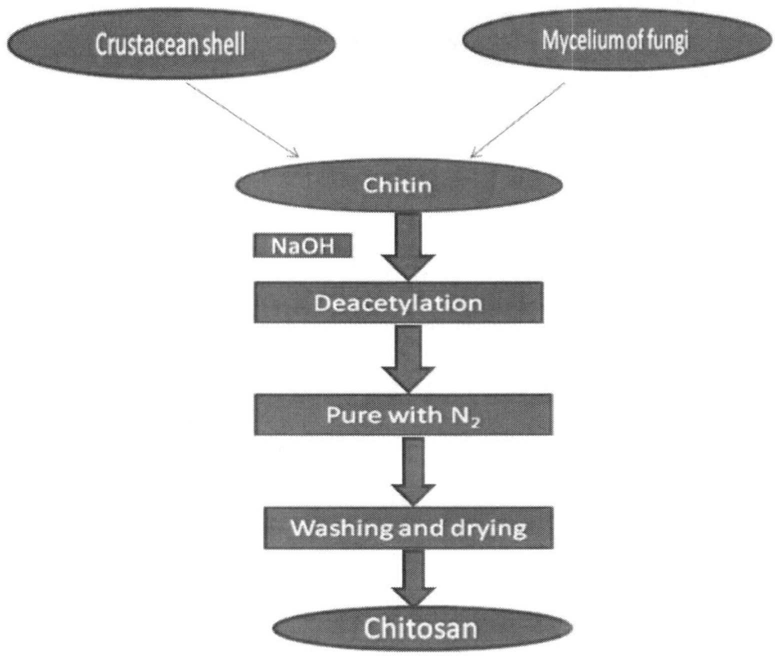

Figure 10. Preparation of Chitosan by Thermochemical Deacetylation.

3.3.1. Heterogeneous Deacetylation

This method is used mainly in industry. This type of deacetylation involves the preferential action in the amorphous regions of the solid chitin so that crystalline native regions remain intact in the parent chitin. In this method, chitin flakes are treated with aqueous solution of NaOH (30-60%) at 80-120 °C for 4-6 h. This method can give highly N-deacetylated products during long treatment time and repeated treatments usually lead to depolymerisation [80].

3.3.2. Homogeneous Deacetylation

Homogeneous deacetylation is carried out on pre-swollen chitin under vaccum in aqueous solution [78]. Alkaline chitin in the form of solution of the sodium salt of chitin is treated with 1.4% NaOH, at 25 °C. The process can produce partially N-deacetylated derivatives of chitosan/chitin, which are soluble in water. However, this method is not very efficient and a random distribution of N-acetyl groups is found in the products.

Availability of chitosan produced by these methods is restricted and seasonal for commercial use [81]. Besides, large amount of concentrated alkaline solution is used in this process which causes environmental pollution. To overcome all these problems several alternate sources of chitin/chitosan have been suggested including insects, Antarctic krill and diatoms which can be utilized as raw materials in industries[82]. Chitosan-producing fungi, particularly zygomycetes which have chitosan in their cell wall, can be cultured [83]. Chitosan has been produced by fermentation of zygomycetes such as *Absidia coerulea* [84,85,86,87], *Mucor rouxii* [19,88,89] and *Rhizopus* oryzae [67]. The advantage of using fungi is the easy handling, harvesting and controlling to produce high quality chitosan. But in the fungal mycelia, chitosan and glucan are covalently linked. The complex of chitosan and glucan causes considerable problems in the extraction of both chitosan and glucan from fungal mycelia [90]. The various major polymeric components present in fungal cell wall are shown in Figure 11.

Taxonomic group	Oomycets	Chytridomycetes	Zygomycetes	Basidiomycetes	Ascomycetes/ Deuteromycetes
Fibrillar polymers	β(1,3), β(1,6)-Glucan, cellulose	Chitin, glucan	Chitin, Chitosan	Chitin, β (1,3)-β (1,6) glucans	Chitin, β (1,3)- β (1,6) glucans
Matrix polymers	Glucan	Glucan	Polyglucuronic acid, glucuronomannoproteins	α(1,3)-glucan, xylomannoproteins	α(1,3)-glucan, glucuronomanno-proteins

Figure 11. Major polymers in Fungal Cell Wall (Latge et al [91]).

3.4. Biosynthesis of Chitosan

Chitosan is synthesized in nature by enzymatic deacetylation of chitin. Chitin deacetylase is the enzyme that catalyzes the conversion of chitin to chitosan by the hydrolysis of acetamido groups (Figure 12) of N-acetyl-D-glucosamine present in chitin. Chitin deacetylase was first identified and particularly purified from extracts of fungus *M.rouxii* by Araki and Ito in 1975. Since then, the presence of this enzyme has been reported in several other microorganisms. Chitin deacetylase is found in marine bacteria, several fungi and a few insects [92]. It catalyzes the deacetylation of chitin to chitosan by recognizing a sequence of four GlcNAc units in the chitin, one of which undergoes deacetylation.

3.5. Production of Chitosan from Fungus

Fungi can be cultivated for chitosan production by both solid state and submerged fermentation. Maghsoodi [93] had produced chitosan by submerged fermentation of *Aspergillus niger BBRC,*20004. Saboroud dextrose broth was used as a culture medium and glucose concentrations were changed in the medium from 2% to 12%. It was found that chitosan yield increased from 0.8455 g/l to 0.9121g/l in the medium containing 8% glucose. Logesh [94] had isolated the four endolichenic fungi *Aspergillus niger,Penicillium citrinum, Rhizopus oryzae* and *Fusarium oxysporum* from the thallus of lichen, *Roccella montagnei* and subjected them for submerged fermentation to produce chitosan. On the basis of yield *Aspergillus niger* was found to be potential candidate for the production of chitosan after 12 days of cultivation.

For production of chitosan by solid state fermentation, different types of substrates were used by different workers. For instance wheat straw, Sweet potato and corn seed residue were used as substrates by an edible mushroom, *Lentinus edodes* [95], a Zygomycetes fungus *Gongronella butleri* USDB 0201 [90] and Ascomycetes fungus *Aspergillus niger BBRC,*20004 [96] respectively. Crestini [95] compared both the solid state and submerged fermentations using *L. edodes* and found that the solid-state fermentation method yields 50 times higher chitosan than submerged fermentation. They further found that chitosan obtained from the fungus *L. edodes* have high-quality chitosan in terms of yields and/ degree of acetylation. The degree of acetylation of chitosan obtained from submerged fermentation increased with incubation time. They have attributed it to more susceptibility of chitin for chitin deacetylase enzyme because chitin is less crystalline during initial stages of growth. However, there was no such effect on degree of acetylation of chitosan obtained from solid-state fermentation. The quality of chitosan is considered better with low degree of acetylation due to high positive charge density and Acetylation degrees of chitosan obtained from *L. edodes* ranged from *5* to 13% and that from commercial crabshell chitosan ranged from 18-22% making fungus as a better contender for good quality chitosan.

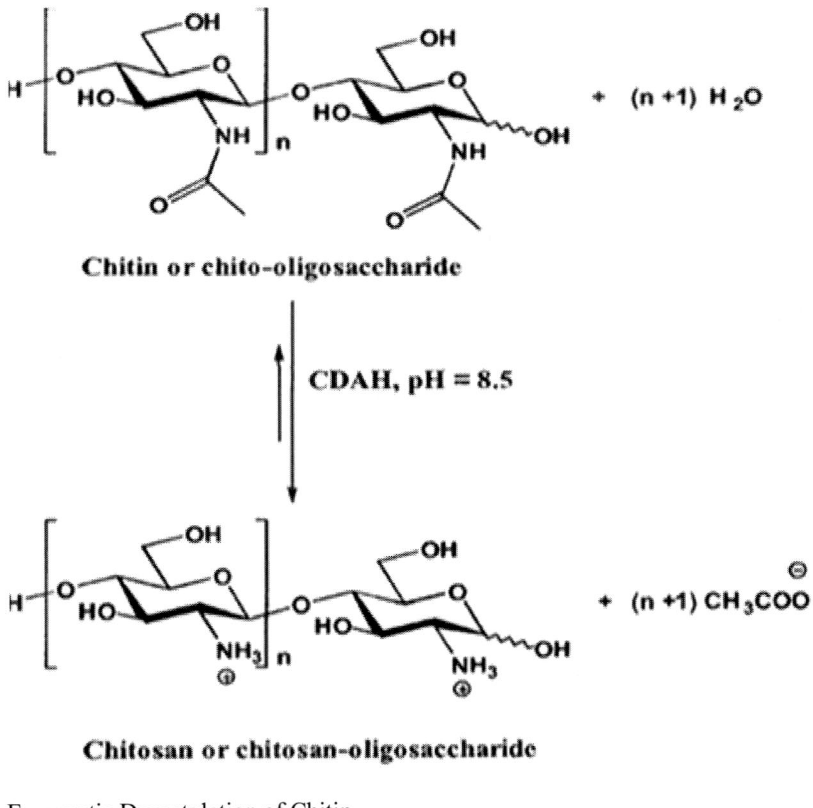

Figure 12. Enzymatic Deacetylation of Chitin.

3.6. Extraction of Chitosan from Fungal Cell Wall

First of all, the cell wall is isolated from the fungal biomass by its alkali treatment (with dilute NaOH solution) at elevated temperature (e.g. 90-120%). At high temperature, the alkali dissolves proteins, lipids, and alkali-soluble carbohydrates and the cell wall remains as alkali insoluble material (AIM). After this, chitosan is separated from AIM by dissolving it in an acid solution (e.g. 2–10% acetic acid at 25–95 °C for 1–24 h). During this step, the other components of cell wall remain as alkali- and acid-insoluble material (AAIM). Finally, fungal chitosan is precipitated by increasing the pH to 9-10 and chitosan is recovered by centrifugation.

In another approach, treatment of cell wall with hot dilute Sulfuric acid solution has resulted in a high yield and more pure form of chitosan. Unlike acetic, citric, lactic and hydrochloric acids, chitosan is not soluble in dilute Sulfuric acid solutions at room temperature [97]. However, it is soluble in hot boiling solution of Sulfuric acid. The temperature dependent Solubility of chitosan in Sulfuric acid solutions is not shared with the other components of cell wall such as chitin and polyphosphates. Therefore, by treating the cell wall with hot dilute Sulfuric acid, chitosan becomes soluble in hot acid and it can be separated from other components of cell wall by filtration. In the next, step chitosan can be recovered from Sulfuric acid by cooling. Zamani reported that extraction of chitosan from cell wall of Zygomycetes Fungi by Sulfuric acid results in a product with higher purity and yield compared to traditional extraction method e.g. by acetic acid [98].

3.7. Properties of Chitosan

Chitosan occurs as flabby powder or flake. It is odorless and its colour varies from yellow to white. The properties of chitosan depend on two factors, degree of deacetylation (DD) and molecular weight (M_w). The term chitosan includes a variety of polymers that differ from each other in their average molecular weight and degree of deacetylation. The DD and M_w of chitosan depend on the reaction conditions during the manufacture of chitosan from chitin. Chitosan is nontoxic, biodegradable as well as antimicrobial polymer. The unique properties of chitosan, are due to its cationic nature and solubility in acidic solutions, compared to other natural polysaccharides which are usually neutral or negatively charged [31]. Chemically chitosan is a linear polyamine having reactive amino and hydroxyl groups. It can form chelates with transition metal ions. A large no. of applications of chitosan are due to its biodegradable, biocompatible, haemostatic, fungistatic, spermicidal, antitumor, antimicrobial, central nervous system depressant and immunoadjuvant character [5].

3.7.1. Molecular Weight

Chitosan is a polymer of high molecular weight. Its molecular weight can be determined by methods of chromatography [99], light scattering [31] and viscometry [100]. Its molecular weight varies with the raw material and method of preparation. Depending on the process and grades of the product, the molecular weight of commercial chitosan was in the range of 100,000 –1,200,000 Daltons. High temperature causes thermal degradation of chitosan, thereby lowering its molecular weight [101].

3.7.2. Solubility

Chitosan is unique because of its polyamine character which makes it soluble at acidic pH. The pH-dependent solubility of chitosan is due to its amino groups (—NH_2), which become protonated upon dissolution at pH 6 or below to form cationic amine groups (—NH_3^+). This causes the increased intermolecular electric repulsion, resulting in a polycationic soluble polysaccharide, with a large number of charged groups. In contrast, chitosan tends to lose its charge at higher pH, and may therefore precipitate from solution due to deprotonation of the amine groups [102]. Chitosan is soluble in dilute acidic solutions below pH 6.0.

Chitosan is soluble in organic acids such as acetic, formic, and lactic acids. The most commonly used solution is 1% acetic acid at about pH 4.0. Solubility of chitosan in inorganic acids is quite limited as it is soluble in 1% hydrochloric acid and insoluble in sulfuric and phosphoric acids. Concentrated acetic acid solutions at high temperature can cause depolymerization of chitosan. However above pH 7.0, solubility of chitosan is poor because chitosan solution forms poly-ion complex with anionic hydrocolloid resulting in the precipitation or gel formation [103].

3.7.3. Cationic Nature

The naturally occurring polysaccharides e.g. agarose, pectin, cellulose, dextrin, alginic acid, agar and carragenans are acidic in nature but chitosan are highly basic polysaccharides [104]. Both, chitosan and cellulose have the β-(1, 4)-anhydroglucosidic bonds but the characteristic properties of chitosan are not shared by cellulose [105]. The presence of free amino groups makes it positively charged, that is why it reacts with many negatively charged polymers.

3.7.4. Biodegradability

Chitosan is liable to degradation by enzymes, such as lysozymes (present in tears, saliva, blood and milk), cellulases, chitinases, chitosanases, hemicellulases, proteases (papain and pronase), lipases and ß-1, 3-glucanases. Chitosanases (EC 3.2.1.132) catalyze the endohydrolysis of *ß*- (1, 4)-glycosidic linkages between D-glucosamine (GlcN-GlcN) residues in partly acetylated chitosan[106]. Their biodegradation leads to the production of non-toxic oligosaccharides of variable length which can be subsequently incorporated to glycosaminoglycans and glycoproteins to metabolite pathways or be excreted [107].

3.7.5. Biocompatibility

The most important biological property of chitosan is biocompatibility i.e. it should not be affected by the host and at the same time should not elicit any undesirable local or systemic effects. Our living tissues, including the skin, nasal epithelium and ocular membranes can easily tolerate chitosan. That is why chitosan is also used in nasal drug delivery and cosmetics [106]. Biocompatibility of chitosan depends on the characteristics such as natural source, method of preparation, molecular weight and degree of deacetylation (DD) [78].

3.7.6. Non-Toxicity

The oral LD50 (median lethal dose) of chitosan in mice was found to be in excess of 16 g/day/kg body weight, which is higher than that of sucrose. The toxicity of chitosan depends on its purity.

4. Applications of Chitin and Chitosan

Chitin and Chitosan have a no. of applications [1,5] such as in wastewater treatment, food industry, medical & pharmaceutical industry, biotechnology, agriculture, photography, paper industry and cosmetics etc. [25,66,108]. It can be formed into fibers, films, gels, sponges, beads or nanoparticles. Due to the biodegradable nature of chitin and chitosan, they are used as carbon source for single cell protein production and biodegradable packaging materials. Chitin and its derivative are also used in removal of metals like mercury and color from wastewater & textile mill effluents respectively.

4.1. Biomedical Applications

Chitosan has antacid and anti-ulcerogenic properties due to which it has protective effect on stomach mucosal line. The Antidiabetic (hypoglycemic) action of chitosan makes it suitable agent in lowering of blood glucose level; increasing glucose tolerance and insulin secretion. It is hypocholesterolaemic; thus prevents fat absorption and helpful in weight loss and reduction of blood lipid levels. Chitosan has antibacterial activity against *vibrio cholerae* and *E.coli* [55]. It also has fungicidal and fungistatic properties. Because of its antimicrobial nature It is used in Wound dressings; products for wound treatment, inhibition of fibroplasia; promotion of tissue regeneration; acceleration of wound-healing with minimal scar formation [70]. It has all the characteristics required for contact lenses such as optical clarity, mechanical stability, sufficient optical correction [15], immunological compatibility, gas permeability, wettability, antimicrobial etc. It is also used in bandage making. Chitosan is used in dentistry and oral medicine because of its unique properties like bioadhesive, viscosity-enhancer, permeabilizer [6], antimicrobial, prolonged drug release in buccal cavity, anti-dental caries treatment of periodontal diseases, oral candidiasis, tooth mobility and reduction of plaque formation. It is used as antioxidant for scavenging radicals and chelation of divalent metals. It also has anticoagulant, hemostatic and anti-inflammatory properties. Chitosan is water soluble and bioadhesive which readily binds to negatively charged surfaces such as mucosal membranes. Chitosan enhances the transport of polar drugs across epithelial surfaces. Purified quantities of chitosans are available for biomedical applications. Chitosan may also help reduce blood levels of both low-density lipoprotein and total cholesterol. Chitosan can be used in direct tablet compression, for the improvement of drug dissolution and for production of controlled release implant systems for delivery of hormones [109].

4.2. Wastewater Treatment

The unique molecular structure of chitosan provide it an extremely high affinity for many classes of dyes, including direct, disperse, reactive, vat, acid, sulfur and naphthol dyes [5]. chitosan has low affinity for basic dyes. Thus it helps in removal of metal ionsflocculant/coagulant, protein dye; amino acids [6, 28,29,30,47,58,98]. This makes it very useful substance in water engineering.

4.3. Food Industry

In food industry it is used for removal of dye, suspended solids, preservative, food stabilizer [11], color stabilization, thickener and gelling agent, animal feed additive, etc. it is also used as food additive because of its nutritional value. It is used for clarification of beverages and fruit juices. It is known as fat magnet as it is good for reducing body weight safely [110]. It has been marketed a dietary supplement that in Japan and Europe for about 20 years membrane industry and pulp and paper industry. Due to their inherent properties, coupled with the ability to form films, alone or in combination with other polymers, chitin, chitosan and oligosaccharides are desirable food packaging materials [48,74,111].

4.4. Agriculture

In agriculture it is used as a Seed coating agent, fertilizer [94,108,112], controlled agrochemical release [5,54,58] etc. Coating of depolymerized chitin, carboxymethyl (CM)-Chitin and hydroxyethyl (HE)-chitin on Soyabeans seeds was found to increase the seed germination rate, pod number, the plant dry weight and the crop yield [15].

4.5. Cosmetics

It is used in Moisturizer, face, hand [80,100,113] and body creams, bath lotion [27,44,92] etc. because of its polycationic nature.Moreover it has fungicidal and fungistatic properties [15].

4.6. Biotechnology

It is used for enzyme immobilization, protein separation, cell recovery and chromatography [92] etc. Chitin and chitosan are used for the preparation of biotechnological materials such as cell stimulating material, as fat trapper, Anti-throbogenic and haemostatic materials [10] etc.

Conclusion

The unusual properties of chitin and chitosan are due to its primary amino group and primary and secondary hydroxyl groups. Thus, the reactivity of these groups can be utilized to find applications of chitin and chitosan in different areas. In this review an attempt has been made to simplify the study of these versatile and promising biopolymers in terms of their characteristics and importance. The progress in study of chitin and chitosan is truly remarkable and attracting worldwide attention. Currently, chitin and chitosan are widely used in food, animal feed, textile, cosmetics, and paper and pulp industries as well as in research and development. These developments together with improved scientific knowledge are expected to pave the way for a remarkable success in the study of chitin and chitosan.

Acknowledgment

Authors are thankful to Prof. (Dr.) Ephraim Cohen, Robert H. Smith Faculty of Agriculture, Food and Environment, The Hebrew University of Jerusalem, Rehovot, Israel for personal communication in providing his research papers and granting permission to use flow diagram from his publication. Authors are thankful to the University Grants Commission (UGC, New Delhi) for financial support.

References

[1] Ghormade V, Kulkarni S, Doiphode N, Rajamohanan PR, Deshpande MV (2010) Chitin deacetylase: A comprehensive account on its role in nature and its biotechnological applications.In: Mendez-Vilas A, editor, Formatex Microbiology Series No.2: *Current research, Technology and Education Topics in Applied Microbiology and Microbial Biotechnology*(vol 2, pp: 1054-1066) Badajoz, spain: Formatex.

[2] Bracannot H (1811) Sur la nature des champignons. *Ann Chim Phys* 79:265-304.

[3] Deshpande MV (2005) Chitosan in fungi. Chitin and chitosan - opportunities and challenges.India, Contai, Midnapur: *SSM International publication* pp: 59-68.

[4] Tsigos I, Martinou A, Kafetzopoulos D, Bouriotis V (2000) Chitin deacetylase: new versatile tools in biotechnology. *Trends Biotechnol* 18:305-312.

[5] Dutta PK, Dutta J, Tripathi VS (2004) Chitin and Chitosan: Chemistry, properties and applications. *J Sci Ind Res* 63:20-31.

[6] Keisuke K (2006) Chitin and chitosan: functional biopolymer from marine crustaceans. *Mar Biotechnol* 8:203-226.

[7] Rinaudo M (2006) *Progress in Polymer Science* 31: 603–632Rane KD, Hoover DG (1993) An evaluation of alkali and acid treatments for chitosan extraction from fungi. *Process Biochem* 28:115-118.

[8] Duchesne LC, Larson DW (1989) Cellulose and the evolution of plant life. *Biosci* 39:238-241.

[9] Gooday GW (1990) Physiology of microbial degradation of chitin and chitosan. *Biodegr* 1:177-190.

[10] Synowiecki J, Al-Khateeb NA (2000) The recovery of protein hydrolysate during enzymatic isolation of chitin from shrimp Crangon crangon processing discards. *Food Chem* 68:147-152.

[11] Cheba BA (2011) Chitin and Chitosan: Marine Biopolymers with Unique Properties and Versatile Applications. *Global J Biotechnol Biochem* 6(3):149-153.

[12] Madhavan P (1992) Chitin, Chitosan and their Novel Applications. *Science Lecture Series,* CIFT, Kochi: 1.

[13] Hirano S (1997) Applications of chitin and chitosan in the ecological and environmental fields. In: Goosen MFA (Ed), Application of chitin and chitosan, *Lancaster, PA: Technomic Publishing Corporation*, Inc, pp:31-54.

[14] Knorr D (1991) Recovery and utilization of chitin and chitosan in food processing waste management. *Food Technol* 45:114-122.

[15] Kumar MNVR (2000) A review of chitin and chitosan applications. *Reac Funct Poly* 46:1-27.
[16] Wang SL, Chang WT (1997) Purification and characterization of two bi-functional chitinase/lysozymes extracellularly produced by *Pseudomonas aeruginosa* K-187 in a shrimp and crab shell powder medium. *Appl Environ Microbiol* 63:380-386.
[17] Rudall KM, Kenchington W (1973) Chitin and its distribution. *Biol Rev* 48:597-602.
[18] Bartinicki-Garcia S (1989) Chitin and Chitosan, Sources, Chemistry, Biochemistry, Physical Properties and Applications. *Elsevier Essex*: 23-35.
[19] White SA, Farina PR, Fulton I (1979) Production and isolation of chitosan from *Mucor rouxii. Appl Env Microbiol* 8:323-328.
[20] Mathur NK, Narang CK (1990) Chitin and chitosan, versatile polysaccharides from marine animals. *J Chem Educ* 67:938–942.
[21] Austin PR, Brine CJ, Castle, JE, Zikakis JP (1981) Chitin: new facets of research. *Sci* 212:749-753.
[22] Cohen E (1987) Chitin biochemistry: synthesis and inhibition. *Annu Rev Entomol* 32:71-93.
[23] Jeuniaux C, Vors-Foucart MF, Poulicek M, Busrem JC (1989) Sources of chitin estimated from new data on chitin Biomass and production. *Chitin and Chitosan, Elsevier Applied Science 3-11.*
[24] Bartnicki-Garcia S and Lippman E (1982) Fungal cell wall composition In: Laskin AJ,Lechevalier HA, editor, *CRC hand book of microbiology*(2nd edition,vol.iv, *Microbial composition: carbohydrates, lipids, minerals* pp 229-252) Boca Raton, FL: CRC Press.
[25] Ruiz-Herrera J (1978) The distribution and quantitative importance of chitin in fungi. Proceedings of the Ist International Conference on Chitin and Chitosan. Muzzarrelli RAA, Pariser ER (Eds.), *MIT Sea Grant Program, Cambridge,* MA.: 11-21.
[26] Yen MT, Mau JL(2006) Preparation of fungal chitin and chitosan from shiitake stipes.Fung Sci 21:1-11.
[27] Blumenthal HJ, Roseman S (1957) Quantitative estimation of chitin in fungi. Rackhant Arthritis Research Unit and Departments of Bacteriology and Biochemistry, University of Michigan, Ann Arbor, Michigan 74:222-224.
[28] Walker AN, Garner RE, Horst MN (1990) Immunocytochemical Detection of Chitin in *Pneumocystis carinii. Infect Immun* 58(2):412-415.
[29] Jeuniaux C (1963) Chitine et chitinolyse, un chapitre de biologie moléculaire. *Masson, Paris* 181.
[30] Jeuniaux C (1982) La chitine dans le regne animal. *Bull Soc Zool France* 107:233-249.
[31] Muzzarelli RAA (1977) Chitin. Elmsford, NY: *Pergamon Press,* Oxford: 89-94.
[32] Delmer DP, Amor Y (1995) Cellulose biosynthesis. *Plant Cell* 7:987-1000.
[33] Suzuki Y, Okamoto Y, Morimoto M, Sashiwa H, Saimoto H, Tanioka SI, Shigemasa Y, Minami S (2000) Influence of physico-chemical properties of chitin and chitosan on complement activation. *Carb Poly* 42:307-310.
[34] Jang MK, Kong BG, Jeong YI, Lee CH, Nah JW (2004) Physicochemical characterization of a-chitin, b-chitin and g-chitin separated from natural resources. *J Poly Sci: Part A: Polymer Chemistry* 42:3423–3432.
[35] Minke R, Blackwell J (1978) Structure of Alpha-Chitin. *J Mol Biol* 120(2):167-181.

[36] Gardner KH, Blackwell J (1975) Refinement of the structure of β-chitin. *Biopoly* 14(8):1581-1595.

[37] Herth W, Barthlott W (1979) The site of β-chitin fibril formation in centric diatoms. *J Ultrastruct Res* 68:6-15.

[38] Lehane MJ (1997) Peritrophic matrix structure and function. *Annu Rev Entomol* 42:525–550.

[39] Blackwell J (1969) Structure of Beta-Chitin or Parallel Chain Systems of Poly-Beta-(1, 4)-N-Acetyl-D-Glucosamine. *Biopol* 7(3):281-286.

[40] Urbanczyk GW, Lipp-Symonowicz B, Szosland I, Jeziorny A, Urbaniak J, Domagala W, Dorau K, Wrzosek H, Sztajnowski S, Kowalska S, Sztajnert E (1997) Chitin filaments from dibutyrylchitin precursor: Fine structure and physical & physicochemical properties. *J Appl Poly Sci* 65(4):807-819.

[41] Simpson BK, Gagne N, Simpson MV (1994) Bioprocessing of chitin and chitosan. Fisheries Processing: *Biotechnol Appl* 7:155-173.

[42] Roberts GAF (1992) Chitin chemistry.3rd edition. London: *Mcmillan Press*, Ltd 85-91.

[43] Shahidi F, Arachchi JKV, Jeon YJ (1999) Food applications of chitin and chitosans. *Trends Food Sci Tech* 10:37-51.

[44] Blackwell J, Parker KD, Rudall KM (1967) Chitin Fibres of Diatoms *Thalassiosira Fluviatilis* and *Cyclotella Cryptica*. *J Mol Biol* 28(2):383-388.

[45] Kozloff E (1990) Invertebrates. *Saunder College Publishing,* New York.

[46] Peter MG, Kegel G, Keller R (1986) Structural studies on sclerotized insect cuticle. Chitin in Nature and Technology, (Eds. Muzarelli RAA, Jeuniaux C, Gooday GW). Press, New York.

[47] Nemtsev S, Zueva O, Khismatullin R, Khismatullin M, Varlamov V (2001) Bees as Potential Source of Chitosan.In: Varlamov VP, Bykova VM, Nemtsev SV, Zueva OY,editor, Proceedings of VI International Conference on new achievements in study of chitin and chitosan, Moscow-Schelkovo on Oct 22-24,2001(pp 398) Moscow-VNIRO Publishing.

[48] Shimahara K, Takiguchi Y, Kobayashi T (1989) Chitin and Chitosan. *Elsevier Applied Science* pp:171-178.

[49] Percot A, Viton C, Domard A (2003) Optimization of chitin extraction from shrimp shells. *Biomacromol* 4:12-18.

[50] Hackman RH (1954) Studies on chitin. 1. Enzymatic degradation of chitin and chitosan esters. *Aust J Bol Sci* 13:568-577.

[51] No HK, Meyers SP, Lee KS (1989) Isolation and characterization of chitin from crawfish shell waste. *J Agri Food Chem* 37(3):575-579.

[52] Sandford PA (1989) Chitosan: commercial uses and potential applications. Chitin and Chitosan. Sources, Chemistry, Biochemistry, Physical Properties and Applications, *Elsevier Applied Science:* 51-70.

[53] Manni L, Ghorbel-Bellaaj O, Jellouli K, Younes I, Nasri M (2010) Extraction and Characterization of Chitin, Chitosan and Protein Hydrolysates Prepared from Shrimp Waste by Treatment with Crude Protease from *Bacillus cereus* SV1. *Appl Biochem Biotechnol* 162:345–357.

[54] Austin PR (1988) Chitin solutions and purification of chitin. *Meth Enzymol* 161:403-407.

[55] Liang TW, Chen YJ, Yen YH, Wang SL (2007) The antitumor activity of the hydrolysates of chitinous materials hydrolyzed by crude enzyme from *Bacillus amyloliquefaciens* V656. *Process Biochem* 42:527-534.

[56] Wang S, Jayaram AS, Hemphala J, Senti KA, Tsarouhas V, Jin H, Samakovlis C (2006). Septate-junction-dependent luminal deposition of chitin deacetylases restricts tube elongation in the *Drosophila* trachea. *Curr Biol* 16:180-185.

[57] Abdou, ES, Nagy, KS, and Elsabee, MZ(2008). Extraction and characterization of chitin and chitosan from local sources.*Biores Technol*_99(5):1359-1367..

[58] Glaser L, Brown DH (1957) The synthesis of chitin in cell free extracts of *Neurospora crassa*. *J Biol Chem* 228:729-742.

[59] Jaworsky E, Wang L, Margo G (1963) Synthesis of chitin in cell-free extracts of *Prodenia eridania*. *Nature* 198: 790.

[60] Cabib E (1981) Chitin: Structure, metabolism and regulation of biosynthesis. Encyclopedia of Plant Physiology N. S. 13B: 395–415.

[61] Gooday GW, Trinci APJ (1980) Wall structure and biosynthesis in fungi. Eukaryotic Microbial Cell. Soc. Gen. Microbiol. Symp., Cambridge. *Cambridge University Press,* Cambridge 30:207–252.

[62] Benson RL, Friedman S (1970) Allosteric control of glucosamine phosphate isomerase from the adult housefly and its role in the synthesis of glucosamine 6-phosphate. *J Biol Chem* 245:2219–2228.

[63] Merzendorfer H (2006) Insect chitin synthases: a review. *J Comp Physiol B-Biochemical Systemic and Environmental Physiology* 176(1):1-15.

[64] Cohen E (2010) Chitin Biochemistry: Synthesis, Hydrolysis and Inhibition. *Adv Insect Physiol* 38:5-74.

[65] Peniston QP, Johnson E (1980) Process for the manufacture of chitosan. *US Patent No. 4, 195*, 175, 5 pp..

[66] Li K, Hwang Y, Tsai T, Chi S (1996) Chelation of Iron Ion and Antioxidative Effect on Cooked Salted Ground Pork by N-carboxymethylchitosan (NCMC). *Food Sci Taiwan* 23:608-616.

[67] Suntornsuk W, Pochanavanich P, Suntornsuk L (2002) Fungal chitosan production on food processing by-products. *Process Biochem* 37:727-729.

[68] Zhao JL, Wang HL (1999) The preparation of chitosan from waste mycelia of *Aspergillus niger*. *Ind Microbiol* (Chinese) 29:33-37.

[69] Synowiecki J, Al-Khateeb NA (1997) Mycelia of *Mucor rouxii* as a source of chitin and chitosan. *Food Chem* 60:605-610.

[70] Wu WT, Huang TK, Wang PM, Chen JW (2001) Cultivation of *Absidia coerulea* from chitosan production in a modified airlift reactor. *J Chin Ins Chem Eng* 32:235-240.

[71] Tan SC, Tan T K, Wang SM, Khor E (1996) The chitosan yield of *Zygomycetes* at their optimum harvesting time. *Carbohyd Poly* 30:239-242.

[72] Aruchami M, Gowri N, Sundara-Rajulu G (1986) Chitin deacetylases in Invertebrates. *Chitin in Nature and Technology, plenum:* 263-265.

[73] Pochanavanich P, Suntornsuk W (2002) Fungal chitosan production and its characterization. *Lett Appl Microbiol* 35:17-21.

[74] Singla AK, Chawla M (2001) Chitosan: Some pharmaceutical and biological aspects- an update. *J Pharm Pharmacol* 53:1047-1067.

[75] Suresh PV, Sachindra NM, Bhaskar N (2011) Solid state fermentation production of chitin deacetylase by *Colletotrichum lindemuthianum* ATCC 56676 using different substrates. *J Food Sci Technol* 48(3):349-356.

[76] Mao JS, Cui YL, Wang XH, Sun Y, Yin YJ, Zhao HM, Yao KD (2004) A preliminary study on chitosan and gelatin polyelectrolyte complex cytocompatibility by cell cycle and apoptosis analysis. *Biomat* 25:3973-3981.

[77] Johnson EL, Peniston QP (1982) Utilization of shellfish waste for chitin and chitosan production. Chemistry and Biochemistry of marine food products; Martin, R.E.,Flick, G.J., Hebard, C.E., Ward, D.R., Eds.; *AVI Publishing: Westport,* CT Chapter 19:415-422.

[78] Aranaz I, Mengibar M, Harris R, Panos I, Miralles B, Acosta N, Galed G, Heras, A (2009) Functional characterization of Chitin and Chitosan. *Curr Chem Biol* 3:203-230.

[79] Sannan T, Kurita K, Iwakura Y (1975) Studies on chitin. 1. Solubility change by alkaline treatment and film casting. *Macromol Chem* 176:1191-1195.

[80] Lamarque G, Chaussard G, Domard A (2007) Thermodynamic Aspects of the Heterogeneous Deacetylation of α-chitin: Reaction Mechanisms. *Biomacromol* 8:1943-1950.

[81] Ashford NA, Hattis D, Murray A (1977) Industrial prospects for chitin and protein from shellfish wastes. Cambridge Mass MIT Sea Grant Program. *Report no. MITSG*: 77-73.

[82] Allan GG, Fox JR, Kong N (1978) A critical evaluation of the potential sources of chitin and chitosan. *Proceedings of the First International Conference on Chitin and Chitosan. MIT Sea Grant Program, Massachussets:* 68–78.

[83] Arcidiacono S, Kaplan DL (1991) Molecular weight distribution of chitosan isolated from *Mucor rouxii* under different culture and processing conditions. *Biotechnol Bioeng* 39:281-286.

[84] Davoust N, Hansson G (1992) Identifying the conditions for development of beneficial mycelium morphology for chitosan- producing *Absidia* spp. *Appl Microbiol Biotechnol* 36:618-620.

[85] Rane KD, Hoover DG (1993) An evaluation of alkali and acid treatments for chitosan extraction from fungi. *Process Biochem* 28:115-118.

[86] Muzzarelli RAA, Ilari P, Petrarulo M (1994) Solubility and structure of N-carboxymethyl chitosan. *Int J Biol Macromol* 16:177-180.

[87] Niederhofer AM, Uller B W (2004) A Method for Direct Preparation of Chitosan with Low Molecular Weight from Fungi. *Eur J Pharmaceut Biopharmaceut* 57: 101-105.

[88] Knorr O, Klein J (1986) Production and conversion of chitosan with cultures of *Mucor rouxii* or Phycomyces blakesleeanus. *Biotechnol Letts* 8:691-694.

[89] Chatterjee S, Adhya M, Guha AK, Chatterjee BP (2005) Chitosan from *Mucor rouxii*: production and physico-chemical characterization. *Process Biochem* 40:395-400.

[90] Nwe N, Stevens WF (2002) Production of Fungal chitosan by solid substrate fermentation followed by enzymatic extraction. *Bioprocess Technol Prog* 24:131-134.

[91] Latgé JP, Calderone R (2006) The Fungal Cell Wall, A Comprehensive Treatise on Fungi as Experimental Systems for Basic and Applied Research, Growth, Differentiation and Sexuality, 2nd Edition, pp 75-85.

[92] Zhao Y, Park RD, Muzzarelli RAA (2010) Chitin deacetylase: Properties and applications. *Mar Drugs* 8:24-46.

[93] Maghsoodi V, Razavi J, Yaghmaei S (2009) Production of chitosan by submerged fermentation from *Aspergillus niger*. *Chem Chem Eng* 16:145-148.
[94] Logesh AR, Thillaimaharani KA, Sharmila K (2012) Production of chitosan from endolichenic fungi isolated from mangrove environment and its antagonistic activity. *Asian Pac J Trop Biomed* 2(2):140-143.
[95] Crestini C, Giovannozzi-Sermanni G (1996) Solid state fermentation of *Lentinus edodes*: A new and efficient approach to chitosan production. In: Muzzarelli RAA (Ed.), Chitin Enzymology, *Atec Edizioni,* pp:595–600.
[96] Maghsoodi V (2008) Influence of Different Nitrogen sources on amount of chitosan production by *Aspergillus niger* in Solid State Fermentation. *Iran J Chem Eng* 27(1):47-52.
[97] Lu Shaojie XS, Deyong C, Yiping C, Kangde Y (2004) Preparation of water soluble chitosan. *J Appl Polym Sci* 91(6):3497-3503.
[98] Zamani A, Edebo L, Sjöström B, Taherzadeh MJ (2007) Extraction and precipitation of chitosan from cell wall of zygomycetes fungi by dilute sulfuric acid. *Biomacromol* 8(12):3786-3790.
[99] Bough WA, Salter WL, Wu ACM, Perkins BE (1978) Influence of manufacturing variables on the characteristic and effectiveness of chitosan products. 1. Chemical compositions, viscosity, and molecular-weight distribution of chitosan products. *Biotechnol Bioeng* 20:1931–1943.
[100] Maghami GG, Roberts GAF (1988) Studies on the adsorption of anionic dyes on Chitosan. *Macromol Chem* 189:2239-2243.
[101] Rout SK (2001) Physicochemical, Functional, and Spectroscopic analysis of crawfish chitin and chitosan as affected by process modification. *Dissertation.*
[102] Fernandes JC, Tiera MJ, Winnik FM (2006) Chitosan nanoparticles for non-viral gene therapy. *ACS Symposium Series* 934:177-200.
[103] Kurita K (1998) Chemistry and application of chitin and chitosan. *Poly Degr Stab* 59:117-120.
[104] Peter MG, Eberlin MN (2010) Applications of mass spectrometry to analysis of structure and bioactivity of chitooligosaccharides. In: Kim SK (Ed.), Chitin, Chitosan, oligosaccharides and their derivatives. *CRC Press, Boca Raton,* USA pp:127-148.
[105] Hudson SM, Smith C (1998) Polysaccharide: Chitin and chitosan: chemistry and technology of their use as structural materials. *Biopol Ren Res.* In: Macromolecular Systems -Materials Approach, Kaplan DL (Ed), *Springer Berlin Heidelberg,* pp:96-118.
[106] Dodane V, Vilivalam V (1998) Pharmaceutical application of chitosan. *Pharm Sci Technol* 1:246-253.
[107] Park H, Park K (1996) Biocompatibility issues of implantable drug delivery systems. *Harm Res* 13:1770-1776.
[108] Pangburn SH, Trescony PV, Heller J (1982) Lysozyme degradation of partially deacetylated chitin, its films and hydrogels. *Biomat* 3(2):105-108.
[109] Rageeb M, Usman M, Jain BV, Pawar SR, Shaikh TY, Potda MB, Patil GA (2013) Many potential applications of chitosan as a pharmaceutical excipient. *Int JA PS.BMS* 2(1):016-020.
[110] Skaugrud O, Sargent G (1990) Chitin and Chitosan: Crustacean Biopolymers with Potential. *International By-products Conference* pp:61-72.

[111] Dutta J, Tripathi S, Dutta PK (2011) Progress in antimicrobial activities of chitin, chitosan and its oligosaccharides: a systematic study needs for food applications. *Food Sci Tech* Int 18:3-34.
[112] Kaur K, Dattajirao V, Shrivastava V, Bhardwaj U (2012) Isolation and characterization of chitosan-producing bacteria from beaches of Chennai, India. In: Palomo JM, editor, Hindawi Publication Corporation, *Enz Res*(vol 2012 Article ID 421683, 6 pages) doi:10.1155/2012/421683.
[113] Merzendorfer H, Zimoch L (2003) Chitin metabolism in insects: structure, function and regulation of chitin synthases and chitinases. *J Exp Biol* 206:4393-4412.

In: Microbes in Process
Editors: Neelam Garg and Abhinav Aeron
ISBN: 978-1-63117-127-7

Chapter 17

Development of Nanocompounds as Antimicrobial Agents

Nayera A. Abdel-Wahed *
Department of Chemistry of Natural and Microbial Products,
Division of Pharmaceutical Industries, National Research Center, Cairo, Egypt

Abstract

Nanomaterials are at the leading edge of the rapidly developing field of nanotechnology. Over the past decade, several delivery vehicles have been designed based on different nanomaterials, such as polymers, dendrimers, liposomes, nanotubes, and nanorods. These nanoparticles have recently emerged as an attractive candidate for delivery as well as for diagnosis. High percentages of bacterial inhibition by nanoparticles were demonstrated against microorganisms, and in some cases were found to completely inhibit growth of a variety of gram-negative and gram-positive bacteria species. The unique size-dependent properties of nanoparticles make these materials superior and indispensable in many areas of human activity. This review tries to summarize the most recent developments in the field of applied nanomaterials, in particular their application in biology and medicine as antimicrobial agent. It also looks at a range of new potential antibacterial therapeutics implements nanotechnology, most of which are in early stages of research but has already shown positive signs of application into nanomedical therapeutics and diagnostics. Nanocarriers have provided a novel platform for target-specific delivery of therapeutic agents.

1. Introduction

Nanotechnology was first introduced by Eric Drexler which is often described as "the founding father of nanotechnology. A basic definition for nanotechnology is the engineering of functional systems at the molecular scale [1]. Nanotechnology is enabling technology that

* Email: niarawahed@yahoo.com.

deals with nano-meter sized objects. It is expected that nanotechnology will be developed at several levels: materials, devices and systems. The nanomaterials level is the most advanced at present, both in scientific knowledge and in commercial applications. A decade ago, nanoparticles were studied because of their size-dependent physical and chemical properties [2]. Now they have entered a commercial exploration period [3]. Living organisms are built of cells that are typically 10 μm across. However, the cell parts are much smaller and are in the sub-micron size domain. Even smaller are the proteins with a typical size of just 5 nm, which is comparable with the dimensions of smallest manmade nanoparticles. This simple size comparison gives an idea of using nanoparticles as very small probes that would allow to spy at the cellular machinery without introducing too much interference. Understanding of biological processes on the nanoscale level is a strong driving force behind development of nanotechnology [4]. Out of plethora of size-dependant physical properties available to someone who is interested in the practical side of nanomaterials, optical [5] and magnetic [6] effects are the most used for biological applications. As mentioned above, the fact that nanoparticles exist in the same size domain as proteins makes nanomaterials suitable for bio tagging or labelling. However, size is just one of many characteristics of nanoparticles that itself is rarely sufficient if one is to use nanoparticles as biological tags. In order to interact with biological target, a biological or molecular coating or layer acting as a bioinorganic interface should be attached to the nanoparticle. Examples of biological coatings may include antibodies, biopolymers like collagen [7], or monolayers of small molecules that make the nanoparticles biocompatible [8]. In addition, as optical detection techniques are wide spread in biological research, nanoparticles should either fluoresce or change their optical properties. The approaches used in constructing nano-biomaterials are schematically presented below (see Figure 1).

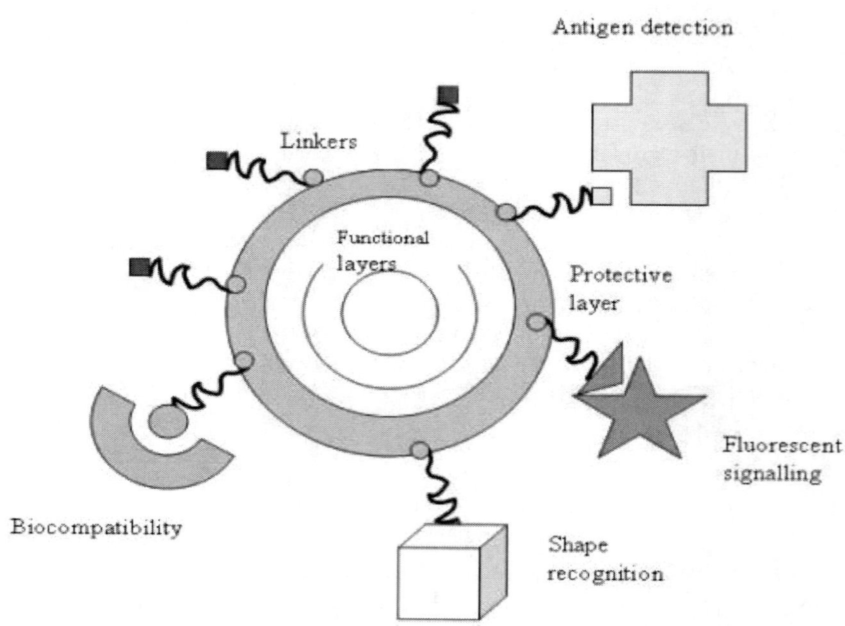

Figure 1. Typical Configurations Utilized in Nano-Bio Materials applied to Medical or Biological Problems.

Nano-particle usually forms the core of nano-biomaterial. It can be used as a convenient surface for molecular assembly, and may be composed of inorganic or polymeric materials. It can also be in the form of nano-vesicle surrounded by a membrane or a layer. The shape is more often spherical but cylindrical, plate-like and other shapes are possible. The size and size distribution might be important in some cases, for example if penetration through a pore structure of a cellular membrane is required. A tight control of the average particle size and a narrow distribution of sizes allow creating very efficient fluorescent probes that emit narrow light in a very wide range of wavelengths. This helps with creating biomarkers with many and well distinguished colors. The core itself might have several layers and be multifunctional. For example, combining magnetic and luminescent layers one can both detect and manipulate the particles. The core particle is often protected by several monolayers of inert material, for example silica. Organic molecules that are adsorbed or chemisorbed on the surface of the particle are also used for this purpose. The same layer might act as a biocompatible material. However, more often an additional layer of linker molecules is required to proceed with further functionalization. This linear linker molecule has reactive groups at both ends. One group is aimed at attaching the linker to the nanoparticle surface and the other is used to bind various moieties like biocompatibles (dextran), antibodies, fluorophores etc., depending on the function required by the application. Pathogenic bacteria remain a major health concern, which are responsible for causing a large number of deaths and hospitalizations each year. Although we have current treatments such as antibiotics, bacteria are gaining resistance to these therapeutics at an alarming rate. That is why new therapeutic and diagnostic treatments are necessary if we want to be prepared against known and unknown pathogenic bacterial infections. The aim of this review is firstly to give reader a historic prospective of nanomaterial application to biology and medicine. Secondly is to overview the most recent developments in this field. Finally is to discuss the antimicrobial activity of these compounds. The review will also give new growing prospective for improving disease management and control in the future.

2. Carbon Nanotubes and Fullerenes

Since the discovery of carbon nanotubes they have flourished and have been applied in a broad range of areas in science and in a way represent the momentum of nanotechnology itself. They have been sourced due to their unique optical, thermal, magnetic, high strength properties as well as the ability to be conductive, almost all of which can be tuned depending on the dimensions of the nanotube (Figure 2). As of late they have been used in medical research in applications such as drug delivery and as components in medical nanodevices. A recent study was performed that looked into the antimicrobial properties of carbon nanotubes and showed that they are strong antibacterial agents [9]. The study conducted by Kang used low metal content, well characterized, narrowly distributed, pristine Single walled nanotubes (SWNTs), which are to observe bactericidal activity on the bacterium *Escherichia coli* K12. The bacteria were incubated with the SWNTs for 60-mins and microbial viability was checked and showed that over 80% of the microbes were killed. The suggested mode of death was the direct interaction of the SWNTs with the bacterial cell membranes causing significant membrane damage [10]. Similar research using nanotubes of different materials such as peptides and lipids also have shown the same mode of toxicity on microbes, suggesting that the nanotubes are able to physically penetrate the cell membranes to their physical dimensions leading to cell death [11]. Fullerene C60 is another carbon structure that has been investigated as an antimicrobial agent, as a means of identifying environmental impacts of fullerene C60. Research was conducted to test the toxicity of the fullerene molecules on two species of bacteria, *E.coli* DH5 and *Bacillus subtilis*, the former being a gram negative and the latter being gram positive [12]. The study showed that the fullerenes did act as bactericidal agents on both bacteria types, but it was also identified that differences in physical dimension of fullerene C60 types have alternative effects therefore more specific testing would be required [12].

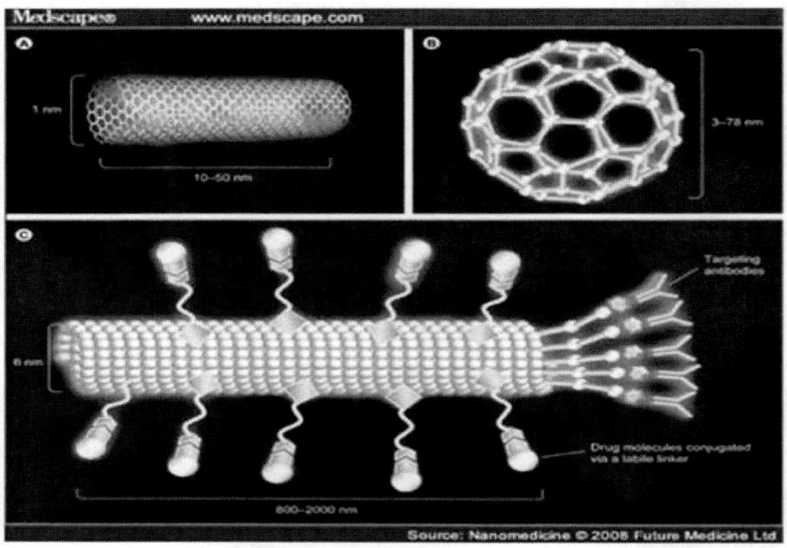

Figure 2. Several Potential Antibacterial Nanomedicines. (A) A Single-Walled Carbon Nanotube (B) A C60 Fullerene (C) A Targeted Drug-Carrying Phage Nanoparticle.

Carbon nanotubes have been applied as diagnostic tools in a range of areas of medicine such as cancer and other human pathogens *in vitro*, but not much has been done in the way of bacterial diagnostics *in vivo*. A study to test the effectiveness of photothermal antimicrobial nanotherapy and nanodiagnostics using self-assembling carbon nanotubes was performed by Kim [13], and has shown as an effective method of being able to achieve both bacterial diagnosis and destruction. The study involved the use of dispersed and shortened carbon nanotubes (dsCNTs) and minimally treated carbon nanotubes clusters that when in solution with bacteria are able to cluster at the bacterial surfaces or alternatively when using the CNT clusters which are able to form bacterial clusters. Once the acterial or CNT clusters have formed they can be used to detect a single bacterium and or can be used to destroy the pathogenic bacteria. The process works by taking advantage of the responsiveness of carbon nanotubes to inherent near infrared lasers, which when targeted and applied causes observed bubbling phenomena that is able to completely destroy bacterial cells. Another advantage of this technique is due to the self-binding affinity of CNTs with bacteria it is unnecessary for any bioconjugation of targeting molecules, although if required for species specific purposes it is possible to do so. The study highlighted the potential application of CNT with pulse lasers as effective photothermal contrast agents that are able to self-deliver to the bacterial infection site and could be applied in the detection of a single bacterial cell in real-time [13]. Another similar study has been completed which used the same techniques of photothermal carbon nanotubes but employs targeted gold nanoparticles to do the job, and was shown to be equally effective [14].

3. Silica Nanoparticles

Magnetic *N*- halamine nanocomposites were synthesized through the encapsulation of the magnetic silica nanoparticles with antibacterial *N*- halamine polymer. The fabricated magnetic *N*-halamine nanocomposites possessed enhanced antibacterial activity against both Gram-positive and Gram-negative bacteria compared with their bulk counterparts. The effect of chlorine content of the magnetic *N*-halamine nanocomposites on the antibacterial activity was investigated. Performances derived from the synergism between magnetic core and antibacterial shell suggest that the magnetic *N*-halamine nanocomposites are qualified for antibacterial applications and separable by the aid of the external magnetic field [15].

4. Silver Nanoparticles

Silver nanoparticles have applications as antimicrobial agents and virucidal agents although not used commercially but have shown great promise among other such agents.

4.1. Silver Nanoparticles As Antimicrobial Agent

Silver nanoparticles show great potential as antimicrobial agents in applications, as they are of low cost and easily synthesized, and could be used to treat material surfaces to provide highly effective antibacterial materials, medical equipment such as in and devices and paints. Silver has long been used in medicine, even in ancient times it was used as a preservative and also to reduce inflammation and prevent against infection of wounds [16]. Various silver-based dressings have been introduced in the past few years and have become the latest and greatest ''innovation'' in wound care products. The ''innovation'' involved in these new wound care products is the simple fact that silver itself is incorporated within the dressing rather than being applied as a separate salt, compound, or solution. The basic issues in choosing a silver-containing dressing can be broadly conceptualized in terms of: (1) the characteristics of the ''carrier'' dressing and (2) the delivery of silver by the dressing to the wound. Keeping these basic issues in mind can help make sense of some of the media marketing blitz accompanying these products [17].

A study was performed to test the effectiveness of silver nanoparticle antimicrobial activity using *E.coli* was undertaken by Sonda [18]. In the study, 10^5 CFU of Gram negative *E. coli* colonies were grown on LB agar plates with different concentrations of Ag-nanoparticles. Ag-NPs at concentrations of 10 $\mu g\ cm^{-3}$ showed inhibition of bacterial growth by up to 70%, and at concentrations of 50-60 μg cm^{-3} there was 100% inhibition of bacterial growth [19]. A SEM and TEM analysis showed the accumulation of the nanoparticles within membranes and some were able to penetrate into the cell, the SEM also highlights the damage caused by the Ag-NPs where visible "pits" can be observed (Figure 3, 4). Silver ions antimicrobial activity is believed to work by impairing DNAs ability to replicate and through the inactivation of key proteins by denaturation when they are bound together [19]. Another study showed that with the use of silver nanoparticles in conjunction with antibiotics (Figure 5) such as penicillin G, amoxicillin, erythromycin, clindamycin and vancomycin, provided an increase in the effectiveness of the antibiotics [20].

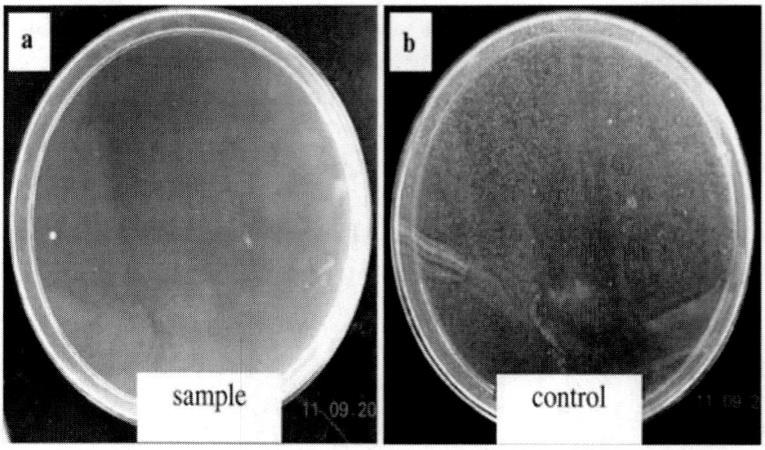

Figure 3. Photographs of Colonies of *E. coli* incubated on Agar Plates obtained from Cultivated Suspensions with (a, sample) and without (b, control) Ag Nanoparticles.

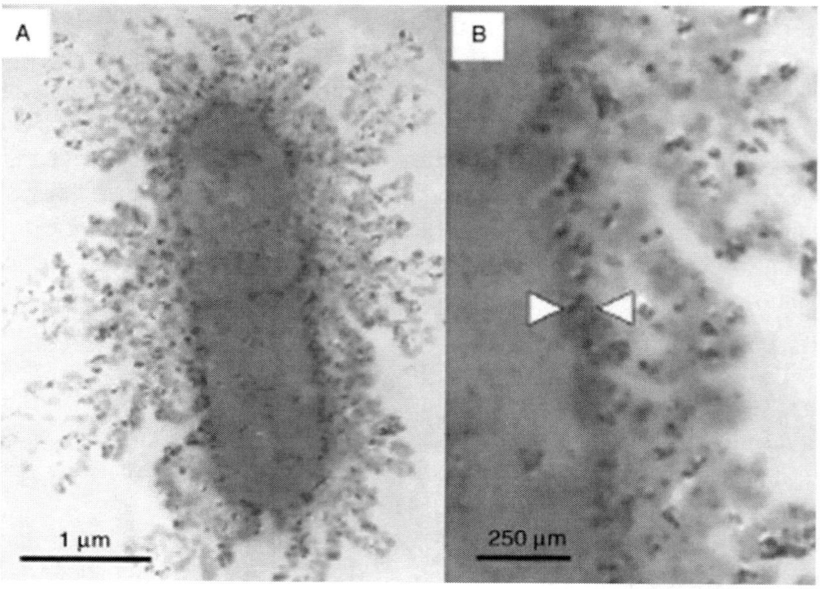

Figure 4. TEM Images of *E. coli* Treated with Silver Nanoparticles (A) and Enlarged View of the Incorporation of Silver Nanoparticle into the Cell Membrane Structure.

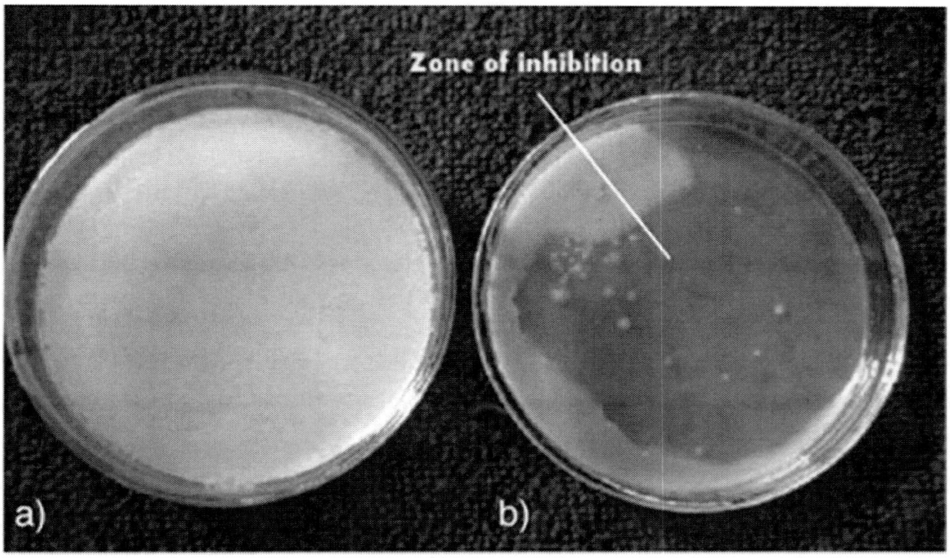

Figure 5. (a) Growth of Bacterial Colonies around Grafted Filter Paper (b) Zone of Inhibition around the AgNPs Loaded Grafted Filter Paper.

Calcium phosphate composite quaternary ammonium dimethacrylate (QADM) and nano-silver (NAg) were incorporated into calcium phosphate composite for the first time. NACP+QADM+NAg were strongly antibacterial and greatly reduced metabolic activity, and acid production of *S. mutans* biofilms, while possessing mechanical properties similar to commercial composites. These nanocomposites are promising to have the double benefits of remineralization and antibacterial capabilities to inhibit dental caries. Tetracalcium phosphate composite contains quaternary ammonium dimethacrylate with antibacterial properties [21].

4.2. Silver Nanoparticles As Virucidal Agents

Virucidal agents differ from virustatic drugs in that they act directly and rapidly by lysing viral membranes on contact or by binding to virus coat proteins. Nevertheless, the interaction of AgNPs with viruses is still an unexplored field. However, the mechanism of action of AgNPs as an antiviral and virucidal has been studied against several enveloped viruses. Recently, it has been suggested that nanoparticles bind with a viral envelope glycoprotein and inhibit the virus by binding to the disulfide bond regions of the CD4 binding domain within the HIV-1 viral envelope glycoprotein gp120, as suggested by Elechiguerra and colleagues [22]. This fusion inhibition was later elegantly demonstrated by Lara and colleagues (Lara in their latest report). The antiviral effects of AgNPs on the hepatitis B virus (HBV) have been reported using a HepAD38 human hepatoma cell line. There has been evidence of high binding affinity of nanoparticles for HBV DNA and extracellular virions with different sizes (10 and 50 nm). Moreover, it has been demonstrated that AgNPs could also inhibit the production of HBV RNA and extracellular virions in vitro, which was determined using a UV-vis absorption titration assay. Further investigation will be needed to determine whether this binding activity prevents HBV virions from entering into host cells or not [23]. In an another report Sun and colleagues showed that AgNPs were superior to gold nanoparticles for cytoprotective activities toward HIV-1-infected Hut/CCR5 cells [24]. It is generally understood that Ag, in various forms, inactivates viruses by denaturing enzymes via reactions with sulfhydra, amino, carboxyl, phosphate, and imidazole groups [25]. However, it is necessary to design studies in vivo to increase therapeutic benefit and minimize adverse effects. Among antiviral activities, the capacity of AgNPs to inhibit an influenza virus was determined in a MDCK cell culture and was demonstrated that with AgNPs at 0.5 μg/ml concentration viral infectivity was reduced. Nanosilver may interfere with the fusion of the viral membrane, inhibiting viral penetration into the host cell [26]. Lara and colleagues further demonstrated that AgNPs inhibited a variety of HIV-1 strains regardless of their tropism, clade and resistance to antiretrovirals [27]. The fact that AgNPs inhibited number of HIV-1 isolates suggests that their mode of action does not depend on cell tropism and that AgNPs are broad spectrum anti-HIV-1 agents.

5. Nanoencapsulation

In recent years, encapsulation of antimicrobial drugs in nanoparticle systems has emerged as an innovative and promising alternative that enhances therapeutic effectiveness and minimizes undesirable side effects of the drugs. Here the current progress and challenges in synthesizing nanoparticle platforms for delivering various antimicrobial drugs are reviewed. Nanoencapsulation is defined as a technology to pack substances in miniature making use of techniques such as nanocomposite, nanoemulsification, and nanoestructuration and provides final product functionality that includes controlled release of the core. Donsi et al. [28] describe the nanoencapsulation of essential oils.

A terpene mixture extracted from *Melaleuca alternifolia* and d-limonene were encapsulated into nanoemulsions. These antimicrobial agents in the nanocapsules were tested by the authors in pear and orange juices against *Lactobacillus delbrueckii, Saccharomyces cerevisiae, Escherichia coli* and complete inactivation of microorganisms with 5.0 g/l terpenes was attained. Nanoencapsulation is, therefore, indicated to include antimicrobial agents in foods, and sensitive ingredients from interaction with the foodmatrix. Silver compounds may be used to coat implantable materials in order to prevent biofilm formation and other complications due to the presence of bacteria.

However, these compounds may be too soluble and even toxic for the host. Encapsulation might be very advantageous in order to increase the stability and biocompatibility of silver drugs. Liposome-encapsulated antibiotics may be very useful against infections by intracytoplasmic pathogens. Targeting of other cells and organs is studied by using monoclonal antibodies (immune-liposomes) or polysaccharide as a specific ligand. Liposome encapsulation decreases the toxicity of some drugs and improves their therapeutic efficacy. Various formulations for amphotericin B have been used clinically for fungal infections in Europe and the United States [29]. According to the structural organization biodegradable nanoparticles are classified as nanocapsule, and nanosphere. The drug molecules are either entrapped inside or adsorbed on the surface [30].

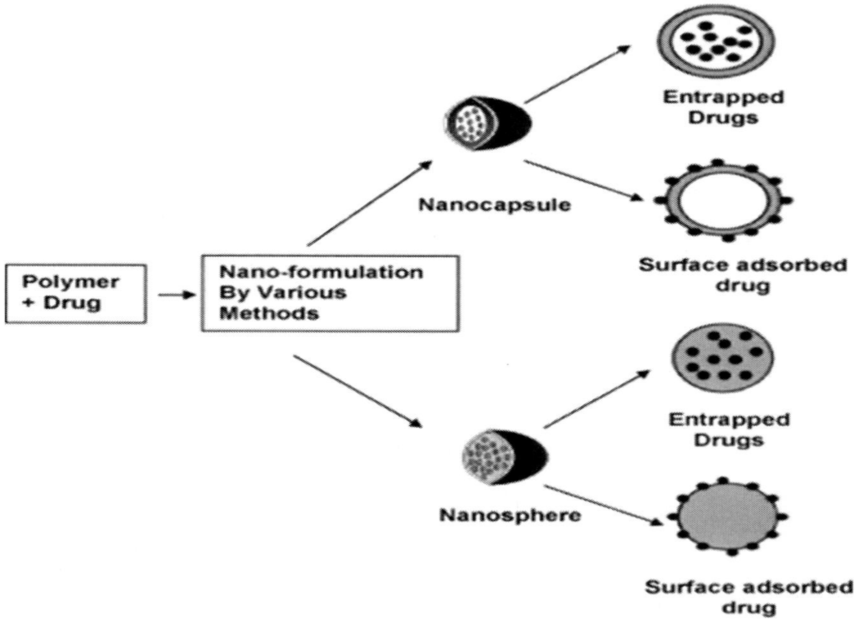

Figure 6. Type of Biodegradable Nanoparticles.

6. Iron Nanoparticles

Zero-valent iron nanoparticles (nano-Fe^0) recently have been shown as an effective alternative to granular Fe^0. The advantages of nano- Fe^0 over microscale granular Fe^0 include its improved mobility and increased reactivity due to its higher surface area [31, 32]. Despite increasing interest in the antimicrobial activity of metal nanoparticles and the widespread use of nano-Fe0 for environmental remediation, little is known about the antimicrobial activity of nano- Fe^0. Previous investigators have reported on inactivation of bacteriophages by other types of iron-based compounds such as iron oxide-coated sands [33] or microscale iron powder [34]. However, the biocidal activity required relatively high doses (several g/L) or long treatment times (for days or weeks), indicating that their biocidal activities are negligible compared to that of nano- Fe^0. A strong bactericidal effect of nano-Fe^0 on *Escherichia coli* (*E. coli*) was reported. This bactericidal effect was found to be a size-related and element-specific property of nano- Fe^0.

Control of pathogenic viruses for water supplies includes not only disinfection for drinking water treatment but also prevention of viral contamination due to untreated wastewater. A comprehensive study on the role of virus interfacial behavior in aquatic environments, to gain fundamental understanding of factors, has influencing the fate and transport of virus in subsurface environment. It was found that hematite nanoparticles with diameters from 5 to 20 nm provide an attractive substrate for the viruses, fouling by natural organic matter substantially reduces virus removal capacity. In addition, the presence of water hardness (i.e., Ca^{2+} and Mg^{2+}) significantly increases virus adsorption because the divalent cations allow virus to adsorb on the tural organic matter (NOM) which adsorbed on the surface of hematite nanoparticles. They also obtained microscopic evidence that adsorption of bacteriophage MS2 to hematite nanoparticles does not destroy the viral capsid. With appropriate solution conditions, they were able to desorb 40% of adsorbed viruses.

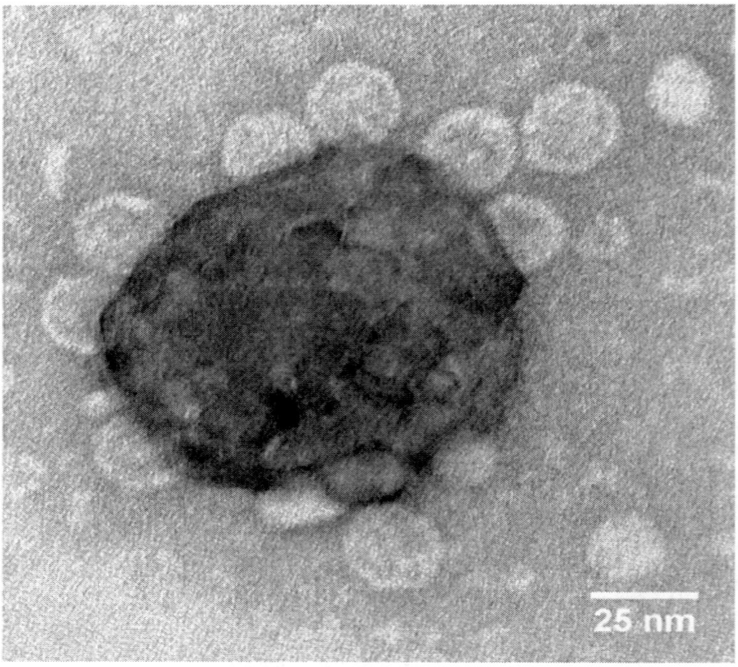

Figure 7. Bacteriophage MS2 adsorb to Hematite Nanoparticles.

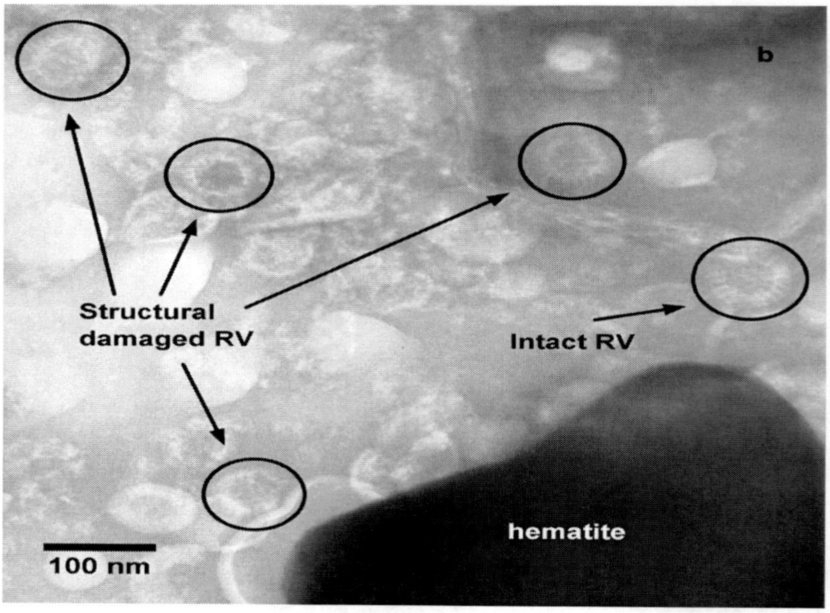

Figure 8. Rotavirus Adsorb to Hematite Nanoparticles.

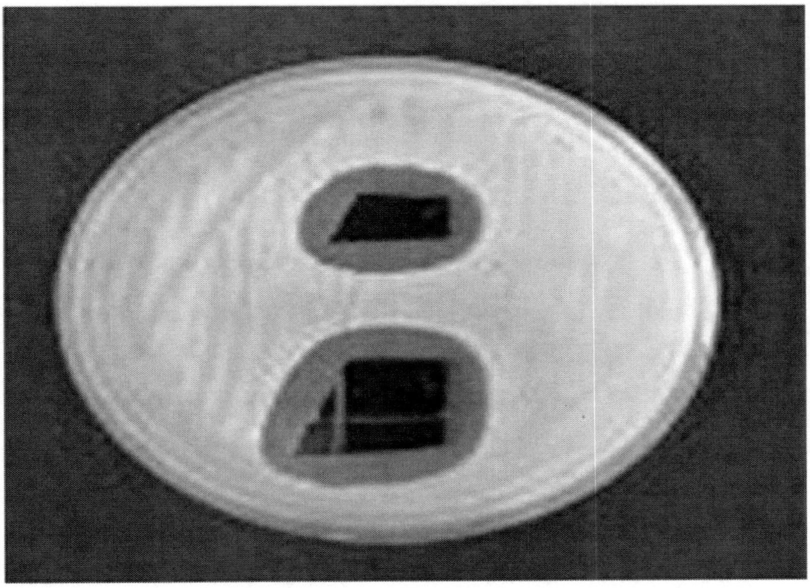

Figure 9. Inhibition Halo of as Deposited Silver Nanocluster-Silica Composite Coatings against *Staphylococcus aureus* strain.

7. Fluoride Nanoparticles

It was reported on the antimicrobial and antibiofilm activities of nanosized magnesium fluoride (MgF_2 nanoparticles (NPs) synthesized in ionic liquid using microwave chemistry a novel water-based synthesis of MgF_2 NPs using sonochemistry. The sonochemical irradiation of an aqueous solution of $[Mg(OAc)_2{\cdot}(H_2O)_4]$ containing acidic HF as the source afforded crystalline well-shaped spherical MgF_2 NPs that showed much improved antibacterial properties against two common bacterial pathogens (*Escherichia coli* and *Staphylococcus aureus*).

It is also able to demonstrate that the antimicrobial activity was dependent on the size of the NPs. Finally, the antimicrobial activity of MgF_2 NPs against established biofilms was also examined. Results highlight the potential to further develop the concept of utilizing these metal fluoride NPs as novel antimicrobial and antibiofilm agents The antibiofilm activity of crystalline MgF_2 nanoparticles against E. coli C600 and S. aureus FRF1169 bacterial pathogens revealed that the $MgF_2{\cdot}$Nps attach and penetrate into the cells causing a disruption in membrane potential, enhanced lipid peroxidation and DNA binding. These processes disrupt the ability of the bacteria to form biofilms.

Further utilization of the microwave synthesis procedure to effectively coat glass surfaces with $MgF_2{\cdot}$Nps showed that these surfaces can restrict biofilm formation of the studied bacteria. Taken together the results of this study emphasize the potential of using metal fluoride nanoparticles for the design of sterile surface coatings that may be useful for various medical applications [35].

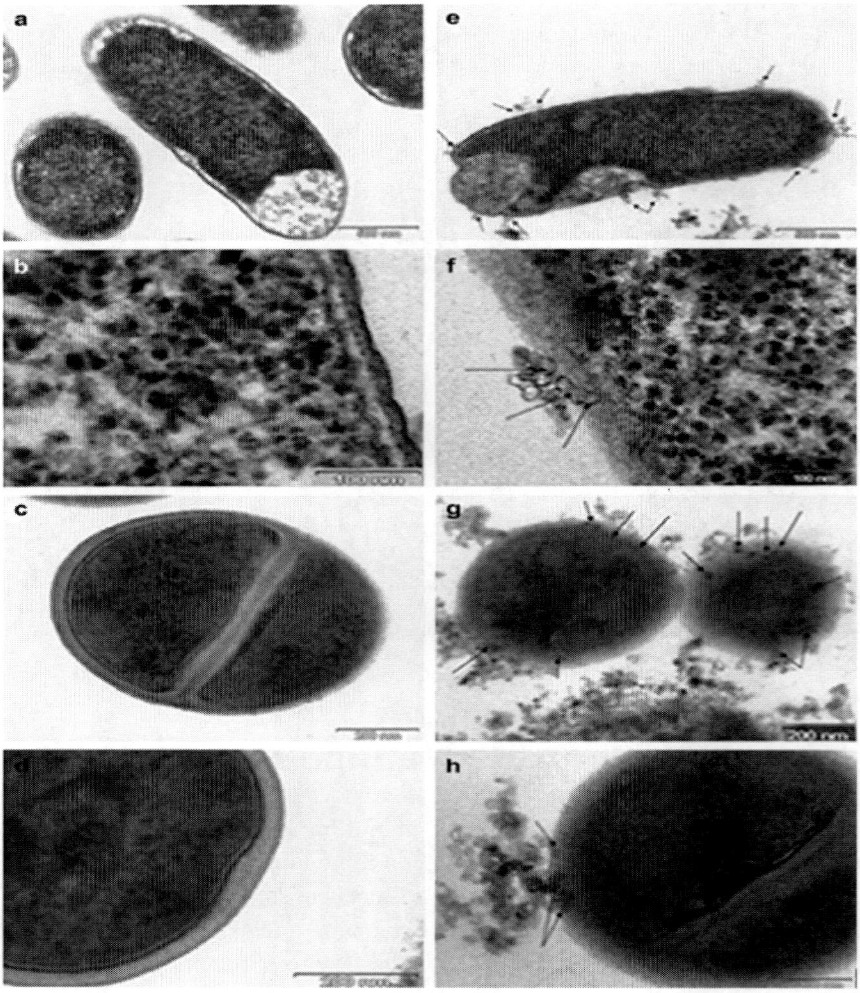

Figure 10. Transmitted Electron Microscopy of MgF_2·Nps Treated and Untreated Cells. TEM Micrographs of *E. coli* and *S. aureus* Thin Sections: Untreated *E. coli* (a, b) and *S. aureus* (c,d); MgF_2·Nps (1 mg/ml) Treated *E. coli* (e,f) and *S. aureus* (g,h). Arrows Indicate MgF_2 Nanoparticles.

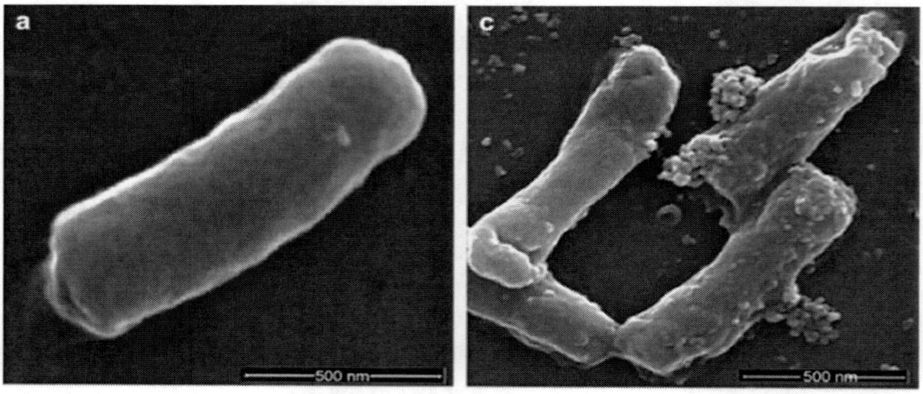

Figure 11. (Continued).

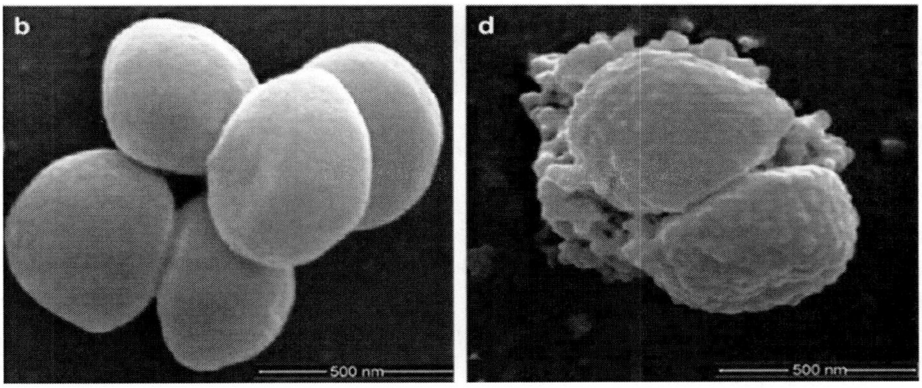

Figure 11. MgF_2·Nps Characterization. (a) Powder X-ray Diffraction (XRD) Patterns of MgF_2·Nps; (b–c) HR-TEM Micrographs and (d) Nanoparticles Size Distribution. The MgF_2·Nps Examined were a Product of a 2 min Microwave Reaction.

8. Magnesium Oxide Nanoparticles

Nano-MgO has been involved in many different areas of research due to its novel properties, particularly in nanocomposite materials but now has been approached as antibacterial agents. MgO was originally known to be an antibacterial agent and it was thought that the damage caused to bacteria was due to the formation of superoxide anions on its surface; this inspired further research into understanding the mode of action that allows nano-MgO to exhibit antibacterial activity [36]. It was found that nano-MgO particles are able to take up halogen gases due to its defective natured surface and its positive charge, which results in a strong interaction with bacteria and spores as they are negatively charged. In a study the nano-MgO particles were produced and bromide and chloride were absorbed into the particles and then tested against *E. coli, B. cereus* and *B. globigii* as well as the spores from the *Bacillus* species. Experimental results showed that nano-MgO in conjunction with the uptake of halogens, in particular chloride, that effective destruction of the bacteria and bacterial spores was observed [36].

Another study was conducted which also covered the antimicrobial properties of nano-MgO and showed highly controlled synthesis of the particles with the ability to control particle size [37].Through this process, using a range of Nano-Mg particle sizes ranging from 70 nm down to around 8nm, it was a observed that there was significant antimicrobial activity that actually increased the with a decrease in particle size, in particular with *B. subtilis* [37]. Nano-MgO has a great potential to be used as an antimicrobial agent, and probably more so than other metal-oxide nanoparticles showing antimicrobial activity due to the lack of required light that is necessary for other metal oxides to function, as well as the benefit of being non-toxic. It has been tested already in Phenylpropyl type interior paints and shown significant antimicrobial activity, this type of application could be applied in many medical facilities as well as households as a means to reduce contamination and pathogenesis to humans [37].

9. Zinc Oxide Nanoparticles

It is known that zinc oxide exhibits antibacterial activity like many other metal oxide groups and like the others only few have been scaled down to the nano size and researched further, such as ZnO nanoparticles. It has been shown that ZnO nanoparticles show significant antibacterial activity over a broad range of bacterial species and in particular against *Staphylococcus aureus* where it out competes five other commonly used metal oxide nanoparticles [38]. Research was carried out by Jones [38] to test the antibacterial activity of nanoparticles as well as the impact of particle size on bactericidal efficiency. In the experiment they used four different bacteria species, which included *Staphylococcus aureus* N315, *Staphylococcus epidermidis* 1487, *Streptococcus pyogenes* NZ131 and *Bacillus subtilis* as well as testing four different ZnO particle concentrations of 0mM, 2mM, 5mM and 10mM. The results showed a significant loss in viable cell numbers of all four bacteria species, with 5mM and 10mM being the most effective and 2mM being the worst, where it provided only a slight improvement from the control [18]. The effect of particle size was also tested using three different sized ZnO particles. Results showed that the smaller particle sizes were able to reduce bacterial growth in some cases up to 99% while the larger particles for the same conditions only managed 40 – 50% inhibition of bacterial growth [38]. There is range of potential antibacterial applications for ZnO nanoparticles to be implemented into such as ointments, surface coatings, disinfectants for dentistry and in medical facilities. In further study, to protect children's feet from being infected by the fungi, a novel nano antimicrobial composite containing nano ZnO and isothiazolinone was prepared and was used to treat the children's insole by spraying. The inhibitory effects of the nano antimicrobial composite with 2% could inhibit the growth more than 97% of the fungi [39].

10. Silicon Dioxide Nanoparticles

SiO2 is used very extensively in material sciences and has also been more recently used in a range of medical research such as drug delivery [40]. Nano-SiO2 is not used directly as an antimicrobial agent itself but acts as a carrier for antimicrobial and bactericidal agents due to its exceptionally porous structure. Due to this porous structure SiO2 is able to absorb ions and organic molecules quite easily, making it highly efficient and a promising carrier for antimicrobial applications. As observed in previous examples in this review, metal-containing inorganic materials has shown to be effective as antimicrobial agents; therefore a study was conducted and has implemented zinc and silver into SiO_2 nanoparticles as a means of creating a high-performance bactericidal agent [41]. In the experiment conducted by Jia [41], they synthesized two different types of SiO_2 particles one that has been loaded with silver cations known as SLS and a co-absorption of zinc-silver loaded into the other, titled SLSZ. The SiO_2 particles were around 20nm ± 5nm in size and composed of 0.5 % Ag+ and 4 -9% of Zn^{2+} in the SLZS with the amount of silver ions evenly distributed in each of both SLS and SLZS. The nanoparticles were tested on two bacteria species being *E. coli* and *Streptococcus faecalis* and were incubated with two singular concentrations of 2.5mg and 5mg for both SLS and SLZS. The results showed that both SLS and SLZS exhibit excellent antibacterial activity, for both *E. coli* and *S. faecalis*. SLS was able to kill all the viable *E. coli* cells after a contact time of 6hrs when 2.5mg was used and was brought down to 4hrs when 5.0mg of the loaded nanoparticles were used. Likewise, the SLZS was able to decrease the contact time needed even further reducing it from 4hrs to 2hrs. Results using *S. faecalis* produced were just as promising, with SLS able to kill all viable cells after an incubation period of 10hrs using 2.5 mg and down to 6hrs using 5.0mg. Respectively for the SLZS particles the incubation time needed to kill all the viable cells was brought down from 6hrs using the 2.5mg to 2hrs using the amount of 5.0mg. SLZS was noticeably the more effective of the two, which is believed to be due to an increase in antibacterial sites enhancing the contact of the particles with the bacteria so the process speed was increased.

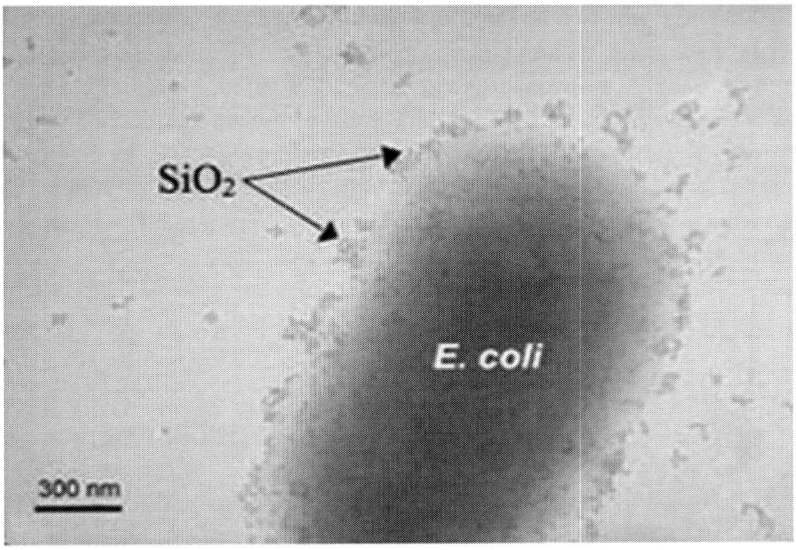

Figure 12. Adhesion of Silicon Dioxide Nanoparticles on the *E. coli* as Observed by TEM.

11. Copper Oxide Nanoparticles

CuO nanoparticles were effective in killing a range of bacterial pathogens involved in hospital-acquired infections. However, in comparison with nano Ag and nano Cu, higher concentrations of nano CuO were required to achieve a bactericidal effect. It has been suggested that the reduced amount (between 3- and 20-fold) of negatively charged peptidoglycans would make Gram-negative bacteria less susceptible to such positively charged antimicrobials [42]. This is supported by the MBC findings with nano CuO in this study for *P. aeruginosa* and *Proteus* spp. However, in the time–kill experiments the Gram-negative strains tested showed a greater susceptibility to nano CuO combined with nano Ag.

Preliminary research on the antiviral (versus influenza A and SARS viruses) capabilities of nanomaterials has shown that nanoparticles such as Ag and Cu release Ag^{+} and $Cu2^{+}$ ions that cause local pH and conductivity changes. This liberation of metal ions into solution will then have the capability to inactivate or kill viruses. Cu^{2+} and Ag^{+} ions are also small enough to disrupt bacterial cell membranes and gain entry in order to disrupt enzyme function. Indirect effects through changes in the surrounding charge environment may also impact on the effectiveness of nanoparticulate metals against microorganisms [18]. Studies to assess the potential of nano CuO embedded within a range of polymer materials have shown a lower contact-killing ability in comparison with release killing ability against MRSA strains which would suggest a release of ions into the local environment is required for optimal antimicrobial activity. Atomic force microscopy (AFM) and TEM are useful techniques to investigate the changes in bacteria upon exposure to antimicrobials. Studies using AFM and TEM with 'aerogel'- generated nano magnesium oxide and *Escherichia coli* have shown that the cell wall of this bacterium is extensively damaged, allowing the contents to leak out and nanoparticles to gain entry [36].

Likewise, Ag nanoparticles have been shown to attach to the microbial cell surface and penetrate inside, where intracellular targets, including respiratory enzymes, are disrupted [18]. The precise mechanism of action of nano CuO is the subject of ongoing investigations. Hospitals and transport are two particular areas that offer opportunities for the use of nanoparticulate metals and metal oxides to prevent the spread of infection. With the increase in air travel and greater mobility in general of people, airborne, vector-borne and zoonotic spread of infectious agents are important public health issues. Within the hospital environment, wall coverings, equipment, clothing and bedding are all potential risk areas for spread of infection [43]. Therefore, in terms of potential use, the incorporation of nanoparticulate metals and metal oxides, including nano CuO, into surfaces and other objects could be envisaged.

12. Iron Oxide Nanoparticles

Iron oxide (IO) has been widely used in biomedical research because of its biocompatibility and magnetic properties [44]. Nanoparticles of IO, with sizes less than 100 nm, have been developed as contrast agents for magnetic resonance imaging [(45] as hyperthermia agents [46] and as carriers for targeted drug delivery to treat several types of cancer [47]. It is further believed that through the use of magnetic nanoparticles, an optimal drug delivery system can be developed by using an external magnetic field to direct such nanoparticles to desirable sites (such as implant infection) for immediate treatment [48].

Along this line, some have hypothesized that reactive oxygen species (ROS) generated by Fe_3O_4 nanoparticles could kill bacteria without harming non-bacterial cells [49]. Specifically, Pareta et al. [48] cultured osteoblasts (bone-forming cells) with IO nanoparticles (at a concentration of 4.25 mg/mL) and found that cell density was greatly enhanced in the presence of IO nanoparticles compared with cells cultured without nanoparticles. In order to study the effects of iron oxide (IO) nanoparticles on *Staphylococcus aureus*, IO nanoparticles were synthesized via a novel matrix-mediated method using polyvinyl alcohol (PVA). The IO nanoparticles were characterized by transmission electron microscopy and dynamic light scattering. Further, *S. aureus* were grown in the presence of three different IO nanoparticle concentrations for four, 12, and 24 hours. Live/dead assays were performed and the results provide evidence that IO/PVA nanoparticles inhibited *S. aureus* growth at the highest concentration (3 mg/mL) at all time [50].

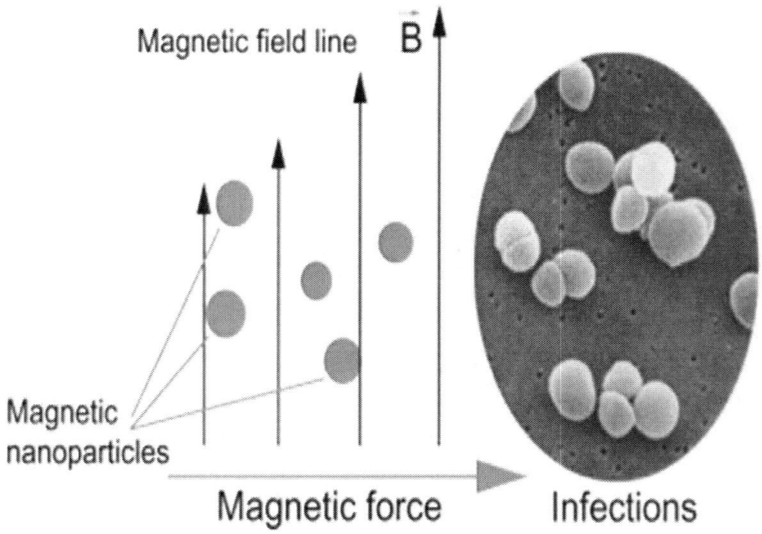

Figure 13. A Simplified Design for a Magnetic Drug Delivery System in which Magnetic Nanoparticles Coated with Drugs are Directed to Infection Sites by an External Magnetic Field.

13. Selenium Nanoparticles

Staphylococcus aureus is a key bacterium commonly found in numerous infections. *S. aureus* infections are difficult to treat due to their biofilm formation and documented antibiotic resistance.

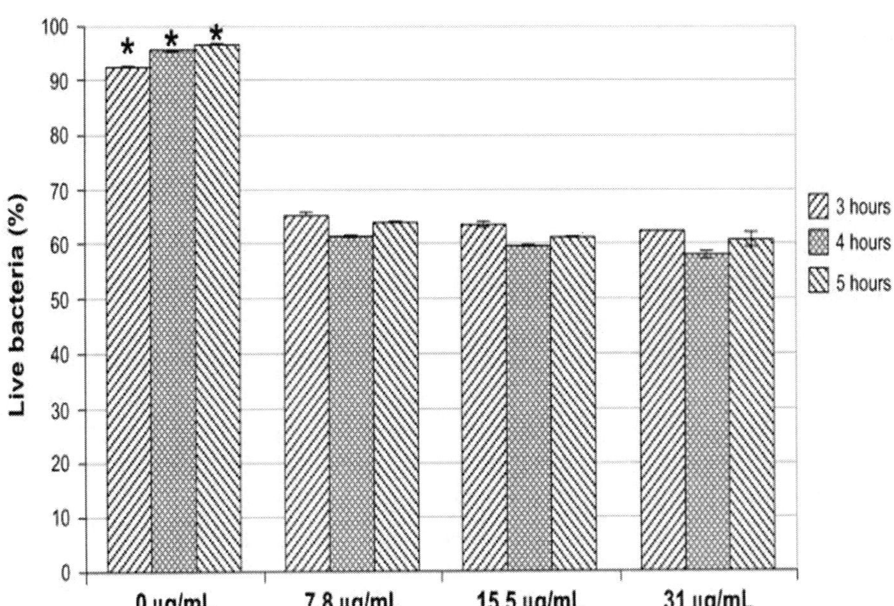

Figure 14. Decreased Percentages of Alive Bacteria in the Presence of Selenium Nanoparticles.

While selenium has been used for a wide range of applications including anticancer applications, the effects of selenium nanoparticles the growth of *S. aureus* provided the first evidence of strongly inhibited growth of *S. aureus* after 3, 4, and 5 hours at 7.8, 15.5, and 31 μg/mL. The percentage of live bacteria also decreased in the presence of selenium nanoparticles. Therefore, selenium nanoparticles may be used to effectively prevent and treat *S. aureus* infections and thus should be further studied for such applications [51].

14. Poly-L-Lactide Nanoparticles

Nisin is an antimicrobial protein produced by *Lactococcus lactis* and is known as a lantinobiotic due to the fact that it contains unusual amino acids such as lanthionine [52]. Nisin has just recently been approved by the FDA for use as a food additive to provide preservation due to its broad spectrum antibacterial activity against gram-positive bacteria. One major problem when implementing nisin as a food additive is due to its weak structural integrity and certain bacteria having a minor tolerance to its effects, hefty amount of nisin is required. But it has now been combined with the use of Poly-Lactide nanoparticles in a study by Salmaso [52]. Poly-L-Lactide was used as it is biodegradable and nontoxic and could be used in certain applications of interest, as well as being easily producible. The nisin loaded poly-L-lactide nanoparticles were produced using a mixed solvent method implementing super critical CO_2, and particles were shown to be of low diameter, low dispersity and were capable of loading high protein quantities. The average particle size was found to be between 250 – 400nm. They were then tested for their antibacterial properties against *Lactobacillus* using two nisin loaded concentrations of 20mg of a 5% and 20mg of 20% nisin loaded particle. Results of the experiment showed a significant improvement in the antibacterial activity time compared with the free nisin, with free nisin being active for around 7 days to an increased 40 days with the use of the nisin loaded in the poly-L-lactide nanoparticles. There is a considerable potential for this technique in protein formulation as shown by the nisin loaded poly-L-lactide nanoparticles, with observed results demonstrating improved protein stabilization and prolonged antibacterial activity, thus allowing such a compound to be applied in food and pharmaceutical preservation [52].

Conclusion

Nanomedicine has provided new avenues for antibacterial therapeutics and diagnostics as well as enhancing current systems through the use of new novel nanomaterials. A variety of potential antibacterial therapeutics has been mentioned in the review from both synthetic and biologically inspired materials that have been engineered to a nanoscale. High percentages of bacterial inhibition by nanoparticles were demonstrated against microorganisms, and in some cases were found to completely inhibit growth of a variety of gram-negative and gram-positive bacteria species. Nanomaterials have a collection of uses such as ointments, sterilizing paints and materials, sterilization of medical equipments and clinical environments.

Acknowledgments

I would like to thank Prof. (Dr.) Shadia M. Abdel-Aziz, Microbial Chemistry Dept., National Research Center, for her support and help in revising and arranging this review article.

Conflict of Interest

The author declares that there is no conflict of interest.

References

[1] Eric D (1981) Molecular engineering: An approach to the development of general capabilities for molecular manipulation._Proceedings of the National Academy of Sciences PNAS September 1, 78:5275-5278.

[2] Murray R, Grammar P, Rodwell V (2000) Harpers Biochemistry. 25th Edn., McGraw-Hill Co., USA.,740-742.

[3] Mazzola L (2003) Commercializing nanotechnology. *Nature Biotechnol.* 21:1137.

[4] Taton T (2002) Nanostructures as tailored biological probes. *Trends Biotechnol.* 20:277-279.

[5] Parak W, Gerion D, Pellegrino T, Zanchet D, Micheel C, Williams C, Boudreau R, Le Gros M, Larabell C, Alivisatos A (2003) Biological applications of colloidal nanocrystals *Nanotechnol.* 14:R15-R27.

[6] Pankhurst Q, Connolly J, Jones S, Dobson J (2003) Applications of magnetic nanoparticles in biomedicine. *J. Phys. D. Appl. Phys.* 36:R167-R181.

[7] Sinani V, Koktysh D, Yun B, Matts R, Pappas T, Motamedi M, Thomas S, Kotov N (2003) Collagen coating promotes biocompatibility of semiconductor nanoparticles in stratified LBL films. *Nano Lett.* 3:1177-1182.

[8] Zhang Y, Kohler N, Zhang M (2002) Surface modification of super paramagnetic magnetite nanoparticles and their intracellular uptake. *Biomater* 23:1553-1561.

[9] Weir E, Lawlor A, Whelan A, Regan F (2008) The use of nanoparticles in anti-microbial materials and their characterization. *Analy* 133:835-845.

[10] Seoktae K (2007) Single-walled carbon nanotubes exhibit strong antimicrobial activity. *Langmuir* 23:8670-8673.

[11] Chipot C, Tarek M (2006) Interaction of a peptide nanotube with a watermembrane interface. *Phys Biol.* 3:S20-25.

[12] Lyon D, Fortner J, Sayes C, Colvin V, Hughe J (2005) Bacterial cell association and antimicrobial activity of a C60 water suspension. *Environ. Toxicol. Chem.* 24:2757-2762.

[13] Kim J, Shashkov E, Galanzha E, Kotagiri N, Zharov V (2007) Photothermal antimicrobial nanotherapy and nanodiagnostics with self-assembling carbon nanotube clusters. *Lasers Surg. Med.* 39:622-634.

[14] Zharov VP (2006) Photothermal nanotherapeutics and nanodiagnostics for selective killing of bacteria targeted with gold nanoparticles. *Biophys. J.* 90(2):619-627.

[15] Alideertu D, Shi L, Jinfeng H, Tao W, Tianyi Z, Xiao L, Wang W, Xin Z, Liu F, Ge G, Chen Y (2011) Modifying Fe_3O_4-Functionalized Nanoparticles with N-Halamine and Their Magnetic/Antibacterial Properties. *ACS Appl. Mater Interfaces* 3:4228-4235.

[16] Moghimi S (2005) Nanomedicine: prospective diagnostic and therapeutic potentials. Asia Pacific Biotech News 9:1072-1077.

[17] Mooney E (2006) Silver dressings [safety and efficacy reports]. *Plast Reconstr Surg.* 117:666-669.

[18] Sondi I, Salopek-Sondi B (2004) Silver nanoparticles as antimicrobial agent: a case study on *E. coli* as a model for Gram-negative bacteria. *J. Coll Inter Sci.* 275:177-182.

[19] Shahverdi AR (2007) Synthesis and effect of silver nanoparticles on the antibacterial activity of different antibiotics against *Staphylococcus aureus* and *Escherichia coli* Nanomedicine. *Nanotechn. Biol. Med.* 3:168-171.

[20] Ahmad R, Sara M, Hamid R, Hossein J, Ashraf-Asadat N (2007) Rapid synthesis of silver nanoparticles using culture supernatants of *Enterobacteria*: A novel biological approach. *Process Biochem.* 42:919-923.

[21] Cheng L, Weir M, Xu H, Antonucci J, Kraigsley A, Lin N, Lin-Gibson S, Zhou C (2012) Antibacterial amorphous calcium phosphate nanocomposites with a quaternary ammonium dimethacrylate and silver nanoparticles. *Dent Mater* Feb 1 (Accepted).

[22] Elechiguerra J, Burt J, Morones J, Camacho-Bragado A, Gao X, Lara H, Yacaman M (2005) Interaction of silver nanoparticles with HIV-1. *J. Nanobiotechn.* 3:6: doi:10.1186/1477-3155-3-6.

[23] Lu L, Sun R, Chen R, Hui C, Ho C, Luk J, Lau G, Che C (2008) Silver nanoparticles inhibit hepatitis B virus replication. *Antivir. Ther.* 13:253-262.

[24] Sun R, Chen R, Chung N, Ho C, Lin C, Che C (2005) Silver nanoparticles fabricated in Hepes buffer exhibit cytoprotective activities toward HIV-1 infected cells. *Chem. Commun.* (Camb), 5059-5061.

[25] Rai M, Yadav A, Gade A (2009) Silver nanoparticles as a new generation of antimicrobials. *Biotechn. Adv.* 27:76-83.

[26] Mehrbod P, Motamed N, Tabatabaian M, Soleimani E, Amini E, Shahidi M (2009) In Vitro Antiviral Effect of "Nanosilver" on Influenza Virus. *DARU*, 1**7:**88-93.

[27] Lara H, Ayala-Nuñez N, Ixtepan-Turrent L, Rodriguez-Padilla C (2010) Bactericidal effect of silver nanoparticles against multidrug-resistant bacteria. *World J. Microbiol. Biotechnol* 26:615-621.

[28] Donsi F, Annunziata M, Sessa M, Ferrari G (2011): Nanoencapsulation of essential oils to enhance their antimicrobial activity in foods. *LWT - Food Sci. Technol.* 44(9):1908-1914.

[29] Shigeru K, Tomono K, Maesaki S (1998) Drug delivery systems for infection: Liposome-incorporating antimicrobial drugs Share this item. *J. Inf. Chem.* 4:159-173.

[30] Avnesh K, Kumar S, Yadav S, Yadav C (2010) Biodegradable polymeric nanoparticles based drug delivery systems. *Coll Surf. B: Biointerfaces* 75:1-18.

[31] Zhang WX (2003) Nanoscale iron particles for environmental remediation: An overview. *J. Nanopart Res.* 5:323-332.

[32] Hydutsky B, Mack E, Beckerman B, Skluzacek J, Mallouk T (2007) Optimization of Nano- and Microiron Transport through Sand Columns Using Polyelectrolyte Mixtures. *Env. Sci. Technol.* 41:6418-6424.

[33] Ryan J, Harvey R, Metge D, Elimelech M, Navigato T, Pieper A (2002) Field and laboratory investigations of inactivation of viruses (PRD1 and MS2) attached to iron oxide-coated quartz sand. *Env. Sci. Technol.* 36:2403-2413.

[34] You Y, Han J, Chiu P, Jin Y (2005) Removal and inactivation of waterborne viruses using zerovalent iron. *Env. Sci. Technol.* 39:9263-9269.

[35] Lellouche J, Friedman A, Lahmi R, Gedanken A, Banin E (2012) Antibiofilm surface functionalization of catheters by magnesium fluoride nanoparticles. *Intr. J. Nanomedicine* 7: 1175-1188.

[36] Stoimenov PK (2002) Metal oxide nanoparticles as bactericidal agents. *Langmuir* 18:6679-86.

[37] Huang C, Cha D, Ismat S (2005) Progress report: short-term chronic toxicity of photocatalytic nanoparticles to bacteria, algae, and zooplankton; 2005. EPA Grant Number: R831721. http://cfpub.epa. gov/ncer_abstracts/index.cfm/fuseaction/ display.abstractDetail/abstract/7384/report/0

[38] Jones N (2008) Antibacterial activity of ZnO nanoparticle suspensions on a broad spectrum of microorganisms *FEMS Microbiol. Lett.* 279:71-76.

[39] Hui L, Jin Z, Rui S, Chen W (2011) Identification of fungi from children's shoes and application of a novel antimicrobial agent on shoe insole. *Afr. J. Biotechnol.* 10:14493-14497.

[40] Anglin EJ (2008) Porous silicon in drug delivery devices and materials. *Adv. Drug Deliv. Rev.* 60:1266-77.

[41] Jia H (2008)The structures and antibacterial properties of nano-SiO_2 supported silver/zinc silver materials. *Dent Mater* 24:244-249.

[42] Kawahara K, Tsuruda K, Morishita M, Uhida M (2000) Antibacterial effect of silverzeolite on oral bacteria under anaerobic conditions. *Dent. Mater* 16:452-455.

[43] Allaker R, Ren G (2008) Potential impact of nanotechnology on the control of infectious diseases. *Trans Roy Soc. Trop Med. Hyg.* **102:**1–2.

[44] Gupta A, Gupta M (2005) Synthesis and surface engineering of iron oxide nanoparticles for biomedical applications. *Biomater* 26:3995-4021.

[45] Babes L, Denizot B, Tanguy G, Le J, Jean J, Jallet P (1999) Synthesis of iron oxide nanoparticles used as MRI contrast agents: A parametric study. *Coll Interf Sci* 212:474-482.

[46] Gonzales-Weimuller M, Zeisberger M, Krishnan K (2009) Size-dependent heating rates of iron oxide nanoparticles for magnetic fluid hyperthermia. *J. Magn. Mag. Mat.* 321:1947-1950.

[47] Kohler N, Sun C, Wang J, Zhang M (2005) Methotrexate-modified superparamagnetic nanoparticles and their intracellular uptake into human cancer cells. *Langmuir* 21(19):8858-8864.

[48] Pareta RA, Taylor E, Webster TJ (2008) Increased osteoblast density in the presence of novel calcium phosphate coated magnetic nanoparticles. *Nanotechnol* 19(26):265101.

[49] Touati D (2000) Iron and oxidative stress in bacteria. *Arch. Biochem. Biophys.* 373:1-6.

[50] Nhiem T, Aparna M, Dhriti M, Sinha A, Nayar S, Thomas J (2010) Bactericidal effect of iron oxide nanoparticles on *Staphylococcus aureus. Int. J. Nanomed.* 5:277-283.

[51] Tran P, Webster T (2011) Selenium nanoparticles inhibit *Staphylococcus aureus* growth. *Intr. J. Nanomed* 6:1553-1558.

[52] Salmaso S (2004) Nisin-loaded poly-l-lactide nano-particles produced by CO_2 anti-solvent precipitation for sustained antimicrobial activity. *Int. J. Pharma* 287:163-173.

Index

#

A

B

C

D

E

F

G

H

I

J

K

L

M

O

P

Q

R

S

T

U

Y

Z

α

β

γ